Cave and Karst Systems of the World

Series Editor

James W. LaMoreaux, P.E.LaMoreaux and Associates, Tuscaloosa, AL, USA

More information about this series at http://www.springer.com/series/11987

Greg A. Brick • E. Calvin Alexander Jr.
Editors

Caves and Karst of the Upper Midwest, USA

Minnesota, Iowa, Illinois, Wisconsin

Springer

Editors
Greg A. Brick
Lands and Minerals Division
Minnesota Department of Natural Resources
St. Paul, MN, USA

E. Calvin Alexander Jr.
Earth and Environmental Sciences
University of Minnesota
Minneapolis, MN, USA

ISSN 2364-4591 ISSN 2364-4605 (electronic)
Cave and Karst Systems of the World
ISBN 978-3-030-54632-8 ISBN 978-3-030-54633-5 (eBook)
https://doi.org/10.1007/978-3-030-54633-5

This Springer imprint is published by the registered company Springer Nature Switzerland AG
The registered company address is: Gewerbestrasse 11, 6330 Cham, Switzerland

The truth of nature lieth hid in certain deep mines and caves.—Democritus

Foreword

It was shrewd of the editors to ask me to write this Foreword, because I have little experience in their home territory and am presumably free of bias. But having spent much time elsewhere in caves and their surrounding landscapes, I can safely offer my congratulations on a job well done. The list of subjects covered by this book promises a level of discussion well beyond that of a simple tourist guide. It explains the basics of how water can pass underground through certain kinds of bedrock, dissolving channels that enlarge into caves, and how the overlying surface responds by developing depressions and sinking streams, to form what is known as a karst landscape. It goes much farther, investigating cave biology, history, archeology, and mining, as well as land management, the effect of past climates, and far more. The authors are among the most accomplished figures in American karst science. But except for them, few specialists in this field have more than a passing acquaintance with the Upper Midwest USA. That alone is justification for this book and its depth of coverage.

Many vacationers drive straight through this area on their way to higher ground, as though drawn by a magnet. In doing so, they are likely to miss the many curiosities that make this region special—small towns, isolated farmhouses, strange and specialized museums, amusing signs—all surrounded by gently rolling prairie and a feeling of peace. Most significantly, this quiet land is far more important to our economy and well-being than any number of majestic mountains. Preserving its fertility is a serious matter, and a significant part of that strategy is to understand the impact of karst. Land subsidence, soil degradation, flooding, and contamination can be major obstacles to productive land use. Proper land management is the only effective remedy, and this book provides the appropriate guidance. One of the most important goals of this book is to stress how to protect the soil and groundwater by adjustments in agricultural practices. The authors have gained much, or most, of their professional experience in the areas included in this book, and their insight extends far beyond its boundaries. Some are known worldwide for their skill at mapping caves and others for their insight into how to use cave deposits to interpret geologic history and ancient climates. The lead author of the section on cave biology began his cave studies in the heart of this region in the late 1950s and is still going strong.

In addition, several early specialists in the field, now long retired, spent large parts of their lives studying the caves and karst of this region. Of special note are the late J Harlen Bretz (no period after the "J"!), internationally famous for his early interpretations of cave origin and co-author of a book on Illinois Caves; James Hedges, prolific author and editor of descriptive cave literature; and David Morehouse, who, in a study of Iowa caves, was among the first to recognize the origin and enlargement of certain caves by sulfuric acid, a by-product of the reaction between oxygen-rich water and local iron-sulfide minerals.

Caves in the Upper Midwest USA have their own special character. Many of them have fissure-like outlines with long straight passage segments that intersect in a zig-zag pattern through the soluble rocks. Many others are truncated remnants of phreatic mazes high on ridge tops, making for mudbaths or painful crawling. Decorative "cave formations" (speleothems) are sparse but intriguing. The total mapped passage lengths of individual caves in this area range up to an impressive 17 miles, although the areas open to tours are limited to the most accessible and attractive sections. Some caves are quite challenging to explore, with deep shafts and cold lakes, but these obstacles are cleverly bypassed in caves open to tours. Minnesota has 490 known caves, two open to the public. Iowa contains 418, including at least five that are open to tours. Wisconsin has 367, one open to the public. Illinois contains 832 known caves, five or six open to tours. The karst areas of Illinois extend southward into younger rocks than those described elsewhere in this book, and they give a glimpse of the differences produced by variations in local geology.

Karst landscapes and caves are intriguing not only to scientists, but also to those who enjoy exploring and adventure. However, they pose a challenge. Without careful land management, there is a tendency for such areas to degrade rapidly by the subsidence and erosion of soil and the spread of contaminants. This is especially true during rainy periods. One of the most important goals of this book is to stress how to facilitate the protection of soil and groundwater by adjustments in agricultural practices.

My wife and I have only limited experience in the Upper Midwest USA. Still, our visits to this area have provided us with warm memories. A few decades ago we were engaged in fieldwork for the State of Minnesota at Mystery Cave, the well-known show cave. Determining how the structure of the limestone affected the origin and development of the cave was a satisfying experience. Twice we had a chance to lead visitors through it when there was a shortage of guides. Pointing out the attractions, we were glad to see how eager the groups were to understand the cave and to know where each passage led. Children were as interested as adults, and even more so where side-passages branched into darkness and we had a chance to turn on the lights and see for ourselves where they went. It helped that the two of us were as curious as the visitors.

One day, while we were mapping the cave's rock layers, we met a lone contractor who was busy with trail reconstruction. He had once worked at Disney World. "I wish I could spend a month working in here," he said—"I could make it look like a **real** cave!" We still laugh about that. Visitors find it an enchanting place, even though its walls are not pink and there are no fairies flapping overhead.

A couple of friends recently bought some property nearby that is surrounded by broad karst lowlands. Showing us around, the husband paused at a viewpoint overlooking the plain several dozen feet below. He quietly confided to us: "Looking out from here, your spirit just **soars**...."

We laughed quietly, but knew that he was right. With that kind of outlook, one can easily fall in love with our Upper Midwest karst!

Arthur N. Palmer
Professor Emeritus
Earth and Atmospheric Sciences/Water Resources
State University of New York (SUNY)
Oneonta, NY, USA

Preface

The Upper Mississippi Valley was recognized as one of the "important karst regions of the Northern Hemisphere" by Herak and Stringfield (1972, p. 470–471). Yet it's been 40 years since the publication of the 1980 NSS Guidebook on the caves of Minnesota, Wisconsin, and Iowa, edited by Alexander (1980), which over time became the most heavily cited speleological work on the part of the United States covered in the present volume. To this roster we now add a fourth state, Illinois, the states arrayed symmetrically around the Upper Mississippi River. While the editors realize there are many differing opinions as to what states constitute the Midwest, let alone the Upper Midwest, these contiguous states will be referred to as the Upper Midwest, or the Upper Mississippi Valley (UMV) region, in this book.

During those 40 years our understanding and stewardship of caves and karst has advanced significantly. There has been the widespread recognition of hypogene processes, the refinement of AMS radiocarbon dating, dye tracing to define springsheds, the use of LiDAR, and the establishment of the Minnesota Cave Preserve and other karst conservancies. With the discovery of many more caves in the interim, the time is ripe for a new overview of the region. In doing so, and in keeping with modern conservation practice, the exact locations of ungated wild caves are not disclosed.

While each state is described in a separate chapter, the introduction gives a seamless overview of the entire region, correlating the separate stratigraphic nomenclatures of the four states. The geologic names provided within Chap. 1 have been reviewed for conformance with the usage of the U.S. Geological Survey and the standards of the North American Stratigraphic Code. Geologic names provided in all other chapters have not been rigorously reviewed and thus are the sole responsibility of the respective authors. (When in doubt about the use of a particular stratigraphic name, we suggest consulting the generalized correlation charts in Figs. 1.6 and 1.9 in Chap. 1.) Special attention is then given to the lead-zinc mining region, biology, agriculture and karst, paleoclimate research, in separate chapters. A significant array of advances in karst science originated largely in the Upper Midwest, and these are described in the context of the various fields concerned, in a separate chapter.

Active caving groups in all four states have found, explored, and mapped new caves. Countless anonymous "sherpas" have contributed to the karst science efforts. Critical masses of research scientists at the State Universities and staff at the State Geological Surveys, Regulatory Agencies, and Soil and Water Conservation Districts in each of the four states recognized the significance of karst phenomena—as did the USGS, USDA, and USEPA. These groups, in a very synergistic and cooperative way, have researched, mapped, monitored, and informed the areas' residents about the impact and limitations imposed by karst phenomena on human activities and vice versa. "Karst" has progressed from an unknown, esoteric geographic term to an important consideration in the everyday lives of people in the area. For example, it is no longer socially acceptable to use sinkholes as convenient garbage and waste disposal sites.

The advances in karst hydrogeologic sciences have been funded by federal, state, local, and private grants to a wide range of individual researchers, organizations, and agencies. That funding has been matched by years of volunteer time and effort by innumerable people. The paleoclimate research has been funded primarily by federal research grants to the research universities. The many individuals and organizations who have supported the karst science advancement are gratefully acknowledged.

St. Paul, MN, USA
Minneapolis, MN, USA

Greg A. Brick
E. Calvin Alexander Jr.

References

Alexander EC Jr (ed) (1980) An introduction to Caves of Minnesota, Iowa, and Wisconsin, Guidebook for the 1980 National Speleological Society Convention, Huntsville, AL, 190p

Herak M, Stringfield VT (1972) Karst: Important karst regions of the northern hemisphere. Elsevier, Amsterdam, 551p

Contents

Contributors

E. Calvin Alexander Jr. Department of Earth and Environmental Sciences, University of Minnesota, Minneapolis, MN, USA

Raymond R. Anderson University of Iowa, Iowa City, IA, USA

Greg A. Brick Lands and Minerals Division, Minnesota Department of Natural Resources, St. Paul, MN, USA

Michael J. Day Department of Geography, University of Wisconsin-Milwaukee, Milwaukee, WI, USA

Joseph A. Devera Illinois State Geological Survey, Champaign, IL, USA

James A. Dockal Department of Earth and Ocean Science, University of North Carolina, Wilmington, NC, USA

Daniel H. Doctor United States Geological Survey, Reston, VA, USA

Jeffrey A. Dorale Department of Earth and Environmental Sciences, University of Iowa, Iowa City, IA, USA

Kevin Erb Division of Extension, Natural Resources Institute, University of Wisconsin-Madison, Green Bay, WI, USA

Patricia N. Kambesis Western Kentucky University, Bowling Green, KY, USA

Michael J. Lace Coastal Cave Survey, West Branch, IA, USA

Julian J. Lewis Biological Consulting LLC, Borden, IN, USA

Donald E. Luman Illinois State Geological Survey, Champaign, IL, USA

Benjamin J. Maas Department of Environmental Science, Buena Vista University, IA, USA

Samuel V. Panno Illinois State Geological Survey, Champaign, IL, USA

Stewart B. Peck Biology Department, Carleton University, Ottawa, ON, Canada; Entomology, Canadian Museum of Nature, Ottawa, ON, Canada

John O. Whitaker ISU Center for Bat Research Outreach and Conservation, Indiana State University, Terre Haute, IN, USA

Karst Geology of the Upper Midwest, USA

1

Daniel H. Doctor and E. Calvin Alexander Jr.

Abstract

Karst in the Upper Midwest occurs within a thick sequence of mixed carbonate and siliciclastic Cambrian through Pennsylvanian sedimentary rocks, with a minor occurrence of karst in Proterozoic sandstone. Deposition of the sediments occurred on a marine epeiric ramp that spanned much of the North American continent through most of the Paleozoic. The Upper Midwest region experienced dramatic changes in sea level over geologic time, resulting in the observed sequence of interbedded carbonate and clastic rocks. The greatest degree of karst development occurs within (1) the Lower Ordovician Prairie du Chien Group below the Sauk-Tippecanoe (Knox) unconformity, (2) the Upper Ordovician Galena Group, (3) the Middle and Upper Devonian Wapsipinicon and Cedar Valley Groups, and (4) the Middle Mississippian Mammoth Cave Group and correlative formations. Uplift and exposure of the rocks likely occurred in the Permian, with some later deposition of Cretaceous terrestrial sediments atop the marine strata. Nearly all the Cenozoic sedimentary units were removed by ice sheets during the Pleistocene; however, pockets of Cretaceous sediments persist on the margins of the Driftless Area, a region of the Upper Mississippi River Valley that remained largely free of ice during the last ice age.

D. H. Doctor (✉)
United States Geological Survey, 12201 Sunrise Valley Drive, MS 926A, Reston, VA 20192, USA
e-mail: dhdoctor@usgs.gov

E. C. Alexander Jr.
Department of Earth and Environmental Sciences, University of Minnesota, 150 Tate Hall, 116 Church Street SE, Minneapolis, MN 55455, USA

1.1 Introduction

Travel due east on US Interstate 90 from Austin, Minnesota, toward the Mississippi River, and you traverse one of the great karst regions of central North America. Moving across the great flat plains of the upland plateau one sees wide expanses of fields planted in corn, soybeans, and alfalfa that mark a seemingly never-ending horizon. Continue east on Minnesota Highway 16 after passing through the village of Spring Valley, and with a sideways glance out the window you might catch sight of the crown of a tree that stands alone in a field, its trunk hidden from view as if pulled vertically down into the Earth. One, two, three more sunken trees appear like as many gophers peering out of their burrows. The trees are sunken reminders of a once wooded plain, now entirely reclaimed for agriculture, except for the natural closed depressions that enclose isolated arbors. Cross the sinkhole plain along a spur of the plateau that is dissected by headwater stream valleys on either side, and you enter the town of Preston and descend off the plateau down into the valley of the Root River, a spring-fed designated trout stream. Bluffs of stratified pale yellow and gray rock peek through the winding, tree-lined river valley. Aided by the cool waters of springs emanating from vertical fissures and bedding plane partings in the bluffs, the Root River has incised its way through layer upon layer of limestone and dolostone, cutting a path to join the Mississippi River just south of La Crosse, Wisconsin. Some of the voids in the bedrock are wide enough that they beckon to be explored, holding the promise of passage into a hidden underground maze.

This is the karst of the Upper Mississippi Valley (UMV), formed across a time span of more than half a billion years by geologic processes that first built up then sculpted down the landscape to its present appearance. The geologic history of the region is tied to the constructive and erosive power of water: water of seas from hundreds of millions of years ago, water of ice from hundreds of thousands of years ago, and

G. A. Brick and E. C. Alexander Jr. (eds.), *Caves and Karst of the Upper Midwest, USA*,
Cave and Karst Systems of the World, https://doi.org/10.1007/978-3-030-54633-5_1

water of the present day that continues to shape the landscape. Water is not only the key to the geologic past, it is also the key to the more recent human history of the UMV region. The present human interaction with the landscape depends on its water resources, and those water resources are the key factor that will determine the future of the region. As in most karst regions, humans were drawn to settle in the UMV because of the abundant and readily available water resources provided by the karst aquifers. Continued economic growth and prosperity of the region rely on sustainable use and management of those aquifers. How might the water resources of the region be impacted by human land use, and how does the geologic history of the region help us to better understand those impacts? To begin to answer those questions, we need to look at what lies beneath the surface.

Figure 1.1 is an artist's rendering of what a block of earth cut out from southeastern Minnesota might look like if lifted out of the ground. At the surface, rolling hills covered in lush vegetation and crops sustain farming communities. The vegetation is rooted in a mantle of soil and sediment that blankets stratified layers of sedimentary rocks, mostly carbonates intercalated with shales, siltstones, and sandstones. The history of the deposition of these layers of sedimentary rock is where we begin to take a step back in time.

1.2 Paleozoic Geology and Paleogeography

Karst in the upper midwestern states of Minnesota, Wisconsin, Iowa, and Illinois is formed within a stratigraphic sequence of carbonate rocks (limestone and dolostone) that are interbedded with siliciclastic deposits of sandstone, siltstone, and shale. The geologic history of these sedimentary rocks began in the early Paleozoic Era, approximately the last 500 million years of Earth history. The oldest rocks that are currently exposed at the surface in the region were deposited as nearshore and shallow marine sediments that accumulated onto a stable foundation of crystalline igneous and metamorphic basement rocks of the North American craton. The marine sediments blanketed the basement rocks and subsequently were buried, compressed, and solidified through the process of diagenesis whereby loose grains of sediment were cemented and transformed into sedimentary rock. These sedimentary rocks were uplifted in the late Paleozoic, and some layers of overlying rock were eroded to reveal those exposed at present.

A simplified geologic map of the region is shown in Fig. 1.2 with the geologic units displayed according to the time at which the predominantly carbonate rocks were deposited. This figure illustrates where shallow, nearshore

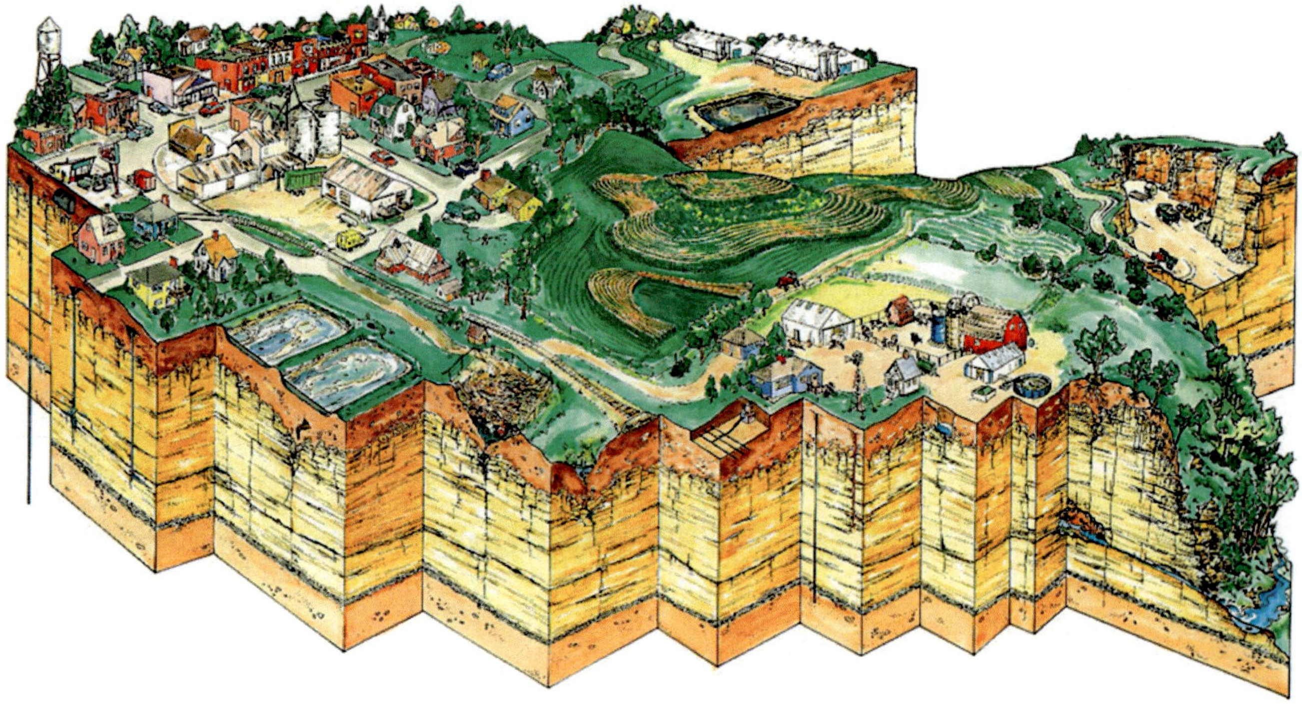

Fig. 1.1 Block diagram of southeastern Minnesota karst surface and subsurface (image used with permission of the Southeast Minnesota Water Resources Board)

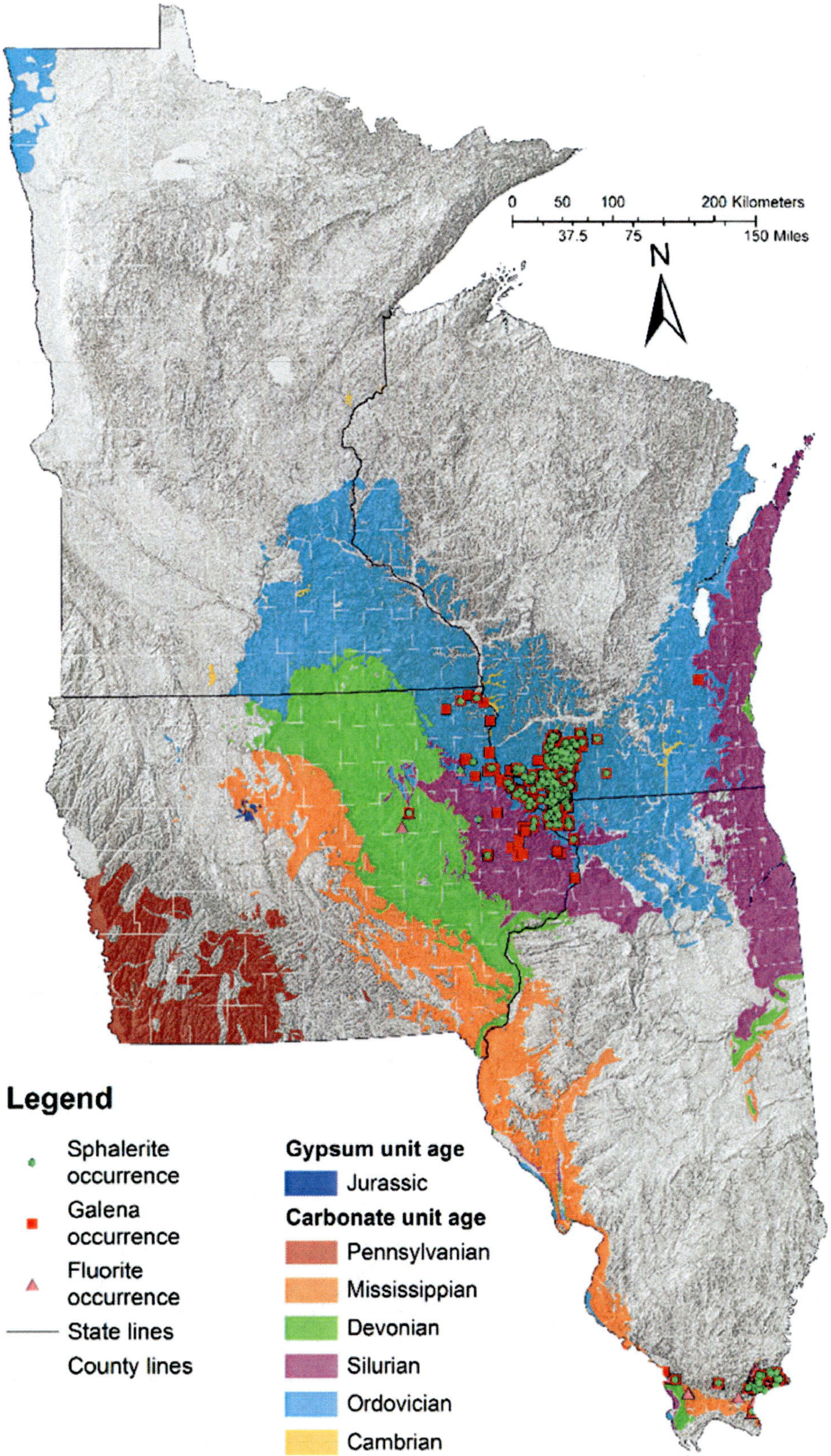

Fig. 1.2 Soluble bedrock geologic map of Iowa, Minnesota, Wisconsin and Illinois showing mineral occurrences. The carbonate rocks range from the Cambrian to the Pennsylvanian. The youngest Paleozoic carbonate strata are exposed only in southwestern Iowa, where karst development is not well expressed at the surface as result of overlying glacial deposits. Much younger Jurassic evaporite deposits occur near Fort Dodge in central Iowa and host gypsum karst

marine to deeper carbonate ramp environments were located at different periods of geologic time. Changes in sea level during sediment deposition and tectonic warping of the initially near-horizontal layering account for the map pattern. Also shown in Fig. 1.2 are the locations of mineralization in the rocks caused by the migration of metal-bearing brines that deposited zinc and lead sulfide minerals and associated fluorite. These ores are called Mississippi Valley Type (MVT) deposits due to their localization in this region and further to the south. See Dockal (2020), Chap. 7, this volume.

During the Middle and Late Cambrian (ca. 513 to 488 Ma) and through the Ordovician (ca. 488 to 444 Ma), the area of the North American continent that was exposed as land above the world ocean was much smaller, roughly half the size of the present-day continent (Fig. 1.3). This paleo- continental land mass is known as Laurentia, the center of which would have been located just south of the equator (Runkel et al. 2012). Laurentia was in a tropical to sub-tropical climatic zone across much of its extent; however, at that time in Earth history vascular land plants had not yet evolved and the land surface would likely have been a sparsely vegetated, barren, desert-like environment of rock, sand, and finer sediments. As a result, much siliciclastic sediment eroded off this landmass and was deposited in the adjacent shallow marine environments (Runkel et al. 2007). Fringing this landmass on its paleo-western side was the Iapetus Ocean, within which an enormous shallow bank of siliciclastic and carbonate sediment built up and covered the crystalline basement rocks of the Proterozoic craton (Fig. 1.3).

Although the North American continent shifted position over millions of years, the basement topographic features of the Transcontinental Arch, Wisconsin Dome, Forest City Basin, Illinois Basin, and Michigan Basin persisted. The processes responsible for the formation of the large regional basins in the surface of the crystalline craton have been attributed to fault-controlled mechanical subsidence and thermal subsidence under tensional stress during rifting, and flexural foreland subsidence resulting from orogenic episodes (Klein and Hsui 1987; Armitage and Allen 2010). The distribution and thickness of sedimentary cover that filled the basins evolved in response to sea level changes, sediment supply from adjacent emergent highlands, and broad tectonic warping of the interior craton. Sediments that filled the basins during submergent episodes record a history of approximately 175 million years, from the Late Cambrian to the Pennsylvanian.

1.2.1 Paleozoic Depositional Environments

In southeastern Minnesota, northern Iowa, northern Illinois, and southwestern Wisconsin, the Paleozoic sediments were deposited during multiple cycles of transgression (sea level rise) and regression (sea level fall) within a shallow marine ramp called the Hollandale Embayment. The embayment was a depression between regional basement highs of the Transcontinental Arch to the west, and the Wisconsin Dome to the east (Fig. 1.3). As a result, a single snapshot in geologic time would capture a position of the marine margin that would likely differ from earlier or later times. Due to the constantly shifting position of the paleoshoreline, a timeline passing through the rock strata of the basin would cross changes in lithologies related to the depositional environment or facies. Such changes in facies during the Late Cambrian are shown schematically in Fig. 1.4 (Runkel et al. 2007).

The Hollandale Embayment was a marine ramp that sloped to the south into the Forest City Basin that underlies most of Iowa. At present, the Paleozoic strata dip in general to the southwest at 1.9–2.9 m per km (Alexander 1980).

Most of the carbonate rocks in the region are a mixture of calcium-carbonate-dominated limestones and magnesium-carbonate-dominated dolostones. Fine siliciclastic sediment is abundant throughout the section, but siliciclastic input varied greatly as the sediments were deposited. The large supply of fine siliciclastic sediment onto the carbonate bank resulted in numerous layers of sand, silt, and clay that interrupted carbonate deposition. As will be discussed later, some of these siliciclastic layers have had an impact on later diagenesis and solutional development within predominantly carbonate rocks.

Deposition resulted in a large buildup of sedimentary rocks that were uplifted and exposed, and that host the karst features of the greater UMV region. The thickness of the carbonate-dominated sequence varies greatly according to the depositional environment and its relation to regional structures (e.g., Kolata 2005), but the stratigraphic section generally thickens toward the south and east from southeastern Minnesota into the interiors of the Forest City Basin and the Illinois Basin. In the Twin Cities of Minnesota, the thickness of the Paleozoic section is approximately 610 m (2,001 ft), while in central Illinois the thickness is more than 4 km (13,123 ft). This increase in sediment thickness is the result of a progressive deepening of the marine environment moving offshore along an epeiric ramp (Fig. 1.5).

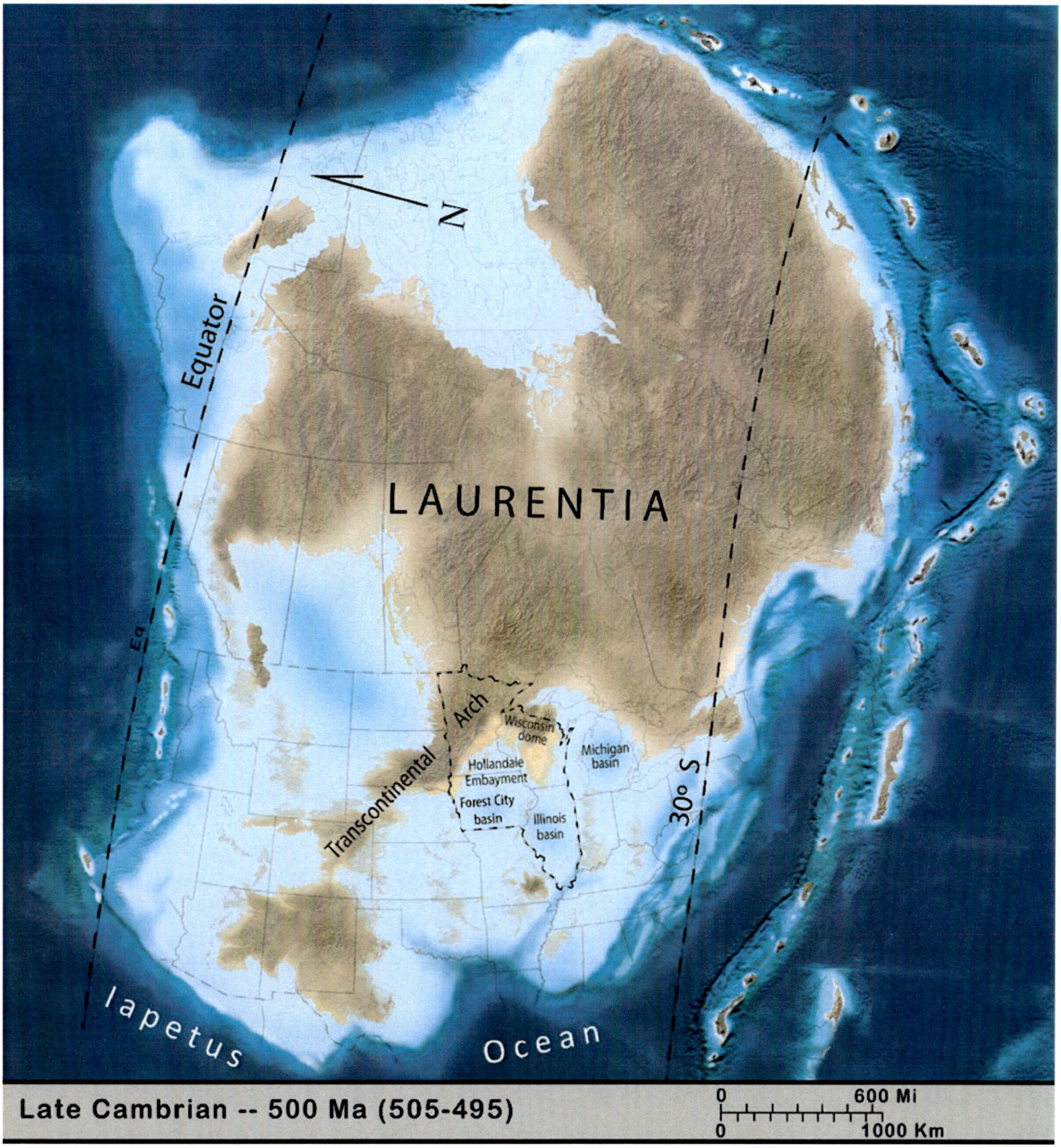

Fig. 1.3 Laurentia in Late Cambrian time (©2013 Colorado Plateau Geosystems Inc.). The land area of the Upper Midwest was covered by a shallow sea that resulted in the thick accumulation of carbonate rocks that host karst features today. The Forest City Basin, Michigan Basin, and Illinois Basin are depressions in the basement rocks of the North American craton where thicker amounts of sedimentary rocks accumulated, yielding economic hydrocarbon and mineral deposits in many areas

Rocks that were once deeply buried have become exposed at the surface today as result of tectonic uplift and prolonged erosion. The rocks are thought to have been buried beneath approximately 600–800 m (1,970–2,620 ft) at the time of MVT ore deposition based on temperature reconstructed from mineral fluid inclusions (Bailey and Cameron 1951). The major tectonic episodes of mountain-building, or orogenies, that affected the rocks of the region occurred along the margins of the North American continent but transferred stresses into the continental interior that warped and deformed the existing sediments. The last major orogeny to impact the eastern and south-central portion of

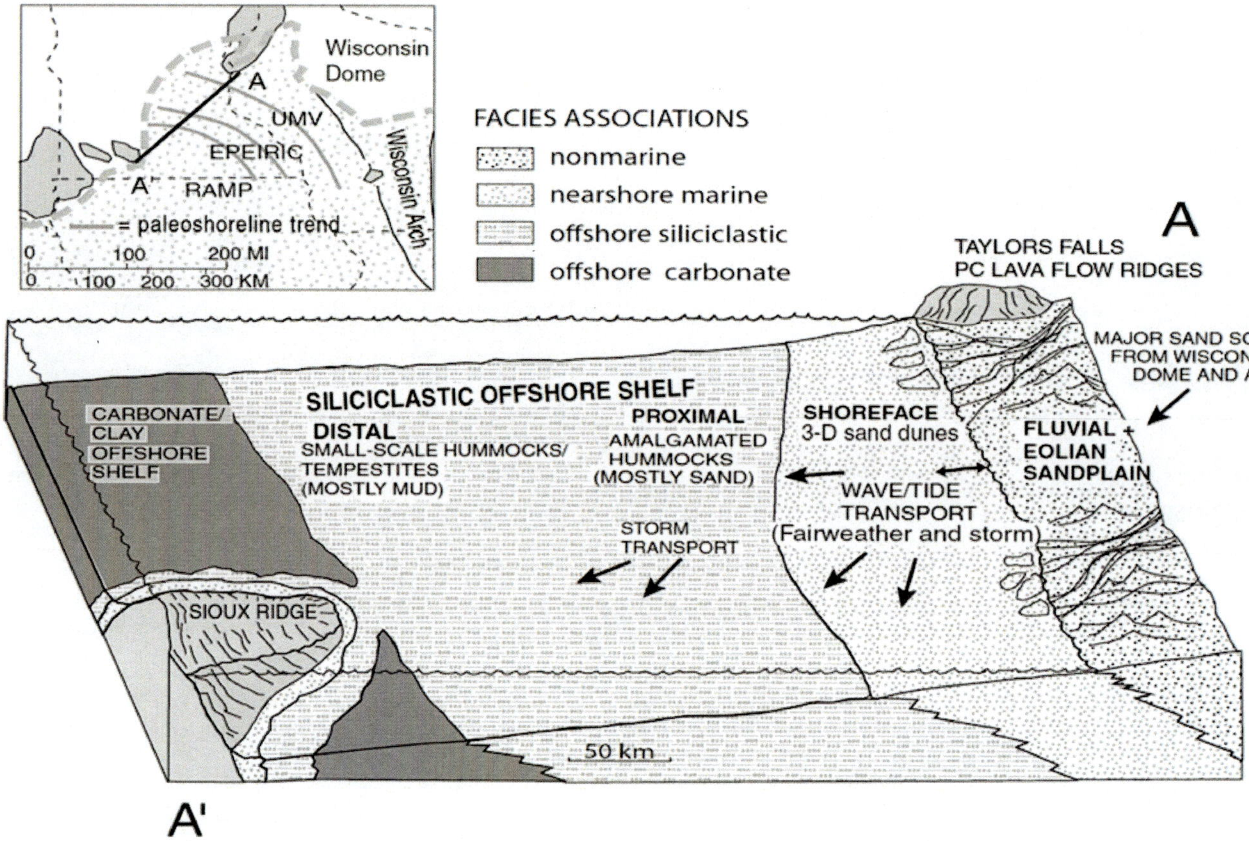

Fig. 1.4 Schematic block diagram illustrating the different depositional environments represented within a snapshot of time in the Late Cambrian (from Runkel et al. 2007)

USA was the Alleghanian orogeny that occurred between approximately 320 to 260 million years ago, during which the Appalachian and Ouachita Mountains were uplifted.

1.3 Stratigraphy of Karst-Bearing Units

Nearly all karst development occurs in carbonate rocks; exceptions include cave development in Proterozoic sandstones in Pine County, Minnesota (Shade 2002; Shade et al. 2015), and caves in other sandstones such as the Ordovician St. Peter Sandstone (see Alexander and Brick 2020, this volume). Although detailed stratigraphic correlations are beyond the scope of this chapter, generalized stratigraphic correlation charts are shown in Fig. 1.6 for the Cambrian through Silurian, and later in Fig. 1.9 for the Devonian through Jurassic to facilitate the comparison of stratigraphic formation names between States and in the subsequent chapters of this volume. The names presented herein conform with formal usage currently accepted by the US Geological Survey but may differ from earlier usage or informal names used in various states.

1.3.1 Cambrian System

The earliest Cambrian sediments covered an interval of Proterozoic clastic sediments derived from the erosion of the underlying crystalline and metamorphic basement rocks. Overlying the Proterozoic clastic rock is the Mount Simon Sandstone, found throughout the region. The Mount Simon serves as a regional aquifer, providing water generally of good quality for domestic use. Above the Mount Simon are a series of sandstones and finer grained siliciclastic units that extend throughout southeastern Minnesota, northern Iowa, and southwestern Wisconsin. These are represented by the Eau Claire, Bonnterre, and Wonewoc Formations, and the Tunnel City Group. Cambrian sandstones in Wisconsin are valuable resources of frac sand, as windblown sands on the shorelines of the Paleozoic seas caused rounding and sorting of sand grains in the deposits, making them ideal for use as fracking proppant.

A prolonged period of carbonate deposition occurred throughout the continent of North America resulting in a thick sequence of Upper Cambrian through Upper Ordovician rocks. The resulting stratigraphic succession is referred

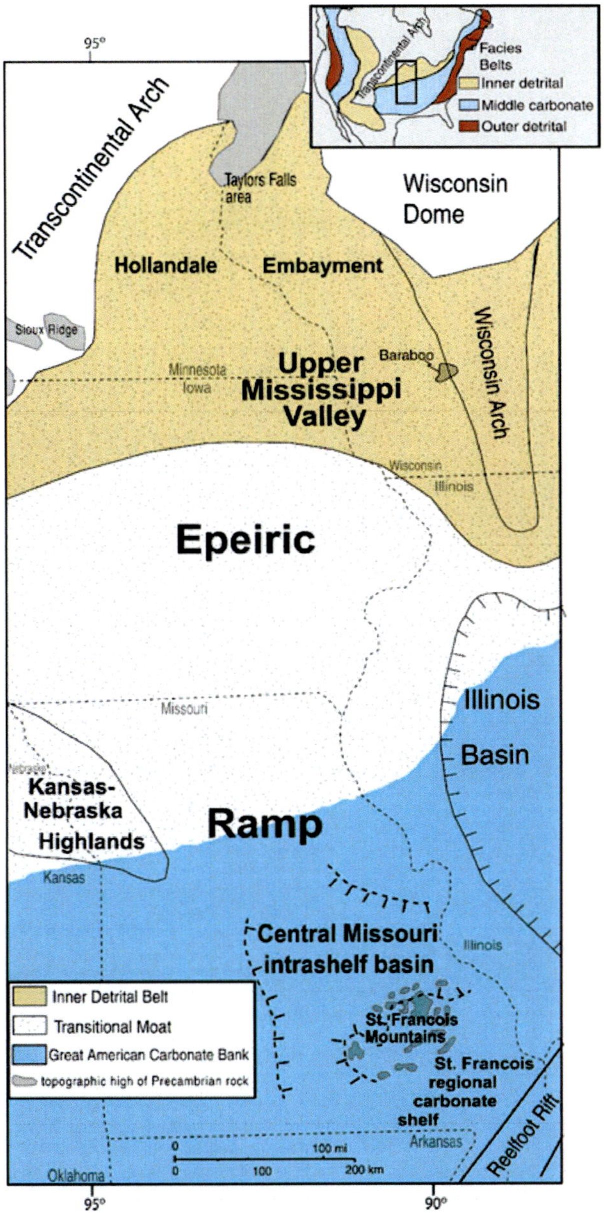

Fig. 1.5 Schematic map illustrating the change in depositional environment across the upper Mississippi valley during the Late Cambrian (from Runkel et al. 2012)

to as the "great American carbonate bank" (Fritz et al. 2012). In the UMV region, the sedimentation of this interval can be divided into two packages: a lower primarily siliciclastic sediment package and a successive package of primarily carbonate sediment (Runkel et al. 2012). The deposition of primarily carbonate sediments began near the Cambrian-Ordovician boundary following the deposition of primarily sandstones and other siliciclastic rocks mixed with some carbonates in the Upper Cambrian.

The area of the Hollandale Embayment was especially shallow in Late Cambrian time, resulting in deposition of mostly sand, silt, and clay mixed with some carbonate, and represented by the upper part of the Tunnel City Group (Franconia Formation), the St. Lawrence Formation, and the Jordan Sandstone (Runkel et al. 2007). While predominantly siliciclastic deposition occurred in the northern part of the UMV in Late Cambrian time, carbonate deposition of the Potosi Dolomite and Eminence Dolomite was occurring further south in Illinois and extending into Missouri. An interval of mixed carbonate and siliciclastic sediments in the Upper Cambrian strata of the northern portion of the UMV region is represented by the St. Lawrence Formation. As a result of the mixture of these sediment types, the St. Lawrence has been described as an "aquitardifer" due to its low vertical transmissivity but high horizontal solutional transmissivity developed at the stratigraphic boundaries between the intercalated siltstone and carbonate layers (Anderson et al. 2011; Runkel et al. 2018). Overlying the St. Lawrence Formation is the Jordan Sandstone, another regionally important sandstone aquifer.

1.3.2 Ordovician System

Above the Cambrian-Ordovician boundary, a relatively brief depositional hiatus occurred (Mossler 2008). The erosional surface was covered by carbonate rocks of the Prairie du Chien Group (Fig. 1.7). The lowest of three main karst systems within the Paleozoic stratigraphy is formed within this stratigraphic interval as illustrated in Fig. 1.8 (Runkel et al. 2014). In Minnesota, Wisconsin, and Iowa, the Prairie du Chien Group is composed of the Oneota Dolomite and the overlying Shakopee Formation. The boundary between the Oneota and the Shakopee is marked by a significant depositional hiatus and is a paleokarst horizon in southeastern Minnesota, with paleokarst features prominent in quarry exposures (Alexander and Wheeler 2015). Crystal Cave, the largest commercial cave in Wisconsin, is hosted within the Prairie du Chien; the upper levels are in the Shakopee Formation, and the lower levels extend into the Oneota Dolomite where spongework passages occur (Day 2009). In northern Illinois and parts of southeastern Wisconsin, the depositional hiatus between the Oneota and Shakopee may be absent, and further to the south and east these units are separated by the New Richmond Sandstone or Roubidoux Formation.

Following deposition of the Prairie du Chien Group, one of several major continental-scale periods of erosion, or depositional unconformities, occurred. This period of erosion divides the Sauk and Tippecanoe major depositional episodes on the interior of North America (Sloss 1963) and may have lasted over 10 million years, from the Early to Middle Ordovician. Evidence is found throughout the continent for paleokarst development in carbonate deposits exposed during this time interval. In the southeastern United States, the carbonate rocks that were deposited prior to being

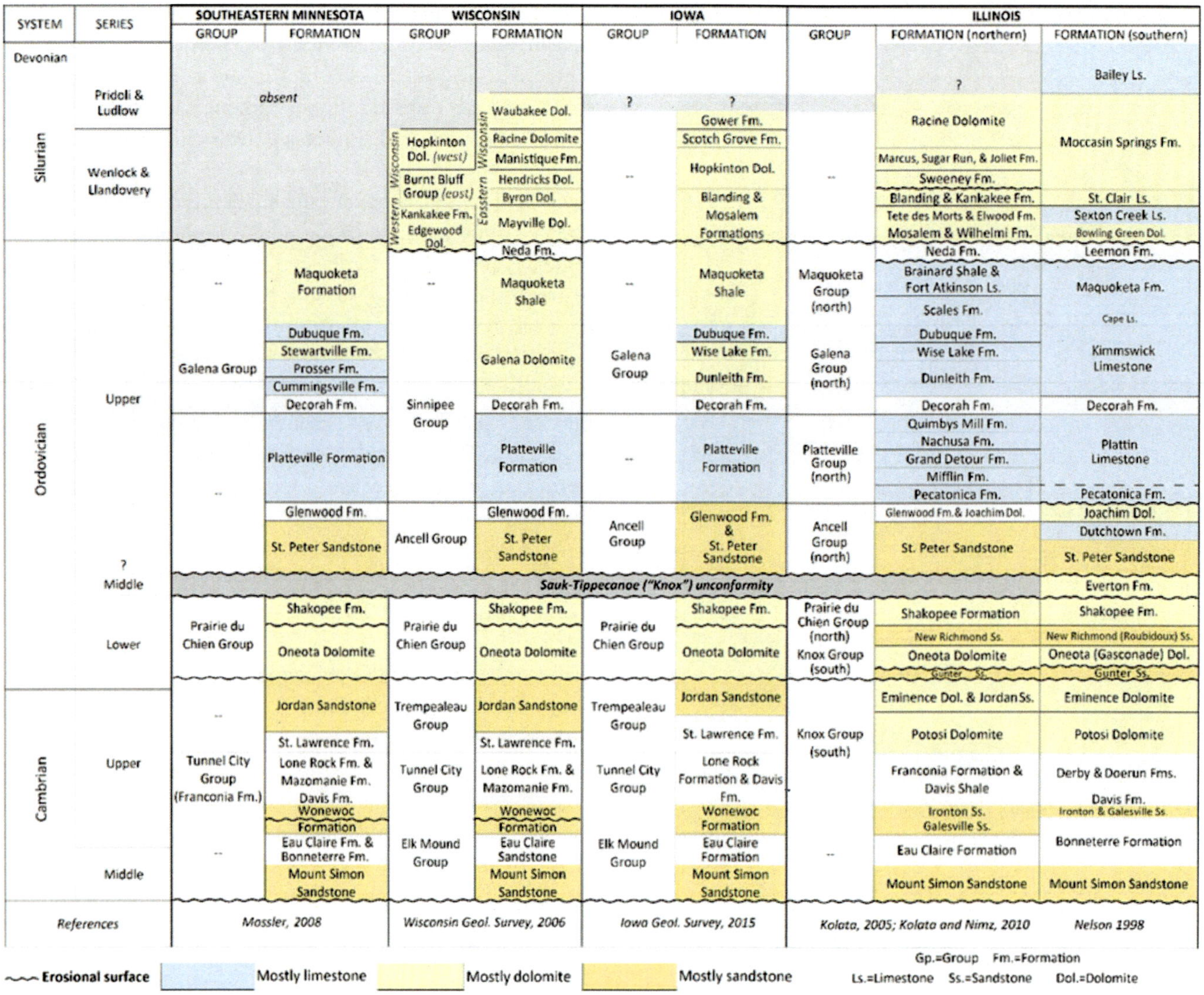

Fig. 1.6 Generalized correlation chart of the lower Paleozoic stratigraphy (Cambrian through Silurian) of Minnesota, Wisconsin, Iowa, and Illinois. References provided in the bottom row of the chart

exposed at this interval belong to the Knox Group, and thus this period of erosion is often referred to as the "Knox unconformity" (Fig. 1.6).

After this prolonged period of erosion, the surface was blanketed by the St. Peter Sandstone, an aeolian and near-shore marine sand that spread across much of the Upper Midwest (Mazzullo and Ehrlich 1987). The St. Peter Sandstone is composed of well-sorted, well-rounded, frosted, and pitted quartz grains that are weakly cemented by silica, carbonate minerals, and some iron oxides. As a result, the sandstone is friable where weathered, and often does not stand up well to erosion. Nonetheless, outcrops of this unit are frequently found, and may form low, rounded hills on the landscape. Breccia pipes have been found in the St. Peter where collapse has occurred into the underlying Prairie du Chien (Barr et al. 2008), possibly as a result of voids formed within the paleokarst surface at the Knox unconformity.

Atop the St. Peter Sandstone, shale of the thin and discontinuous Glenwood Formation was deposited, and above this, the Platteville Formation. Flaggy limestone layers of the Platteville Formation are conspicuous along the bluffs of the Mississippi River in the Twin Cities. Although the Platteville hosts few caves, several large caves in the cities of Minneapolis and St. Paul have formed within the St. Peter Sandstone and are capped with broad, flat ceilings held up by the Platteville (Brick 1997, 2009). In northern Illinois, the Platteville Group is subdivided into several formations based on the relative degree of argillaceous limestone and calcareous shale contained in the rock units and includes the Pecatonica, Mifflin, Grand Detour, Nachusa, and Quimbys Mill Formations. The Platteville is generally finer grained and thinner bedded than the rocks of the overlying Galena Group; however, the Pecatonica and Nachusa Formations resemble some parts of the

Fig. 1.7 Lower portion of the Shakopee Formation of the Prairie du Chien Group exposed in a roadcut near Chatfield, Minnesota (photo by D.H. Doctor)

Galena Group and may be mistaken for those younger rocks (Willman and Kolata 1978).

Overlying the Platteville is the Decorah Formation, the basal unit of the Galena Group. The Decorah is fossiliferous hosting well-preserved brachiopods, crinoids, bryozoans, and other fossils in a matrix of shale and fine-grained argillaceous carbonate. The Decorah Formation contains the well-known Deicke and Millbrig potassium-bearing bentonite beds that permit radiometric age determination and correlation with other rocks of this age across North America (Kolata et al. 1998). The Deicke has been dated using the $^{40}Ar/^{39}Ar$ technique on single grain biotite phenocrysts to 449 ± 2.3 Ma, and the Millbrig to 448 ± 2.0 Ma (Min et al. 2001).

The majority of caves in the UMV region, particularly in Minnesota, Iowa, and Wisconsin, are hosted by the carbonate rocks of the Galena Group that overlie the Decorah Formation. These rocks are generally more massive carbonate rocks than the older units. Significant zones of solutional development occurring along tensional fractures (joints) and bedding planes in these rocks have been documented throughout the UMV region. In Minnesota, solutional porosity tends to be concentrated at the contact between the Cummingsville and overlying Prosser Formations of the Galena Group (Runkel et al. 2003). Larger caves are also found in the Stewartville Formation of the Galena Group, including Mystery and Niagara Caves, two developed show caves in Minnesota. The Stewartville is primarily a dolostone and has distinctive *Thalassinoides* trace fossil burrows. Interbedded limestone and shale of the Dubuque Formation overlie the Stewartville in southeastern Minnesota, and these grade into thinner beds of shaly limestone.

Above the Galena Group is the Maquoketa Formation, a shaly dolostone in Minnesota and Iowa, mostly shale in Wisconsin, and shaly limestone in Illinois. Although some karst features are known in the Maquoketa in Minnesota, the unit becomes increasingly shale-dominated toward the east and is essentially a shale in Wisconsin such that it forms a regional aquitard.

1.3.3 Silurian System

Silurian rocks are generally poorly exposed throughout much of the UMV region due to glacial cover. The most significant karst within Silurian units occurs in the Door Peninsula of northeastern Wisconsin, parts of northeastern Iowa, and parts of northeastern Illinois (Fig. 1.2). The outcrops of Silurian strata occur on the edges of two large regional basins; thus, the Silurian strata are not well correlated between Iowa, Illinois, and Wisconsin. In eastern Wisconsin and northeastern Illinois, the Silurian rocks outcrop in an arcuate pattern that defines the western edge of the Michigan Basin (Fig. 1.2). In Iowa and northwestern Illinois, the Silurian rocks are found along the northwestern edge of the Illinois Basin. The Silurian strata were extensively dolomitized during diagenesis.

In Door County, Wisconsin, the Silurian is primarily represented by the rocks of the Burnt Bluff Group. Soderman and Carozzi (1963) reported that "the Burnt Bluff Group of

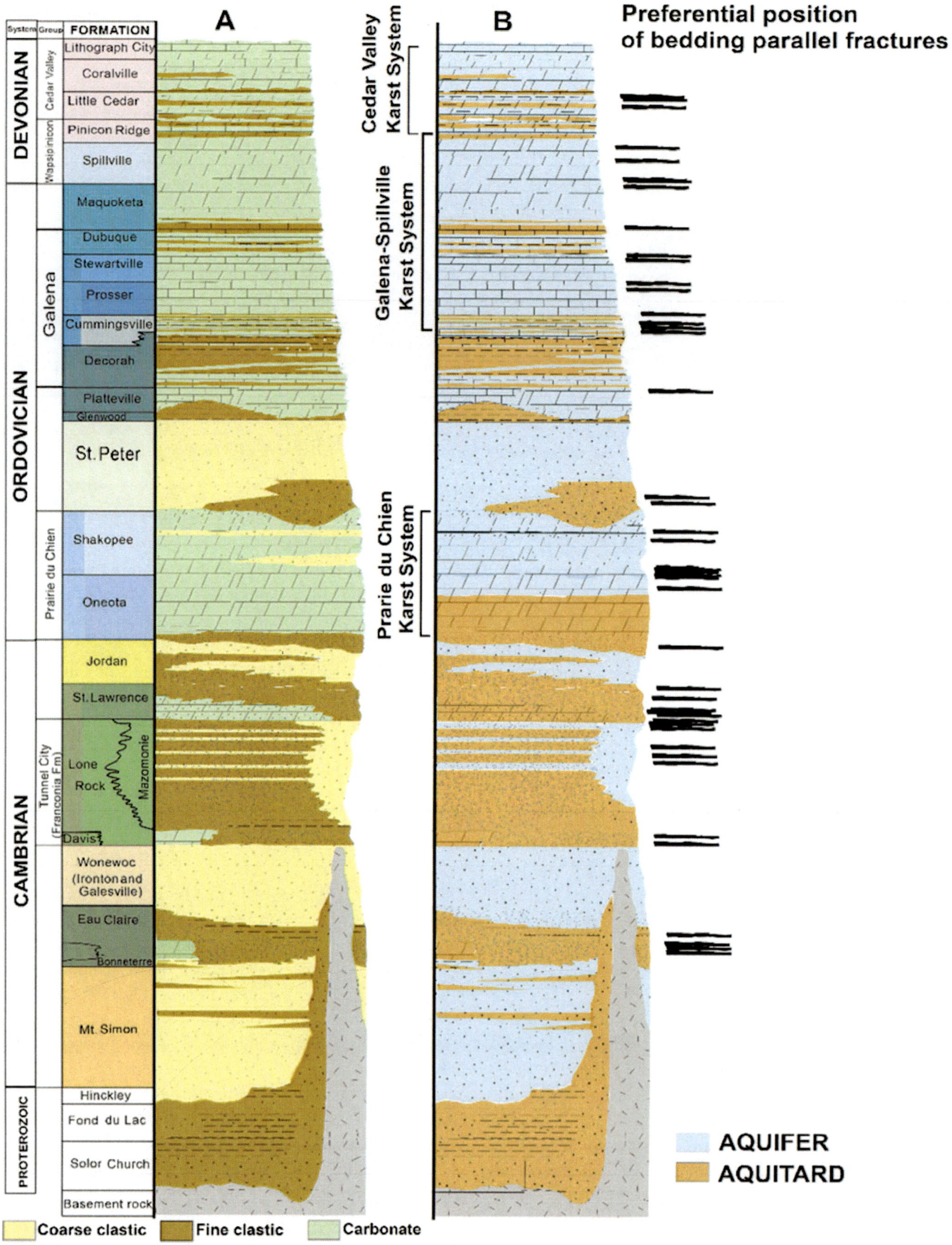

Fig. 1.8 Paleozoic bedrock stratigraphy in southeastern Minnesota. Column A illustrates the lithostratigraphic components and column B illustrates the hydrogeologic units. Also shown are stratigraphic positions of the three major karst systems and preferential positions of bedding parallel fracture networks (from Runkel et al. 2014)

Wisconsin consists of approximately 30 m (100 ft) of predominantly thin-bedded dolomite and some partings and thin beds of clay, is apparently conformable with the underlying Mayville Dolomite and the overlying Manistique Group, and correlates approximately with the Burnt Bluff Group of Michigan and the central part of the Kankakee Dolomite of Illinois" and that "subdivision of the Burnt Bluff Group in Michigan is based on paleontological data (Ehlers and Kesling 1957), but a lack of fossils in the sections studied in Wisconsin precludes the use of this method."

Karst features in the peninsula of Door County, Wisconsin, occur in several formations (Johnson and Stieglitz 1990) but primarily on the western side of the peninsula within the Byron and Hendricks Formations of the Burnt Bluff Group (Rosen and Day 1990). The Maquoketa Shale underlies the Silurian dolostones and serves as a regional aquitard (Rayne et al. 2001).

Tecumseh Cave, one of the longest known caves in Wisconsin with 740 m (2,428 ft) of explored passages, is a solutional cave on the Door Peninsula; the second longest cave on the peninsula, Dorchester Cave, is formed in the Manistique Formation, but is thought to have formed through gliding of a large rock block under glacial stresses (Johnson and Stieglitz 1990). Fracture orientations in Door County are likely controlled by the present-day stress field or past stress fields associated with the Alleghanian and Ouachita orogenies (Underwood et al. 2003).

Silurian strata in the Lincoln Hills region of southern Illinois host karst features primarily within the Sexton Creek Limestone (Panno et al. 1997, 2020 this volume). To the northwest in the Driftless Area of Illinois and Iowa, the Silurian units that host karst features include the Mosalem, Tete des Morts, Blanding, and Hopkinton Formations, with most sinkholes occurring in the dolomitic Hopkinton Formation (Panno et al. 1997; Brick 2004). In northeastern Illinois, most of the bedrock at the surface is Silurian in age and is located on the northeast flank of the Kankakee Arch, which forms the division between the Illinois Basin and the Michigan Basin. Numerous karst features occur within Silurian rocks in this part of Illinois, particularly within the Racine Dolomite. Much of the bedrock surface is buried beneath glacial till and other sediments but would seem to host an intrastratal karst. Bretz (1940) reported a cave intersected by quarry activities in the Joliet Dolomite as having slumped fill of Pennsylvanian shale from an overlying unit, and concluded "that not all buried 'sink holes' at the contact of a limestone and an overlying shale formation are relicts of a former karst topography, that some such 'sinks' may develop de novo or may become greatly enlarged by solutional removal of limestone after the shale was deposited" (Bretz 1940, p. 1) (Fig. 1.9).

1.3.4 Devonian System

Rocks of the Lower Devonian are only encountered in southern Illinois; lack of deposition or erosion caused an absence of Lower Devonian rocks in Minnesota, Iowa, and Wisconsin. The Lower Devonian rocks are sometimes referred to as the "Helderbergian" sequence in Illinois (Willman et al. 1975) and correlate to the well-known karst-bearing rocks of the Helderberg Group in the eastern states of New York, New Jersey, Pennsylvania, Maryland, Virginia, and West Virginia. In the Shawnee Hills region of southern Illinois, a few sinkholes and caves are found within the Lower Devonian Bailey and Backbone Limestones, and in the Middle Devonian Grand Tower and Lingle Limestones of the western Shawnee Hills (Panno et al. 1997, 2020, this volume).

Middle Devonian carbonates that host karst features occur primarily in strata of the Wapsipinicon and Cedar Valley Groups in southeastern Minnesota, northeastern Iowa, and northwestern Illinois. In Wisconsin, Devonian carbonates of the Thiensville and Milwaukee Formations crop out at the surface adjacent to Lake Michigan in only small areas (Fig. 1.2).

In Minnesota, the Devonian carbonates and shales of the Wapsipinicon and Cedar Valley Groups are the first bedrock under the southwest corner of Fillmore County and most of Mower County in Minnesota (Mossler 1998a). However, in a broad northwest/southeast band across Mower County and the southwestern corner of Fillmore County, the Devonian bedrock is covered by more than 15 m (50 ft) of pre-late Wisconsinan glacial drift and younger alluvium (Mossler and Hobbs 1995; Mossler 1998b; Meyer and Knaeble 1998).

The Middle Devonian Spillville Formation lies above a regional unconformity that sits atop the Upper Ordovician Maquoketa Formation. Together with the Galena Group rocks, the Spillville forms a regionally continuous horizon of karst development, the Galena-Spillville karst (Runkel et al. 2014; Fig. 1.8). Surface karst features are present in Devonian bedrock in western Fillmore County (Witthuhn and Alexander 1995), as well as along the eastern edge of Mower County, especially in and around Le Roy (in the Lithograph City Formation) in the southeastern corner along the Upper Iowa River and in Mower County along the Cedar River and its tributaries (in the Little Cedar Formation) around and south of Austin, Minnesota (Green et al. 2002). The surface karst features include sinkholes, springs, caves, and a blind valley. For example, Junk Hole Cave is located just north of Cherry Grove in Fillmore County, in the lower part of the Spillville Formation (Brick 1989). The largest blind valley in Minnesota, the York Blind Valley, in Sect. 21 of York Township, is also in the Spillville Formation.

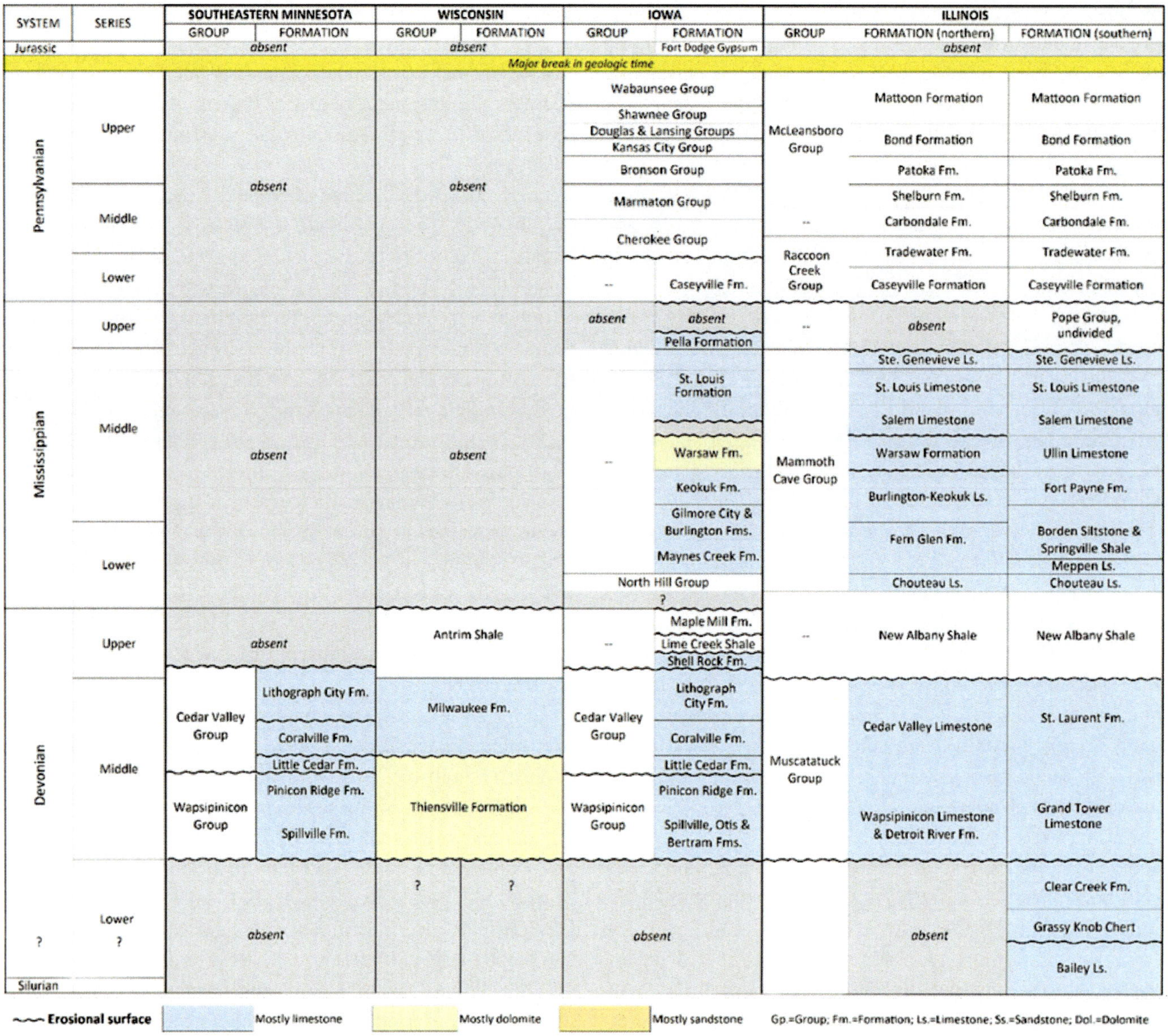

Fig. 1.9 Generalized correlation chart of the upper portion of the Paleozoic (Devonian through Pennsylvanian) stratigraphy in Minnesota, Wisconsin, Iowa, and Illinois. References provided in the bottom row of the chart in Fig. 1.6

The topography over Devonian bedrock in Minnesota is relatively flat and few natural outcrops exist. Several aggregate quarries mine Devonian carbonates and the walls of those quarries reveal the carbonates to be highly jointed with well-developed horizontal bedding partings. While the land surface is relatively flat, the alluvium/bedrock surface is highly irregular at the few meter scales with solution-produced cutters and grikes (Green et al. 2002). The sinkholes and springs in Devonian strata in western Mower County around and south of Austin are the northern end of a larger array of sinkholes extending south-southeast across Iowa. The sinkholes and springs around Le Roy in the southeastern corner of Mower County are the northern end of a second large array of sinkholes extending south into Iowa.

1.3.5 Mississippian System

The Mississippian strata of North America host the richest development of caves and karst in the conterminous United States (Palmer and Palmer 2009). However, in the UMV region, rocks of Mississippian age are exposed only in southwestern Illinois and in Iowa. These rocks host the majority of well-developed surface karst and caves in Illinois, particularly within the Shawnee Hills karst area (Panno et al. 1997).

Underlying the Mississippian strata in Illinois and Iowa is the Devonian New Albany Shale. These shales were rich in organic matter and provided the source rock for petroleum reservoirs contained within the overlying carbonate units.

Gypsum-bearing evaporite deposits are known within both the Illinois Basin and Forest City Basin (Fig. 1.3).

Several units bearing sinkholes and caves in Illinois and Iowa include the Salem, St. Louis, and Ste. Genevieve Limestones, which are the same units that host Mammoth Cave in Kentucky. Sinkholes are also found within the Glen Dean and Menard Limestones of the overlying Pope Group. For a detailed discussion of karst development in the Mississippian strata in Illinois, see Panno et al. 2020 (this volume).

1.3.6 Pennsylvanian System

Pennsylvanian strata of south-central Iowa commonly exhibit repetitive depositional cycles consisting of limited amounts of carbonate units intercalated with siliciclastic and coal-bearing sediments. Carbonates are known to occur throughout much of the Pennsylvanian section, but particularly within the Bronson and Kansas City Groups.

Due to the thin and often shaly nature of these beds, carbonates in the Pennsylvanian strata are unlikely to form caves, or even to result in sinkholes at the land surface. However, solutional voids within these units are known to contain paleokarst features that may become excavated and convey groundwater at elevated velocities. Therefore, these karst features may be of interest for water resource management and protection.

For example, a 22 foot (6.7 m) thick section of the Bethany Falls Member of the Swope Formation of the Bronson Group was measured along the south high wall at the Decatur City Quarry, Decatur County and was described as being "heavily karsted," possibly with solutionally formed root holes (Pope and Marshall 2010). Apparently, this was a paleokarst surface, as the unit is described as having an irregular contact with the overlying Galesburg Shale (Fig. 1.10).

1.4 Paleokarst

The development of karst within the Paleozoic strata is partially controlled by periods of surface exposure and erosion that interrupted the deposition of the sediments. These periods are gaps in the geologic depositional record when sediments were either not deposited, or deposited and subsequently removed by erosion before the overlying sediments were deposited. More importantly, these unconformities and lesser disconformities within the stratigraphy of the UMV region have permitted development of paleokarst horizons in the geologic past. The paleokarst resulted from the exposure of the carbonate formations at the land surface, before, during, and after the formations were buried by overlying sediments. Karst processes have been episodically modifying these Paleozoic strata throughout their geologic history (Hedges and Alexander 1985; Kluessendorf et al. 1988).

A major episode of paleokarst formation is the Sauk-Tippecanoe megasequence boundary, represented by the St. Peter Sandstone overlying the carbonate rocks of the Prairie du Chien Group. This sequence boundary is correlated to the Knox unconformity of eastern North America. Paleokarst at this boundary is visible in outcrops below Bridal Veil Falls at Pikes Peak State Park in Clayton County, Iowa where the St. Peter Sandstone fills a paleokarst basin in the underlying dolomite of the Prairie du Chien Group (McKay 1996).

Lower in the Ordovician Prairie du Chien (OPDC) section near the boundary between the Oneota Dolomite and the overlying Shakopee Formation is an extensive, regionally mappable karst weathering zone (Fig. 1.11). This "OPDC High Transmissivity Zone" (Tipping et al. 2006) is a major conduit for long distance, rapid groundwater movement in the Prairie du Chien (Fig. 1.9). Many of the sinkholes, caves, and other karst features are reactivating and modifying much older paleokarst features. For example, three of the roughly 20 wastewater treatment lagoons built in the 1970s and 1980s to handle municipal waste from small towns in southeastern Minnesota catastrophically failed when sinkholes opened and drained the lagoons. All three of the lagoon failures, Altura in 1974 and 1976 (Alexander and Book 1984), Lewiston in 1991 (Jannik et al. 1992) and Bellechester in 1992 (Alexander et al. 1993) were constructed directly above the OPDC High Transmissivity Zone. The sinkholes apparently drained into the paleokarst horizon.

In Illinois, several zones of paleokarst development have been documented in rocks of varying ages (Illinois Paleokarst fills: https://www.kenig-ogl.org/paleokarsts-and-sediment-fills.html, accessed Feb. 20 2020). Paleokarst in Illinois contains fossil pollen and plant remains of Pennsylvanian age (Plotnick et al. 2009). The caves have been discovered during quarry operations in the northeastern part of the state, primarily in Silurian and Devonian strata. Bretz (1940) described solutional cavities in the Silurian Niagara Dolomite in a quarry near to Joliet, Illinois. In addition, a significant and widespread paleokarst surface has been recognized at the top of the Silurian Kankakee Formation in northeastern Illinois, and extending into southeastern Wisconsin over an area of ~10,000 km^2 (~3,860 mi^2) (Kluessendorf and Mikulic 1996).

Evidence for a Pennsylvanian karst surface on top of Silurian dolostone in northeastern Illinois has been described by Bretz (1940). Sediment-filled paleokarst caves were discovered in 2004 in the Central Limestone Company Quarry, Kendall County, Illinois (Plotnick et al. 2008). The host

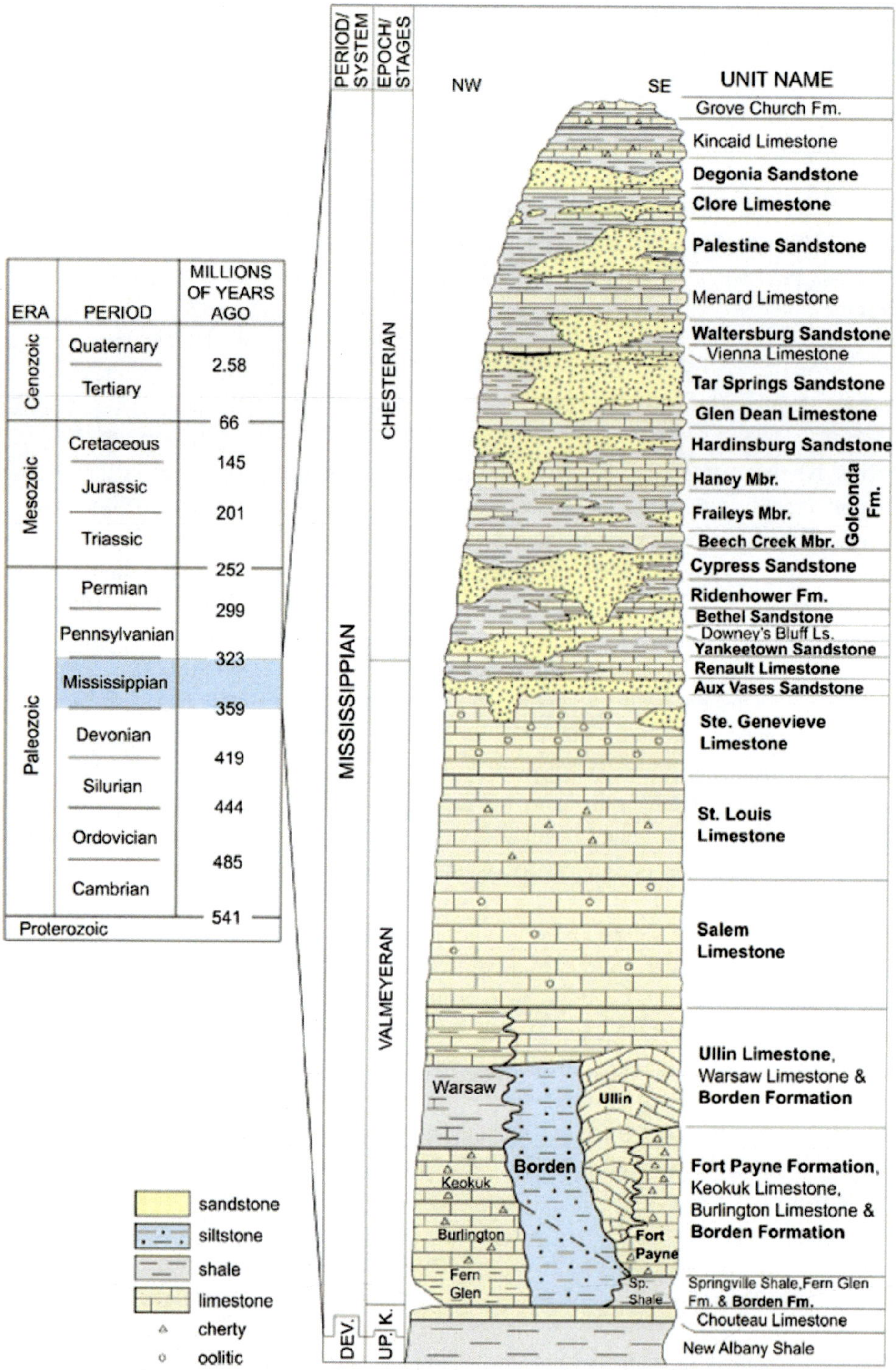

Fig. 1.10 Mississippian stratigraphy of southern and western Illinois. Units that have produced oil are shown in bold type. The base of the Fort Payne Formation is approximately equivalent in time to the base of Keokuk Limestone (dashed line) (Kolata and Nimz 2010). DEV, UP = Upper Devonian, and K = Kinderhookian (from the Illinois State Geological Survey: https://isgs.illinois.edu/outreach/geology-resources/mississippian-rocks-illinois)

rocks are Upper Ordovician limestones of the Dunleith Formation and Galena Group. These caves contain sediments of Pennsylvanian age. The sediment fills apparently represent a speleogenetic process via paragenesis, whereby the sediments in-filled passages nearly contemporaneously with the passage formation. Fossil pollen and plant remains of early North American conifers are contained in the fills, and remains of a scorpion were also found (Plotnick et al. 2015). These paleokarst features appear to be correlative to paleokarst that formed after deposition of the Kaskaskia, and thus likely formed in the latest Mississippian or earliest Pennsylvanian (Plotnick et al. 2008).

In northwestern Illinois, paleokarst features are known from Devonian limestones of the Rock Island area (Leary 1981; Leary and Trask 1985). A Devonian paleokarst surface has also been recognized at the top of the Thiensville Formation in eastern Wisconsin (Kluessendorf et al. 1988).

In Iowa, paleokarst occurrences have been documented in several portions of the eastern part of the state. For example, the Linwood Mine in Scott County contains mineralized

Fig. 1.11 OPDC high transmissivity zone: large solution cavities aligned along the Oneota Dolomite-Shakopee Formation contact, Wabasha County, Minnesota. Person in lower left for scale (Runkel et al. 2014)

zones of paleokarst cavities (Garvin 1995). In addition, numerous paleokarst features with fills of Pennsylvanian sediment have been documented within the Devonian Cedar Valley Group in the Klein and Conklin quarries near Iowa City (Marshall and Witzke 2010).

1.5 MVT Ore Mineralization

Mississippi Valley Type (MVT) lead and zinc ore deposits are associated with the Paleozoic carbonates of southwestern Wisconsin, northwestern Illinois, and northeastern Iowa. The most comprehensive work describing the geologic context of the deposits was by Heyl et al. (1959). The ore minerals within the deposits are galena (PbS) and sphalerite (ZnS), with lesser amounts of iron and copper sulfides; fluorite, barite, calcite, and dolomite are gangue minerals found in abundance. The ore deposits of the UMV are hosted by Ordovician rocks, primarily immediately overlying the St. Peter Sandstone. The Ordovician Platteville and Galena Groups are the main intervals of rocks that host the deposits, although some mineralization also occurs in the Prairie du Chien Group and in the lower Silurian rocks in northeastern Iowa (Heyl et al. 1959). For example, the Thompson-Temperly mine located in Lafayette County, Wisconsin, yielded abundant ore from the Quimbys Mill Member of the Platteville Formation, the Decorah Formation, and the lower part of the Galena Group (Heyl et al. 1959; Hatch et al. 1986).

Several studies have shown that the ore fluids were derived mainly from evaporated seawater (Leach et al. 2010). The prevailing model for ore emplacement is the movement of basinal brines through the platform carbonates by tectonic forcing during major orogenic events. Age dating of sphalerite using $^{87}Rb/^{86}Sr$ radiometric methods has supported this idea, and some sphalerite ores in Wisconsin have been dated to 270 ± 4 Ma and supported by paleomagnetic data (Pannalal et al. 2004). This age places ore emplacement during the Alleghanian orogeny, when the Appalachian and Ouachita Mountains were uplifted. Uplift of the Ouachita Mountains is thought to have been a potential driver for migration of deep basinal ore-bearing brine fluids to the north into the UMV where economic minerals were ultimately deposited (Appold and Garven 1999; Appold and Nunn 2005).

Deposition of lead and zinc sulfide ores is accompanied by carbonate mineral dissolution and precipitation of associated accessory and gangue minerals. Certain accessory minerals such as fluorite and barite are often economically recovered in addition to the lead and zinc sulfides; in 1965, fluorite mining was a multi-million dollar per year industry in Illinois and fluorite became the state mineral of Illinois (Illinois Geological Survey 2019).

The dissolution of the carbonate rock is evident in some caves intercepted during mining activities that contain wall coatings of these ore minerals (e.g., Garvin and Ludvigson 1993). A detailed description of the history of mining of these ores in the UMV region and associated karst development is provided by Dockal (2020), Chap. 7, this volume.

1.6 Evaporite Deposits

Deeply buried basin sediments host evaporites in the form of gypsum and anhydrite in the subsurface of Iowa and Illinois. In southern Iowa, these deposits occur within the Forest City Basin in the Silurian and Devonian Cedar Valley and Wapsipinicon Groups (Witzke et al. 1988). In addition to these deposits occurring in the same units within the Illinois Basin, gypsum occurs in the Mississippian strata within the St. Louis Limestone (Saxby and Lamar 1957).

The Fort Dodge Gypsum is perhaps the most well-known evaporite of the UMV region. Initially thought to be Cretaceous (Keyes 1893; Wilder 1903), the age of these deposits is now known to be Jurassic (Clark 2014). The thickness of the gypsum deposits is highly variable and ranges from 3 of 4 feet to more than 30 feet, with the average thickness being about 16 feet (Keyes 1893). The gypsum beds are unconformably overlain by glacial till and rest unconformably upon Pennsylvanian coal measures. The Fort Dodge Gypsum has been mined for many decades, and some quarries are still active (Clark 2014). A comprehensive discussion of the Fort Dodge Gypsum is provided by Lace et al. (2020) (this volume) on the karst of Iowa.

1.7 Pleistocene Glaciation

Several episodes of glacial ice sheet advance and retreat occurred in the Upper Midwest during the Pleistocene Epoch, from 2.58 million years to about 11,700 years ago, after which the earth warmed to its present state of interglacial climate (Fig. 1.12). The geologic record of these glacial episodes is contained in the sediments left behind once the ice sheets retreated.

Most, but not all, of the areas in the four states of the UMV region have been repeatedly covered by continental glaciation during the Pleistocene (Fig. 1.12). The geochronology of the most recent late Wisconsinan ice sheets is well established by various age dating techniques. The southern extent of the Wisconsinan ice cover is shown by the red line in Fig. 1.13. Multiple glacial cycles have profoundly modified the pre-Pleistocene karst landscapes and subsurfaces. Sinkholes and some caves were filled with glacial till. The upper portions of the epikarst were scraped off by the ice during the glacial advances leaving relatively level, flat carbonate rock surfaces in many areas. Enormous volumes of carbonate and non-soluble tills were left on top of the karst landscapes in other areas during the retreat of the glacial ice. Many cubic kilometers of meltwater flowed over and through the karst areas. The direction and magnitude of groundwater gradients in the karst repeatedly changed as the glaciers advanced and

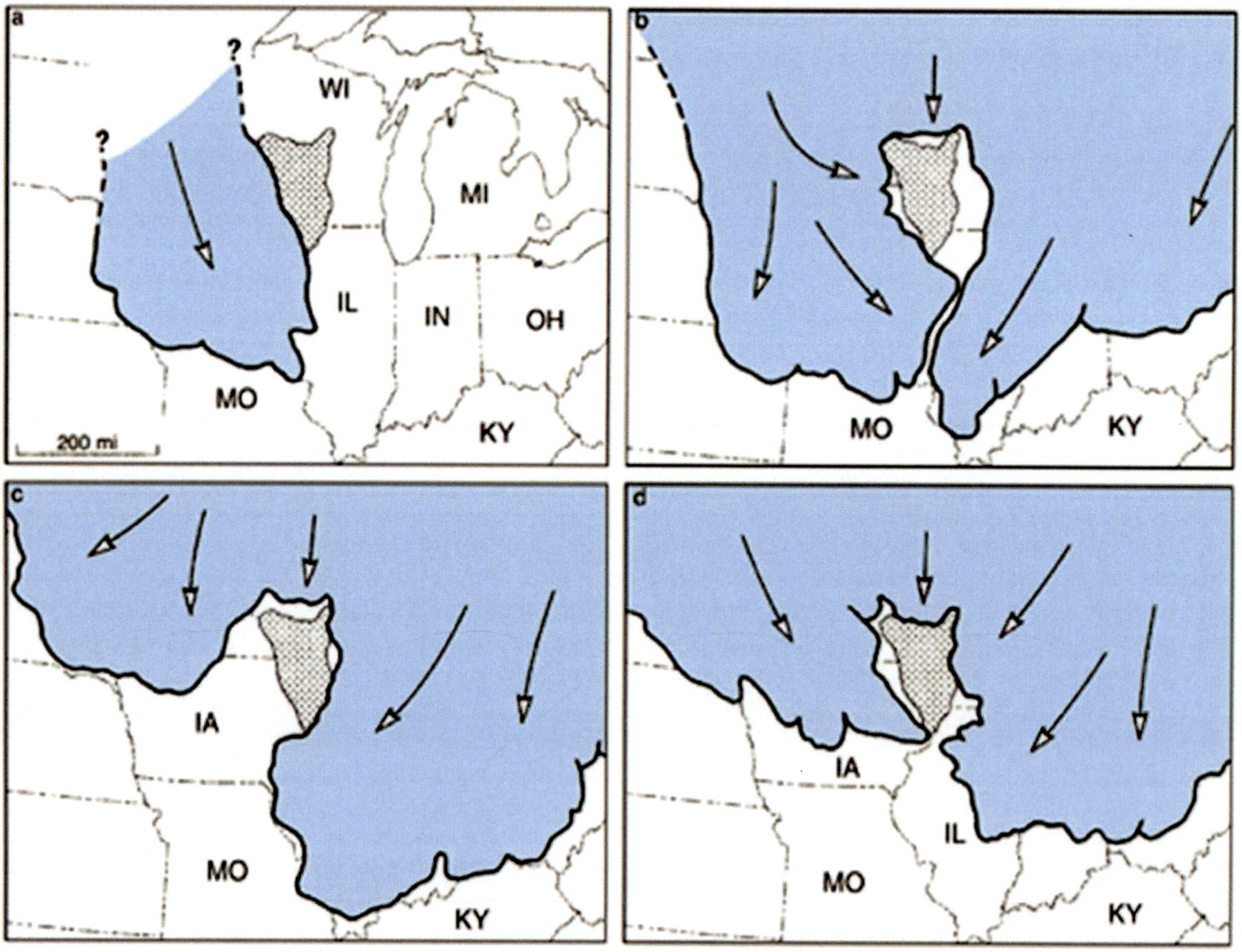

Fig. 1.12 Maximum extent of **a** early Pre-Illinois glacial episode (~1,000,000 years age); Driftless Area shown by stippled pattern; arrow indicates direction of ice movement; **b** late Pre-Illinois glacial episode (~600,000 years ago); **c** Illinois glacial episode (~250,000 years age); **d** late Wisconsin glacial episode (~22,000 years ago) (from Illinois Geological Survey: https://isgs.illinois.edu/outreach/geology-resources/glaciers-smooth-surface, accessed Feb 20 2020)

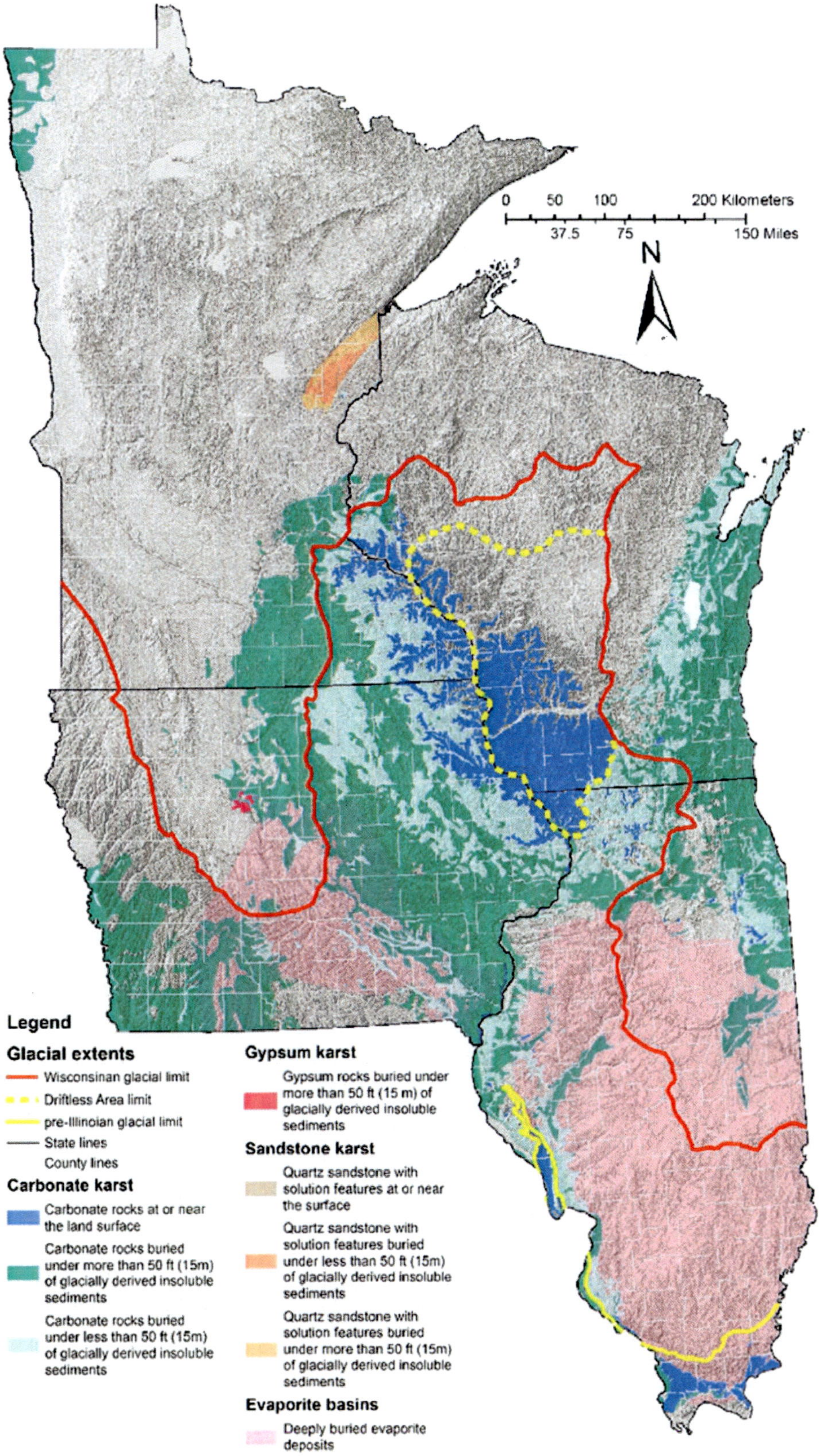

Fig. 1.13 Soluble bedrock units of Minnesota, Iowa, Illinois, and Wisconsin shown according to rock type and thickness of glacial sediment cover. Evaporite rocks deeply buried within basin are also shown. The limit of glacial ice advance during the last (Wisconsinan) ice age is depicted with the red line. Data from Weary and Doctor (2014)

retreated. The meltwater and those changing gradients moved enormous volumes of outwash through the systems, resulting in what have been termed glacially "deranged" karst systems (Ford and Williams 2007).

Deep bedrock valleys incised by ancient rivers during ice sheet retreat lie buried beneath glacial sedimentary cover that may be hundreds of meters deep. The major exception is the Driftless Area enclosed by the dashed yellow line in

Fig. 1.14 An outcrop of the Galena Group in southeastern Minnesota located to the west of the Driftless Area. Note the lack of well-developed epikarst; despite the lack of glacial sediments atop the bedrock, this area was impacted by glaciation in the Pleistocene. North-northwest dipping systematic joints cutting across near-horizontal bedding partings; these fractures likely resulted from stresses applied during the Alleghanian orogeny and were likely enhanced by stress-release after retreat of glacial ice. Photo taken along County Hwy 5 between the villages of Fillmore and Wyckoff in Fillmore County (photo by D.H. Doctor)

Fig. 1.13 in southwestern Wisconsin and the extreme northwestern corner of Illinois. This area has apparently never been covered with glaciers (the southern tip of Illinois may also have never been covered by glaciers). This Driftless Area has been influenced by the Pleistocene glacial advances and retreats. During the glacial maximum, permafrost, loess deposition, and other periglacial processes affected the Driftless Area. During ice melt-backs, glacio-fluvial sediments washed through parts of the Driftless Area.

Hobbs (1999) speculated: "The Driftless Area occupies the eastern part of the Paleozoic Plateau, a relatively high area of Paleozoic sedimentary bedrock that is generally permeable and is deeply dissected by the Mississippi River and its tributaries. Bedrock of the Paleozoic Plateau acted as a giant sieve that was able to dewater the base of advancing ice. The exposed bedrock created pinning points that inhibited ice advance across the Paleozoic Plateau. Ice therefore flowed around the eastern and western margins of the Driftless Area and continued its advance as far as southern Illinois and central Missouri." See also Iannicelli (2010) and references therein.

The area between the eastern lateral moraine of the late Wisconsinan Des Moines lobe and the Mississippi River in southeastern Minnesota and northeastern Iowa is too often referred to as part of the Driftless Area in scientific and popular usage (Fig. 1.13). However, the bedrock surface had been scraped off by ice such that there is little to no development of a thick epikarst horizon (Fig. 1.14). Discontinuous sheets of multiple weathered tills are also mapped in the surficial geology of the area. Meter-scale glacial erratics of igneous and metamorphic cobbles abound in the landscape indicating that the area was glaciated in the pre-Illinoisan Pleistocene (Fig. 1.15).

1.8 Age of the Karst Landscape

"How old is the cave?" is a question one often hears while exploring underground passages. The answer is usually not simple, nor easily determined. While an upper bound on the age of a cave is determined by the age of the bedrock and a lower bound by the age of the oldest deposits within the cave, the actual age of a cave is somewhere in between. The timing of the creation of a cave void is determined by the processes that control the overall evolution of the karst landscape that hosts the cave.

Similar to caves, we gain clues about the age of a karst landscape based on the oldest deposits that can be found within the "pockets" of the karst itself, the sinkholes. Sinkholes can be excellent repositories of geologic information, often containing deposits that have survived intact and in place for thousands of years, and occasionally for millions of years. In the Upper Midwest, the processes that formed the karst span millions of years, and some sinkhole deposits are similarly as old. Although the preservation of such old deposits is quite sparse, when discovered they provide invaluable insight into the age of the landscape.

For example, Sloan (1964) described the informal Iron Hill Member of the Cretaceous Windrow Formation in southeastern Minnesota as being a "heterogenous deposit of limonite (dominantly goethite) and admixed chert fragments, silt and clay. It forms bouldery, irregular masses, without noticeable bedding or other sedimentary structures, and contains local residual blocks of older carbonate beds. Typically, it occurs on a karst topography, primarily as fillings in enlarged joints and sinkholes or caves." The age of the formation is inferred from fossils of leaf imprints, carbonized wood, and shark teeth. Sloan (1964) also reported the presence of a Cretaceous fossil shark tooth found in place within a sinkhole pit that had been mined for clay and iron ore at Bellechester. These findings led to acknowledgement of preservation of some karst features within the landscape dating back to the Cretaceous.

In Missouri, a remarkable find in Bollinger County was unearthed in the early 1940s by the Chronister family while they were digging for a cistern in clay deposits that had filled a sinkhole. A geologist from the Missouri Geological Survey, Dan R. Stewart, who happened to be interested in such deposits was led to the site. The family had found and saved several bones, which were later described by Charles Gilmore of the Smithsonian Museum of Natural History as being

Fig. 1.15 Entrance to Goliath Cave in southeastern Minnesota. People are standing at the contact where loess deposits cover the bedrock exposures at the site (photo by D.H. Doctor)

vertebrae from the tail of a dinosaur, *Hypsibema missouriensis* (Gilmore and Stewart 1945). Since then, excavations at the site in the 1990s have resulted in numerous additional fossil finds, including specimens of other dinosaurs, fish, turtles, and plants. The Chronister site sits atop limestone likely of Mesozoic age but is surrounded by the Roubidoux sandstone of Ordovician age, and faulting is thought to have juxtaposed the formations (Gilmore and Stewart 1945). Although in Missouri, this site is at the same latitude and located within correlative stratigraphic rocks as the region covered by the other states in this volume where glaciation has obliterated evidence of earlier karst development. This find confirmed the presence of a Cretaceous-age erosional surface within the karst of Missouri, portions of which extend up the Mississippi Valley into southeastern Minnesota.

References

Alexander EC Jr (1980) Geology field trip. In: Alexander EC Jr (ed) An introduction to caves of Minnesota, Iowa and Wisconsin, Guidebook for the 1980 NSS Conv. Minn Geol Survey, St. Paul, pp 143–188

Alexander EC Jr, Brick GA (2020) this volume

Alexander EC Jr, Book PR (1984) Altura Minnesota lagoon collapses. In: Beck B (ed) Sinkholes: their geology, engineering and environmental impact, Balkema, Rotterdam, pp 311–318

Alexander EC Jr, Broberg JS, Kehren AR (1993) Bellechester Minnesota Lagoon collapses. Environ Geol 22(4):353–361

Alexander EC Jr, Wheeler BJ (2015) A Proposed hypogenic origin of iron ore deposits in southeast Minnesota Karst. In: Doctor DH, Land L, Stephenson JB (eds) National Cave & Karst research institute symposium 5, Sinkholes and the engineering and environmental impacts of Karst, Proceedings of the 14th multidisciplinary conference, Oct 5–9 2015, Rochester, MN, NCKRI, Carlsbad, NM, pp 167–175

Anderson JR, Runkel AC, Tipping RG (2011) Hydrostratigraphy of a fractured urban aquitard. Geol Soc Am Field Guides 24:457–475

Appold MS, Garven G (1999) The hydrology of ore formation in the Southeast Missouri district; numerical models of topography-driven fluid flow during the Ouachita Orogeny. Econ Geol 94(6):913–935

Appold MS, Nunn JA (2005) Hydrology of the Western Arkoma basin and Ozark platform during the Ouachita orogeny: implications for Mississippi valley-type ore formation in the Tri-State Zn–Pb district. Geofluids 5(4):308–325

Armitage JJ, Allen PA (2010) Cratonic basins and the long-term subsidence history of continental interiors. J Geol Soc 167(1):61–70

Bailey SW, Cameron EN (1951) Temperatures of mineral formation in bottom-run lead-zinc deposits of the upper Mississippi Valley, as indicated by liquid inclusions. Econ Geol 46:626–651

Barr K, Alexander EC Jr, Klimchouk A (2008) Hypogenic Karst and Its implications for Minnesota hydrogeology. In: Yuhr LB, Alexander EC, Beck B (eds) Proceedings of the eleventh multidisciplinary conference on sinkholes and the engineering and environmental impacts of Karst, Tallahassee, Florida, 22-26 Sept, pp 43–53

Bretz JH (1940) Solution cavities in the Joliet limestone of northeastern Illinois. J Geol 48(4):337–384

Brick GA (1989) Junkhole Junket: New Year's Day 1989. Minnesota Speleol Mon 21(2):20–21

Brick GA (1997) Classification and morphology of St. Peter sandstone piping caves in Minnesota. J Cave Karst Stud 59(3):168

Brick GA (2004) Iowa Underground: A Guide to the State's Subterranean Treasures. Trails Books, Black Earth, WI, 223 p

Brick GA (2009) Subterranean Twin Cities. University of Minnesota Press, Minneapolis, 226 p

Clark RJ (2014) Highlights of the unique geology of the Fort Dodge, Iowa, area. Iowa City: Geol Soc of Iowa Guidebook 92, 18 pp

Day M (2009) Wisconsin. In: Palmer AN, Palmer MV (eds) Caves and Karst of the USA, glaciated central lowlands. National Speleological Society, Huntsville, pp 143–145

Dockal JA (2020) Lead-Zinc crevices of Iowa, Illinois, and Wisconsin. Chapter 7, this volume

Ehlers GN, Kesling RV (1957) Silurian rocks of the northern peninsula of Michigan: Mich Geol Soc, Guidebook for Annual Geological Excursion (June), 63 p

Ford D, Williams PD (2007) Karst geomorphology and hydrology. John Wiley & Sons, London, 601 p

Fritz RD, Morgan WA, Longacre S (2012) Introduction. In: Derby JR, Fritz RD, Longacre SA (eds) The great American carbonate bank: the geology and economic resources of the Cambrian-Ordovician Sauk megasequence of Laurentia. AAPG Memoir 98, pp 1–3

Garvin PL, Ludvigson GA (1993) Epigenetic sulfide mineralization associated with Pennsylvanian paleokarst in eastern Iowa, USA. Chem Geol 105(4):271–290

Garvin PL (1995) Paleokarst and associated mineralization at the Linwood Mine, Scott County Iowa. J Iowa Acad Sci 102(1–2):1–21

Gilmore CW, Stewart DR (1945) A new sauropod dinosaur from the Upper Cretaceous of Missouri. J Paleontol: 23–29

Green JA, Alexander EC Jr, Marken WJ, Alexander SC (2002) Karst Hydrogeomorphic Units, Geol Atlas of Mower Co, MN, Co Atlas C-11, Part B, MN Dept Nat Res, St. Paul, MN, Plate 10

Hatch JR, Heyl AV, King JD (1986) Organic geochemistry of hydrothermal alteration, basal shale and limestone beds, middle Ordovician Quimbys Mill Member, Platteville Formation, Thompson-Temperly zinc-lead mine, Lafayette County, Wisconsin. In: Dean WE (ed) Symposium on organic matter and ore deposits. Denver Regional Exploration Geologists Society, Denver, pp 93–104

Hedges J, Alexander EC Jr (1985) Karst-related features of the upper Mississippi valley region. Stud Speleol 6:41–49

Heyl AV Jr, Agnew AF, Lyons EJ, Behre CH Jr, Flint AE (1959) The geology of the Upper Mississippi Valley zinc-lead district: US Geol Surv Prof Paper 309, 310 p

Hobbs H (1999) Origin of the driftless area by sub-glacial drainage–a new hypothesis. In: Mickelson DM, Attig JW (eds) Glacial processes past and present, Boulder CO, GSA Special Paper 337, pp 93–102

Iannicelli M (2010) Evolution of the driftless area and contiguous regions of midwestern USA through Pleistocene periglacial processes. Open Geol J 4:35–54

Illinois State Geological Survey (2019) Fluorite: Illinois' state mineral. https://isgs.illinois.edu/outreach/geology-resources/fluorite-illinois-state-mineral. Accessed 02 Feb 2020

Iowa Geological Survey (2015) Stratigraphic Column of Iowa, University of Iowa, 1 plate, https://www.iihr.uiowa.edu/igs/files/2015/09/stratcolumn-2015.pdf. Accessed 20 Feb 2020

Jannik NO, Alexander EC Jr, Landherr LJ (1992) The sinkhole collapse of the Lewiston, MN waste water treatment facility Lagoon. In: Quinlan J, Stanley A (eds) Proceedings of 3rd conference on hydrogeology, ecology, monitoring of ground water in Karst Terranes. NGWA, Dublin, OH, pp 715–724

Johnson SB, Stieglitz RD (1990) Karst features of a glaciated dolomite peninsula, door county, Wisconsin. Geomorphology 4(1):37–54

Keyes CR (1893) Gypsum deposits of Iowa. Iowa Geol Surv Annl Rpt 3:257–304 https://doi.org/10.17077/2160-5270.1313

Klein GD, Hsui AT (1987) Origin of cratonic basins. Geology 15 (12):1094–1098

Kluessendorf J, Mikulic DG (1996) An early Silurian sequence boundary in Illinois and Wisconsin. Geol Soc Amer Spec Paper 306:177–185

Kluessendorf JJ, Mikulic DG, Carman MR (1988) Distribution and depositional environments of the westernmost Devonian rocks in the Michigan Basin. In: McMillan NJ, Embry AF, Glass DJ (eds) Devonian of the World; proceedings of the 2nd international symposium on the Devonian system; Volume I, Regional synthesis: Canadian Society of Petroleum Geologists Memoir, 14, pp 251–263

Kolata DR (2005) Bedrock geology of Illinois. Illinois State Geological Survey, Map, p 14

Kolata DR, Huff WD, Bergström SM (1998) Nature and regional significance of unconformities associated with the Middle Ordovician Hagan K-bentonite complex in the North America midcontinent. Geol Soc Amer Bull 110:723–739

Kolata DR, Nimz C (eds) (2010) Geology of Illinois. University of Illinois at Urbana-Champaign, Institute of Natural Resource Sustainability, Illinois State Geological Survey, 530 pp

Lace et al (2020) this volume

Leach DL, Taylor RD, Fey DL (2010) A deposit model for Mississippi Valley-type lead-zinc ores. chap. A of Mineral deposit models for resource assessment: US Geological Survey Scientific Investigations Report 2010–5070–A, 52 p

Leary RL (1981) Early Pennsylvanian geology and paleobotany of the Rock Island County, Illinois, area: Part 1, Geology: Illinois State Museum. Rep Investig 37:1–88

Leary RL, Trask CB (1985) Early Pennsylvanian paleotopography and depositional environments, Rock Island County, Illinois: Illinois State Geological Survey, Guidebook 18, 42 pp

Marshall TR, Witzke BJ (2010) The Pennsylvanian strata of Johnson County, Iowa: channel and karst fills in Klein and Conklin Quarries. In: The Geology of Klein and Conklin Quarries, Johnson County, Iowa, Geol Soc Iowa Guidebook 87, p 117

Mazzullo JM, Ehrlich R (1987) The St. Peter Sandstone of southeastern Minnesota: mode of deposition. Minnesota Geol Surv Rep Investig 35:44–51

McKay RM (1996) Paleokarst in Iowa. Iowa. Geology 21:21–23

Meyer GN, Knaeble AR (1998) Surficial geology. Geol Atlas of Mower Co, MN, Co Atlas C-11, Part A, MN Geol Surv, St. Paul, MN, Plate 3

Min K, Renne PR, Huff WD (2001) 40Ar/39Ar dating of Ordovician K-bentonites in Laurentia and Baltoscandia. Earth Planet Sci Lett 185(1–2):121–134

Mossler JH (1998a) Bedrock Geology. Geol Atlas of Mower Co, MN, Co Atlas C-11, Part A, MN Geol Surv, St. Paul, MN, Plate 2

Mossler JH (1998b) Depth to the pre-cretaceous bedrock surface. Geol Atlas of Mower Co, MN, Co Atlas C-11, Part A, MN Geol Surv, St. Paul, MN, Plate 5

Mossler JH (2008) Paleozoic lithostratigraphic nomenclature for Minnesota. Minnesota Geological Survey Report of Investigations 65, St. Paul, 83 p

Mossler JH, Hobbs HC (1995) C-08 Geologic atlas of Fillmore County, Minnesota [Parts A and C]: Minnesota Geological Survey, 5 plates, 1:100,000 scale. https://hdl.handle.net/11299/58513

Nelson, WJ (1998) Bedrock geology of the Paducah 1 degree X 2 degree CUSMAP quadrangle, Illinois, Indiana, Kentucky, and Missouri, US Geological Survey Bulletin 2150-B, 1 plate, 1:250,000 scale

Palmer AN, Palmer MV (2009) Caves and Karst of the USA. National Speleological Society, Huntsville, AL, USA, p 445

Pannalal SJ, Symons DTA, Sangster DF (2004) Paleomagnetic dating of Upper Mississippi Valley zinc–lead mineralisation, WI, USA. Jour App Geophys 56(2):135–153

Panno SV, Weibel CP, Li W (1997) Karst regions of Illinois. Illinois State Geological Survey Open File1997-02, 42 p

Panno et al (2020) this volume

Plotnick RE, Kenig F, Scott A, Glasspool I (2008) Central Quarry: an exceptionally well-preserved Pennsylvanian paleocave and fill. In: Curry B (ed) Deglacial history and Paleoenvironments of Northeastern Illinois, 54th midwest friends of the pleistocene field conference, May 16–18 2008. DeKalb, Illinois, pp 63–70

Plotnick RE, Kenig F, Scott AC, Glasspool IJ, Eble CF, Lang WJ (2009) Pennsylvanian paleokarst and cave fills from northern Illinois, USA: a window into late Carboniferous environments and landscapes. Palaios 24(10):627–637

Plotnick RE, Kenig F, Scott AC (2015) Using the voids to fill the gaps: caves, time, and stratigraphy. Geol Soc London, Spec Pub 404 (1):233–250

Pope JP, Marshall TR (2010) Pennsylvanian geology of Decatur City and Thayer Quarries. In: The Pennsylvanian Geology of Southcentral Iowa, Guidebook 86, Geol Soc Iowa, pp 1–26

Rayne TW, Bradbury KR, Muldoon MA (2001) Delineation of capture zones for municipal wells in fractured dolomite, Sturgeon Bay, Wisconsin, USA. Hydrogeology J 9(5):432–450

Rosen CJ, Day MJ (1990) Glaciated Karst Terrain in the Door Peninsula of Wisconsin. Trans Wisconsin Acad Sci Arts Lett 78:39–44

Runkel AC, McKay RM, Cowan CA, Miller JF, Taylor JF (2012) The Sauk megasequence in the cratonic interior of North America: Interplay between a fully developed inner detrital belt and the central great American carbonate bank. In: Derby RD et al (eds) The great American carbonate bank: The geology and economic resources of the Cambrian-Ordovician Sauk megasequence of Laurentia, AAPG Memoir 98, pp 1001–1011

Runkel AC, Miller JF, McKay RM, Palmer AR, Taylor JF (2007) High-resolution sequence stratigraphy of lower Paleozoic sheet sandstones in central North America: the role of special conditions of cratonic interiors in development of stratal architecture. Geol Soc Am Bull 119(7–8):860–881

Runkel AC, Steenberg JR, Tipping RG, Retzler AJ (2014) Geologic controls on groundwater and surface water flow in southeastern Minnesota and its impact on nitrate concentrations in streams. Minn Geol Surv OFR14–02, St. Paul, 154 p https://hdl.handle.net/11299/162612

Runkel AC, Tipping RG, Alexander EC Jr, Green JA, Mossler JH, Alexander SC (2003) Hydrogeology of the Paleozoic bedrock in southeastern Minnesota: Minnesota Geological Survey Report of Investigations 61, 105 p, 2 plates

Runkel AC, Tipping RG, Meyer JR, Steenberg JR, Retzler AJ, Parker BL, Green JA, Barry JD, Jones PM (2018) A multidisciplinary based conceptual model of a fractured bedrock aquitard: improved prediction of aquitard integrity. Hydrogeol J 26(7):2133–2159

Saxby DB, Lamar JE (1957) Gypsum and anhydrite in Illinois. Circular No. 226

Shade BL (2002) The genesis and hydrogeology of a sandstone Karst in Pine County, Minnesota. MS thesis, U of MN, 171 p

Shade BL, Alexander EC Jr, Alexander SC (2015) The Sandstone Karst of Pine County, Minnesota. In: Doctor DH, Land L, Stephenson JB (eds) NCRI symposium 5, Sinkholes and the engineering and environmental impacts of Karst, Proc 14th Multidisciplinary Conf, Oct 5–9 2015. Rochester, MN, NCKRI, Carlsbad, MN, pp 157–166

Sloan RE (1964) The Cretaceous System in Minnesota. Report of Investigations 5, Minnesota Geological Survey, St. Paul, MN, 64 p

Sloss LL (1963) Sequences in the cratonic interior of North America. Geol Soc Am Bull 74(2):93–114

Soderman JW, Carozzi AV (1963) Petrography of algal bioherms in Burnt Bluff Group (Silurian), Wisconsin. AAPG Bull 47(9):1682–1708

Tipping RG, Runkel AC, Alexander EC Jr, Alexander SC, Green JA (2006) Evidence for hydraulic heterogeneity and anisotropy in the mostly carbonate Prairie du Chien Group, southeastern Minnesota, USA. Sed Geol 184:305–330

Underwood CA, Cooke ML, Simo JA, Muldoon MA (2003) Stratigraphic controls on vertical fracture patterns in Silurian dolomite, northeastern Wisconsin. AAPG Bull 87(1):21–142

Weary DJ, Doctor DH (2014) Karst in the United States: A digital map compilation and database. US Geological Survey Open-File Report 2014–1156 23 p https://dx.doi.org/10.3133/ofr20141156.

Wilder FA (1903) The age and origin of the gypsum of central Iowa. J of Geol 11(8):23–748

Willman HB, Kolata DR (1978) The Platteville and Galena Groups in Northern Illinois, Illinois Geological Survey Circular 502/1978, 75 p

Willman HB, Atherton E, Buschbach TC, Collinson CW, Frye JC, Hopkins ME, Lineback JA, Simon JA (1975) Handbook of Illinois Stratigraphy: Illinois State Geological Survey Bulletin 95, 261 p

Wisconsin Geological Survey (2006) Bedrock Stratigraphic Units in Wisconsin (Large poster). Open-file report WOFR2006–06LG, 1 plate

Witthuhn KM, Alexander EC Jr (1995) Sinkholes and Sinkhole Probability. Geol Atlas of Fillmore Co, Co Atlas C-8, Part B, MN Dept Natl Res, St. Paul, MN, Plate 8

Witzke BJ, Bunker BJ, Rogers FS (1988) Eifelian through lower Frasnian stratigraphy and deposition in the Iowa area, central midcontinent, USA. IN: Devonian of the World: Proceedings of the 2nd international symposium on the Devonian system—Memoir 14, Volume I: Regional Syntheses, pp 221–250

2 Exploration and Institutional History of Caves of the Upper Midwest, USA

Greg A. Brick

Abstract

Cave history among the states comprising the Upper Midwest, USA, involves three periods: (1) European exploration of the region, usually before 1850; (2) Euroamerican settlement in the century from 1850 to 1950; and (3) modern exploration by cave clubs or NSS grottoes, usually following World War II. The earliest mentions of caves in this region are often tied to French explorers along rivers in the seventeenth and eighteenth centuries. Some of the most famous show caves had their beginnings in the subsequent, settlement period. Finally, the project caving of dedicated clubs and the "pulsed paradigm" of exploration of the longer cave systems have continued to the present time. Ironically—contrasting with caves like Mammoth in Kentucky—the smaller the caves the larger their history: Cave in Rock (Illinois), Carver's Cave and Fountain Cave (Minnesota), Decorah Ice Cave (Iowa), and the caves of the Dells (Wisconsin) are good examples.

2.1 Introduction

Cave history among the states comprising the Upper Midwest, USA, involves three periods: (1) European exploration of the region, usually before 1850; (2) Euroamerican settlement in the century from 1850 to 1950; and (3) modern exploration by cave clubs or NSS grottoes, usually following World War II. Each state in turn will be reviewed from this tripartite perspective.

The earliest mentions of caves in the Upper Mississippi Valley Region are often tied to French exploration along rivers in the seventeenth and eighteenth centuries. Marquette and Joliet paddled major Midwestern rivers in 1673 but do not mention any caves, although they would have seen much St. Peter Sandstone en route. The French fur trader Pierre-Charles Le Sueur, in his ascent of the Mississippi River in 1700, described caves in his Journal. Caves were noted along the Ohio River by French explorers a generation later.

2.2 Minnesota

Much of the information for this section is derived from *Minnesota Caves, History and Lore* (Brick 2017) unless otherwise noted.

2.2.1 Explorers

Apart from purely mythological stories, such as the Dakota god of the underworld (Brick 2009a), the **Lake Pepin Saltpeter Caves** are the first mention of caves in Minnesota, and cave saltpeter in North America (Shaw 1992, p 52). The caves were described by the French fur trader Le Sueur in 1700 (Brick 2013). Various copper and lead mines are described on the resulting map (Brick 2016).

Carver's Cave, the baptismal font of Midwestern caving, was visited in 1766/67 by British colonial explorer Jonathan Carver, who made the 4,800-mile (7,723 km) roundtrip out from Boston. This sandstone cave, in what is now the city of St. Paul, was described in his journal and published in *Travels through the Interior Parts of North America*, a bestseller of the day (Fig. 2.1).

The most dramatic exploration narrative connected with Carver's Cave occurred in 1913. John H. Colwell, President of the Mounds Park Improvement Association, was appointed to its "exploration committee" and promptly set about relocating the cave, which had been "lost thirty years" owing to the accumulation of talus at the foot of the bluff.

G. A. Brick (✉)
Lands and Minerals Division, Minnesota Department of Natural Resources, 500 Lafayette Road, St. Paul, Minnesota 55155, USA
e-mail: drgregbrick@gmail.com

G. A. Brick and E. C. Alexander Jr. (eds.), *Caves and Karst of the Upper Midwest, USA*,
Cave and Karst Systems of the World, https://doi.org/10.1007/978-3-030-54633-5_2

Fig. 2.1 Interior of Carver's Cave showing the subterranean lake described by Jonathan Carver in 1766. Stereograph image by W.H. Illingworth, courtesy of the Trustees of the Boston Public Library

Colwell's goal was to commercialize the cave, stringing lights and building a flight of stairs down the bluffs from Short Street. To be sure, there had been earlier, ineffectual, attempts to make Carver's Cave into a show cave (Brick 2007b). But first, Colwell had to drain the 12-foot (3.7 m) deep lake inside Carver's Cave and explore it to the very end. This would not be an easy task, as others have found. While the lake water drained away, more was pouring into the rear of the cave through hidden conduits. It was difficult to find a spot low enough to drain the lake without deluging the nearby railroad tracks.

On December 15, 1913, Frank Koalaska, Colwell's erstwhile foreman—but now a rival—claimed to have found the "4th Carver's Cave." "The innermost chamber is 50 feet [15 m] high at one place," a newspaper reported. "The roots of trees growing on the bluff penetrate the walls, and there is a 20-foot [6.1 m] fall of pure water in it. A piece of clay pottery bearing Indian hieroglyphic inscriptions was found underneath the sand on the floor of the cavern." The room was dubbed "the most beautiful of any so far discovered." Perhaps because of this bitter feud, Colwell's original plans never came to fruition. A decade later, Colwell authored a series of eight articles, "The Story of Dayton's Bluff," which appeared in the *Minneapolis Tribune* in late 1924. "Carver's Cave," he concluded somewhat mysteriously, given all the media hoopla, "has never been officially explored."

A journalist, Charles T. Burnley, drafted a conjectural map of the alleged discoveries in 1913, and in his crude cartography, Carver's Cave resembled the stomach of a cow with its various chambers. The Burnley map would be the starting point for others many years later. Getting into those rooms—especially that elusive and spectacular waterfall room at the very back—was quite a draw for later generations (Brick 2000).

Carver's Cave was repeatedly lost to sight over the years owing to detritus sloughing down from the bluffs above. On September 16, 1977, the cave was again relocated and opened with a backhoe as part of an official city bicentennial project, and Native Americans visited it the next day. Double steel doors were erected, which in the coming decades were themselves buried by a debris fan of the sloughed detritus. In 2005, Carver's Cave was incorporated into the new Bruce Vento Nature Sanctuary, named for the congressman who represented the East Side of St. Paul for many years.

In 2016, the 250th anniversary of Carver's first visit to the cave, remotely controlled floating rovers with lights and video recorders were used to produce YouTube footage of the remote corners of the cave. Human explorers, whose feet stirred up the fine silt on the bottom of the subterranean lake—thus clouding the water—could not get this clarity of footage. Old graffiti could be read below the water line, and it was hoped that petroglyphs, too, would be found.

The U.S. Army also got into the business of cave exploration. On July 16, 1817, Major Stephen H. Long, of the newly created U.S. Corps of Topographical Engineers, disembarked from his "six-oared skiff" in what is now St. Paul to explore **Fountain Cave** (Brick 1995) (Fig. 2.2). Other explorers soon followed. Henry Rowe Schoolcraft (1793–1864) visited Fountain Cave in 1820, recording his observations in the *Narrative Journal of Travels*, published the following year. Mistakenly assuming that he had found Carver's Cave, Schoolcraft was understandably puzzled by the bizarre metamorphosis the cave seemed to have undergone in the half century since Carver had explored it. The French émigré scientist Joseph N. Nicollet visited Fountain Cave in 1837. It is marked "New Cave" on his famous 1843 map, *Hydrographic Basin of the Upper Mississippi River*—the so-called mother map of Minnesota.

The 1837 treaty with the Ojibway having opened for settlement the triangle of land between the St. Croix and Mississippi Rivers, Pierre "Pig's Eye" Parrant staked a claim at this cave because it was the nearest point to Fort Snelling that was not actually on the military reservation, thus shortening the distance for the soldiers to whom he sold whiskey. But since the platting of the city of St. Paul actually began in 1849 with "St. Paul Proper," in what is now the downtown area, and not at Fountain Cave, the traditional claim that Parrant founded the city is dubious. The geologist David Dale Owen (1807–1860) visited Fountain Cave in 1848. Having been educated in Switzerland, Owen poetically compared the snowy whiteness of the St. Peter Sandstone cave to a glacier cave. It was Owen, indeed, who coined the very term "St. Peter Sandstone," based on his study of rock outcrops along St. Peter's River—now the Minnesota River—at Fort Snelling.

The years from 1850 to 1880 were Fountain Cave's golden age. It became a fashionable Victorian cave—the first commercial or show cave in the upper Midwest. An 1856 letter described "the torch of birch-bark which your guide manufactures for the occasion."

The oldest known graphic depiction of a Minnesota cave is a pencil and watercolor of Fountain Cave by an unknown artist, looking into the entrance, about 1850. The most elaborate literary account of Fountain Cave at this time was

Fig. 2.2 Fountain Cave, named for the spring water flowing from its entrance, was the supposed birthplace of the city of St. Paul, MN. Stereograph image by W.H. Illingworth, courtesy of the Trustees of the Boston Public Library

presented in E.S. Seymour's *Sketches of Minnesota, the New England of the West,* published in 1850. Seymour establishes that the cave was basically an unbranched tube, wholly in the sandstone layer. Apart from widenings of this passage, called rooms, much of the passage was crawlway. There were four rooms successively decreasing in size upstream, of which he gave the dimensions. The third room back was the only named feature in the cave, called "Cascade Parlor," because it contained a waterfall two feet (0.6 m) high. He did not go beyond the fourth room, having traveled an estimated distance of 60 rods (302 m), but stated that he could hear a second waterfall in the distance.

By 1960, Fountain Cave lay directly in the path of a new highway, providing the engineers with a good place to dump "surplus excavated material" and seal the cave. The Minnesota Historical Society erected a historical marker for Fountain Cave over the buried ravine in 1963 (Brick 1995).

For discussion of the speleogenesis of Carver's and Fountain Caves, see Brick (2020), Chap. 8, this volume.

2.2.2 Settlers

Canton's Bear Cave. A famous bear story from the 1850s, as recorded in the biblically voluminous county histories of yesteryear, is noteworthy as the oldest known cave reference for Fillmore County that can be associated with a presently known karst feature (Kehret 1974). For chronological perspective, note that the first settler in Fillmore County arrived in 1851, the year that the Treaty of Traverse des Sioux opened southern Minnesota to Euroamerican settlement. Considering that the first geological survey of Fillmore County (1876) does not even mention caves, the next earliest known cave reference, after this 1850s bear story, is "George Bacon's Cave" (aka Petrified Indian Cave) with a record dating back to 1885.

There are two different published versions of the bear story. Neill's *History of Fillmore County* (1882) is less detailed but more readable, so it was the obvious choice for inclusion in Kehret's *Minnesota Caves of History and Legend* (1974). The cave's location is given as a mile and a half (2.4 km) north of the town of Boomer (Canton) and the date of the occurrence as 1856. Based on existing clues, the Bear Cave sinkhole was relocated (Brick 2005).

Petrified Indian Cave, or merely Indian Cave, crams into its short, tight passages, the kinds of archeological and historical associations that the promoters of Mystery Cave, its far larger neighbor, probably would have liked to have had.

Overlooking the South Branch of the Root River, in Fillmore County, the large, frost-shattered mouth of the cave narrows down abruptly to a tight corridor running back into the Galena limestone bluff for 240 feet (73 m). It received its name from the story told by settler George Bacon of how he went far into the cave as a youth in 1885 and saw a petrified Indian woman standing against the wall.

Upon exploring the cave in 1969, however, modern cavers, squeezing sideways, came across a flowstone deposit on the wall in the suspected location of the supposed Indian. They concluded that what Bacon had seen was a stalagmitic column fused up against the wall. The validity of the petrified Indian story notwithstanding, the cave does contain genuine artifacts. In 1970, archeologist Jerry Oothoudt found a hearth while excavating the entrance, but the dark zone of the cave does not appear to have been used. As early as the 1920s, archeologist Albert Jenks had recognized that the unglaciated southeastern corner of Minnesota, with its caves and rock shelters, had archeological potential, like the Les Eyzies Region of France, where so many art and bone caves had been discovered, and Oothoudt's work built on this insight (Brick 2017, pp 27–28).

Chute's Cave, on the banks of the Mississippi River near St. Anthony Falls, was the first and only show cave in the city of Minneapolis, giving boat tours from 1876 to 1880 (Brick and Petersen 2004) (Fig. 2.3a). About 60 m long and 30 m wide, a floor-to-ceiling collapse mound fills the greater part of the cave (Fig. 2.3b). Adelbert Russell Moore, engineer for the St. Anthony Falls Water Power Company, explored Chute's Cave in 1909 and drafted a map, describing the mound as consisting of "fallen lime rock in a very irregular shape, but on the whole somewhat resembling a huge fountain built up in tiers, over which water trickled, the water being impregnated with iron had colored the stone to almost a jet black giving it an extremely beautiful appearance." The water-laid mineral deposits described by Moore are well known to cavers as flowstone, and Moore's mound is assuredly what had been referred to as the "Tower of St. Anthony" in the late 1800s Nesmith Cave hoax, which was based on this cave.

2.2.3 Modern Exploration

Mystery Cave was first conceived in the mind of state geologist N.H. Winchell in 1876, based on where water entered the ground, and where it came back out again at springs. But in the days before organized caving clubs after World War II, such startlingly obvious cues were not often followed up on. It was left to the gravedigger Joe Petty to stumble across the cave quite by accident, more than 60 years later, in February 1937, when he saw vapor coming out of the hillside on a cold winter day. There are no legends associated with the cave, nor are there any prehistoric artifacts found within the cave, contrasting with the nearby, but much smaller, Petrified Indian Cave, suggesting that Petty's find was truly original (Brick 2018).

(a)
(b)

Fig. 2.3 **a** Chute's Cave was the first and only show cave in the city of Minneapolis with boat tours from 1876 to 1880. Stereograph image circa 1876 courtesy of Vic Thorstensen. **b** Chute's Cave showing flowstone-coated slabs of Platteville Limestone on the collapse pile filling much of the cave, and a fallen support timber. Photo by Greg Brick, 1993

Since 1918, there had been a picnic ground at Mystery Cave Park, on the South Branch of the Root River, in what is now Forestville State Park, south of the town of Wykoff. The namesake cave featured a ceiling spring, and a rock basin was built to catch the water, which was piped out to the picnic grounds. But the much longer cave lurked nearby, undiscovered for years. Once the longer cave was discovered, the name *Mystery Cave* was transferred to it from the first one, which then became Old Mystery Cave, nowadays a bustling raccoon den whose occupants were only too happy to take over the spring basin.

Mystery Cave is the longest known natural limestone cave in Minnesota and was the first to show the phased pattern of human discovery, afterward seen at Spring Valley Caverns. Thus, we have Mystery I, II, and III. Not surprisingly, the cave, with its 12 miles of passages, became the focus of an early cave club, the **Minnesota Speleological Survey (MSS)**, organized in 1962 by Clarence "Slim" Prohaska, the owner of the cave (Alexander 2012).

Mystery I, the original discovery, where owner Clarence Prohaska began tours in 1947, has passages in the Dubuque layer of the Galena Group. It features Cathedral Domes, Turquoise Lake, and the Bomb Shelter, among other named features. Mystery II was discovered in 1958 during the longest marathon trip inside a Minnesota Cave, lasting 84 h. During that trip, a new exit to the surface was dug at what is now the Mystery II entrance. Tours began at this second entrance to the system under the moniker Minnesota Caverns, in 1960, following passages in the Stewartville layer of the Galena Group. Fourth and Fifth Avenues and Garden of the Gods—a creamy rich welter of flowstone—are highlighted on these tours. Mystery I and II are connected by passages, called the Door-to-Door Route, which would not be welcome to the casual visitor but are a delight to cavers themselves. In 1967, Mystery III was discovered, with its Dragon's Jaw Lake and Eureka passage. Mystery Cave is unusual in that it went from private to public ownership (Fig. 2.4). In 1988, the cave was acquired by the Minnesota Department of Natural Resources and became part of Forestville State Park—which was renamed Forestville/Mystery Cave State Park. The old Mystery I tour route was refurbished, made handicap accessible, and a new visitor's center constructed.

The 1924 discovery of **Niagara Cave**, on the Minnesota-Iowa border near the town of Harmony, has a familiar structure—the lost animal narrative. Three boys found the cave while in pursuit of three pigs, which had fallen into a sinkhole leading to the cave. In 1932, three investors established the commercial tour, which has become the longest continuously privately operated show cave in Minnesota history. The tour leads the visitor down a long flight of steps, past the 60-foot (18.3 m) waterfall for which the cave is named. The name Niagara has historically been associated with weddings, which entitles this cave, with its Crystal Chapel, to be the most dramatic venue for cave weddings in Minnesota (Fig. 2.5).

Beyond the sump at the lower end of Niagara Cave, cavers clad in wetsuits dug mightily for years to find more passages. They even built narrow-gauge wooden trackways to haul off the sediment—trackways that floated away each time the cave flooded.

As compared with Mystery Cave, which has many passages off-limits to the casual tourist, most of Niagara Cave is visible during the tours. Accordingly, these two great caves provide a contrast in how cave features are named. Many of the caver supplied names found off the tourist routes in Mystery Cave relate to earth materials (mud, sand), the types of movements required to traverse cave passages (crawls, straddles), or stages of exploration (base camp, discovery route). By contrast, most of the names in Niagara Cave are aimed squarely at tourists, as indeed they are along the avowedly commercial routes in Mystery Cave.

Niagara Cave was considered interesting enough to have inspired the formation of its own cave club, the **Niagara**

Fig. 2.4 Scene in Mystery Cave, Minnesota's largest show cave, postcard from author's collection

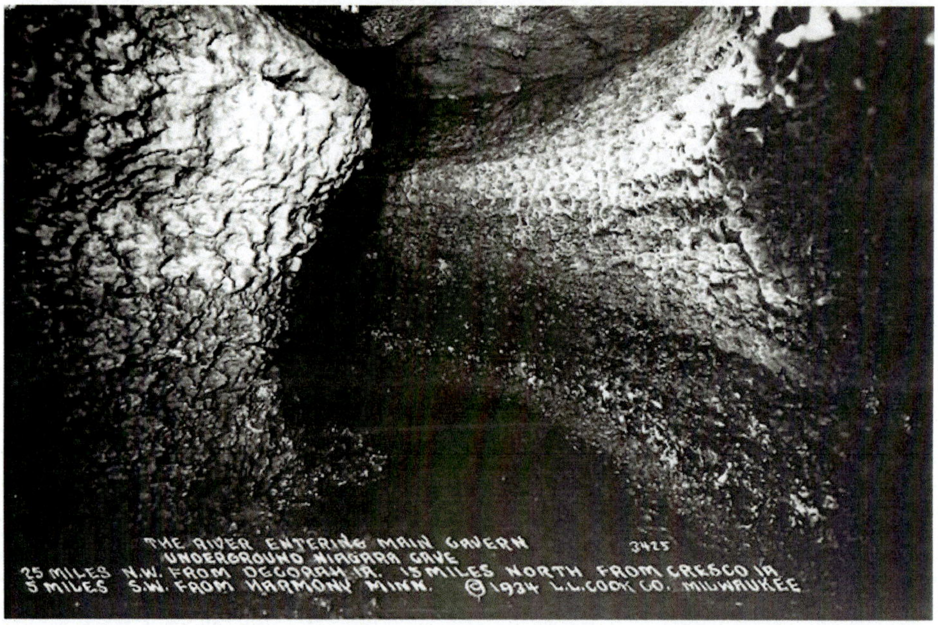

Fig. 2.5 Niagara Cave with its subterranean waterfall is Minnesota's best known cave for weddings, postcard from author's collection

Cavers (Brick 2007a). While now extinct, this club applied a whole separate nomenclature to Minnesota caves than that used by the current caving clubs.

Spring Valley Caverns is perhaps the best example among Minnesota Caves of what persistent weekend caving by organized cavers can accomplish in expanding a cave that was declared played out by bystanders. The cave jumped from one half mile (0.8 km) to more than 5.5 miles (8.8 km) of known passages within several years.

John Latcham found Spring Valley Caverns in the narrow ridge of Galena limestone separating Bear and Deer Creeks just north of the town of Spring Valley, Minnesota, in 1966, while searching for a lost calf. But according to neighbors it had been known locally before then. Called Latcham's Cave it was developed as a show cave, operating for two seasons (1968–69), under the name International Caverns and then Spring Valley Caverns. The Frozen Falls (a flowstone cascade on the cave walls) and the Medusa (a jellyfish shaped formation) are two noteworthy features advertised in early brochures for the cave. But the cave was too far off major highways and the property was eventually foreclosed upon. It sat vacant for decades thereafter.

The original Quonset hut over the sinkhole, whose floor had collapsed, was replaced by a sturdy stone building by caver John Ackerman, who acquired the property in 1989. Starting from the original tourist nucleus with a half mile

(0.8 km) of known passages, Ackerman pushed every lead in the cave by following wind and water with "Houdini-type tricks" and judicious application of explosives. I should know, as I helped to wire some of the charges, often while suspended precariously over some frightful gulf. The discoveries fit the pulsed paradigm, in five episodes, leading to the designations Spring Valley Caverns I through V for various portions of the cave. Ackerman installed culverts with ladders into man-made entry shafts that lead into the cave for the ease of accessing distant parts of the cave system. This cave is the usual Minnesota network maze on the prevailing intersecting joint pattern.

Spring Valley Caverns is now the show piece of Ackerman's Cave Farm. Ackerman harvested a crop of three dozen caves on this land over the decades, often using a trackhoe. The Cave Farm in turn is part of his Minnesota Cave Preserve, which includes caves scattered across Fillmore County. This was the foundation of Ackerman's **Minnesota Caving Club**, established in 2012 (Alexander 2012).

One of these other caves is **Tyson Spring Cave**, captured in stereopticon views and postcards from the early days, when it was a popular picnic spot. Spring water gushes from the entrance at 50 gallons per minute (0.16 cubic meters per second) at the base of a 120-foot (36.6 m) cliff of the Galena limestone. A commercialization attempt was made in the 1930s, but the cave was not thoroughly explored until the Argonaut Society and others came along with scuba gear in the 1970s, allowing explorers to get past the sump, or water-filled passages, to more walking passages beyond. Some of the ledges in the passages were lined with frogs, staring blankly at the determined explorers as they marched past. Ackerman installed a culvert to allow dry access beyond the sump to the estimated three miles of passages in this branchwork cave.

2.3 Wisconsin

This section has greatly benefitted from the assistance of Wisconsin speleohistorian Gary K. Soule.

2.3.1 Explorers

The earliest caves mentioned in Wisconsin are those of its perimeter, along the Great Lakes and rivers surrounding the state. Perhaps the earliest mention of a cave in Wisconsin is in the Journal of Le Sueur, in 1700, when he stopped at **Snake Cave** in 1700, notorious as a rattlesnake den, while ascending the Mississippi River (Reeder and Day 1990).

Joseph LeConte was a well-known nineteenth-century geologist. A 17-page typescript by LeConte, dated 1899, recounts a summer vacation spent among the Apostle Islands and its caves in 1844, which he took to be a textbook example of wave action (Brick 2002).

2.3.2 Settlers

In 1896, Tecumseh Cave, or **Horseshoe Bay Cave**, the longest undeveloped wild cave in the state, was discovered in the Niagara Escarpment of Door County. At more than 3,100 feet (945 m), and requiring the diving of sumps, it is now part a Wisconsin Department of Natural Resources bat hibernaculum, off-limits to most caving.

The so-called Driftless Area is the heart of Wisconsin cave country, however, as first described by Lange (1909) and Martin (1932), with caves such as Bear, Eagle, Blue Mounds, and John Gray. The Wisconsin River neatly divides the area, with caves in Prairie du Chien carbonates north of the river, from caves in the Galena Group, south of that river. **Bogus Bluff Cave** was the supposed lair of a vanished Indian race (a common theme in American cave folklore, see Brick 2009b) as well as counterfeiters. **Castle Rock Cave** was described in the visit of an early Wisconsin governor (Blaine 1919). Soule (2002) has documented an 1850 Wisconsin cave hoax that involves a cave in basalt with copper and lead deposits, which does not match the geology of that vicinity.

Several show caves sprang up in the Driftless Area and beyond. Small, romanticized sandstone caves attracted attention early at the Dells of the Wisconsin River (Brick 2007c) (Fig. 2.6). **Eagle Cave**, originally used for apple storage and long a favorite for Boy Scout camping, was discovered in 1849 and tours began in 1938, making it the first show cave in the state. **Cave of the Mounds**, 25 miles (40 km) west of Madison, was discovered by a blast in the Brigham Quarry—a longtime fossil collecting spot for geology students—and tours began the next year, 1940. As documented by Soule (1999), this cave has a rich tradition of postcards (214 varieties known) and brochures. **Kickapoo Indian Caverns**, on the crooked Kickapoo River in the heart of the Driftless, supposedly contained Indian artifacts, as noted by early lead miners. Tours began in 1947 until just recently. Crystal Cave, at the edge of the Driftless, was discovered near Spring Valley in 1881 but tours did not begin until 1942. While containing a mile (1.6 km) of passages, many of these are tight crawlways through the Oneota dolomite and only accessed during dig projects or spelunking tours.

2.3.3 Modern Exploration

The **Wisconsin Speleological Society (WSS)** was founded by Dick Kuhlen in 1960. Some of its stellar projects include

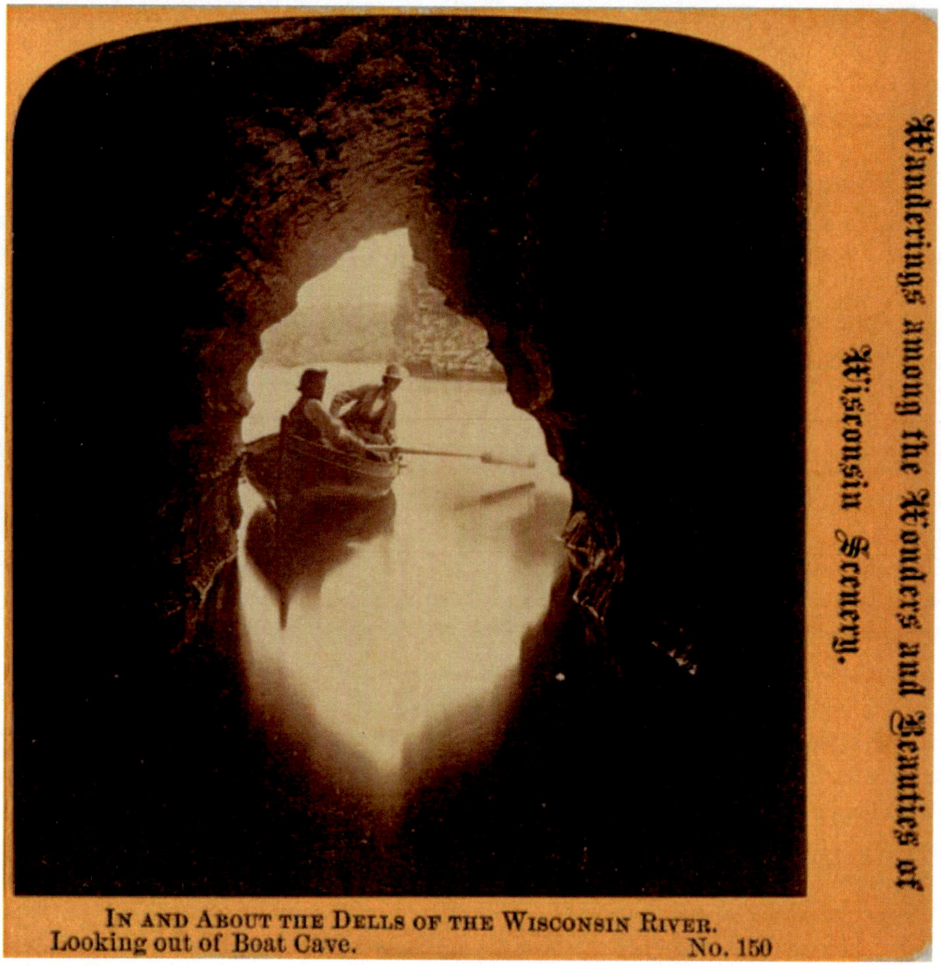

Fig. 2.6 The sandstone caves of the Wisconsin Dells inspired writers in neighboring states. Stereograph image courtesy of the Rhode Island School of Design Museum

the discovery by Gary Soule of **Paradise Pit Cave** in 1968, an excavation project that lasted for years, involving blasting and truck-mounted winches. The **Carolyn's Caverns System** and **Montgomery Cave** at Ledge View Nature Center have provisions for guided public touring (Rudy 1998). At Cherney Maribel Caves County Park, electrically lighted and guided walking tours of **Maribel New Hope Cave**, as well as crawling tours in the historic **Tartarus Cave System**, are periodically opened and offered to the public. Maribel was the location of a springwater spa in the first quarter of the last century, complete with a hotel, while the historic Tartarus Cave was the subject of a special study by Soule (2013). Other Wisconsin cave clubs were short lived, informal, or merely a post office box drop (Soule 2016).

Archeological caves have been a distinct thread in Wisconsin cave history since 1879, when a study of **Samuel's Cave**, near La Crosse, with its famous shaman petroglyph, included a unique investigation of the stratigraphy (Brown 1926). Archeologists have also been active in the documentation of pictograph caves such as **Arnold/Tainter Cave** and **Gottschall Rock Shelter** (Boszhardt 2003).

In its most philosophical aspect, William Cronon (1991), a Professor of environmental history at Yale, and longtime Wisconsin caver, has written about caves as providing a sense of place and stories about the landscape.

2.4 Iowa

Much of the information for this section is derived from *Iowa Underground* (Brick 2004) unless otherwise noted.

2.4.1 Explorers

On the Mississippi River, the lead mines in the Dubuque area were mentioned by Le Sueur in his ascent of the river in 1700. Lewis and Clark described a fish-fossil-containing cave on the banks of the Missouri River in what is now Iowa or Nebraska (Brick 2010).

2.4.2 Settlers

While hunting for bees and killing rattlesnakes one day in 1848, Lorenzo Dutton found a cave. Along with Falling

Springs, located on the opposite side of the town of West Union, it became a favorite spot for local picnickers. Reportedly, **Dutton's Cave** contained a lake on which people could row boats.

Early accounts described "a small wonderland" known as Burt's Caves, later dubbed Morehead's Caves. Early visitors mentioned the "milk-white stalactites" that adorned the caves, long since snatched away by greedy souvenir hunters. The area was a popular picnicking spot long before it became a state park, in 1921, at which time it was renamed **Maquoketa Caves**. The whole park was added to the National Register of Historic Places in 1991. The caves in the park were used by Native Americans as indicated by archeological finds. The recorded history of the caves began in 1837, when hunters chased a herd of deer up the Raccoon Creek Valley from the Maquoketa River. When they saw the deer enter what is now the lower entrance of Dancehall Cave, they concluded the deer were trapped in the cave and could be taken at leisure. Later, however, they found that the deer had long since just walked out the other side—the upper entrance of the cave.

James Sherman Minott was an American Civil War veteran who made his home in **Minott's Cave** along the Cedar River. He lived by fishing, hunting, and trapping, and was fond of giving bizarre names to the natural features around him—names like Blow Out Hollow and Screeching Sands Hollow. In the late 1890s, he bought land on the river and built a general store, hotel, restaurant, and boat livery. He sold lots for summer cottages. He played a key role in making the Palisades of the Cedar River a popular recreation area. It wasn't until 1922, however, a decade after Minott's death, that the Palisades became a state park, to which the Louis Kepler Memorial Area was added a few years later. The poet Carl Sandburg and his Cornell College students frequently picnicked in the park.

Decorah Ice Cave, in Decorah, Iowa, was known to the international scientific community in the late nineteenth century for its strange behavior: the formation of ice in spring, yet the cave was devoid of ice in winter. After an on-going discussion in *Scientific American* and other journals, Alois Kovarik, an Instructor at the Decorah Institute (now Luther College) finally explained the phenomenon along currently accepted lines. Few American caves have had such an influence on the development of cave meteorology (see Brick 2020, Chap. 8, this volume). The cave thereafter drew large crowds and was commercialized (see below), later being acquired as a city park (Hedges and Knudson 1975) (Fig. 2.7).

Fig. 2.7 Decorah Ice Cave, the celebrated "Cave of Paradox," photo by Greg Brick

At present, Iowa has two operating show caves, **Crystal Lake Cave** (commercialized in 1932) and **Spook Cave** (commercialized in 1955). However, at least five other Iowa show caves were open to the public in the past. The earliest was **Timmen's Cave**, located in Union Park, Dubuque. A 1909 postcard depicts tourists on an electrically lighted stairway inside the cave. Indeed, Union Park was initially developed by a trolley company to showcase the use of electricity, but it never really recovered from a devastating flood in 1919, and the cave was blasted shut.

The 1930s were the golden decade of show caves in Iowa. **Decorah Ice Cave** was commercialized by Stanley Scarvie in 1929 and operated until 1941. Scarvie strung lights in the cave and built a zoo nearby. He also operated **Glenwood Cave**, nine miles (14.5 km) east of Decorah, from 1931 to 1935. It featured a 1,200-foot (366 m) tour by boat—foreshadowing the boat tours at Spook Cave. In later years, under the moniker "Kegger Cave," Glenwood became notorious as a major party spot for local college students. In 1936, Scarvie opened **Wonder Cave**, just outside Decorah, which remained in operation until 1976. The cave featured a Petrified Forest, a floodlit Rock of Ages, and the "World's Largest Known Stalactite" (in fact, it was nothing of the sort). In 1937, Gerald Mielke, who would later develop nearby Spook Cave, opened **Wompi Cave**, billed as "Iowa's

Deepest," to the public, but it closed during World War II when he could no longer obtain gasoline for the electrical generator that powered the lights. The cave contained a bizarre formation resembling an elephant's head.

Before the discovery of Coldwater Cave in 1967, **Crystal Lake Cave** was the largest known cave in Iowa, and it remains the state's largest show cave today. James Rice discovered the cave in 1868 while digging for lead. He sank a 40-foot (12.2 m) shaft on a hillside south of Dubuque and struck a cave passage. The cave proved barren of ore, however. It was called Rice's Cave at first, and then, in the usual fashion, took the name of subsequent owners, becoming Linden's Cave. It wasn't commercialized until 1932 when Bernard Markus, one of the original miners, christened it Crystal Lake Cave and built the pretty white latticework cage around the facade of the entrance building that you see today. Formerly obscure, Crystal Lake Cave came to the attention of the scientific world in 1938 when geologist J Harlen Bretz, from the University of Chicago, published a classic scientific study of caves in the Galena Formation (Bretz 1938).

2.4.3 Modern Exploration

The **Iowa Grotto** (originally, the Iowa City Grotto) was established in 1949, and re-established in 1957 with James Hedges' *Iowa Cave Book*, which was succeeded by *The Intercom*, as its chief publication. The early grotto was focused on the lead mines and caves of the Dubuque area, where the Iowa climbing cam was developed as a practical mechanical ascender for exploring the lead-mining pits, in 1964.

The discovery of **Coldwater Cave** changed Iowa caving forever. On September 17, 1967, Steve Barnett, a University of Iowa geology student, performed a free dive at Coldwater Spring and swam up into what is now called the First Room. He was acting on a tip from a farmer who claimed that during Prohibition he had set up a still in a room just behind the cliff-face. The room was small, and there was a sump at the far end. This second sump, it turned out, was far too long to attempt another free dive. Assisted by David Jagnow, he began a series of dives with full scuba gear through a quarter-mile (0.4 km) succession of sumps before finally emerging into the air-filled main passage that we now call Coldwater Cave, in 1968.

By 1974, the Iowa Grotto hosted the NSS Convention at Decorah. After Coldwater Cave became available for exploration, the Rock River Speleological Society began the tradition of third weekends of every month for survey trips, and they greatly improved the facilities available, which the Iowa Grotto continued (Burkhead 1999).

2.5 Illinois

This section has greatly benefitted from the assistance of Illinois caver Larry Cohen.

2.5.1 Explorers

The earliest reference to Illinois caves is believed to be that of the French engineer M. de Lery in 1729, for **Cave in Rock** on the Ohio River at the southern tip of the state (Speece 1981). Subsequently marked as "Caverne dans le Roc" on Bellin's map, which accompanied Charlevoix's *History of New France* (1744)—as it was at that time part of French colonial lands—nothing is again recorded until more than a half century later when river pirates made the cave the scene of their depredations, beginning in 1797 (Rothert 1924). Cave-in-Rock State Park was established in 1929 (Fig. 2.8).

2.5.2 Settlers

Cave in Rock is located in the Shawnee Hills, an extension of the Ozarks. These hills are also famous for saltpeter caves. The cave sediments were removed and lixiviated to extract nitrate, which could then be boiled down with lye to extract potassium nitrate (saltpeter), which when combined with carbon and sulfur, produced gunpowder.

2.5.3 Modern Exploration

Early work on Illinois cave and karst geology was conducted by J Harlen Bretz, a University of Chicago Professor and world authority on cave formation. Bretz (1938) explored **Smith Cave**, an old lead miner's cave near the town of Mt. Carroll. He concluded that the cave was totally phreatic in origin and lacked a later vadose phase, to use the terminology that he promoted in his *Caves of Illinois* (Bretz and Harris 1961).

The story of cave exploration in Illinois differs from that of neighboring Wisconsin. Illinois' part of the Driftless Area occupies its northwest corner and is not the heart of its cave country, which is southeast of East St. Louis on the **Salem Plateau** around the town of Waterloo in Monroe County. The Plateau is a crescent-shaped segment of the Ozark Dome separated from the Missouri side by the Mississippi River. The lithology of this area differs from that in much of Wisconsin's Driftless, where maze caves in Oneota dolomite are hydrologically unrelated to the present-day landscape. Project caving in Illinois has been more focused on pushing

(a)

(b)

Fig. 2.8 **a** Cave in Rock on the Ohio River at the southern tip of Illinois, as depicted by artist John J. Egan circa 1850, courtesy of the Eliza McMillan Trust, St. Louis Art Museum. **b** Cave in Rock today, photo by Greg A. Brick, with the author's wife for scale

and surveying lengthy caves rather than excavating glacial sediments. Most Monroe County caves are active sink/spring systems within Mississippian age carbonates, primarily St. Louis limestone. The Plateau itself contains extensive karstic features with hundreds of rounded, shallow sinkholes within a relatively small area.

The "big three" cave systems found on the Plateau from north to south are Krueger-Dry Run, Illinois Caverns, and Fogelpole (Frasz 1983a; Moss 2009). All three have lengthy stream passages, multiple entrances, close proximity to the land surface, and a gradual dip toward the center of the Illinois Basin. The springs resurge along Horse Creek, a tributary of the Mississippi River, a factor which has meant a later discovery than had they been present on the river itself, the chief travel route in pioneer days.

Illinois Caverns, three miles in surveyed length, is the only cave in Illinois ever operated commercially, during the St. Louis World's Fair of 1904 (Cohen 1984). It faded from view thereafter, becoming an unofficial "show cave" with ladders and walkways under the supervision of Armin Krueger (1914–1996), a soybean farmer who spent his entire life within a few miles of the cave entrances. Although subsequently placed under state ownership, efforts to create a viable Illinois State Heritage Park have not succeeded.

Early cave exploration in Monroe County was led by Waterloo native Father Paul Wightman in the 1950s and 1960s. Father Wightman also performed some of the earliest dye traces in Illinois (Wightman 1969). His efforts demonstrated a hydrological connection between the three cave systems and springs along Horse Creek. Wightman led early surveys of the longest cave in Illinois, **Fogelpole Cave**, over 12 miles (19.3 km) in length. Because of its biological heritage, this cave has been, over time, the most protected of the three cave systems.

The **Krueger-Dry Run Cave System** which includes Half Mile Cave is over four miles (6.4 km) in length. Glacial fill plugs passages in both Half Mile Cave and Illinois Caverns that otherwise would connect the two cave systems. The main downstream entrance to the system is blocked by Big Sink from the hydrologically linked, 1.2-mile (1.9 km) long Kelly Spring Cave system. Cave diving efforts at Big Sink to tie the two systems together have so far proved fruitless.

Much of exploration of Monroe County caves was conducted by the **Windy City Grotto (WCG)** of the National Speleological Society. Founded in 1961, WCG cavers were active in cave surveys throughout the United States. The **Little Egypt Grotto (LEG)** based in Carbondale, at the southern tip of the state, a region long known as "Little Egypt." Caving groups from Missouri, including **St. Louis University Grotto (SLUG)** and the **Meramec Valley Grotto (MVG)**, were also active in mapping the known caves of the area.

The heyday of caver-led surveying in Monroe County occurred in the 1970s and 1980s. The Windy City Grotto discovered and surveyed numerous caves throughout Monroe County and lengthened the maps of existing cave systems. Some of the surveys led to connections between known caves creating longer cave systems. During the 1990s and early 2000s, additional cave exploration was led by Philip Moss on the north and northwestern side of the Plateau.

Crawling in stream passages or occasionally swimming is common. Among the significant caves in Monroe County, however, is **Saltpeter (Fults) Cave**. Located 150 feet (46 m) up a bluff overlooking the Mississippi River plain to the west, Saltpeter Cave is unlike any other Monroe County cave. It is an ancient paleokarst window that consists of dry, hard-packed clay, and shows no recent geologic development.

Paleontologists have taken an interest in Monroe County Caves such as **Meyers Cave** (Parmalee 1967) and **Couch Cave** (Frasz 1983b). The mastodon remains found in Babar's Bone Room were delivered to the Field Museum of Chicago.

Acknowledgments The author would like to thank Larry Cohen, Windy City Grotto; Ed Klausner, Iowa Grotto; and Gary Soule, Wisconsin Speleological Society, who have contributed generously to this chapter.

References

Alexander EC Jr (2012) The history of caving in Minnesota. Minnesota Speleol Mon 44(8):3–4
Blaine JJ (1919) Blaine Papers. State Historical Society, Madison, WI
Boszhardt RF (2003) Deep cave rock art in the upper Mississippi valley. Prairie Smoke Press, St. Paul, MN, 94 p
Bretz JH (1938) Caves in the Galena formation. J Geol 46(6):828–841
Bretz JH, Harris SE Jr (1961) Caves of Illinois. Report of Investigations 215. Illinois State Geological Survey, Urbana, IL, 87 p
Brick GA (1995) What happened to Fountain Cave—The real birthplace of St. Paul? Ramsey County History 29(4):4–15
Brick GA (2000) Beyond the sump: the Burnley map of Carver's Cave. J Cave Karst Stud 62(1):40
Brick GA (2002) LeConte's account of the caves of the Apostle Islands in 1844. Minnesota Speleol Mon 34(8): 89, 91
Brick GA (2004) Iowa Underground: A Guide to the State's Subterranean Treasures. Trails Books, Black Earth, WI, 223 p
Brick GA (2005) Canton's Bear Cave: An investigation of Fillmore County's oldest cave reference. Minnesota Speleol Mon 37(3):25, 27–28
Brick GA (2007a) The Niagara Cavers at Big Room Cave. Minnesota Speleol Mon 39(3):28–29
Brick GA (2007b) "Charon Trimmed His Lamp": Carver's Cave and the definition of a show cave. J Spelean History 41(1):23–26
Brick GA (2007c) Recycled imagery: the Minnesota alter ego of a Wisconsin cave. Hollow Earth News 14(7):4
Brick GA (2009a) The serpent god of the iron cave: A native American cave myth from the Upper Midwest. J Spelean Hist 43(1):10–14

Brick GA (2009b) The myth of the American cave man. Proceedings of the 15th International Congress of Speleology, Kerrville, TX 3:1955–1956

Brick GA (2010) Lewis and Clark's reference to a cavern on the Missouri River, 1804. J Spelean Hist 44(2):8–9

Brick GA (2013) The nitrate deposits of rock crevices in the Upper Mississippi Valley. Ph.D. thesis, University of Minnesota, Minneapolis, 165 p. University of Minnesota Digital Conservancy. https://hdl.handle.net/11299/157750

Brick GA (2016) Early French mining in Minnesota. In: Labine MA (ed) They spoke french: French heritage in Minnesota. French-American Heritage Foundation, Arden Hills, MN, pp 65–71

Brick GA (2017) Minnesota Caves: History and Lore. Arcadia/The History Press, Mount Pleasant, SC, 144 p

Brick GA (2018) "Well-known subterranean passages": The prehistory of Mystery Cave, Minnesota. J Spelean Hist 52(1):4–5

Brick GA (2020) Paradigmatic studies in Midwestern cave science. Chapter 8, this volume

Brick GA, Petersen PA (2004) Ten cent tour: The story of Chute's Cave. Hennepin Hist 63(2):4–25

Brown CE (1926) Indian caves in Wisconsin. Wisconsin Archeol 5 (1):7–29

Burkhead L (1999) History of the Iowa Grotto. Intercom 35(3):2–8

Cohen L (1984) Illinois Caverns: A history. J Spelean Hist 18(1):17–21

Cronon W (1991) Landscape and home: Environmental traditions in Wisconsin. Wisconsin Mag Hist 74(2):82–105

Frasz GB (1983a) Twenty miles under Illinois: The land and the caves. NSS News 41(2):72–76

Frasz GB (1983b) Couch Cave, Monroe County, Illinois. Windy City Speleonews 23:50–53

Hedges JA, Knudson G (1975) Additional notes on the Decorah Ice Cave. J Spelean Hist 8(3–4):23–26

Kehret R (1974) Minnesota Caves of History and Legend. Superior Printing, Chatfield, MN, 28 p

Lange EG (1909) Caves of the driftless area of southwestern Wisconsin. University of Wisconsin thesis, 81 p

Martin L (1932) The Physical Geography of Wisconsin, 2nd edn. University of Wisconsin Press, Madison, WI, 608 p

Moss PL (2009) Illinois. In: Palmer AN, Palmer MV (eds) Caves and Karst of the USA. Huntsville, AL, pp 132–135

Parmalee PW (1967) A recent cave bone deposit in southwestern Illinois. NSS Bull 29(4):119–147

Reeder PP, Day MJ (1990) Lead mining from the caves in the upper Mississippi valley: An historical perspective. J Spelean Hist 24 (1):14–22

Rothert OA (1924) The outlaws of cave-in-rock: Historical accounts of the famous highwaymen and river pirates who operated in pioneer days upon the Ohio and Mississippi rivers and over the old Natchez Trace. Arthur H. Clark Company, Cleveland, OH, 364 p

Rudy C (1998) Ledge View's Caves. Calumet Nature Studies Inc, Chilton, WI, 28 p

Shaw TR (1992) History of cave science: The exploration and study of limestone caves, to 1900, 2nd edn. Sydney Speleological Society, New South Wales, Australia, 338 p

Soule GK (1999) A history of the postcards of Cave of the Mounds, Blue Mounds, Wisconsin. J Spelean Hist 33(2):23–41

Soule GK (2002) A great Wisconsin Cave Hoax? J Spelean Hist 36 (1):9–18

Soule GK (2013) Tartarus Cave. Spelean History Series No. 13. Speece Productions, Altoona, PA, 24 p

Soule GK (2016) Founder of the Wisconsin speleological society. www.wisconsincaves.org. Accessed 29 Sep 2016

Speece JH (1981) Early American speleological writings. Proceedings of the Eighth International Congress of Speleology, Bowling Green, KY, p 183–184

Wightman P (1969) Underground stream tracing, Monroe County, Illinois. Unpublished typescript report, 5 p

3 Minnesota Caves and Karst

E. Calvin Alexander Jr. and Greg A. Brick

Abstract

Carbonate caves and karst lands underlie most of southeastern Minnesota. The karst features are developed in relatively flat-lying lower Paleozoic limestones, dolostones, and sandstones. These rocks are major bedrock aquifers and contain about three-quarters of the groundwater resource in Minnesota. Over half of the people in Minnesota live on the karst lands and most depend on groundwater for drinking, agriculture, industrial, and recreational water supplies. Karst phenomena, aquifers, landscapes, and caves are fundamental boundary conditions, often unrecognized, on human activities in these areas. The Minneapolis/St. Paul metropolitan area contains a wide array of caves, natural and artificial, in the St. Peter Sandstone. The natural caves are developed by groundwater piping processes and are usually found near river valleys. This sandstone was mined for silica for use in glass making and for mortar and foundry sands. The resulting artificial caves served for mushroom growing, cheese ripening, entertainment, and other repurposings. Three electronically accessible databases contain information about Minnesota's karst features. The Minnesota Karst Features Database (KFD 2020) lists the locations of sinkholes, stream sinks, stream sieves, and miscellaneous other karst features. Tipping et al. (2015) describe the development of the KFD over several decades. Information on the locations of springs and seeps was originally part of the KFD but was then used as the start of a state-wide Minnesota Spring Inventory (MSI 2020). Brick (2017c) provides the essential guidance document for MSI with extensive background information. Information on dye tracing and springshed delineation is available in the Minnesota Groundwater Tracing Database (MGTD 2020). Green et al. (2018) describe the development and use of the MGTD.

3.1 Bedrock Geology

The southeastern Minnesota carbonate karst lands (Fig. 3.1) are developed in and on nearly flat-lying sedimentary rocks of Cambrian, Ordovician, and Devonian age. These strata were deposited in transgressive and regressive cycles as a succession of seas that alternatingly flooded and exposed the Hollandale Embayment. The Hollandale Embayment was a shallow depression located between the Wisconsin Dome to the northeast and the Transcontinental Arch to the northwest in lower Paleozoic time (Austin 1972). The local bedrock consists of Cambrian, Ordovician, and Devonian sandstones, shales, limestones, and dolostones. A conventional lithostratigraphic column for the southeastern Minnesota sedimentary rocks shows the relative thickness of the various units (Fig. 3.2). The Cambrian units, the Mt. Simon Sandstone up through the Jordan Sandstone, are predominantly sandstones with shale and thin carbonate beds. These Cambrian units are major regional aquifers but contain few caves or karst features.

These Paleozoic rocks are exposed at the surface mainly along the Mississippi River valley and the major tributary rivers feeding the Mississippi. The Lower Ordovician Prairie du Chien Group (OPDC), the Oneota Dolomite, and the overlying Shakopee Formation are the lowest major carbonate units. These units, and the overlying St. Peter Sandstone, are major regional aquifers and host numerous caves and sinkholes.

The Upper Ordovician Glenwood and Platteville Formations and Decorah Shale function as a major regional aquitard. The Upper Ordovician Cummingsville, Prosser,

E. C. Alexander Jr. (✉)
Department of Earth & Environmental Sciences, University of Minnesota, 150 Tate Hall, 116 Church St. SE, Minneapolis, MN 55455, USA
e-mail: alexa001@umn.edu

G. A. Brick
Lands and Minerals, Department of Natural Resources, 500 Lafayette Road, St. Paul, Minnesota, 55155, USA

G. A. Brick and E. C. Alexander Jr. (eds.), *Caves and Karst of the Upper Midwest, USA*,
Cave and Karst Systems of the World, https://doi.org/10.1007/978-3-030-54633-5_3

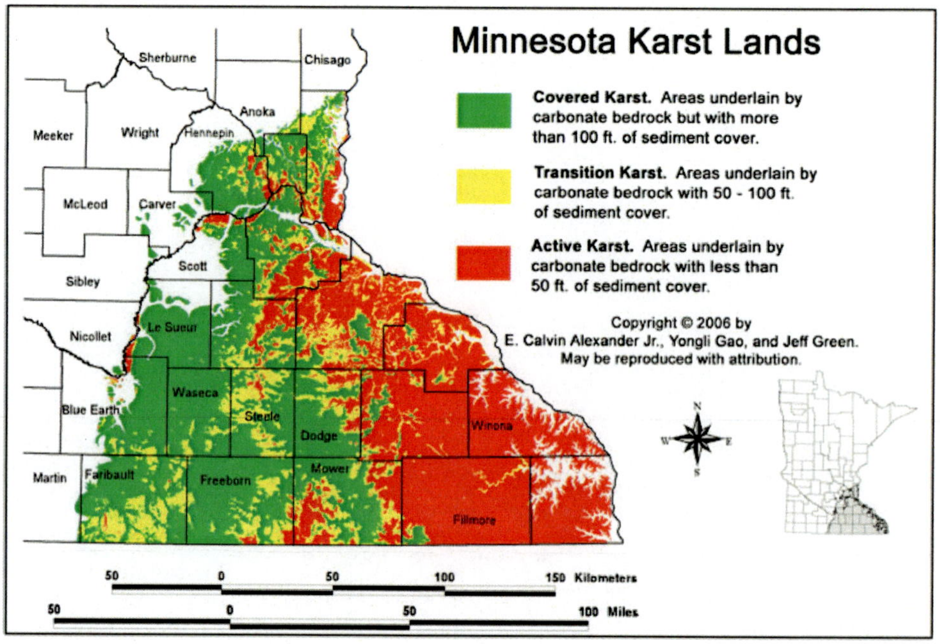

Fig. 3.1 The 2006 Minnesota karst lands map. The colored areas are the portion of Minnesota underlain by carbonate bedrock

Stewartville, Dubuque, and Maquoketa Formations and the overlying Devonian Spillville Formation are primarily carbonates and host most of the caves, sinkholes, blind valleys, and other karst features in Minnesota. The rest of the Devonian rocks, while predominately carbonates, are largely covered with thick glacial sediments and only locally contain sinkholes in Minnesota, primarily in Mower County.

Figure 3.3 shows the same information from Fig. 3.2 but the vertical scale is linear time. The gray areas represent time periods when there was no deposition or when deposits were removed by subsequent erosion. Although much of Minnesota was covered briefly by a Cretaceous sea, there are essentially no Mesozoic marine sediments in southeastern Minnesota; however, non-marine clastic sediments exhibit patchy occurrence (Sloan 1964). In most of southeastern Minnesota the eroded Paleozoic bedrock surface is overlain by Pleistocene and Holocene sediments—an additional 350–550 million-year unconformity. During each of these unconformities, karst solutional processes created and then modified generations of karst features.

The present-day sinkholes, caves, and other karst features, in many cases, are reactivated buried, interstratal paleokarst features (Hedges and Alexander 1985).

Although southeastern Minnesota is often referred to as the "Driftless Area," all of it has been glaciated several times. Scattered patches of several pre-Illinoian tills, several loess layers, glacial erratics, and extensive glacio-fluvial deposits occur in southeastern Minnesota. The multiple pre-Illinoian glacial advances "bulldozed" off much of the pre-existing epikarst, filled the deeper sinkholes and karst depressions, and flushed enormous volumes of sediments into and through the karst systems. When the glacial and recent sediments are systematically removed from the Paleozoic bedrock surfaces, as carbonate aggregate quarries are developed (for example), a relatively smooth, flat bedrock surface is usually revealed. That flat surface typically contains filled vertical joints and filled paleo sinkholes.

The western portions of Paleozoic carbonate rocks are covered by thick layers, more than 30 m, of Des Moines Lobe Wisconsinan till. Although there are few surface karst features in these areas, well drilling records show numerous karst voids in the covered Paleozoic aquifers.

The Twin Cities Metropolitan Area was covered first by Late Wisconsinan Superior Lobe deposits and then by Des Moines Lobe deposits. But in much of the area the glacial deposits are thin enough that surface karst features exist, and the bedrock aquifers are vulnerable to rapid surface contamination. The true Driftless Area karst is on the east side of the Mississippi river in southwestern Wisconsin and extreme northwestern Illinois (see Figs. 1.12 and 1.13 in Doctor and Alexander (2020), Chap. 1, this volume).

Outside the Twin Cities Metro Area and its suburbs, most of southeastern Minnesota is intensively farmed. Row crop agriculture and various types of confined animal feeding operations are common. The common land uses are agriculture and municipal/residential. There are scattered patches of forest particularly along the steep sides of the incised river valleys. There are numerous local quarries that produce primarily aggregate for road and other construction uses. Historically carbonate bedrock was quarried for dimensional stone for buildings and foundations. In the last decade a small, controversial mining of the St. Peter and Jordan Sandstones to produce proppant sands for hydraulic fracturing use has developed.

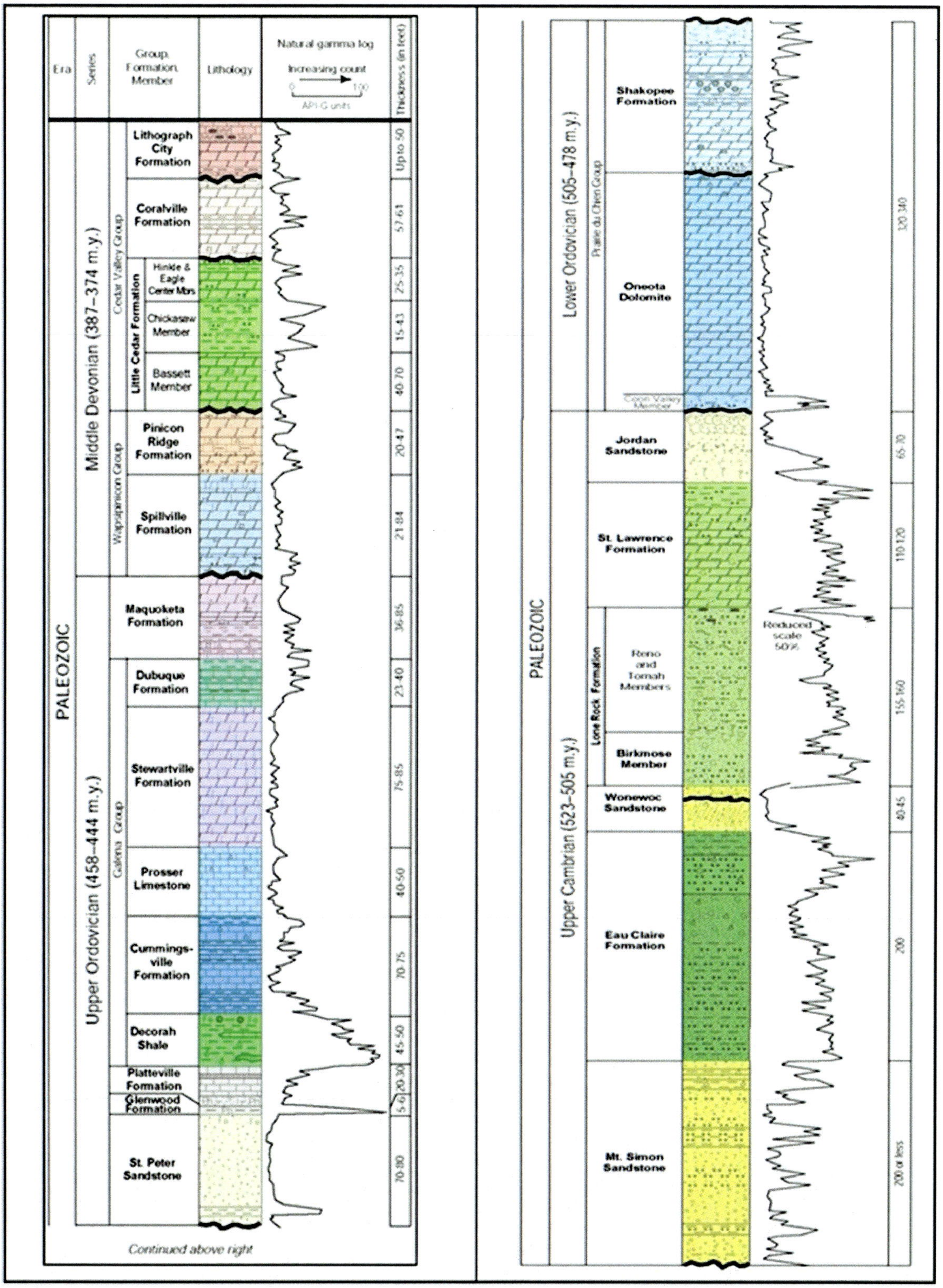

Fig. 3.2 Conventional Paleozoic lithostratigraphic column for southeastern Minnesota. The vertical scale is thickness. The heavy black lines are major unconformities. Modified from Mossler (2008)

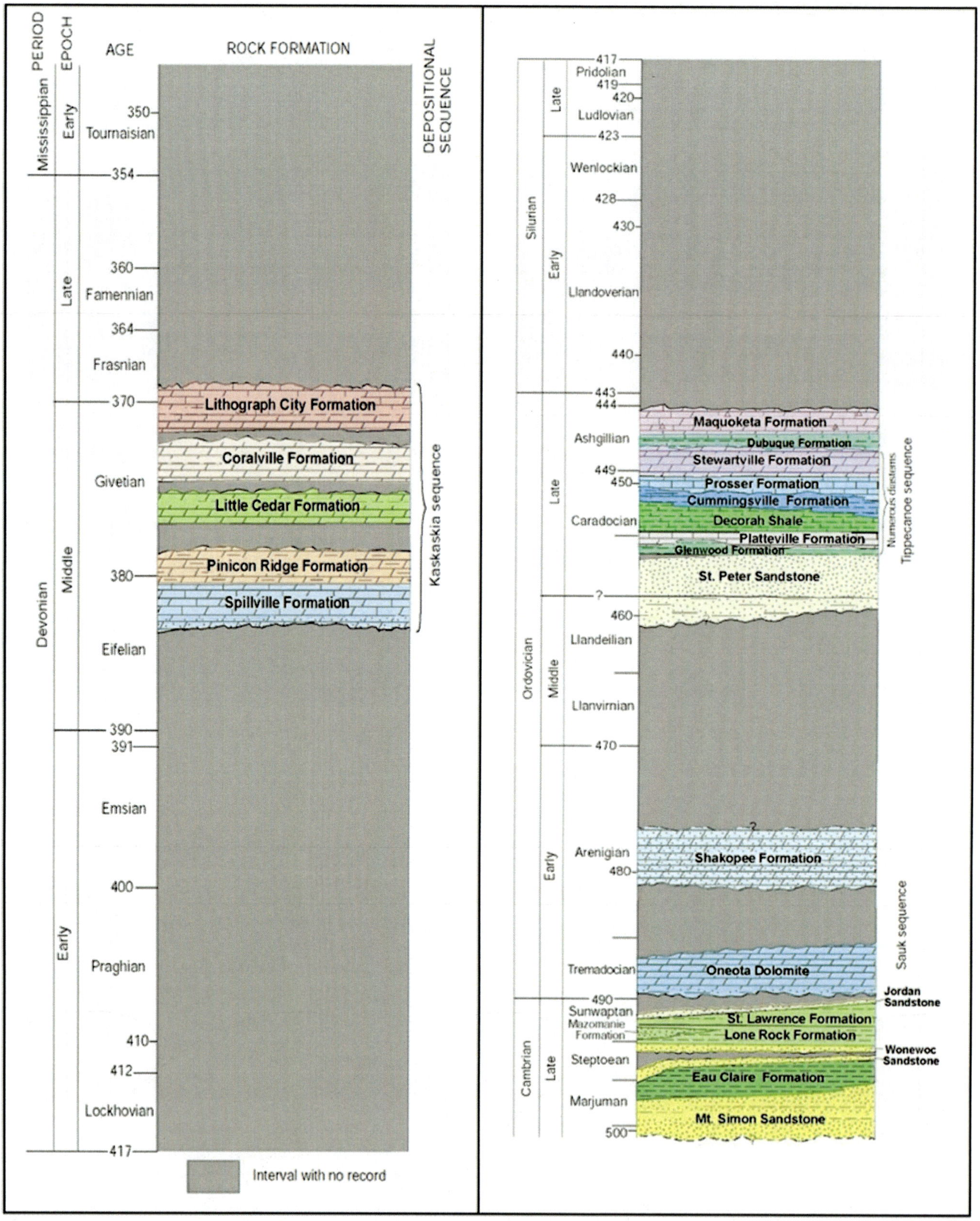

Fig. 3.3 Chronostratigraphic Paleozoic lithostratigraphic column for southeastern Minnesota. The vertical scale is time. The gray areas show the time intervals with no records—either no deposition or subsequent erosion of previously deposited sediments. Modified from Mossler (2008)

3.2 Minnesota Caves

Minnesota has 490 listed caves and over 15,000 karst features. Ownership and control of access to caves and karst features is complex: (1) Several significant caves and karst features are in the Minnesota Department of Natural Resources' (MNDNR) State Parks, Scientific and Natural Area (SNA) Preserves, State Forests, and other kinds of publicly owned property. (2) Many of the other small caves, sinkholes, springs, and karst features are on privately owned land. (3) John Ackerman's Minnesota Cave Preserve (MCP 2020), a non-profit organization, "owns access to 43 caves in Minnesota and Coldwater Cave in northern Iowa."

Although there are caves and karst features in Minnesota from the shores of Lake Superior to the Iowa border, Fillmore County is the heart of caves and karst in Minnesota. A 38.6 by 57.9 km (4 townships by 6 townships) rectangular county on the Iowa border, Fillmore County contains over 10,000 mapped sinkholes—more than are mapped in all the rest of Minnesota combined. Fillmore County contains more kilometers of mapped caves than the rest of Minnesota combined. It is the home of the two largest show caves in Minnesota—Mystery Cave and Niagara Cave. Fillmore County contains the largest karst springs, blind valleys, and sinking streams in Minnesota.

The history of cave exploration in Minnesota is covered in Brick (2020b), Chap. 2, of this volume. An up-to-date general guidebook for visiting publicly accessible caves and karst features in Minnesota is Green and Brick (2019).

3.2.1 The Galena Group Caves

In Minnesota six caves longer than 1.6 km and Niagara Cave form a northwest to southeast, roughly 50 km long, linear array across Fillmore County. Coldwater Cave just south of the Minnesota border in Iowa forms the southeastern end of this array and is described in Lace (2020), Chap. 4, of this volume. These caves are developed in the Upper Ordovician Galena Group, the Dubuque and Stewartville formations, the Prosser Limestone, and the Cummingsville Formation. The caves are currently epigene systems with water moving downward from surface recharge to resurgent springs. But many of the caves contain evidence of earlier, hypogene phases with artesian flow moving upward in their speleogenesis. The caves can be divided into three groups: maze caves, river caves, and compound caves.

3.2.1.1 Maze Caves

The maze caves are rectilinear, joint controlled caves in plan view. Their natural entrances are typically sinkholes or stream sinks in the Dubuque Formation. The caves form in the Dubuque and Stewartville formations near the top of the Galena Group. The caves typically contain upper, older, relatively dry levels, and lower levels. The upper level passages end in breakdown, sediment plugs, and/or pinch off. The lower levels of the caves contain active streams that ultimately sump forming the downstream ends of the explorable cave. Dye tracing has documented that the cave streams resurge in springs typically a kilometer or more beyond the terminal sumps in the caves. These caves do not contain dome pits. Dye tracing has also documented substantial springsheds that feed subsurface flow into these caves in addition to surface watersheds that feed the entrance stream sinks or sinkholes.

Mystery Cave

Mystery Cave, at 20.6 km and the longest cave in Minnesota (Fig. 3.4), is part of the Forestville/Mystery Cave State Park (FMCSP 2020) in south central Fillmore County and offers a range of tours. The cave was discovered in three phases: Mystery I, the southwestern part of the cave, contains the main commercial tour and the original, now modified, entrance. Mystery II, the northeastern part of the cave named "Minnesota Caverns," contains the artificial entrance. Mystery III is the northwestern part of the cave.

Mystery Cave contains three types of passages. In cavers' parlance the first two types are referred to as "Big Dubuque Passages" and "Stewartville Crevices" and Mystery II contains a "big Stewartville" passage, the east end of the Fifth Avenue passage. The Big Dubuque Passages are typically dry and 3–6 m wide and tall. These passages are rectangular in cross-section and are clearly formed by breakdown migration of passages upwards into the Dubuque Formation. The Dubuque consists of alternating roughly 0.3 m thick bands of limestone with centimeter thick beds of shale. The ceilings of the Dubuque passages are typically flat-bedding plane partings. The walls are typically angular broken faces and the floors are covered with breakdown blocks. In some areas the solution processes subsequently have sculpted sections of the ceilings and/or walls.

The Mystery I commercial tour route is mainly in the Dubuque Formation. In places the breakdown floors of the Dubuque passages have been covered by stream deposits. One such stream deposit, the Door-to-Door gravels, can be traced from the near the Mystery I entrance to near the eastern end of Mystery II. The deposit fines along its length —from 100 cm-scaled, imbricated limestone fragments and glacial cobbles in Mystery I to cm-scaled pebbles near its end in Mystery II. The deposit clearly records an episode of surface sediments being washed into the cave since the gravels include a wide variety of igneous and metamorphic rocks that have been transported by glaciers to southeastern Minnesota. At three different places along the length of the

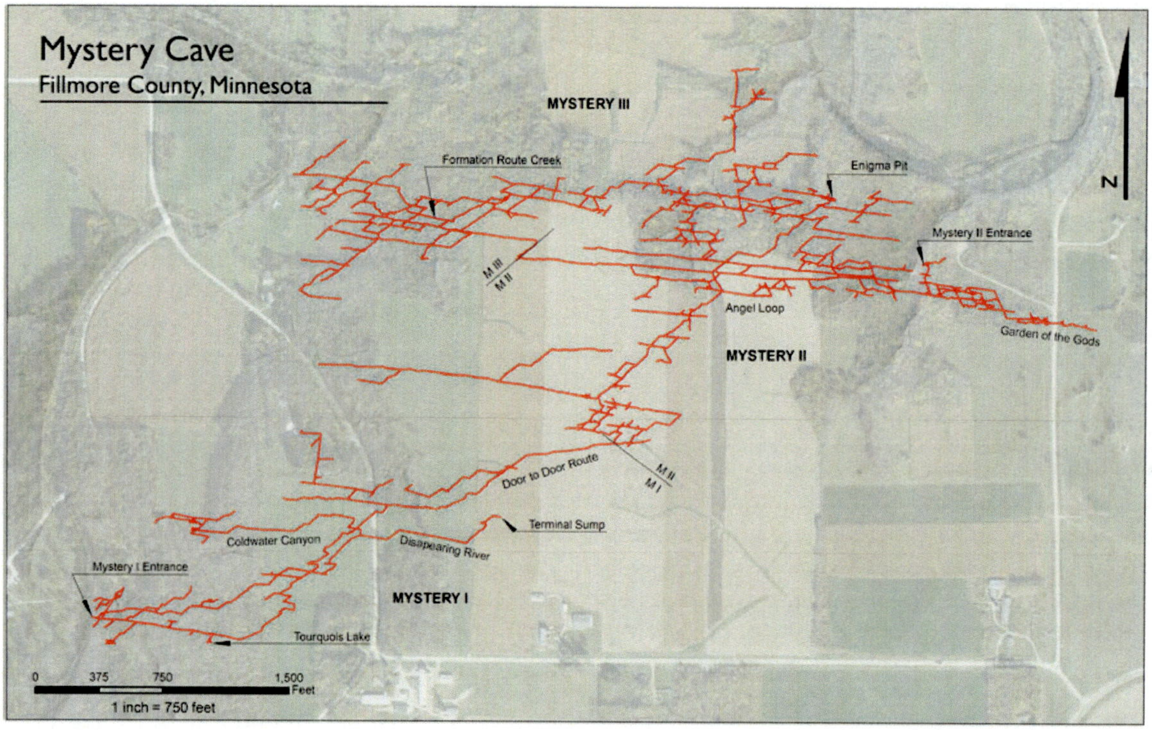

Fig. 3.4 The 1984 MSS Survey stick-line map of Mystery Cave superimposed on an air photo. At this scale most of the passages are narrower than the lines. Graphic by Martin Larsen [1,500 ft = 457 m]

Door-to-Door gravels speleothems growing on top of the gravels have U/Th basal ages of 12,200 years (Milske et al. 1983).

The "Stewartville Crevices" form the lower levels of Mystery Cave. They are clearly solutionally enlarged vertical joints in the Stewartville Formation. These joints can be over 10 m tall, up to kilometers long, and are systematically oriented roughly east/west and northeast/southwest.

In the eastern part of Mystery II, cave passages are developed in three sets of joints in the Stewartville. In the rest of the cave passage development is concentrated in only two joint sets. Careful leveling work by Art and Peggy Palmer (Palmer and Palmer 1993a) documented that the change from two to three sets of solutionally enlarge joints corresponds to a monoclinal change in the bedrock dip. The Stewartville crevices often subtly increase in size upward and terminate at the bottom of the Dubuque. In many places a horizontal, roughly circular tube has developed at the top of the crevice producing a characteristic "keyhole" cross-section. The Stewartville crevices range in width from a meter to just wide enough for a caver to move through—to not quite wide enough for a caver to negotiate. The "Big Stewartville" passage, the east end of Fifth Avenue in Mystery II, has solutionally sculpted walls with occasional, apparently recent, large breakdown blocks.

During the deposition of the Stewartville Limestone, the carbonate mud on the floor of the tropical Ordovician seas was extensively burrowed. These pervasive, centimeter-scale burrow networks were back-filled with additional carbonate muds—but muds with different calcium to magnesium ratios. These two types of carbonates tend to solutionally weather at different rates. The result is that the vertical Stewartville crevice walls are often rough at the centimeter scale. This differential weathering characteristically forms (in caver parlance) "the knobblies" which greatly complicate movement in tight passages.

Occasional breakdown blocks occur in the Stewartville crevices, but many of the crevices are filled with a massive, finely laminated reddish-brown silt. The fill is not the classic residual terra rossa clay common to many karst regions. The silt seems to be most closely related to glacial loess. Many of the side crevices branching off the main passages and the active lower level stream passages are filled almost to or to their tops with this silt. The silt deposition appears to have been a phreatic phenomenon. Where the contacts between the silt and the underlying or adjacent bedrock have been exposed, the silt is always resting on a solution produced surface in the bedrock. There are no known locations where the silt sits on speleothems, but the silt is often capped by breakdown blocks or speleothems.

Hydraulically, Mystery Cave functions as an epigenic underground meander cutoff for the South Branch of the Root River (SBRR) (Fig. 3.5). The SBRR in the vicinity of Mystery Cave flows through two, complex, entrenched bedrock meanders. Under low-to-moderate flows the SBRR progressively vanishes in a long series of stream sinks and

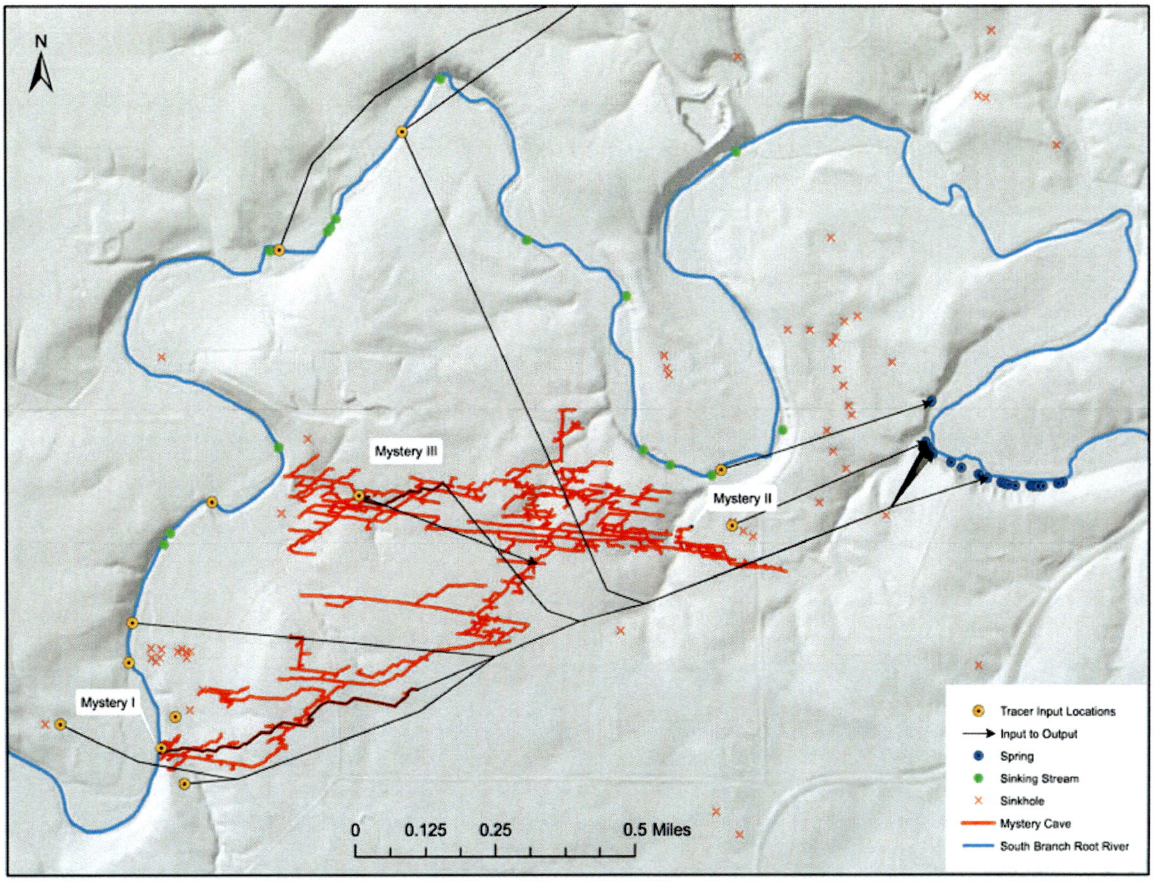

Fig. 3.5 A simplified diagram of the sinking reach of the South Branch of the Root River illustrating how Mystery Cave functions as an underground meander cutoff between Mystery I and the resurgence of the river at Seven, Saxifrage, and Crayfish Springs. Background is a LiDAR shaded relief DEM [0.5 miles = 0.8 km]

then resurges at the Seven (23A1) [see Note 1 in KFD (2020)], Saxifrage (23A113), and Crayfish Springs (23A80) (Fig. 3.6). Under extreme low flow/drought conditions, the terminal sink of the SBRR can be just downstream of the Mystery I entrance. Under low flow conditions, about 8 km of river bed in the two meanders is dry. The straight-line distance from the stream sinks at near the Mystery I entrance to Seven Springs is about 2.4 km and the total river elevation drop is about 18 m, and about 14 m of that drop occurs at or near the stream sinks at the Mystery I entrance. Each of the individual sinking points has a limited flow capacity, and as the flow in the SBRR increases the terminal sink moves further downstream. Under flood conditions part of the flow remains on the surface all the way to Seven, Saxifrage, and Crayfish Springs which are the source springs for the perennial reach of the SBRR and a state-designated trout stream.

Dye traces have documented that the SBRR water that sinks (23B1) near the Mystery I entrance flows through the lower level Stewartville passages via the Disappearing River to Seven and Saxifrage Springs. SBRR water that sinks a few hundred meters further downstream (23B47) flows through the Coldwater Canyon passage and then joins the Disappearing River. SBRR water that sinks further down the surface valley just beyond the west end of Mystery III (23B44) flows though Formation Route Creek in the Stewartville and joins the Disappearing River flow somewhere downstream of the terminal sump of the Disappearing River. SBRR water that sinks much further down the surface valley north of Mystery II (23B64) flows to Crayfish and S1 and S2 of Seven Springs. Crayfish Spring is ephemeral and dry during low flow conditions.

Systematic mapping of Mystery Cave was a major, ongoing project of the Minnesota Speleological Survey (MSS) during the 1970s and early 1980s. The survey data were entered onto punch cards and processed on a University mainframe computer. Although the surveyors made detailed sketches of the walls and cross-sections as the compass and tape surveys were conducted, the main products of the MSS's efforts were survey-line maps. The 1:2250 scale Mystery Cave map in 1980 NSS Convention Guide book (MSS 1980, Plate (1) is an example. In 1984, the surveyed length of Mystery was about 18.98 km. Maps based on the MSS survey data set have served as the basis for much of the

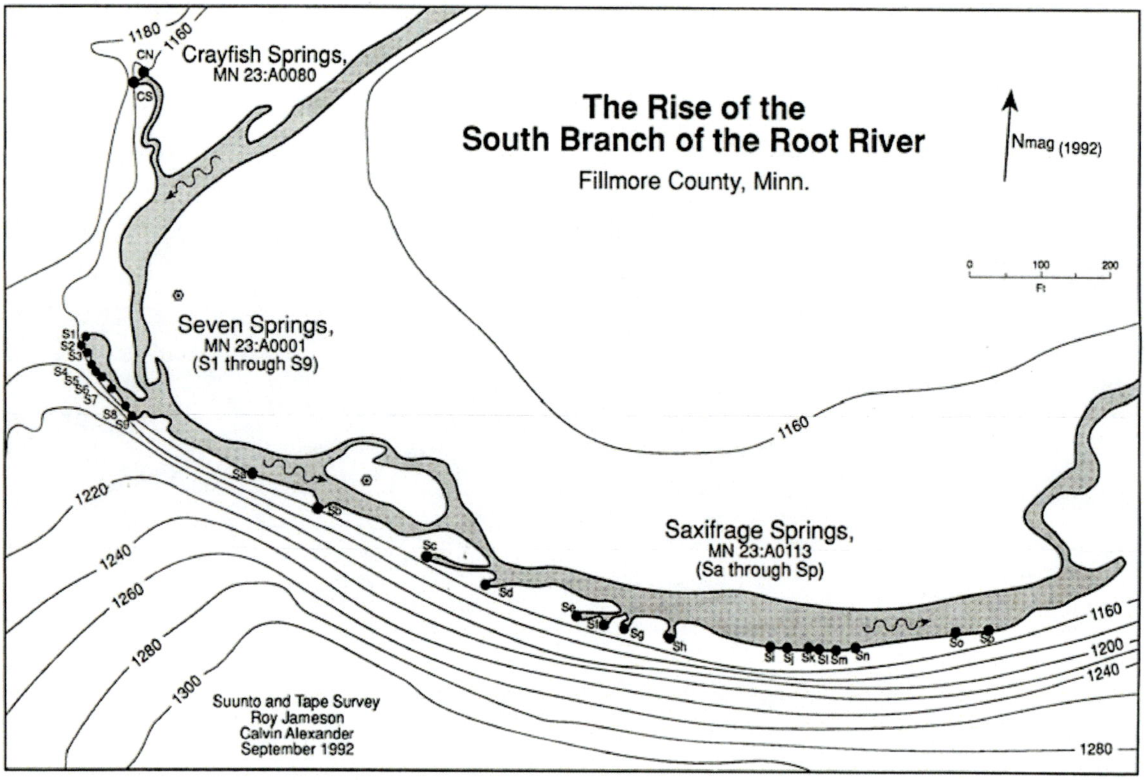

Fig. 3.6 The Seven, Crayfish, and Saxifrage Springs at the rise of the South Branch of the Root River. The source of perennial flow in the SBRR and the head of the trout stream portion of the SBRR [200 ft = 61 m]

subsequent scientific research and environmental management of Mystery Cave. When Mystery Cave was purchased by the State of Minnesota and became part of the state park system, the MSS transferred their survey data to the State.

Between 1991 and 1993 Art and Peggy Palmer conducted high-precision leveling surveys and created a set of 20 profiles of the major passages in Mystery Cave. The profiles have a horizontal scale of 1:960 and 5 × vertical exaggeration (Palmer and Palmer 1993a), and the vertical scale is accurate to a few centimeters. These elegant profiles revealed a wealth of information about Mystery Cave, its relationship to the SBRR and to the local bedrock stratigraphy. In 2005, MNDNR staff initiated a modern, detailed resurvey of Mystery Cave with a major goal to inventory cave resources along the survey. The two-decade-old MSS survey data proved to be very difficult to convert to modern cave survey software formats, as they contained too many survey closure errors and missing sections of survey and did not include resource inventory. Figure 3.7 is a small section of the new map illustrating the detail recorded in the remapping project. Progress on this new survey has been slowed by the need to avoid the commercial tours during the summer season and the need to avoid disturbing the significant bat population in Mystery during the winter hibernation. The latter concerns have been intensified as White Nose Syndrome has begun to impact the bat populations in Minnesota in the past few years. The current remapping effort has covered about 8 km of the cave and extended the total mapped cave to 20.6 km (Dawn Ryan, pers. comm. 2019).

Mystery Cave has served as a natural laboratory for cave and karst research for generations of researchers. Clarence Prohaska, who developed, owned, and operated Mystery from the late 1940s to the 1970s, actively encouraged and facilitated the exploration and mapping of Mystery. In the 1970s and early 1980s, Neil Davies owned and operated Mystery. He also actively encouraged and facilitated exploration, mapping, and research in the cave. Since Mystery became part of the state park system, the MNDNR has built a large, new visitor's center and upgraded commercial tour route. The MNDNR staff has encouraged and facilitated a wide variety of academic and applied research projects, classroom field trips as well as the ongoing modern remapping and inventory project.

In 1978, Richard Lively established an alpha counting U/Th dating laboratory at the Minnesota Geological Survey and began a systematic study of the ages of speleothems in Mystery Cave and surrounding areas (Lively and Alexander 1980; Lively 1983). Milske (1982) and Milske et al. (1983) did the first systematic study of the sediments in Mystery Cave. Lively's U/Th analyses provided the first chronologic

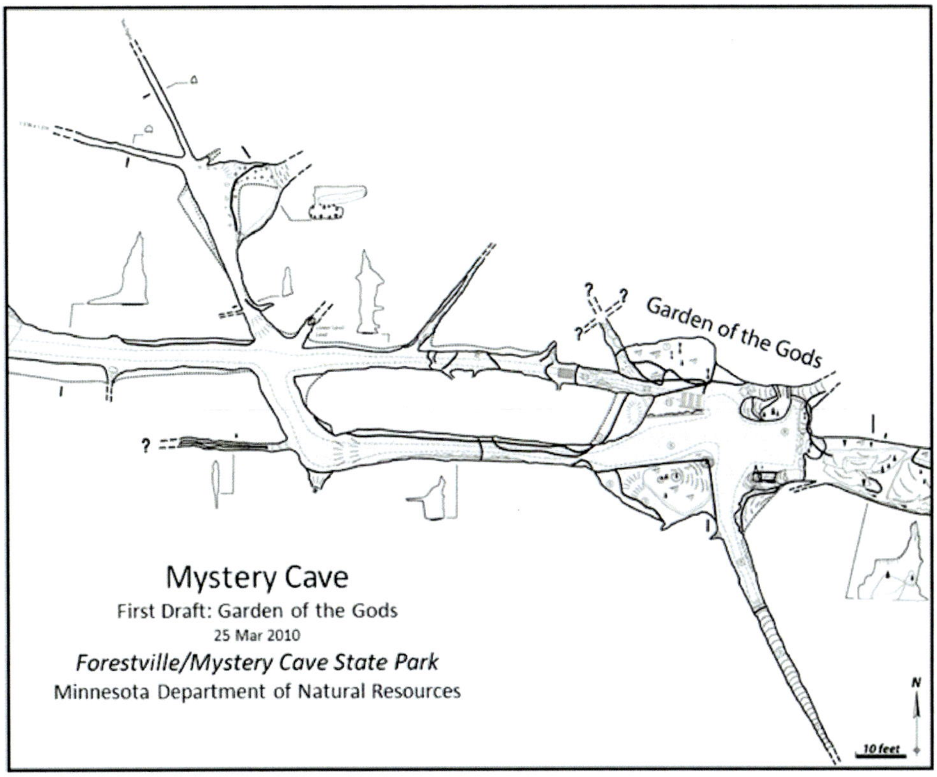

Fig. 3.7 Draft map of garden of the gods area near the east end of Mystery II. Image provided by the MNDNR. [10 ft = 3.0 m]

control for Milske's work in Mystery Cave and other researchers' subsequent geologic investigations.

Dye tracing research began to document the various groundwater basins in and around Mystery Cave and adjacent Fillmore County springsheds (Mohring 1983; Mohring and Alexander 1986). Dye tracing research at Mystery Cave and in Minnesota generally has continued to the present. Grow (1986) focused on the major dissolved ion water chemistry of the surface and spring waters in and around Mystery Cave and did pioneering work in the presence of the herbicide atrazine in Minnesota's karst waters (Adams et al. 1985).

Mystery Cave was purchased by the State of Minnesota in February 1988 and dedicated as part of Forestville/Mystery Cave State Park in August 1988. The Legislative Commission on Minnesota Resources funded a major research project on the Resource Evaluation of Mystery Caves in 1991–1993 biennium. That research project had three major components. Art and Peggy Palmer researched the geology and origin of Mystery Cave (Palmer and Palmer 1993a, b). Calvin Alexander and Roy Jameson studied about the waters of Mystery Cave (Alexander and Jameson 1994; Jameson and Alexander 1995). Rich Lively and Brian Krafthefer investigated the radon concentrations, activities of radon decay products, meteorological conditions, and ventilation in Mystery Cave (Lively and Krafthefer 1993a, b, 1995).

The Palmers interpreted their results to indicate "the cave originated as an underground bypass for the South Branch of Root River, and that the passages developed in several stages: (1) an early west-to-east series of passages, including Fourth and Fifth Avenues; (2) passages in Mystery I and the Door-to-Door Route, with a general northeasterly trend; (3) lower levels that formed as the river level cut downward in its channel, allowing deeper cave development. This evolution did not take place in discrete stages, as they overlap in time, and many of the original paths are still active in the lower levels, as well as periodically at all levels during floods. We view the cave as a dynamic floodwater cave that is intimately tied to the entrenchment history of the South Branch valley, rather than an enlarged remnant of a region-wide system of solutional fissures, although early solutional enlargement previous to the entrenchment of the river probably did contribute to the initial enlargement of some fractures. Entrenchment of the South Branch, and therefore the origin of Mystery Cave, depended on the entrenchment of the Mississippi River into the low-relief pre-glacial landscape" (Palmer and Palmer 1993a, p 29).

Alexander Klimchouk later visited Mystery Cave and observed several features he interpreted to be indicators of hypogene speleogenesis (Klimchouk 2007, Plate 1E, p 41 and Plate 14H-A, p 52). Klimchouk's observations initiated an ongoing scholarly debate on the origins of maze caves generally, and Mystery Cave specifically (Palmer 2007, 2011).

Perhaps relevant to this debate are two factors that neither Klimchouk nor the Palmers discuss. (1) On the land surface, almost on top of Mystery III there is a worked-out iron ore strip mine that was part of an extensive iron ore mining operation in western Fillmore County in the 1950s. These iron ores were mined from paleokarst sinkholes and depressions. Field relations indicate they are of pre-glacial age. The iron ores appear to have been deposited by hypogenic processes (Alexander and Wheeler 2015) and if so the strip mine over Mystery III documents that hypogenic processes were active at Mystery Cave's location before the Pleistocene. (2) The advance and melt-back of glaciers during the Pleistocene repeatedly induced enormous changes in the water table gradients in the various bedrock aquifers containing Mystery Cave, created large changes at the local and regional base levels, and flushed many cubic kilometers of fresh water over and through the karst aquifers.

Mystery Cave is currently functioning as an epigenic meander cutoff of the SBRR. It has functioned similarly in the past. Epigenic process has clearly enlarged and/or overprinted many of the Mystery Caves passages. However, Mystery is developed in carbonate bedrock that has been episodically subjected to karst processes for hundreds of millions of years—long before the Mississippi River drainage, much less the SBRR entrenched meanders existed. Evidence of paleo solution voids is ubiquitous in these carbonate bedrocks, including in wells penetrating the carbonate many hundreds of meters below the current land surface. It is possible, if not probable, that hypogenic processes contributed to the initial development of Mystery Cave. This debate is reminiscent of the great "vadose origin versus phreatic origin" for cave formation debate in the early twentieth century. That debate was resolved by the recognition that both processes occur and are important. Perhaps the same will prove to be true for Mystery Cave.

In the late 1990s, Roy Jameson and Dan Doctor established two drip water monitoring stations in Mystery II—at Coon Lake Drips and in the Garden of the Gods. These stations used data loggers recording continuous records of drip rate, temperature, and conductivity supplemented with periodic water chemistry and ^{18}O and ^{2}H stable isotope analyses of the waters. A series of presentations and papers used the results of this monitoring effort to demonstrate how stable isotope data could complement and strengthen the interpretation or results from data logging cave drip waters (Doctor and Alexander 1998, 2005; Doctor et al. 2001, 2015). The development of commercial cavity ring-down stable isotope analyzer systems has dropped the cost of an individual isotope analysis of water by an order of magnitude or more from the cost of analogous analyses on mass spectrometers. This technical advance has made the cost of high-frequency isotope analyses much more feasible.

Spring Valley Caverns

Spring Valley Caverns (SVC) is the north-west end of the array of big caves in Fillmore County. SVC is a joint-controlled maze cave and the second longest cave in Minnesota with 9.09 km of mapped passages (Fig. 3.8). The larger, dryer passages are in the Dubuque Formation. The lower stream passages are developed in the Stewartville Formation. SVC was briefly a show cave in 1968 and 1969.

Preservation and protection of SVC was the founding rational for John Ackerman's Cave Farm and the Minnesota Cave Preserve. "John Ackerman established Minnesota Cave Preserve in 1989 to preserve, study, protect and to promote conscientious exploration and conservation of our unique underground wilderness. From the beginning, John recognized the need to protect the fragile and timeless environment, both above and underground. Conservation has always been a prime tenet of the Minnesota Cave Preserve. The Minnesota Cave Preserve owns eight preserves in S.E. Minnesota and Northern Iowa, which provides access to 36 miles [58 km] of cave passages. The properties encompass 43 caves, 715.7 surface acres [289.6 ha] and 1,300.33 acres [526.2 ha] of additional subterranean cave rights" (MCP 2020).

SVC is well documented with timelines, photos, and maps on the website (SVC 2020, MCP 2020). Briefly, SVC was discovered by John Latcham in 1966 although it had probably been entered previously. The initial mapping of the cave by Ron Spong revealed about 0.8 km of cave, now known as SVC I. In 1983, the property was acquired by James Manggaard who allowed the MSS to explore the cave. Between 1984 and 1987, Dave Gerboth took a special interest in the cave and SVC I was resurveyed by the MSS. In 1987, Dave introduced John Ackerman to the Manggaard Farm and they discovered several other smaller caves in the 35–40 sinkholes on the property. In 1988, John Ackerman visited SVC I for the first time. John purchased roughly half of the Manggaard Farm outright and underground rights to the other half of the Farm.

In April 1990, John and Dave, using explosives, widened a long narrow crevice at the "dead end" of an SVC I passage and broke through into what became SVC II, SVC III, SVC IV, and SVC V. In 1991, John purchased additional land from neighbors to create new entrances to remote parts of SVC III, SVC IV, and SVC V. New shaft entrances were installed to SVC III and SVC IV. Between 1991 and 1997, more caves on the property were discovered and SVC grew to almost 9.1 km in length.

In 1997–1998, the original Quonset entrance building was demolished; the entrance sinkhole excavated down to bedrock and a new, largely underground earth sheltered building was constructed over the SVC I entrance. The SVC I entrance building is a unique structure—with a very

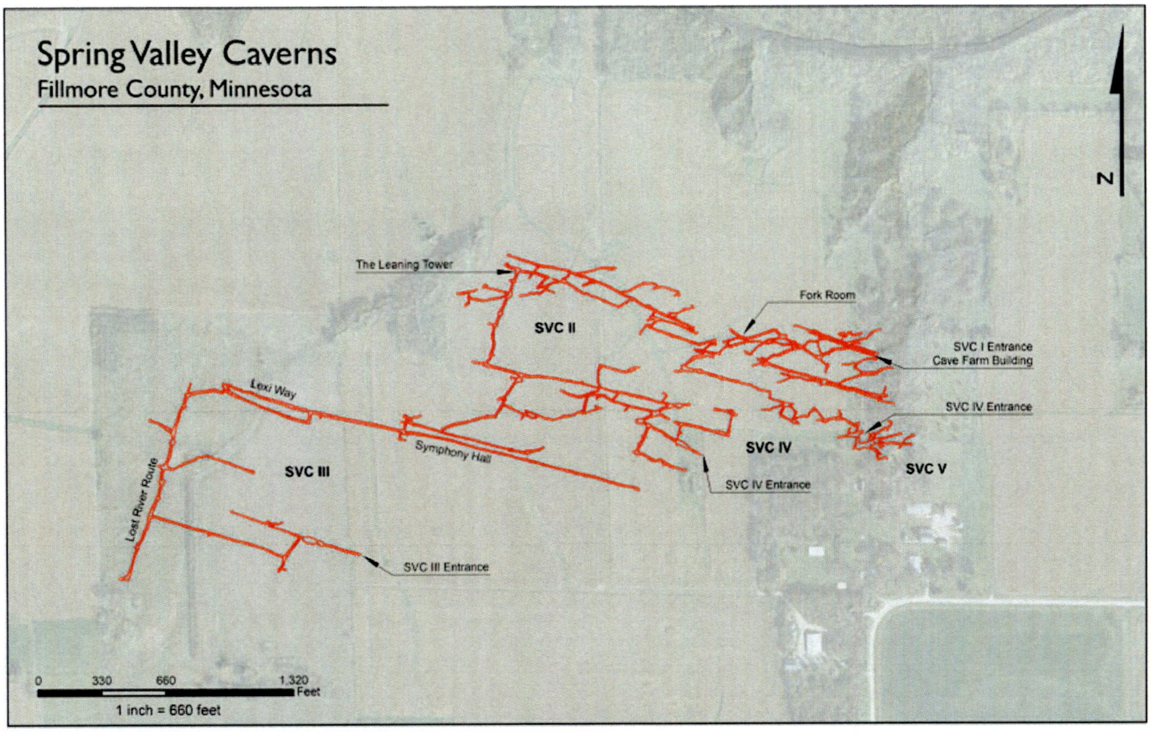

Fig. 3.8 Spring Valley Caverns map on air photo/LiDAR shaded relief DEM background. Mapped by the Minnesota Cave Preserve. Graphic by Martin Larsen [1,200 ft = 366 m]

extensive "basement." Between 1999 and 2012, John purchased five additional properties adjacent to the Cave Farm and added them to the Minnesota Cave Preserve.

As the map of SVC expanded, John realized that a 1940s vintage Amoco-BP oil pipeline crossed over major parts of SVC. In the Eilder Blind Valley (23B159) on the west edge of SVC III, a perennial stream sinks (23B58) directly on top of the pipeline under the blind valley. If the pipeline developed a leak, the spill would follow the water flowing into SVC, contaminate the cave, and contaminate Bly's Spring (23A127) 3.2 km east of the pipeline leak. In 2013, after over 20 years of constant badgering and legal threats, BP finally "stabilized" the sagging portion of the pipeline under the blind valley—instead of replacing or better rerouting that section of the pipeline away from the cave and blind valley.

The Minnesota Cave Preserve supports and encourages the scientific study of its caves, sinkholes, and karst landscapes. A series of dye traces has defined the groundwater flow around and through SVC (Fig. 3.9). SVC developed in the relatively flat interfluve between Bear Creek to the north and Dear Creek to the south (Fig. 3.8). The elevation of Bear Creek is about 15 m lower than the elevation of Deer Creek. Water flowing into sinkholes south and southwest of the cave flows through SVC III, SVC II and then east-northeast about 2.6 km to Bly's Spring (23A127) on the south side of Bear Creek. Water sinking at the Eilder Blind Valley in 23B58, directly on top of the BP pipeline, flows through SVC II and then to Bly's Spring (Costello and Alexander 2006).

Many of the sinkholes in southeastern Minnesota, including those on the Cave Farm, had been used as waste dumps from the initial European settlement in the 1850s up through the 1960s. Volunteer cavers have been systematically excavating the sinkholes on the Cave Farm, both to clean out the dump sites and to search for caves below the sinkholes. Initially these excavations were conducted with hand tools, which proved to be laborious and dangerous.

In 2002, John Ackerman bought a trackhoe and had the main arms rebuilt to enable deeper sinkholes to be cleaned out. The trackhoe, named the "Cave Finder," proved useful for many cave-related projects. The sinkholes typically prove to be bigger and deeper than their surface expression and partially filled with a wide variety of natural and anthropogenic materials. In most cases the excavations reach an enlarged joint or small cave in the bedrock bottom. A total of 35 caves, in addition to SVC, have been found under sinkholes on the Cave Farm. Two of the more interesting and educational sinkhole projects were 1991 Appliance Sink clean out and the 2003 Artifact Sink excavation. Both projects are extensively photo documented in MCP (2020).

Appliance Sink (23D60) had been a major dump site and was overfilled with a wide variety of unwanted materials. As

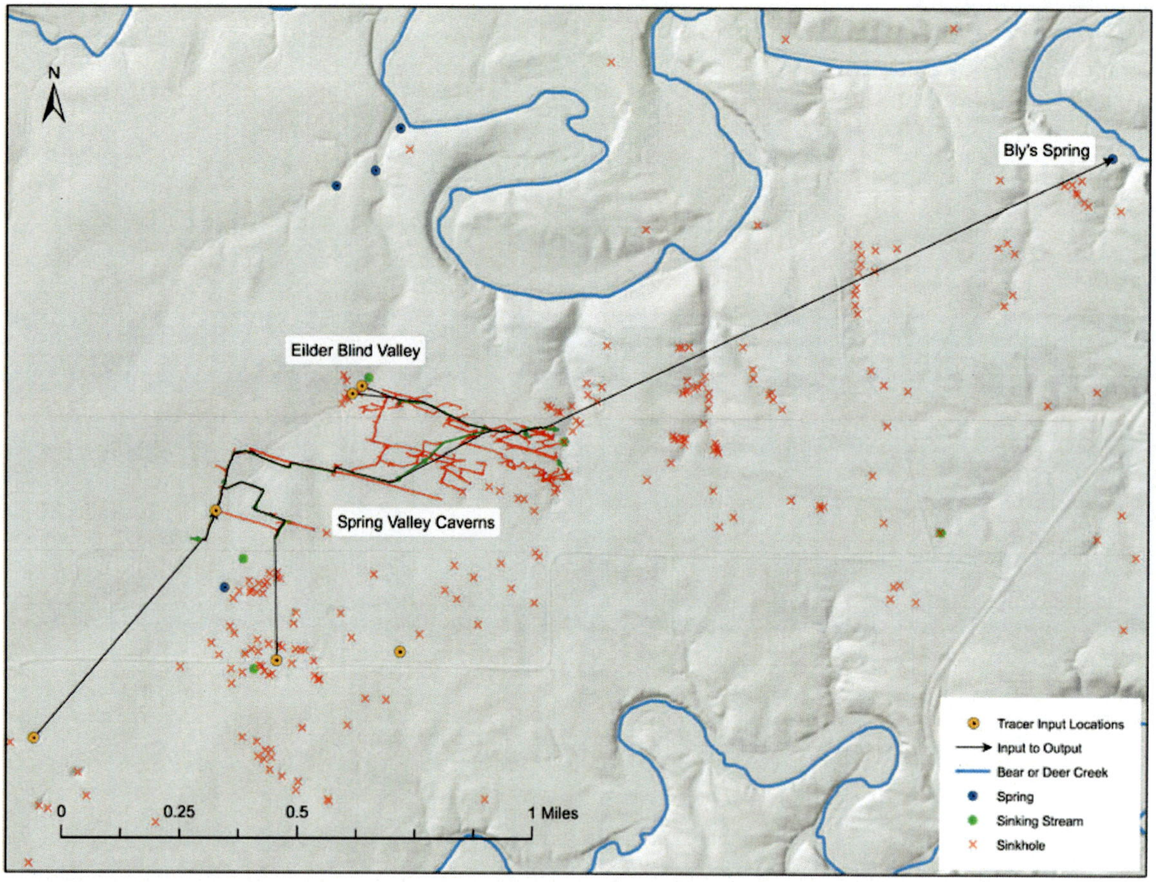

Fig. 3.9 A simplified representation of the of dye tracing through Spring Valley Caverns to Bly's Spring on Bear Creek. Bear Creek flows to the east along the north side of this figure. Deer Creek flows east along the south side of this figure. Background is a LiDAR shaded relief DEM [1 mile = 1.6 km]

the clean out progressed tons of metal objects were hauled to metal recyclers, scrap wood was burned, and household trash transported to licensed landfills. Dozens of decomposing batteries and leaching farm chemical barrels were uncovered and removed to hazardous waste disposal facilities. A cave passage in SVC IV was found to connect directly to the bottom of Appliance Sink. Water flushing through Appliance Sink had deposited a trail of trash (tires, bottles, cans, bones, etc.) hundreds of meters through the cave before dropping into a series of pits into the local water table. This walking passage is readily accessible from the SVC IV entrance and has proven to be a very effective tutorial for visiting environmental managers. A trip through the cave passage vividly illustrates than sizable solid objects, in addition to soluble pollutants, dumped into sinkholes that can and do move into the underlying aquifers.

Artifact Sink (23D1766) appeared to be a typical Cave Farm sinkhole—a clump of trees in the corn field (Fig. 3.10). The sink had been used as a household dump and contained a lot of modern trash. The top 3 m of the fill under the modern trash contained hundreds of late 1800s glass bottles and dozens of stoneware crocks and jugs.

Below the anthropogenic layer, the sinkhole proved to be a 10 m deep, elongated pit in the bedrock (Fig. 3.11). The bottom of the pit was an approximately 4–6 m diameter horizontal, roughly east/west solution conduit. The conduit was almost filled with sediment in both directions. The fill had apparently compacted slightly as it dewatered and there was a few centimeter air gap between the top of the sediment fill and the top of the conduit.

A small garden tractor with a front-end loader was lowered into the pit and used to remove the sediment fill in the conduit (Fig. 3.12). "Sediment removal reached the 100-foot [30 m] mark with no end in sight. A tiny camera was snaked ahead almost 30' [10 m] through a slender air space to reveal that the passage continued forward. Unfortunately, it was almost totally plugged with sediment. This project was subsequently terminated, and the sinkhole was restored" (SVC 2020).

No cave passage of this size and shape is known from SVC or any other cave in southeastern Minnesota. However, a conduit of this size and shape, also filled with sediment, was exposed by quarrying in the west wall of the Big Spring Quarry northwest of the town of Harmony in the 1980s. The

Fig. 3.10 Artifact sinkhole before excavation. Photo from John Ackerman

Fig. 3.11 Excavated artifact sinkhole. The "Cave Finder" at the top edge of the bedrock pit. Photo from John Ackerman

fills in these two large conduits may be related to the Pleistocene glaciations or earlier phases of karst processes that have modified these Ordovician carbonates. Unfortunately, neither of these conduits or their filling sediment were analyzed, and both are no longer accessible for the study.

As is discussed in more detail by Dorale (2020), in Chap. 11 of this volume, SVC has played a major role in paleoclimate research in Minnesota. SVC speleothems contain annual growth bands analogous to tree rings which can be counted under a microscope. These annual bands allow the measurement of paleoclimate records with annual resolution. In addition to the stable isotope records of temperature and vegetation changes on the land surface above SVC, much of SVC floods during the largest recharge events. These large

Fig. 3.12 Sediment removal from conduit below artifact sink. Photo from John Ackerman

floods leave thin layers of mud among the annual growth bands. These have permitted a high-resolution record of big floods at SVC for the past 3,000 years (Dasgupta 2008; Dasgupta et al. 2010; Feinberg et al. 2019).

3.2.1.2 River Caves

The river caves are dendritic to linear joint influenced caves in plan view. Their natural entrances are springs near the bottom of the Galena Group in the Cummingsville Formation. The entrance springs typically sump near the entrance and access to the main air-filled cave passages required cave diving techniques until artificial entrances were installed upstream of the entrance sumps. The cave passages typically contain running streams and require wet suits to avoid hypothermia even when accessed through the artificial entrances. The upstream ends of the passages are formed by unenterable sumps, breakdown or sediment plugs, or the passages pinching down to unenterable sizes. Dye tracing has documented substantial surface springsheds that feed water into these river caves. These caves contain dome pits.

Three large river caves are part of the array of large caves in southeastern Minnesota and northernmost Iowa. Two river caves, Tyson Spring Cave (TSC) and Bat River Cave (BRC) are part of the Minnesota Cave Preserve and are described below. The spring entrances of these two caves are north of the Village of Fillmore. Tyson Spring (23A122) is on the east side of the Middle Branch of the Root River (MBRR). Bat River Spring (23A13) is on the west side of the MBRR. Coldwater Cave, south of Harmony Minnesota in northern Iowa, is the third example of a river cave. Coldwater Cave is the longest of the caves in the northwest/southeast array of Upper Carbonate caves and forms the southeast end of that array.

Tyson Spring Cave

Tyson Spring Cave (TSC) is on the northern edge of the array of big caves in Fillmore County. All the 5.57 km of mapped passages in TSC are stream passages. Although a curvilinear dendritic pattern in plan view (Fig. 3.13), the passages are strongly joint controlled. The main stream passages are developed in the Cummingsville Formation but some of the taller dome pits probably extend up into the overlying Prosser Formation. The cave's history, exploration, and paleontology finds are extensively photo documented at MCP (2020).

A major stream flows out of the picturesque, remote entrance of Tyson Spring Cave at the base of a towering limestone cliff. In 1862, the U.S. Government assigned the land containing Tyson Spring to the widow of a War of 1812 veteran. She deeded the property to Harper Tyson in 1868. In the 1870s, a stereoscope image pair of photographs was taken of the TSC entrance by a professional photographer from the nearby town of Chatfield (Fig. 3.14).

Local tradition relates that, in 1930, residents would take wooden boat trips in the first 275 m of Tyson Spring Cave to the first sump. In the 1960s and 1970s, Ron Spong of the MSS led an effort to free dive the sump and established that it was too long for free diving. In the 1980s, Roger Kehret, a

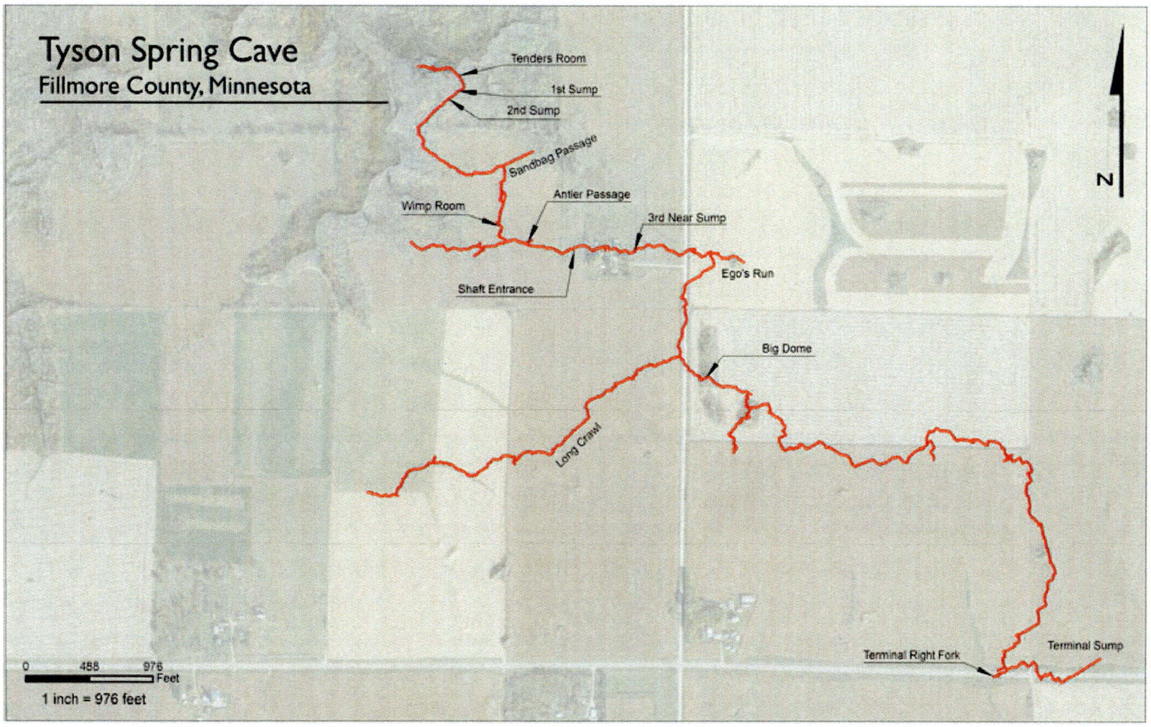

Fig. 3.13 Tyson Spring Cave map on air photo/LiDAR shaded relief DEM background. Mapped by the Minnesota Cave Preserve. The natural entrance, Tyson Spring (23A122) is at the northern end of the cave. Graphic by Martin Larsen [976 ft = 297 m]

Fig. 3.14 The 1870s stereopticon view of Tyson Spring. The man in the photo could be the eponymous Harper Tyson

caver who lived in nearby Chatfield, theorized that rock fall talus from the cliff at the entrance to Tysons Spring had damned the spring raising the water level in the entrance. Roger began what turned into a multi-year effort, to trench through the talus pile and thereby lower the water level in the entrance pool. He hoped that lowering the water level sufficiently might open an air passage above the sump.

In 1985, MSS members Larry Laine and Steve Porter used SCUBA to swim through the sump. They reported that the cave passage suddenly became large and they explored the main passage for over a 1.6 km upstream. In 1987, Steve and Larry returned to the cave and explored another 1.6 km upstream before they returned to the entrance. They turned around in a large main stream passage that showed no signs of diminishing in size.

Between 1988 and 2006, the main cave could only be accessed during low water conditions, combined with a dry forecast and very few cavers could or were interested in risking their lives to explore the cave system. Only a small number of trips were made into the cave by various cavers over the years. In 2006, John Ackerman purchased 0.4 ha of land above the cave and an additional 57.9 ha of subterranean rights, which included Tyson Spring and three outlying caves. After John made numerous trips deep into the cave to compile the necessary data needed to create a safe artificial entrance, a 38 m deep, 0.76 m diameter entrance shaft was completed.

The shaft entrance allowed systematic exploration, mapping, and scientific studies to begin. In 2007, Harper Tyson's grave was located in a local Pioneer Cemetery and a copy of the original land deed was obtained from the National

Archives. A gate was installed at the natural entrance and an informational sign posted. Tami Thomsen dove the far upstream sump and followed the submerged passage for 72.5 m until she encountered a constriction. One very significant scientific result was the serendipitous discovery that TSC contained a wealth of Pleistocene large mammal fossils.

In 2008, a Stag-moose antler was found in TSC (Fig. 3.15). The Stag-moose (*Cervalces* sp.) was a solitary browser and had never before been reported this far north. This was a Minnesota first. Once they realized that the cave contained fossil bones, John and his fellow cavers began to look for the bones—and began finding more. In 2008, Chris Widga examined the site where the Stag-moose antler had been found and collected sediment samples. More bones were found (Fig. 3.16). Back at the Illinois State Museum, Chris and his colleagues analyzed the samples and arranged for ^{14}C age determination and DNA analysis of the cat skull fragment. They reported "A scimitar-toothed cat (*Homotherium serum*) and stag moose (*Cervalces* sp.) are described from Tyson Spring Cave, Fillmore County. These specimens represent the first records of both species in the state, and the first record for *H. serum* the Great Lakes region. Although the *Cervalces* specimen remains undated, it shares features with pre-Wisconsin specimens from the eastern Great Plains. The *H. serum* individual dates to c. 26.9 ka, when the Wisconsin ice margin was less than 60 km away" (Widga et al. 2012).

Bat River Cave

Bat River Cave (BRC) is across the Middle Branch of the Root River from TSC, north of Fillmore village. All 4.57 km of mapped passages in BRC are stream passages. Although a curvilinear dendritic pattern in plan view (Fig. 3.17), these passages are strongly joint controlled. The main stream passages are developed in the Cummingsville Formation but some of the taller passages and domes may extend up into the overlying Prosser Formation. The cave's history, exploration, bats, and paleontology finds are extensively photo documented at MCP (2020).

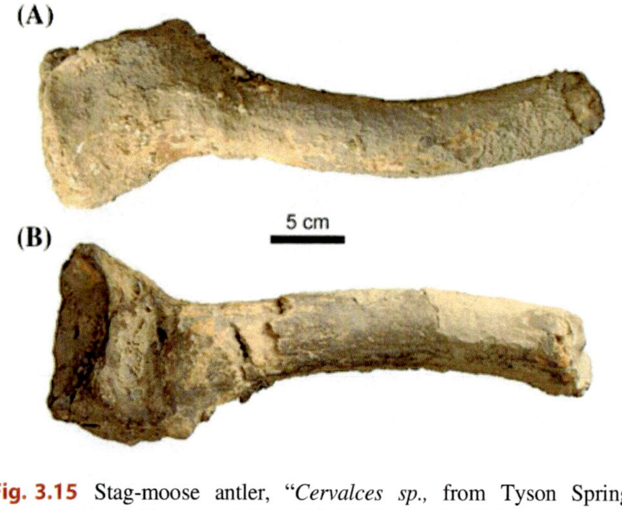

Fig. 3.15 Stag-moose antler, "*Cervalces sp.*, from Tyson Spring Cave. **A** dorsal, **B** ventral." (Fig. 3.3; Widga et al. 2012, p 4)

In July 2006, cave divers dove through the sump and surfaced on the other side. Thousands of bats were discovered hibernating along the walls (Fig. 3.18). The cave was well named! The next year, John Ackerman purchased 0.4 ha of surface and 93.5 ha of subterranean rights to Bat River Cave from the land owner. A new 20 m deep, 0.76 m diameter entrance shaft was drilled. During the fall and winter of 2007–2008 mapping and exploration resumed, new cave segments were discovered and pushed (Fig. 3.18). In December 2007 a bat count documented 4,132 hibernating bats (Fig. 3.19). Bat River Cave is the largest known natural cave bat hibernaculum in the tri-state area.

The bad news: the December 22, 2007 survey counted 4,132 bats in BRC. On February 5, 2011, a second bat survey was conducted, which prompted by the desire to see if caving activities impacted the bat population and to establish a baseline before the spreading White Nose Syndrome (WNS) reached Minnesota. This second survey counted 4,112 bats—a number statistically indistinguishable from the 2007 result. On February 11, 2017, after WNS had reached Minnesota, the third bat survey in BSC counted 2,836 bats. "Sadly, dozens of bats were found dead on the walls and floating in the water in the historic section. We witnessed numerous bats exiting the historic entrance, flying around the snow-covered landscape in a bewildered state. Approximately seventy percent of the bats we viewed in the cave exhibited signs of White Nose Syndrome. Bats were also seen flying around the entrances of other major caves in SE Minnesota, some found dead in the snow." The fourth survey on February 9, 2019 found 82 bats. Almost all of them were infected with WNS (Ackerman 2019). Only time will tell if this bat colony will ever recover.

3.2.1.3 Compound Caves

The compound caves are joint controlled maze and/or rectilinear caves in plan view. Their natural entrances are typically sinkholes or stream sinks in the Dubuque Formation or in the case of Holy Grail Cave the overlying Maquoketa Formation. The caves form in the Dubuque and Stewartville formations near the top of the Galena Group. The caves typically contain upper, relatively dry, levels and lower levels. The lower levels of the caves contain active streams that ultimately sump forming the downstream ends of the explorable cave. Dye tracing has documented that the cave streams resurge in springs typically a kilometer or more beyond the terminal sumps in the caves.

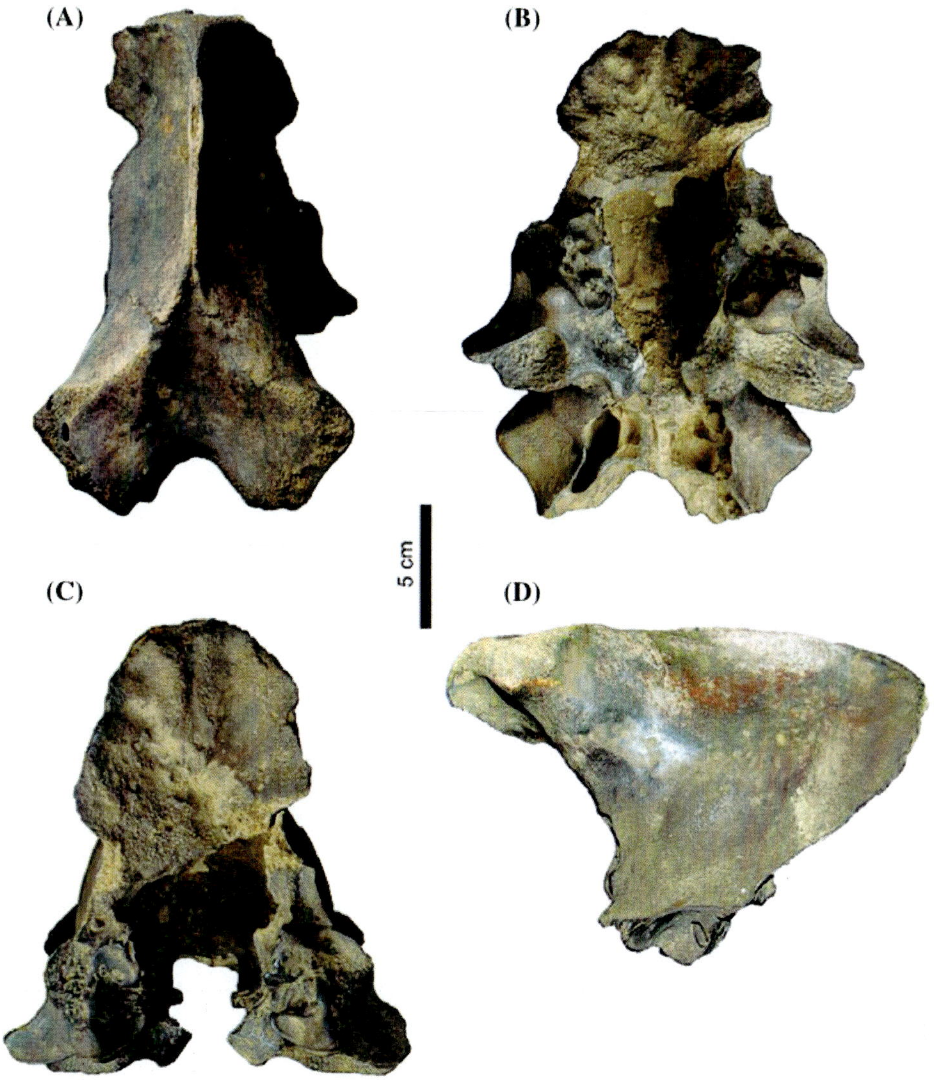

Fig. 3.16 Scimitar-toothed cat skull, "*H. serum* cranium from Tyson Spring Cave. **A** dorsal, **B** ventral, **C** posterior, **D** left side." (Fig. 3.6; Widga et al. 2012, p 6)

Goliath's Cave

The 16 ha, mostly wooded property in Sect. 3 of York Township is a blind valley. Cavers called the property Jessies Grove in memory of Jessie McCracken (1863–1947) who was born and lived her long life in the vicinity. A perennial stream Jessies Kill flows into the southwest corner of Jessies Grove and sinks into several stream sinks/sinkholes in the woods depending on flow. One sinkhole/stream sink (23D4986/23B140) is the natural entrance to the 4.26 km long Goliath's Cave (Fig. 3.20). The section of Goliath's Cave under Jessies Grove appears, in plan view, to be a flood water maze developed along joints in the Dubuque and Stewartville formations. The water flowing through the maze converges into a single underground stream, "The Rubicon," which flows eastward for more than a kilometer through one long joint trend in the Stewartville Formation. The stream drops over a 4.6 m waterfall and then sumps under the Canfield Creek surface valley. This eastern section appears to resemble the river caves. The morphology of the cave passages, however, reveals a complex of older perhaps hypogenic cave segments integrated by currently active epigenic passages (Alexander et al. 2015). Dye traces have documented that the water flows underground from the terminal sump about 3 km northeast to Canfield Big Spring (23A33) and Black Rock Spring (23A34). Those two large springs are the headwaters of the Canfield Creek trout fishery. Jessies Grove is now the MNDNR's Cherry Grove Blind Valley Scientific and Natural Area (CGBV SNA). The part of Goliath's Cave east of 193rd Street, including the artificial shaft entrance, is now part of the Minnesota Cave Preserve. A series of events that produced this division were reported in the book *Opening Goliath* (Griffith 2009) and documented, from the Minnesota Cave Preserve's perspective, in MCP (2020).

The following history of the discovery and exploration of Goliath's Cave is modified and condensed from MCP (2020) and Spong (2006). In the mid-1950s southeast Minnesota cavers discovered a significant cave in a sinkhole near Cherry Grove. Shortly thereafter a massive talus fall blocked the entrance. During the next 20+ years cavers occasionally

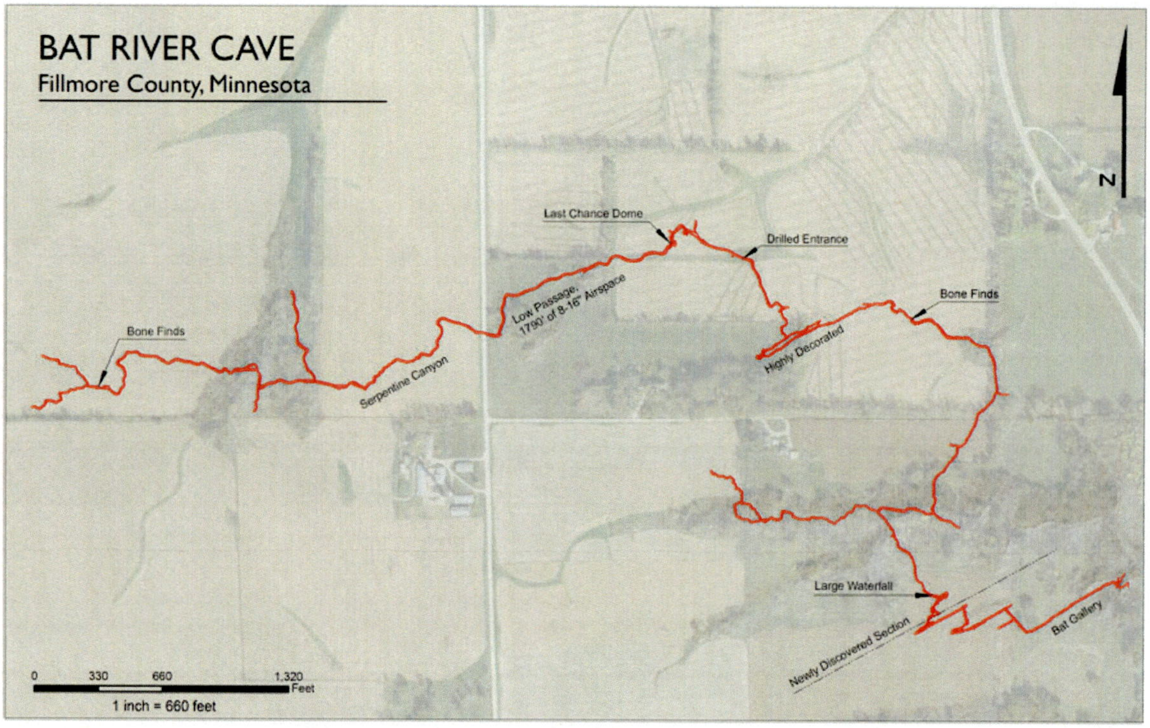

Fig. 3.17 Bat River Cave map on air photo/LiDAR shaded relief DEM background. The natural entrance to the cave, Bat River Spring (23A13) is at the east end of the cave. Mapped by the Minnesota Cave Preserve. Graphic by Martin Larsen [1320 ft = 402 m]

Fig. 3.18 Main passage in Bat River Cave. Photo from John Ackerman

tried to dig through the talus pile with no success. In 1980, the flood from an 18 cm rain removed loose rock and soil from the cave entrance (Fig. 3.21). Ron Spong managed to enter the cave and negotiated an entrance crawl with two near-sumps and walked through the Dubuque upper level of the cave but retreated due to a flood threat. Ron named the discovery Coon Cave.

In 1984, Jim Magnusson, the then MSS president, removed rocks and mud from the sinkhole entrance. Jim explored and surveyed the extensive Dubuque upper level of

Fig. 3.19 Bats in Bat River Cave. Photo from John Ackerman

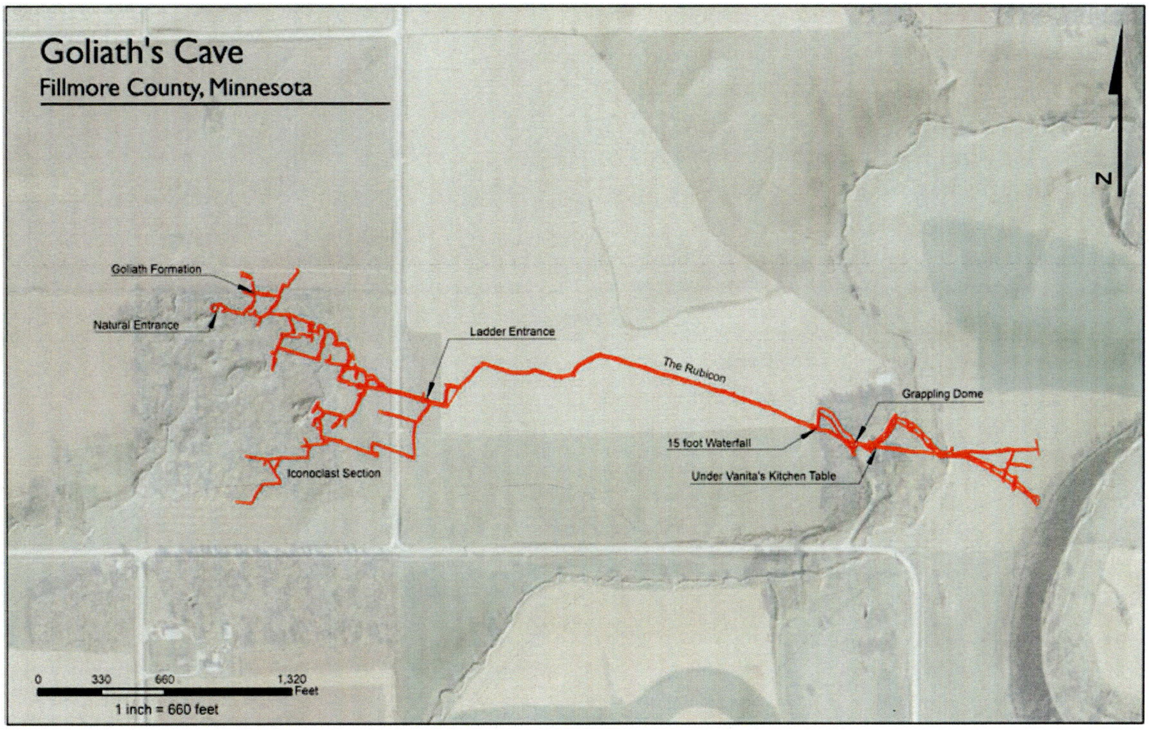

Fig. 3.20 Goliath's Cave on air photo/LiDAR shaded relief DEM background. Mapped by the Minnesota Cave Preserve. Graphic by Martin Larsen [1320 ft = 402 m]

Fig. 3.21 Goliath's Cave natural entrance at the start of the September 23, 2010 flood. An hour later the entire blind valley was filled and overflowed. Photo by Charlie Graling

the cave. A massive stalagmite flowstone formation immediately past the entrance crawl was named Goliath and the cave's name became Goliath's Cave. Subsequent cave visits revealed that the twin dips in the entrance crawl were usually filled with water or sumped. Access to the cave was impossible except in very dry periods. In 1985, Jim Magnusson, Mike Lilja, and Dave Madison traversed the near sumps, made their way through the upper level, and dropped down into a Stewartville lower stream level. They discovered the entire eastern portion of Goliath's Cave and walked about 1.5 km east to a 4.6 m waterfall. They named this long stream passage "The Rubicon" (Fig. 3.22). Climbing down the waterfall, they found that the tall main passage abruptly dropped into a flat, wide, low passage with limited air space. They reported that the passage continued ahead. In 1986, Steve Porter, Larry Laine, and Dave Gerboth surveyed the lower level to the waterfall and Steve moved forward about 120 m beyond the waterfall. He reported extremely low air spaces, interrupted by periodic tall domes in the low, wide stream passage.

In 1986, Jim Magnusson began to excavate an artificial entrance to Goliath's to facilitate year-round access. A site was chosen beyond the entrance near sumps and digging started from both the top, the ground surface, and the bottom, within the cave. A small shed was constructed in the woods near the dig site to make overnight weekend trips feasible. Solid bedrock, rather than the anticipated filled sinkhole, was encountered 2 m down. A cave radio was used

Fig. 3.22 The Rubicon River passage. Photo by Art Palmer

Fig. 3.23 Drilling David's Entrance to Goliath's Cave, 2004. Photo from John Ackerman

to find a better location for the excavation of a different shaft entrance. Excavation of the second entrance began in 1987 and continued into the spring. Just as the 6.4 m level was reached, the owner of the cave notified Magnusson that, due to liability concerns, cavers were no longer welcome on the property.

In 1998, the Kapper family requested for a conditional use permit from Fillmore County to turn Jessies Grove into a limestone quarry for construction aggregate. Local neighbors attending a County Board meeting opposed the requested permit. The MNDNR expressed concern about the obvious potential contamination of Jessies Kill that sinks, flows through the cave, and resurges in a state park. To permanently preclude a quarry from destroying Jessies Grove and Goliath's Cave, and potentially contaminating the state park, the MNDNR purchased the 16 ha in 1999.

The original concept was that Jessies Grove would become part of Forestville State Park. After all, dye tracing had documented that Jessies Grove fed directly into the state park's major springs. However, when it proved, for political reasons, too difficult to add Jessies Grove to the state park, the MNDNR designated the property as a SNA and renamed it the Cherry Grove Blind Valley Scientific and Natural Area (CGBV SNA 2020). The head of the SNA program then met with the MSS and stated that recreational caving would not be permitted in the CGBV SNA, and that the scientific research and mapping could proceed only under a permit system and in accord with the MNDNR's needs and priorities. The natural entrance was promptly gated, and the SNA posted.

In 2004, John purchased 145 ha of underground rights and a 0.8 ha of the surface over the Cave immediately east of 193rd Street. John then drilled a 23 m deep, 0.76 m diameter entrance into Goliath's which was named David's Entrance (Fig. 3.23). The eastern section of Goliath's Cave was reopened for exploration, mapping, scientific studies, and recreational caving.

From 2005 to 2009 a series of digs, climbs, and SCUBA dives found, explored, and mapped major new sections of Goliath's Cave via David's Entrance. Cave radio techniques documented that Venita Sikkink's house, more than 1.6 km east of the natural entrance, is about 46 m directly above the largest body of water in Goliath's. A new, large, epigenic section of Goliath's, named the "Iconoclast" section, was discovered by SCUBA diving through a sump (Fig. 3.24). The Iconoclast section, probably the geologically newest section of Goliath's Cave, drains surface water from Downwater Sink (23B20) and other sinkholes and stream sinks in the southern part of the CGBV SNA to the Rubicon River passage. A climb up a narrow crevice near the east end of the cave discovered a 400 m long upper level walking passage over the Rubicon River passage. A series of SCUBA dives pushed the downstream sump, the eastern end of the cave, about 200 m to a major underwater collapse. "Since the water in this stream passage has been traced to Forestville State Park, it is estimated that a huge cave system awaits discovery along this subterranean route. At this time, it is highly doubtful that a connection can be made through the Goliath's Cave system unless the large terminal sump opens, allowing us access to the remaining stream passage."

David's Entrance allowed safe, easy access to Goliath's Cave for scientific studies. Dye traces using autosamplers to create breakthrough curves in the cave became feasible. Several years of continuous temperature, conductivity, and water level records in the Rubicon at the bottom of David's Entrance documented very rapid changes in all three parameters in response to recharge events (Fig. 3.25). This figure is discussed further in the Springs Section later in this chapter.

Fig. 3.24 Cave diver John Preston in the Iconoclast section of Goliath's Cave. Note the pronounced "knobblies" walls, ceiling and floors of this geologically young, epigenic passage. Photo from John Ackerman

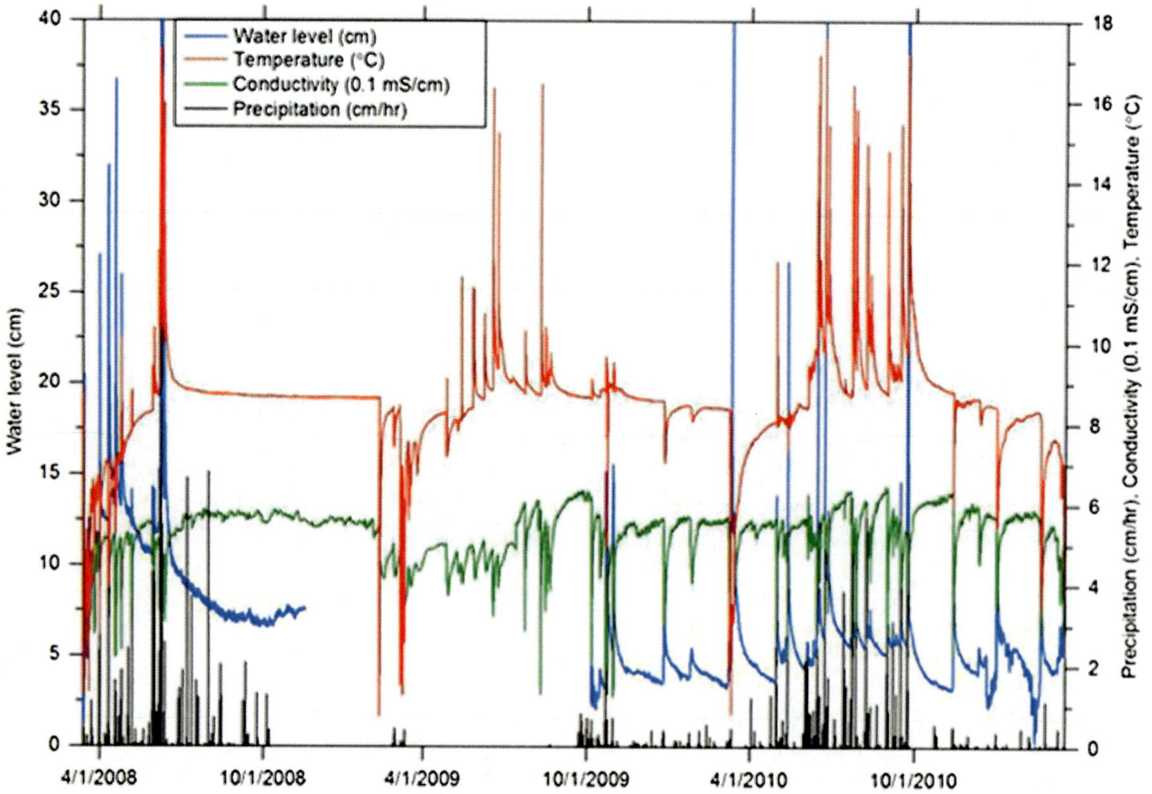

Fig. 3.25 Water level, temperature, and conductivity record from the Rubicon River at the bottom of David's Entrance in Goliath's Cave (Alexander et al. 2015). The largest event was the June 6–7, 2008 flood that opened Holy Grail Cave

York Blind Valley/Odessa Springs Underground River System

The largest known underground river system in Minnesota connects the York Blind Valley (YBV) (23B36) to Odessa Springs (23A20 and 23A240) and Border Uprising Spring (23A138). The YBV in Sect. 21 of York Township of Fillmore County is the largest known blind valley in Minnesota. Runoff from 30.4 km^2 of the (nominal) head of the Canfield Creek watershed feeds a perennial stream that enters the northwest corner of Sect. 21. That segment of Canfield Creek sinks in one or more of about a dozen major stream sinks in the blind valley. The "nominal" in the preceding sentence is because under all but the largest floods, those 30.4 km^2 are part of the drainage basin of the Upper Iowa River via the underground river. Only the largest flood events, flood the YBV and contribute part of that flow downstream through the Canfield Creek valley to the SBRR. The streamsinks in the YBV are at about a 390 m elevation in the Spillville Formation. There are two abandoned iron ore strip mines in the same section, on both sides of the YBV, which may indicate that hypogene processes were active in this area in pre-Pleistocene times (Alexander and Wheeler 2015).

Odessa Spring emerges from the base of a limestone cliff on the north side of the Upper Iowa River in Sect. 31 of Harmony Township, 16 km east of the York Blind Valley. It is the largest known spring in Minnesota with a flow that varies from a base flow of about 1.3 m^3/s to a high flow of over 5.7 m^3/s (UIRWP 2020) (Fig. 3.26). Odessa Spring emerges from the lower Prosser Formation at about a 334 m elevation—56 m lower and 15.75 km east of the sinks in YBV. Dye traces beginning about 1990 have documented the connection between York Blind Valley and Odessa Spring.

Figure 3.27 shows the positions along the southern border of Fillmore County with northern Iowa, of YBV, Odessa and Border Uprising Springs, and Niagara and Holy Grail Caves with the results of the dye traces indicated by the black arrows.

In the late 1880s, "speculating entrepreneurs attempt to gain access to an immense cave they felt existed behind the largest spring resurgence in Minnesota. The Odessa Spring project was so involved that a shanty town named 'Odessa' sprouted up in front of the spring. Odessa Spring is located at the base of a towering bluff along the Upper Iowa River, which is the border between Minnesota and Iowa. The speculators followed the spring into the bluff by blasting a large tunnel, but eventually gave up after running out of funds. The town quickly died with the project" (MCP 2020). The rectangular, west opening (23A20) of Odessa Spring is the remains of that 1880s effort (Fig. 3.26).

For the past several decades cavers had been searching for entrances into what must be a huge cave system. There are scattered sinkholes between YBV and Odessa Spring. They tend to be further apart but larger than most sinkholes in Fillmore County. The largest known individual sinkhole,

Fig. 3.26 Odessa Spring looking north. The left spring is 23A20. The right spring is 23A240. Photo by John Barry

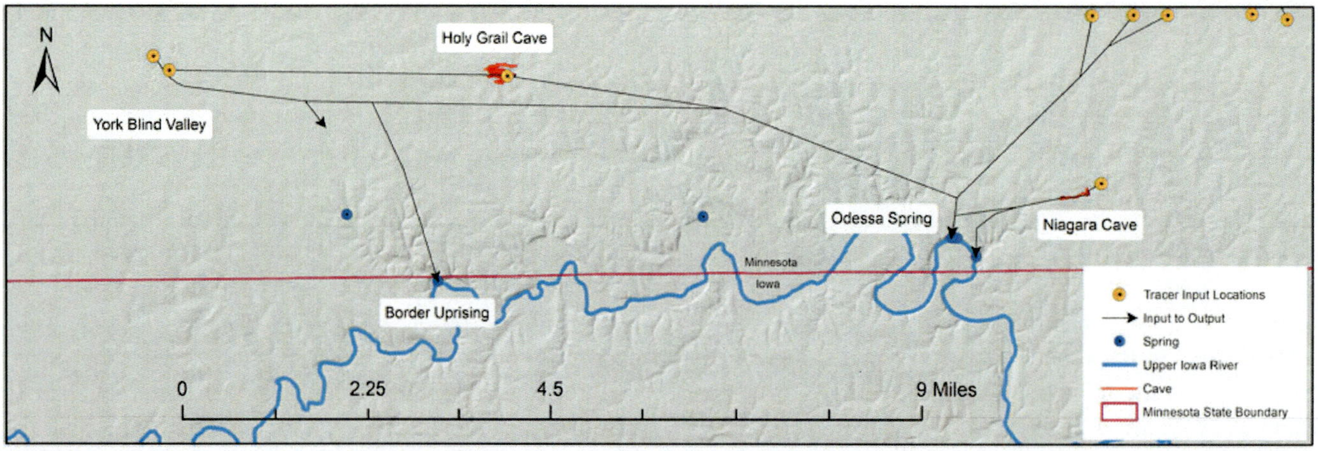

Fig. 3.27 York Blind Valley/Holy Grail Cave/Odessa Spring/Niagara Cave connections. Background LiDAR shaded relief DEM [9 miles = 14.5 km]

Fig. 3.28 Newly formed sinkhole entrance to Holy Grail Cave in 2008. Source: Photo from John Ackerman

Jesse James Sinkhole (23D6744), is a few kilometers northwest of Odessa Spring along the potential flow path. But no entrances were found in any of the existing sinkholes. The stream sinks in the YBV are unenterable crevices and bedding plane partings in the Spillville Formation. The presence of Niagara Cave, a show cave described below, on the eastern edge of Odessa Spring's springshed and the fact that Coldwater Cave was only a few kilometers away in Iowa, all indicated a potential for truly big caves between YBV and Odessa Spring.

Holy Grail Cave

In June 2008, 28 cm of rain hit southern Fillmore County. The resulting flood runoff opened a new sinkhole (23D5160) in Sect. 19 of Bristol Township. Figure 3.28 shows the sinkhole entrance not long after it formed. At the bottom of that sinkhole a 15 m deep bedrock shaft led to what has proved to be a 7.4 km long maze cave (Fig. 3.29). This was the "Holy Grail" for which the cavers had been searching—and that became the newly opened cave's name.

The immediate problem was that Holy Grail Cave (HGC) needed to be protected. The sinkhole entrance was dangerous, both to anyone who might try to enter the cave without equipment and training to safely negotiate the entrance shaft and cave, and to the cave itself. The sinkhole had formed in a grassed drainage way and future runoff would flood runoff and sediment into the cave. Although the cave was on private property, the Minnesota Cave Preserve

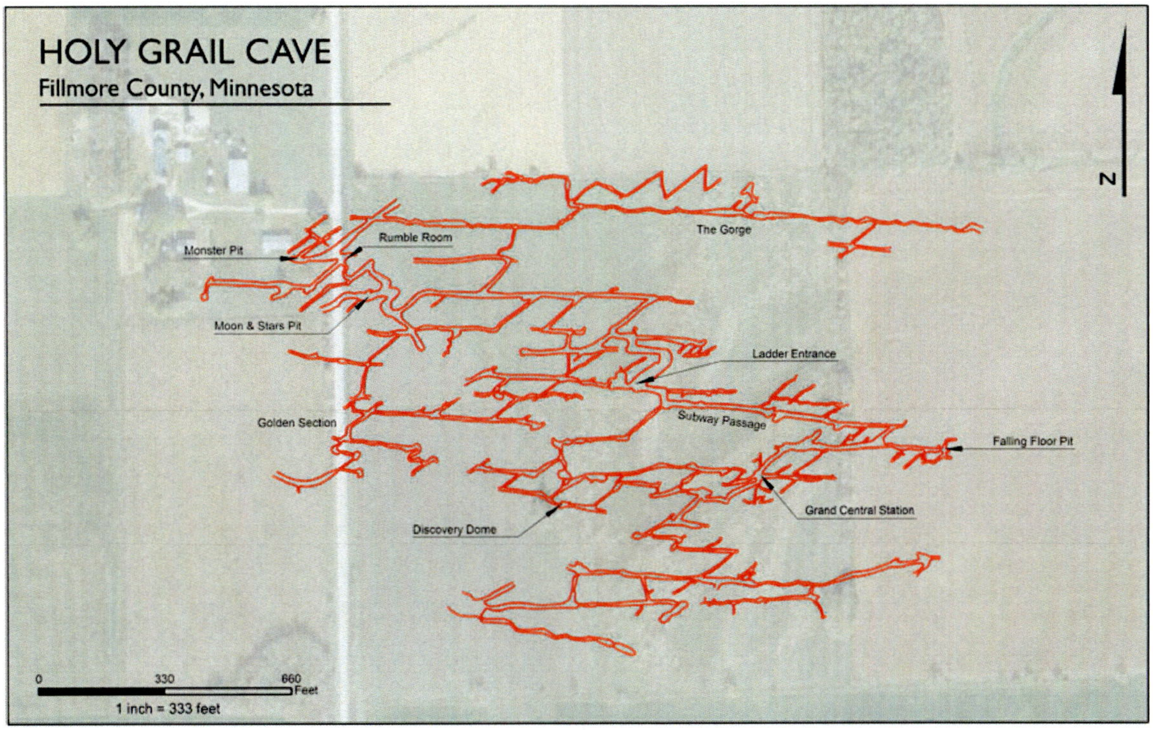

Fig. 3.29 Holy Grail Cave on air photo/LiDAR shaded relief DEM background. Mapped by the Minnesota Cave Preserve. Graphic by Martin Larsen [660 ft = 201 m]

volunteered to cap and seal the sinkhole entrance. The property owners agreed. They were supportive of John Ackerman's quest to preserve, study, and protect Holy Grail Cave. They eventually sold John 4.4 surface ha and 60 ha of subsurface cave rights and HGC became part of the Minnesota Cave Preserve. The sealing of the sinkhole entrance, the subsequent drilling of a 20 m deep, 0.76 m diameter permanent shaft entrance, and the cave's simultaneous exploration and mapping are photo documented in MCP (2020).

HGC is a maze cave in plan view (Fig. 3.29). The cave is developed predominantly along east/west and intersecting northeast/southwest joint sets. But HGC is anomalous in several ways. The cave has no perennial surface water input. There are no stream sinks near the cave. Before the June 2008 flood created the new sinkhole entrance, there were no sinkholes over the cave. The cave is located high in the local landscape without a large surface drainage input. HGC has, by Minnesota caving standards, enormous passages with exceptional vertical development (Fig. 3.30). The upper level passages are in the Maquoketa Formation. Many of the passages and rooms have deep pits that extend down through the Dubuque to a water table in the Stewartville Formation. Although that water table appears to be at relatively constant level across the cave at any one time, the water table level varies by many meters from visit to visit. During flood events on the surface, HGC floods to the ceiling and well up the artificial entrance shaft. Note the deposits of recent, black top soil on ledges all the way to the ceiling in Fig. 3.30.

HGC floods from below. In high water events, water rises through the floor pits, flows short distances horizontally, roughly west to east, through the cave, and then sinks back down through other floor pits. In 2009, a dye trace from ceiling drips at Grand Central Station on the west side of the cave that went to Odessa Spring. In 2017, a dye trace from the terminal sink in York Blind Valley (23X319) was positive to the Falling Floor Pit stream in HGC (Larsen et al. 2018). The underground river from YBV to Odessa runs through HGC. Holy Grail Cave may be a vertical flood water maze cave, that is, a more vertical analog of Art Palmer's (1975) classic flood water maze model. Some sort of restriction in the main conduit system, in the Stewartville Formation, perhaps a breakdown collapse, forces the water to enlarge joints (vertically in this case) to flow around the blockage.

In 2016, a caver entered the Moon and Stars Pit in HGC, looked up and saw "a light up above" that had not been there before. A new sinkhole had opened in the ceiling of the cave (Fig. 3.31). A search on the surface found the new sinkhole (23X61). The sinkhole was about 1 m in diameter and 38 m deep. That new sinkhole was sealed to prevent a potential accident.

Fig. 3.30 Main passage and pit in Holy Grail Cave. Photo from John Ackerman

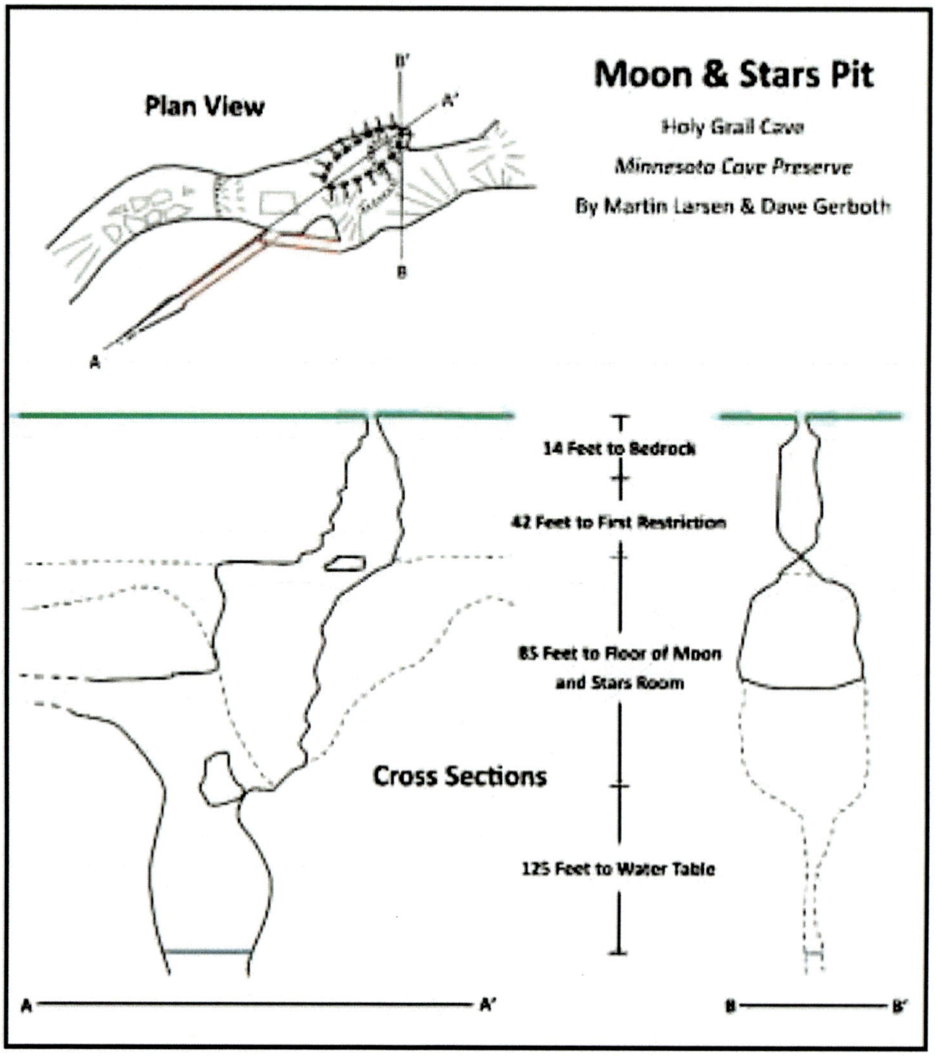

Fig. 3.31 Plan and cross-section of second Holy Grail Cave sinkhole entrance. Graphic by Martin Larsen [125 ft = 38 m]

Fig. 3.32 Water Fall in Niagara Cave. Photo provided by Niagara Cave

Niagara Cave

Niagara Cave is a show cave located southwest of the town of Harmony (Niagara Cave 2020). Figure 3.32 shows the water fall that is the source of the cave's name. Figure 3.33 shows the plan and cross-section maps of the cave. Niagara Cave was discovered in 1924; it was explored and named by cavers Al Cremer, Leo Tekippe, and Joe Flynn in 1932; opened as a show cave in 1934 and described by J Harlen Bretz (1938). Niagara is one of the deepest caves in Minnesota. The cave is currently owned and operated by the Bishop family. The mapped cave is 1.37 km long and 52 m deep. The commercial tour is about 46 m deep, descends, and then climbs 275 stair steps, and is a round-trip distance of about 1.6 km.

The entrance of Niagara is a modified sinkhole in the Dubuque Formation under the entrance building. The tour descends through the Dubuque to a junction in the Stewartville Formation. To the right a perennial cave stream flows from a conduit (Fig. 3.34) into the Wishing Well pool, then down a steep-gradient passage and forms an 11 m waterfall down the side of the "Water Fall Dome" pit. The water fall emerges at about the middle of the wall of Water Fall Dome. An artificial, dead end bridge extends across the dome pit and allows a better view of the dome pit and waterfall (Fig. 3.32).

There is a half tube in the flat ceiling of the passage between the Wishing Well Pool and the top of the water fall. That half tube records the flow path of the original conduit that carried water from the Wishing Well to the side of Water Fall Dome. The floor of that passage is being eroded by the stream producing a steep gradient to the top of the waterfall. The stream at the bottom of the Water Fall Dome flows into a network of low, joint-controlled, water-filled passages which quickly sump. That water appears again as the stream flowing through lowest level at the downstream end of the cave.

Turning left from the junction at the bottom of the entrance stairs, one starts down the main passage of Niagara Cave (Fig. 3.35). The main passage is an abandoned stream passage with solutionally sculpted walls that cut across the relatively flat-lying bedrock. The ceiling of this passage remains relatively level while the floor slopes downward at an initially low gradient. The main passage developed in alternating segments of the roughly east/west and northeast/southwest joint sets is common to most caves in the Upper Carbonates of southeastern Minnesota.

A small Wedding Chapel occupies a room next to the Echo Chamber, which functioned geologically as a short siphon, now dry, in the original stream that formed the main passage. About two-thirds of the way along the main passage is one long, straight cave passage, named the Grand Canyon. The downward gradient of the floor steepens at the same time as the ceiling rises forming the tallest passage in the cave. The floor of the cave drops down into the Prosser Formation.

At the end of the Grand Canyon section, the passage jogs over to the bottom of the Cathedral Dome. The passage then descends down through Stalactite Room. An artificial platform in Stalactite Room is the end of the commercial tour and is in an abandoned lower level passage. From the end of the platform one can look down about 5 m to a cave stream that is the continuation of the water flow from the bottom of the Waterfall Dome and the current lowest level in the cave.

In the 1970s, the lower level room beyond the end of the platform was almost filled with black topsoil that had been washed into the cave after settlers had converted the watershed feeding the cave to row crop agriculture in the 1850s and 1860s. That black topsoil has been largely eroded out of the cave, thanks to the ongoing soil conservation efforts that have significantly decreased the soil washing into the cave. The cave is cleaning out a century and a half accumulation of excess erosion filling the lower levels with sediments.

Niagara Cave is a tutorial on the ongoing evolution of karst drainage. A clear evolutionary history is visible in the

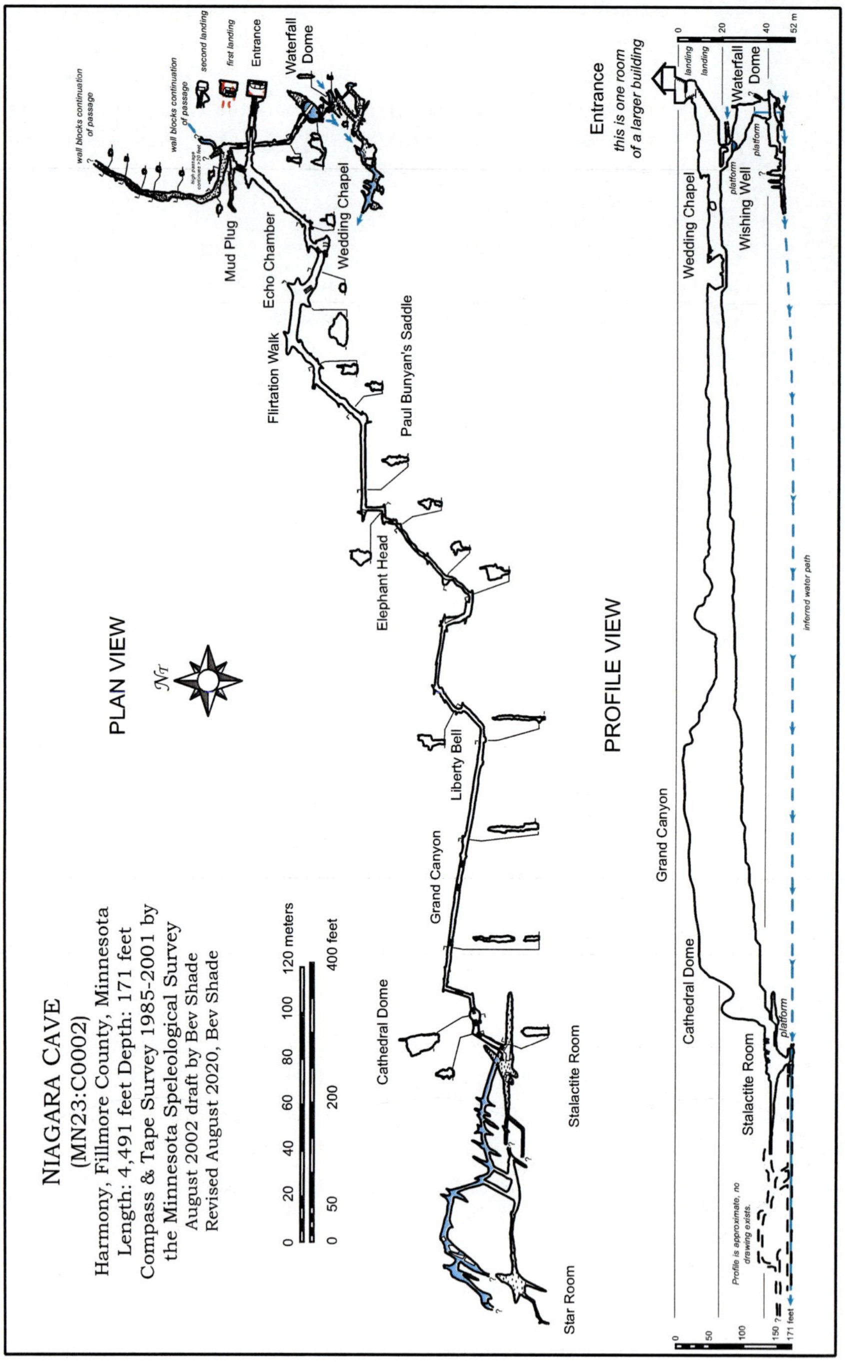

Fig. 3.33 Plan and cross-section maps of Niagara Cave. Image provided by Niagara Cave

Fig. 3.34 Stream entering upper end of Niagara Cave. Photo provided by Niagara Cave

Fig. 3.35 Main passage in Niagara Cave. Photo provided by Niagara Cave

cave's morphology. The main cave passage formed early. Surface runoff sank in the entrance sinkhole, and perhaps, other nearby sinkholes formed a paleo stream. That stream flowed along a sub horizontal bedding parting in the Stewartville, along joints, to the top of the Cathedral Dome. That paleo cave stream eroded down through the bedrock wall of the Dome and the floor of the main passage forming the current downward sloping floor the length of the main passage. Sinkholes upstream of the entrance sinkhole then pirated the flow into the currently active underground stream inlet (Fig. 3.34). That stream may have fed the main passage paleo stream for some time, but eventually dissolved a horizontal conduit to the much nearer Waterfall Dome. The record of that shortcut is the half tube in the flat ceiling of the current stream passage from the Wishing Well to the top of Waterfall Dome. That segment of the current stream is in the process of eroding down the side of the Waterfall Dome. The underground karst drainage system migrates up stream by successive pirating of the flow via dome pits.

Niagara Cave is a part of the springshed of Odessa Spring (Fig. 3.27). Multiple dye traces from the cave stream flowing through Niagara Cave have confirmed that the cave stream feeds Odessa Spring and Hawkeye Spring (23A22). Surface runoff from about a 10 km^2 watershed to the north of Niagara Cave feeds into sinkhole/stream sink 23B70/23D6555 across the road, northeast of the cave. A tile drain outlet also flows continuously into that sinkhole/stream sink and is a major component of the flow through Niagara Cave.

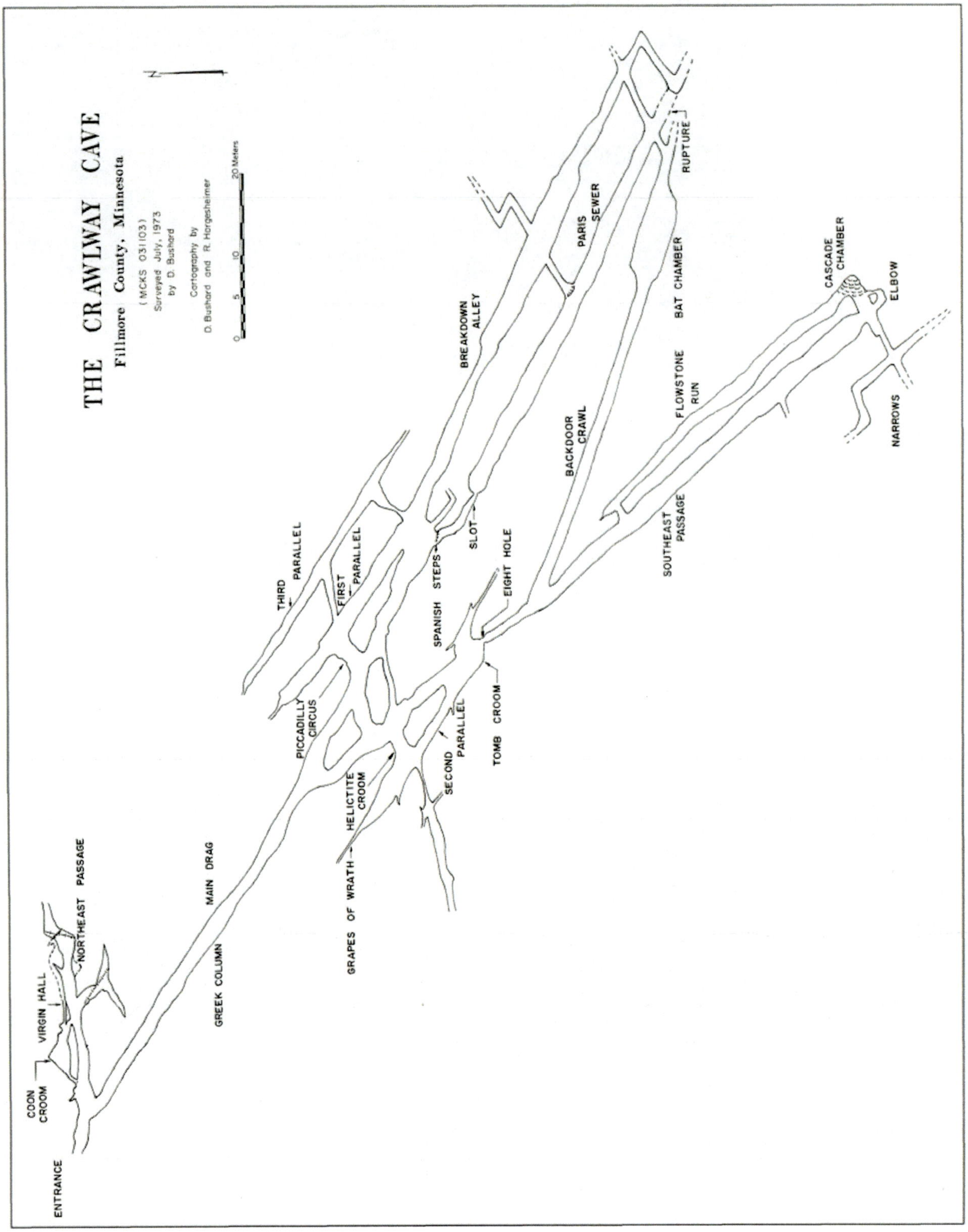

Fig. 3.36 Crawlway Cave map (Spong 1980a)

3.2.1.4 Smaller Upper Carbonate Caves

The Galena Carbonate bedrocks are the first, near-surface layers in a northwest trending band from the Iowa border in Fillmore and western Houston counties, through Olmsted and eastern Dodge County to the Cannon River in Goodhue County. These Upper Carbonate rocks host hundreds of small caves ranging from a few meters to many tens of meters in length. There are six small "outlier caves" around Mystery Cave, **Old Mystery** and **Old Still Caves** (Scobie 1980), **Copeman's**, **Two Skull**, and **Windcourse Caves** (Saylor 1980), and **Petrified Indian Cave** (Oothoudt 1980). The CGBV SNA contains four small caves, **Woodchuck Cave**, **Downwater Cave, Upper Downwater Cave**, and **Wobegone Cave** in addition to the natural entrance of Goliath's Cave (Spong 2006). The Minnesota Cave Preserve lists 35 smaller caves in addition to Spring Valley Caverns, found by excavating sinkholes on the Cave Farm.

In southern Olmsted County there are two small spring entrance caves, **McConnell Cave** (23A9) and **Old Jug Cave** (23A6) along Kinney Creek. Along Bear Creek, southeast of Rochester there is a short, enterable cave: **Bear Overflow Spring Cave** (23A572). All three of these caves fill completely under high flow conditions.

There are a wide variety of smaller caves in the Upper Carbonates. The systematic vertical joint sets seen throughout the Upper Carbonates influence most of the smaller caves to a greater or lesser degree. Some of the caves appear to be recent epigenetic solution caves actively participating in the current drainage systems. Some of the caves appear to be paleo epigenetic solution caves, no longer part of the currently active drainage systems. Some of the caves appear to be reactivation of much older filled karst features. There are a few small caves that appear to have been formed by physical movement of blocks of bedrock.

From a caver's perspective a basic question is: "What fraction of the many sinkholes in the Upper Carbonates have caves under them?" The Cave Farm has provided a significant data set. "I have dug in 51 sinkholes at the Cave Farm and have discovered 38 caves at the bottom of them. Some were worthy of permanent entrances and some I resealed" (John Ackerman, pers. comm. 2019). For the Cave Farm the apparent answer to the basic question is "about 75%." It remains to be seen if the Cave Farm data set is representative. The 2008 opening of the HGC sinkhole entrance suggests that there are many more caves than are currently known.

3.2.2 Platteville Formation Caves

The three formations stratigraphically below the Cummingsvillle Formation—the Decorah Shale, the Platteville Formation, and the Glennwood Formation (Figs. 3.2 and 3.3)—are often lumped together as the aquitard between the Upper Carbonates and the underlying St. Peter Sandstone. The Decorah and Glennwood are clay-rich shales and act as leaky aquitards. However, the Decorah is jointed and fractured and hosts seeps and small springs in numerous outcrops. The Platteville is a complex carbonate unit (Mossler 2008). It is the first bedrock under much of the Twin Cities and forms the ledge over which the St. Anthony Falls of the Mississippi River and Minnehaha Falls formed. Platteville limestone was locally available and used extensively in the 1800s as dimension stone for foundations, buildings, and other construction purposes. The use of the Platteville for construction purposes has been largely abandoned because the Platteville decomposes and crumbles when exposed to surface temperature and moisture fluctuations.

The Platteville acts as an aquitard vertically but as an aquifer horizontally (Runkel 2007). Numerous springs and seeps, emerging from the Platteville along the Mississippi River Gorge downstream from St. Anthony Falls, attest to the Platteville's high horizontal transmissivity (Brick 1997a). One significant cave, **Crawlway Cave**, has been documented in the Platteville Formation west of the town of Fountain. The next four paragraphs and Fig. 3.36 are from Spong (1980a, p 41, 42):

> "The Crawlway is the most extensive solution cave in Minnesota reported in the Platteville Formation. The cave is developed almost wholly in the McGregor Member of the Platteville which locally is a massive-bedded argillaceous dolomitic limestone with shaley partings and is about 3.7 m thick. Slumping in Virgin Hall has exposed the upper 0.8–0.9 m of the Pecatonica Member a 2.7-m-thick, nodular, thin-bedded limestone with shale. The prevalence of copious deposits of sediments (cave fill) obscures the true vertical extent of the cave. The overlying Carimona Member, a 3-m sequence of shale and interbedded limestone is partially exposed with the Decorah Shale above in the roadcut near the Crawlway's entrance. The McGregor and Pecatonica Members crop out below the cave entrance but are normally covered by colluvium.Morphologically, the Crawlway is strictly joint controlled and reticulated (Fig. 3.36). Northeast, cross-joint crevices intersect the major, southeast passages, but sediment has either partially or totally filled countless crevices thereby restricting access to the major passages. Original passage cross sections are nearly obscured by sediment; except for Virgin Hall, only the crowns of the original passages are visible. Crowns are intact throughout most of the cave although is some areas ceilings have weathered to thin plates because of the hydroscopic expansion of shale and have spalled. Extensive slab breakdown occurs at many wide-joint intersections where ceiling beams have failed resulting in progressive collapse upward into tension domes. Cantilevers, remnants of ceiling beams are common in these areas. The thickness and extent of the sediment are unknown, but the slumping of fill which created Virgin Hall exhibits a vaulted cross section that may be typical of other partially filled passages. A 3-m high profile of stratified sediments is exposed by the slumping in Virgin Hall and is comprised chiefly of silt and sand with smaller amounts of clay and gravel. The Crawlway was developed in the phreatic zone, as most speleogens observed (network pattern, stratigraphic control, scalloping and tubular conduits) are phreatic indicators.

> Subsequent rechanneling of the sediment-filled passages by a free surface stream was selective. For example, the Eight Hole was completely choked with fill before it was excavated open. Drip stone and flowstone speleothems are primarily composed of calcite, with some occurrences of aragonite and gypsum. The Platteville is pyritiferous which accounts for the presence of gypsum, limonite and siderite speleothems.There is no conclusive evidence that the Crawlway has ever been associated with the contemporaneous karst drainage net of the overlying sinkhole plane. Groundwater is recharged by both infiltration and the capture of surface runoff by sinkhole and is collected in conduits at the Galena-Decorah contact. Vertical shafts in the Galena caprock are commonly formed in the vadose zone below intermittent sinkhole ponds but usually lack horizontal development unless they intersect antecedent conduits. Gravity springs and seeps are found in the valley scarps where the contact is breached. Discharge is focused by cave springs, integrated conduits which tap the karst reservoirs.The Platteville Formation is a confined aquifer below the Decorah Shale and I hypothesize that the Crawlway was develop by an artesian system initiated when the youthful Sugar Creek breached the Decorah. The supposition that a sulfuric acid mechanism in combination with the carbonic acid mechanism accelerated dissolution is supported by the prevalence of pyrite and marcasite in the Platteville. The preglacial depth of Sugar Creek is unknown, but it was probably 12 m below the cave entrance. As glacio-fluvial deposits filled the preglacial valley, the free surface stream in the Crawlway aggraded gradually filling the cave. Post-Kansan drainage re-established many preglacial valleys including Sugar Creek and selective rechanneling through the sediments began when Sugar Creek entrenched below the cave level. Aeration and speleothem deposition followed, and the natural entrance was eventually sealed."

The last two paragraphs above are careful, detailed observations interpreted in a speleogenic model of what, in current nomenclature, would be called hypogene speleogenesis. "Hypogene" was not part of many American karst scientist's lexicon in 1980. Ron Spong's interpretation was ahead of its time.

3.2.3 St. Peter Sandstone Caves

Much of the information for this section is derived from *Subterranean Twin Cities* (Brick 2009a). Other karst features in the Minneapolis-St. Paul metropolitan area are described by Brick (1997a).

The St. Paul area was known to the Dakota Indians as "White Rocks" because of the snowy white sandstone, exposed in its river bluffs, and geologist David Dale Owen (1807–1860) officially named this rock in 1852 for outcrops near Fort Snelling, along the then St. Peter's River (now Minnesota River). The St. Peter layer has an average thickness of about 30 m regionally. However, it's about 45 m thick at its type section at Fort Snelling. That leaves plenty of three-dimensional volume for caves and tunnels.

The St. Peter Sandstone is very extensive for a single formation, underlying 650,000 km^2 in the Midwest. It is a "sheet sand," meaning that it was laid down flat, like a sheet, over large areas, by a warm, shallow, equatorial sea that invaded the continent from the south. It was the last major sandstone layer to be deposited in the Upper Mississippi Valley, in Middle Ordovician times. In unlined utility tunnels you can sometimes see cross bedding in the walls, representing the old sand dunes of the shoreline beaches.

The St. Peter has an almost saintly purity throughout most of this range, suggesting that it has been recycled from older sandstones, geologically winnowed of its impurities, leaving it 99.44% pure silica. In glass recipes, even small amounts of contaminants are injurious, so this degree of purity was considered crucial. Locally, the sand was used for making windshield glass, as at the former Ford Motor Company plant in St. Paul. Other uses were for foundry sand and mortar sand.

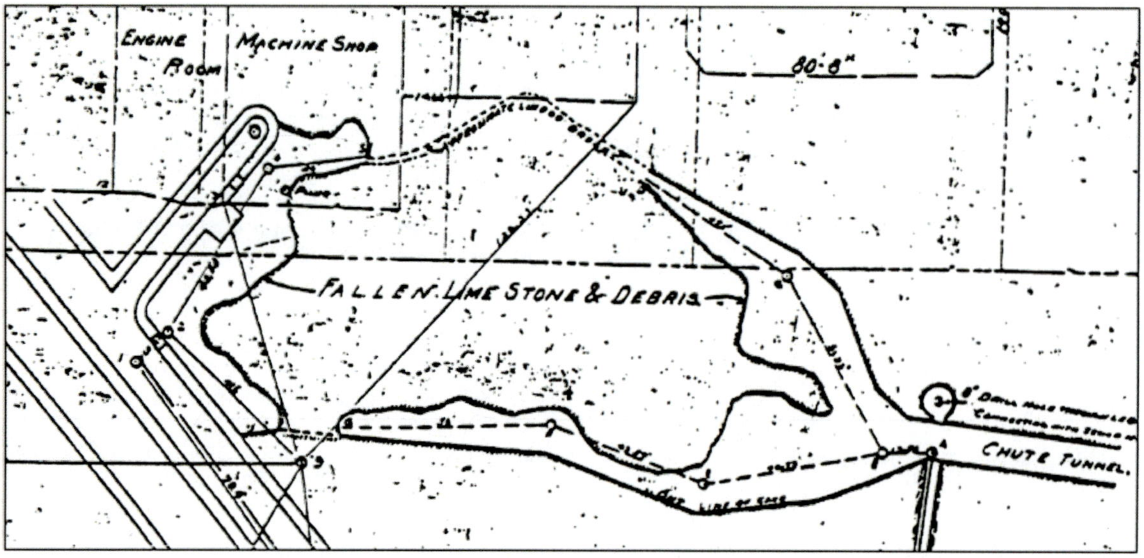

Fig. 3.37 Chute's Cave, 1909 survey. Courtesy of Minneapolis Public Works

Most importantly, the St. Peter Sandstone, in the Minnesota part of its range, lacks natural cementation, hence the individual sand grains are easily excavated by natural and artificial means. Natural caves form in the St. Peter by a process known as "piping," a form of erosion caused by flowing groundwater. The term piping was derived from the pipe-shaped voids created by water flowing underground. Piping forms two different kinds of caves in the St. Peter: tubular caves, best exemplified by Fountain Cave in St. Paul, and maze caves, best seen in Schieks Cave under downtown Minneapolis (Brick 1997b). One layer within the St. Peter, called the "Cave Unit," is more susceptible to piping and thus more favorable to cave formation than others. Five significant St. Peter caves are known in the Twin Cities. The two St. Paul caves, **Carver's** and **Fountain**, much more famous than the rest, are covered at length in Brick (2020b, c), Chaps. 2 and 8 of this volume, respectively.

3.2.3.1 Caves Under Minneapolis

Three significant caves have been discovered under Minneapolis by workers boring mill or sewer tunnels in the St. Peter Sandstone: Chute's, Schieks, and Channel Rock. None of the three big Minneapolis caves are easily accessible to the casual explorer as are the caves in neighboring St. Paul.

Chute's Cave is situated in the St. Peter Sandstone with a ceiling formed by the overlying Platteville Limestone. The cave is on the northeast side of the Mississippi River near St. Anthony Falls. While not the largest cave under Minneapolis, the cave is about 30 m wide and 60 m long and collapsed in the center. The floor-to-ceiling collapse mound fills the greater part of the cave. Adelbert Russell Moore, engineer for the St. Anthony Falls Water Power Company, explored Chute's Cave in 1909 and drafted a map (Fig. 3.37), describing the mound as consisting of "fallen lime rock in a very irregular shape, but on the whole somewhat resembling a huge fountain built up in tiers, over which water trickled, the water being impregnated with iron had colored the stone to almost a jet black giving it an extremely beautiful appearance." The water-laid mineral deposits described by Moore are well known to cavers as flowstone, and Moore's mound is assuredly what had been referred to as the "Tower of St. Anthony" in the late 1800s Nesmith Cave hoax, which was based on this cave (Brick and Petersen 2004).

By 1990, when Chute's Cave was rediscovered by the Minnesota Speleological Survey, the accessible portion consisted of the void space around the perimeter of this enormous mound, and the big room itself. Moore himself called the big room "very wonderful." The triangular chamber, about 15 m on a side and 4.5 m high, contained a primeval forest of decayed timber, at one time meant to hold up the overlying street. Giant slabs of limestone, the size of pontoon boats, littered the floor. Some of them are coffin-shaped—the result of the characteristic rhomboidal jointing pattern of the Platteville Limestone layer, from which they had detached themselves. Another slab jutted out into the huge room like a rocky pulpit, or the prow of a ship (Brick and Petersen 2004).

The origin of Chute's Cave has long been obscure. Was it, like other natural caves in the St. Peter Sandstone, created by groundwater washing away sand grains? Or is it artificial, merely a widened segment of Chute's Tunnel? Or is it partly both? The claim that Chute's Cave is a natural cave was made by Moore in 1931 and became the basis for the account presented in Lucille Kane's classic book, *The Falls of St. Anthony*. But Jeff Dorale (see Chap. 11, this volume)

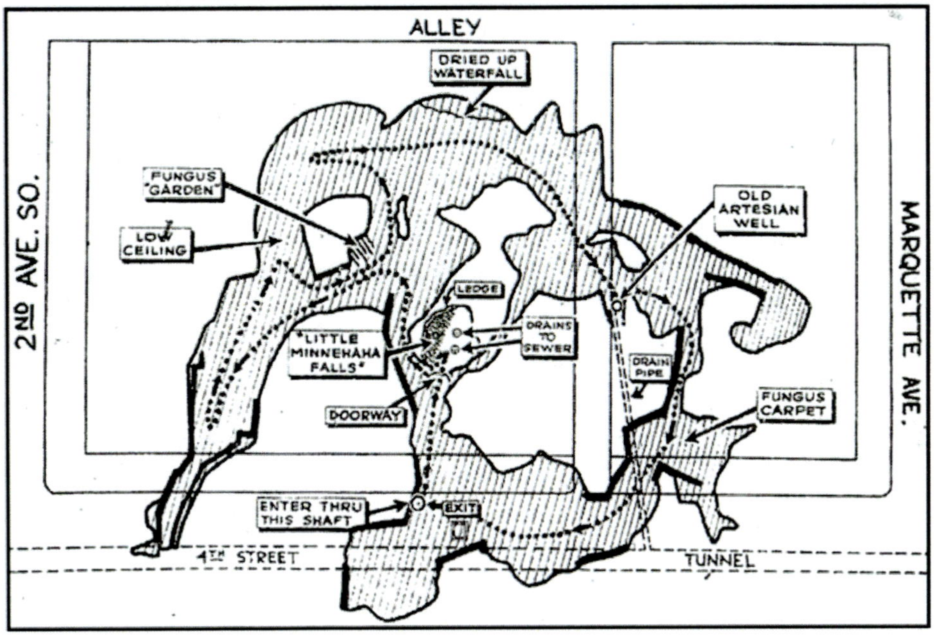

Fig. 3.38 Map of Schieks Cave as visited by David Dornberg (1939), a simplified version of the 1929 Lawton map

took samples of flowstone in the year 2000 from the "Tower of St. Anthony" in Chute's Cave for radiometric dating and found that the flowstone post-dates 1880 (Brick and Petersen 2004).

Schieks Cave, the largest of an archipelago of sewer voids below downtown Minneapolis (Brick 2002), extends for a city block through the St. Peter Sandstone. Carl J. Illstrup, city sewer engineer, who discovered the cave in 1904, described it as a "cave shaped like an inverted bowl," a description that seems puzzling to anyone who has visited the cave. The earliest known documentation of Schieks Cave is Lund's 1904 survey, which notes the former "creeks" and "lakes" of the cave. Reportedly, the cave was kept a secret for several years because city officials feared the public would think downtown Minneapolis was built on a thin shell that would plunge into a hole in the Earth. Another concern was that burglars might have worked undetected and bored directly into the bank's treasure vaults.

In 1907, journalist Jack Longnecker penned an atmospheric discovery narrative titled, "In Caverns of Eternal Night" (Longnecker 1907). By 1929, a Minneapolis Sewer Department press release reported that "the entire business portion of the city is built over a series of subterranean lakes and caverns as mysterious and baffling as the Mammoth caves of Kentucky or catacombs of Rome."

In 1929, when Lawton resurveyed the cave, he labeled his map the Farmers & Mechanics Bank Cave, with reference to the bank building directly above the cave. Lawton depicts the cave extensively modified by the construction of piers, walls, and drainage structures. The Lawton map also shows the 23 m entrance shaft on Fourth Street, the access point for sewer crews today. The original 1904 and 1929 maps are in the Minneapolis Public Works Department files and reproduced as Plate 3 in Kress and Alexander (1980).

In 1931, journalist, Robert Fitzsimmons, waxed poetic about "the beauties of the sewer system" and described Illstrup as "the ruler of this fantastic world." The discovery of the cave in 1904, during the excavation of the North Minneapolis Tunnel, when the crews braved "the lethal breath of deadly gases," was presented as the highpoint of Illstrup's life (Fitzsimmons 1931; Brick 2004).

In 1939, the *Minneapolis Journal* photographer David Dornberg went on a "Camera Safari," as he called it, through this murky cavern (Dornberg 1939). He described it as "a 'lost world,' weird and spooky—the darkest spot for adventure into which my four years as a Journal cameraman ever led me" (Fig. 3.38). One of his most notable observations was "Little Minnehaha Falls," a natural ceiling spring inside the cave. Many years later this groundwater was found to have a temperature far above normal (Brick 2006, 2009a, p 199, 2014).

Schieks Cave's current name derives from Schieks Palace Royale, a gentlemen's club that has occupied the Farmers & Mechanics Bank building for many years. Journalist Kay Miller once described the nightclub as "topless above, bottomless below" because of the underlying cave, but there's no humanly passable physical connection between them.

In 1952, the Twin Cities Grotto, local chapter of the National Speleological Society, visited Schieks Cave, and it was written up for the Sunday supplements. The late David Gebhard, author of architectural guidebooks, was a member of this club. In 1983, the Minnesota Speleological Survey visited the cave, leaving their account of the gloomy cave whose walls were black with cockroaches and the roaring sewer that ran under it (Downes 1983).

The earliest published account of someone visiting Schieks Cave without official sanction was by Roger Kehret. In a 1974 story titled "The Minnesota Rovers Great Manhole Cave Expedition," Kehret revealed how the Rovers, an outing club, dressed as sewer workers, surrounded the entrance manhole with barricades, and used a truck-mounted winch to remove the heavy hexagonal lid (which in recent years has been welded shut). Getting there must have been more than half the fun, however, because Kehret said little about the cave itself, except that it was "almost filled with sand, concrete, pipes and other debris" (Kehret 1974).

Kress and Alexander (1980) discuss the history and speleogenesis of Schieks Cave. Schieks Cave appears to have been formed by the mechanical erosion of the soft St. Peter Sandstone by running water, known as piping. However, when the cave was formed is an open question. One view is "that the cave was formed 10,000–15,000 years ago…a relic of the Ice Age." A contrasting theory, based on the testimony of sewer engineer Illstrup, is that the cave "may have been formed by water escaping from an abandoned artesian well and washing the sand into the sewer," that is, that Schieks Cave may have formed quite recently. The sewer referred to is doubtless the North Minneapolis Tunnel, on which construction began in 1889. Brick (2009a) made several trips to Schieks Cave by hiking through the sewers in the year 2000. Brick observed that the "Old Artesian Well," marked on old maps, does not exist. Nor did it ever exist, because even if it had rusted away completely and the hole had become buried with debris, there would still be a corresponding drill hole through the ceiling, which there was not. Little Minnehaha Falls, or something like it, could have provided the source of water for eroding the cave, once a sink for the sediments had been established by construction of the adjoining tunnel.

No swarms of cockroaches were seen during Brick's 2000 visits to Schieks Cave. They had been displaced by the sort of fly-and-worm ecosystem that has been found in other polluted urban caves around the country (Holsinger 1966). The Schieks Cave biota is basically a guanophile (excrement-loving) community, with "fungus gardens" fed upon by swarms of fungus gnats which in turn supported the spiders in the cave (Brick 2009b).

Fig. 3.39 Channel Rock Cavern, near south end looking back up the main passage. The floor is spoils sand from the 1935 sewer interceptor tunnel excavation. 2008 photo by John Lovaas

Channel Rock Cavern is located immediately west of the Mississippi River gorge under East 34th Street between West River Road and the Seven Oaks Oval Park in Minneapolis (Spong 1980c; Barr and Alexander 2009). The Seven Oaks Oval Park is a large double sinkhole or uvala (85 by 150 m by 7.6 m deep). Channel Rock Cavern was discovered in 1935 during the construction of a sanitary sewer interceptor tunnel in the St. Peter Sandstone about 15 m below the ground surface. The construction came to an abrupt halt when the ceiling of the advancing tunnel collapsed revealing the huge cavern, over 240 m long. The first half of the cave runs west under East 34th Street, then making a right angle turn to the south and extending another 120 m to near the edge of the uvala. The passage is 9–18 m wide and was 6–9 m high (Fig. 3.39).

The cave is developed at the top of the St. Peter Sandstone. The flat ceiling of the main passage is the bottom of the Platteville Limestone. There are a couple of small side passages in the St. Peter walls of the main passage, but they pinch out quickly. The cave ends in a pool of water that is about the same level as the adjacent Mississippi River elevation.

The 1935 map surveyed by the City of Minneapolis shortly after the discovery of Channel Rock Cavern has been modified as shown in Fig. 3.40. Celebrated in the local press at the time, exploration, mapping, and visitation of the cave were facilitated by the excavation of an adit out to the river bluff

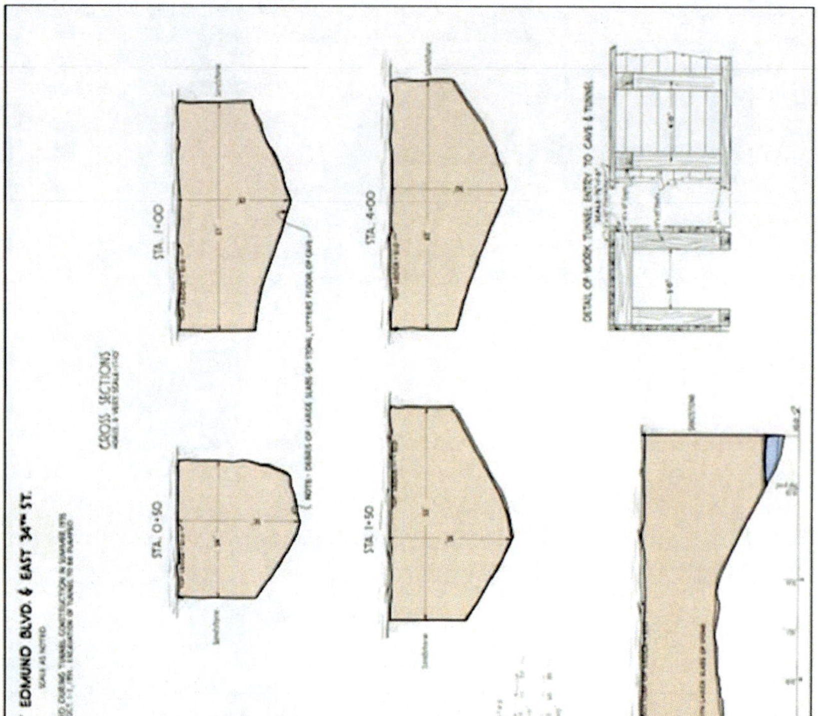

Fig. 3.40 Channel Rock Cavern Map, base map surveyed Oct 1–2, 1935 by City of Minneapolis

Fig. 3.41 A potential hypogenic feeder tube along a joint in the St. Peter Sandstone in Channel Rock Cavern. 2008 photo by John Lovaas

and an access shaft to the surface. The adit became an attractive nuisance and was sealed by the City of Minneapolis. Access to the cave is now via a 15.8 m deep access shaft which is covered by double hexagonal manhole covers in West River Road. As noted on the 1935 map the initial floor of the cave was covered by slabs of breakdown from the Platteville Limestone ceiling. Also noted on the 1935 map is the phrase "excavation from the tunnel to be pumped into the cave." Based on a comparison of the 1935 map with his 1979 map of the cave, Spong (1980c) estimated that the sand back fill ranges from 1 to 4.5 m in depth. The floor of the main passage is now buried with the sewer sand spoils (as seen in Fig. 3.39). Near the east end of the cave, that sand floor has a patina of sewage debris due to back flooding of the sewer interceptor tunnel during storm events. Both Spong (1980c) and Barr and Alexander (2009) propose that Channel Rock Cavern formed by hypogene processes as the upstream retreat of St. Anthony Falls along the Mississippi River passed the cave. The current river gorge from St. Anthony Falls downstream to the junction of the Mississippi and Minnesota Rivers at Ft. Snelling is a recent, post-glacial feature. The gorge has been created by upstream migration of the falls during the last 11,000 years. The falls moved past Channel Rock Cavern about 5,000–7,000 years ago (Wright 1972).

St. Anthony Falls represents about a 30 m drop in the local groundwater base level. The waterfall migration induces a large, transient zone of artesian overpressure, not only in the near surface aquifer, but in the entire stack of lower aquifers. Barr and Alexander (2009, p 57, Fig. 3.10a, b) suggest that the linear arrangement of the Seven Oaks collapse feature, Channel Rock Cavern, and the Shadow Falls ravine on the east side of the Mississippi, indicate a local zone of structural weakness that may have focused the artesian flow and produced the cave.

Along the sides of the main passage there are several open, enlarged joints that may have served as hypogenic feeder tubes for rising waters (Fig. 3.41). However, documentation of the speleogenesis of Channel Rock Cavern is difficult because the backfilling of the cave by sand from tunnel construction has covered the original floor. Moreover, the pool at the end of the cave appears to be controlled by the water level in the adjacent Mississippi River. In the adjacent Mississippi River an artificially high water is maintained at a high level to facilitate barge traffic.

3.2.3.2 Other St. Peter Caves

The known St. Peter caves south of Minneapolis and St. Paul are all small. In Fillmore County **Tennis Shoes' Cave** is one example of a small St. Peter cave (Fig. 3.42). In Olmsted County **Troy Sink** is another example. Troy Sink's entrance is a collapse feature through the Platteville Formation. A map of Troy Sink may be found on the Sinkholes and Sinkhole Probability Plate of the Olmsted County Geologic Atlas (Alexander and Maki 1988).

There are numerous stream sieves and sinks in the St. Peter. A common geometry is for surface runoff from areas higher than the St. Peter along with springs and seeps from the Decorah and Platteville to sink when those small streams reach the St. Peter subcrop. That water may feed springs lower in the St. Peter or drain underground to the Prairie du Chien aquifer.

3.2.3.3 St. Peter Sand Mines and Tunnels

Mushroom Valley

The information and quotations in this section are derived from and referenced in Brick (2009a).

The city of St. Paul—more so than its twin, Minneapolis—is known for its artificial sandstone caves, and the densest cluster of them is surely Mushroom Valley, across the Mississippi River from downtown St. Paul. In this regard, St. Paul's true geological twin city would be Nottingham, UK, famous for its many artificial sandstone caves, as described by Waltham (2008).

Mushroom Valley boasted that it was the largest mushroom-growing center west of Pennsylvania, or

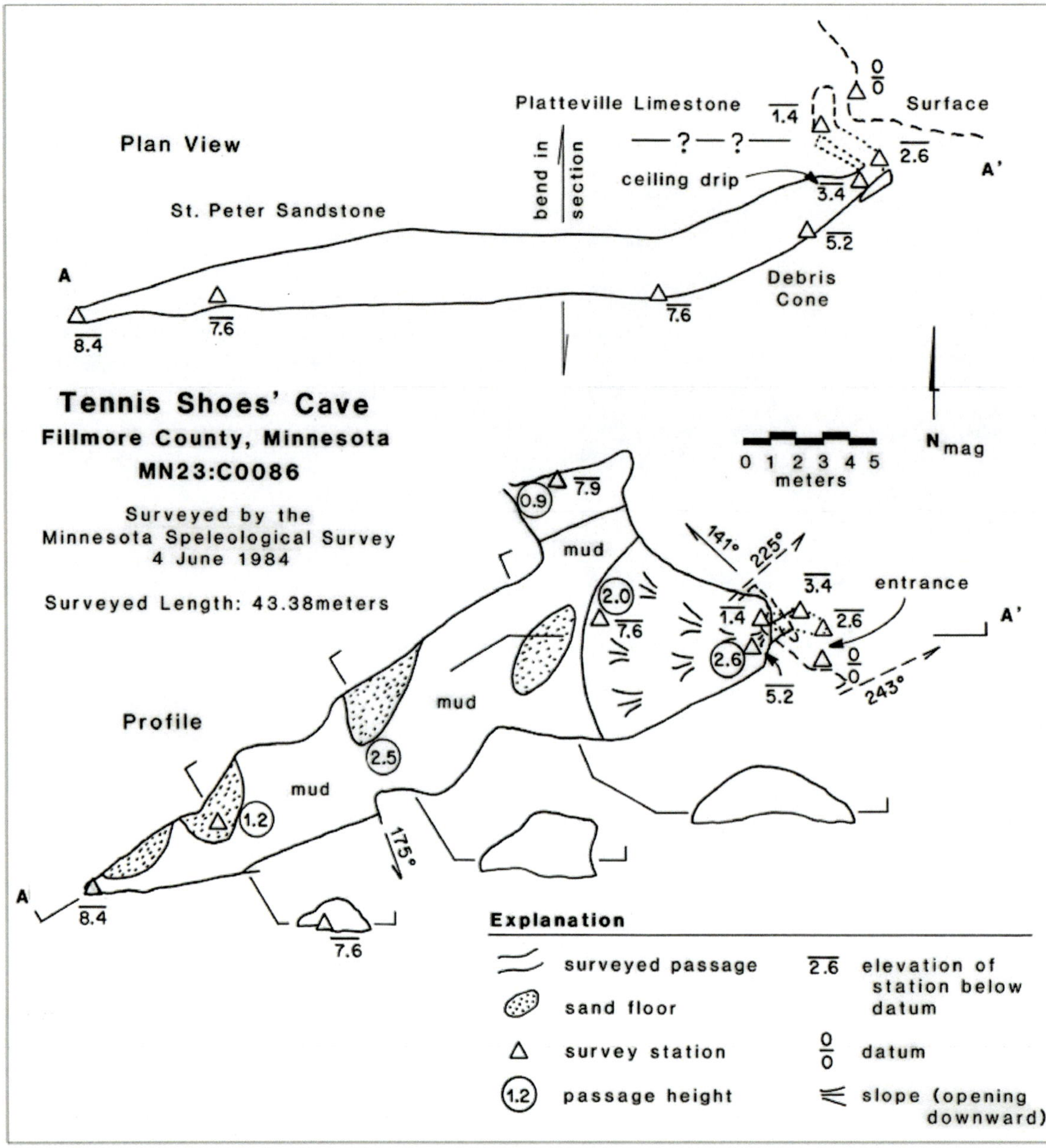

Fig. 3.42 Tennis Shoes' Cave map

alternatively, west of Chicago. Sometimes it was called the Mushroom Capital of the Midwest. The mushrooms were grown in more than 50 sandstone caves that punctuated the bluffs. Although called caves, they were artificial. Begun as silica mines, these caves were subsequently used for mushroom growing and other purposes. According to newspaper columnist Oliver Towne (Gareth Hiebert), "A whole economy and countless legends lie locked from view inside those rustic cliffs." He dubbed them the "Ivory Cliffs" owing to the snowy whiteness of the St. Peter Sandstone.

Used in the broadest sense, Mushroom Valley is divided into three distinct segments: Plato Boulevard, Water Street, and Joy Avenue. Each segment has its own distinct flavor. The Plato segment, incorporating what had been the cave-riddled Channel Street before a 1970 replatting reset the street grid, is capped by Prospect Terrace with its historic houses and magnificent views of the city. The Water Street segment, running along the river and under the High Bridge, had by far the largest caves, forming a labyrinth extending under Cherokee Park. The Joy Avenue segment (now vacated) can still be seen where an unmarked dirt road runs through the woods in Lilydale Regional Park. While most of these caves are very short—root cellars and such—Joy Avenue is anchored by large caves: Mystic Caverns at its

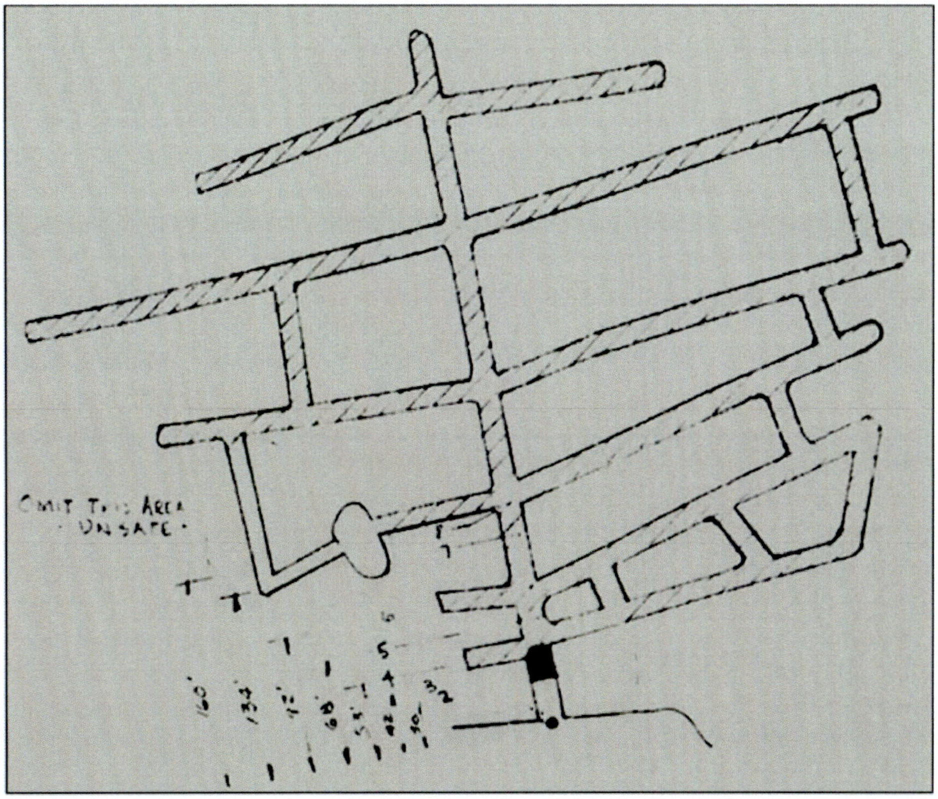

Fig. 3.43 Becker Sand & Mushroom Company cave, drafted 1962 by TKDA for the City of St. Paul

eastern end and Echo Cave at its western end. Named after its namesake acoustic effects, Echo Cave served as brick drying tunnels for the St. Paul Brick Company and was gated as a bat hibernaculum by the MNDNR in 1989. Beyond that are the brick company's clay pits in the Decorah Shale, a bizarre industrial landscape but the city's premier ice-climbing venue because groundwater seeping from the Galena limestone cliffs freezes to form gigantic ice formations in winter.

The Mushroom Valley caves had been considered for bomb shelters during World War II, even before the Pearl Harbor attack. In the early 1960s, they were surveyed by a local firm, TKDA, for suitability as nuclear fallout shelters, producing the only maps that we have today. Generalizing from the TKDA survey, the typical cave is a straight, horizontal passage about 45 m long, often connected by cross-cuts to similar caves on either side, creating network mazes with multiple entrances. A cave operated by the Becker Sand & Mushroom Company was the largest of all, with 10 m ceilings and nearly 1.6 km of passages, its wonderful hybrid name capturing the chief dual usage seen throughout the valley (Fig. 3.43).

The 1962 Civil Defense maps have limited usefulness to the would-be explorer, however, as the surveyors only included passages large enough to accommodate citizens. Crawlways, stoopways, and windows into other passages were deliberately omitted, sometimes leaving out half the passages in the cave. Even if too small for human beings to traverse, such connections often have a major influence on air flow, the hibernation of bats, and so forth.

Not all the former sand mines were used for mushroom growing. Examination of city directories, insurance atlases, and real estate plats allows us to reconstruct a fuller picture of the diversity of people and businesses that inhabited Mushroom Valley. These sources reveal what each of the caves was used for, as it easy to correlate each street address with an individual cave entrance. This discussion will focus on the three chief uses of the Mushroom Valley caves: mushroom gardening, cheese ripening, and as places of entertainment, especially nightclubs.

Mushroom Gardening

The information and quotations in this section are derived from and referenced in Brick (2003b).

The Greeks and Romans were fond of eating mushrooms collected in woods and meadows but it was not until about 1650, in Paris, that one species, the white mushroom (*Agaricus bisporus*), was actually domesticated, or cultivated. Other species had been cultivated in the Orient centuries earlier. The white mushroom thrived on horse manure but not as well on the manure of other animals. About 1800, Parisians found that mushrooms could be grown in the dark in the subterranean stone quarries that honeycombed their city and which maintained a constant year-round temperature. Mushroom cultivation did not reach the United States until 1865. In the 1880s, there was an abortive attempt by the Mammoth Cave Mushroom Company in Kentucky to raise mushrooms in that cave, the product being served up at the Mammoth Cave Hotel and shipped to eastern cities.

The original mushroom farmers in St. Paul were Frenchmen who "had seen mushrooms growing in the caves under the sewers of Paris." Interviewed on several occasions by Oliver Towne, the St. Paul mushroom farmers stated that their predecessors began the local industry in the 1880s. The Pracna Mushroom Caves under Nicollet Island were a much smaller operation in Minneapolis that supplied nearby restaurants.

In the early days, it was often necessary to abandon a cave after growing mushrooms in it for just a few years due to the accumulation of diseases and insects. About 1890, however, a method for the direct germination of mushroom spores was developed at the Pasteur Institute in France. A pure spawn industry began selling disease-free inoculum, grown in milk bottles, to mushroom farmers.

An article in the *St. Paul Pioneer Press* on May 27, 1923, titled "St. Paul's Caves Eclipse Backlot for Gardening, Except for Crop Foes," by Jay W. Ludden, offers a unique glimpse of mushroom farming in the St. Paul caves. Ludden was clearly awed by the sheer size of St. Paul's mushroom caves: "These caverns have cathedral-like arches and looking into them through the dusk that conceals details and accentuates the big lines, one is reminded of etchings of the interiors of medieval temples. This impression is strengthened when at the distant end of the cave the workmen's lamps give light as from an altar."

"As with all gardening," Ludden mused, "the more one goes into it, the more one is disillusioned as regards its simplicity. Pests and blights and molds confront one, and remedies are more or less difficult to apply." By 1923, the pure spawn technique had been adopted by local growers: "Spawn culture is a big industry of one St. Paul company, which has had at one time on the racks used for the purpose, 125,000 milk bottles containing spawn."

Further technological changes after 1923, however, improved the lot of the mushroom farmer. The adoption of the "tray system" in the 1930s, and the consequent disappearance of the old floor beds, was a big step forward in controlling mushroom pests. No longer could the pests seek refuge in the underlying soil during pasteurization, only to later re-infest the beds. Indeed, the remains of these wooden trays form the chief diagnostic artifact of former mushroom caves in St. Paul.

A more recent newspaper article, "Mushroom Farming Is Family Tradition," in the *St. Paul Pioneer Press* of March 28, 1976, painted a sad portrait of Mushroom Valley in its final days. "Mushroom growing," one farmer pointed out, "remains hard and backbreaking work because some things simply cannot be mechanized—including the picking of mushrooms." Although more than a dozen families once engaged in the work, few members of the younger generation seemed willing to adopt the manure-based lifestyle involved. In contrast, William Lehmann, known locally as the "Mushroom King," had already moved his operation to "the world-renowned cement-block caves of Lake Elmo" in 1965. Presumably the more rural setting at Lake Elmo, east of St. Paul, made for cheaper horse manure than in the heart of the motorized city. And specially designed aboveground facilities, as at Lake Elmo, while initially more expensive than caves, allowed for finer tuning of environmental conditions, including the control of pests and diseases. The last cave ceased production in the 1980s during the creation of Lilydale Regional Park. The Mushroom Century had drawn to a close.

Cheese Ripening

The information and quotations in this section are derived from and referenced in Brick (2003a).

The Villaume Box & Lumber Company, founded in 1882, was a well-known St. Paul business, "one of the nation's leaders in the manufacture of custom millwork, shipping cases and boxes," according to a 1940 promotional brochure. The brochure continued:

> Villaume has on its own property, 14 hillside caves with surface level entrances. Each cave has a ceiling height of 12 feet [3.7 m] and is 20 feet [6 m] wide. The 14 caves contain a total of 50,000 square feet [4,600 m^2] of floor space, usable for manufacturing, storage, or as shelters in event of air raids.

From 1933 to the 1950s, the University of Minnesota rented one of the "V Caves," as they were known, and produced a domestic Roquefort cheese—subsequently named Minnesota Blue—an event that would have international repercussions. St. Paul was acclaimed the "Blue Cheese Capital of the World" during World War II.

Professor Willes Barnes Combs, a native of Missouri, was appointed professor of dairy industry at the University of Minnesota in 1925. He soon discovered "a queer local fact. There are dozens of sandstone caves in St. Paul." In the late 1920s, while shopping for mushrooms at a cave in St. Paul, he noticed that a lantern in the cave was covered with rust. The mushroom grower informed him that the atmosphere of the caves was extremely moist. Combs conjectured that the caves might have a combination of temperature and humidity similar to the celebrated Roquefort caves of France, where Roquefort cheese is produced. A crucial problem was keeping the temperature low while maintaining high humidity—an almost paradoxical combination.

Professor Combs met with spectacular success. "Million Yearly Cheese Trade Seen Here," crowed the *St. Paul Pioneer Press* on January 6, 1935, reporting how "the dairy chief," Combs, "disclosed that nearly 10,000 lb [4,500 kg] of Roquefort-type cheese, the flavor of which amazed epicures, was cured this year in a small, experimental cave, within a few minutes' walking distance from St. Paul's City Hall." *Popular Science* magazine featured the university's cave in its April 1935 issue under this heading: "Caves for Cheese Making Discovered in America."

Having begun with such fanfare, Combs's Roquefort project dropped out of sight until the fall of France to invading German armies in June 1940, which cut off French imports decisively. "City's Million-Dollar Cheese Industry Gets Off with Bang," trumpeted the *Pioneer Press* on December 15, 1940. In the autumn of 1940, the Kraft Cheese Company of Chicago rented "one big cave, 150 feet [46 m] deep" from Villaume, and its "K-men" began marketing the ROKA brand of blue cheese, while the Land O'Lakes Company—its rival just down the bluff—rented "two caves, 100 feet [30 m] deep," at the former Castle Royal nightclub, which had gone bankrupt in 1940. In January 1941, the first commercial blue cheese from St. Paul's caves hit the market.

The University Cave featured in the *New York Times* on June 18, 1942. One reporter declared that "St. Paul is well on its way to become the blue cheese capital of the world." By 1945, it was reported that the production of blue cheese ranked second only to cheddar in Minnesota. But concern arose about what would happen to the fledgling blue cheese industry after the war. Fortunately, these fears proved unfounded. In 1949, it was reported that "Imports [of Roquefort] have dropped to practically nothing." The "Land O'Lakes Cheese Cave" was listed at 6 West Channel Street in city directories until 1959. Yet a local caver recalls having seen a derelict sign pointing to the University Cave in the late 1960s. Villaume relocated in 1970, leaving the entire stretch of caves vacant, to the delight of local explorers.

In the town of Faribault, a new cheese plant was established in 1936 to exploit Combs's recipe. Thus originated the Treasure Cave label, now under Canadian ownership. The original sandstone cheese caves go by the name of Caves of Faribault.

Entertainment

Some of the abandoned caves were repurposed as places of entertainment. Examples are felsenkellers used as bowling alleys, and the two great nightclub caves that opened for business during the twilight of Prohibition: Mystic Caverns and Castle Royal. While the former has been lost to history, the latter is still with us as the Wabasha Street Caves. The geologically unique feature of Mystic Caverns was its "seismogram" walls: bluish-black clay-sized sediments in jagged horizontal lines running through a 10 m thickness inside the cave.

Lagering Caves

Prior to 1840, according to historians of the subject, there were no breweries in America producing the German-style lager beer. Lager beer differed from the prevalent English and American beers, such as ale, in that the lager yeast fermented at the bottom of the vat, rather than the top, and the beer required lagering, or storage, for several months at lower temperatures. In the old days, lager beer could only be brewed during the winter months, when cellar temperatures were sufficiently low. But in northern states, such as Minnesota, where natural ice was readily available, ice cakes could be harvested from nearby lakes and rivers in winter and stacked in caves, allowing brewing year-round to meet the growing demand. Examples are Yoerg, Banholzer, Stahlmann, in St. Paul and Heinrich in Minneapolis (Brick 2013b) but they are widely scattered around Minnesota.

The Ford Sand Mines

In the early 1920s, the Ford Motor Company built a Model-T assembly plant in the Highland Park neighborhood of St. Paul. In addition to using hydroelectricity from the adjacent Ford Dam, coal was used to generate steam. The coal was offloaded from trains at the assembly plant and then transported to the steam plant in the river gorge by being lowered down a shaft.

Until 1932, the plant's glass manufacturing facility used silica shipped in from the Minnesota River Valley to make automobile windshields. Ford had not realized that its Highland Park plant sat right over the best silica deposit of all, the St. Peter Sandstone. From a geologist's viewpoint, whoever recognized this fact is an unsung hero of the obvious because the sand had already been much in demand locally for glass making since the late 1800s.

Beginning in the late 1930s, Ford, in their quest for silica, mined out 4.0 km of passages in the St. Peter Sandstone below their Highland Park plant, leaving the main caves, and another 2.4 km below Shepard Road, leaving the Marina Caves. The operation ceased in 1952 when it became more economical to obtain glass elsewhere.

In 1982, the Jaycees (Junior Chamber of Commerce) offered their first Tunnel of Terror event during Halloween, using the eastern third of the Marina Caves. The increasingly elaborate event ran for a quarter century, until concern for cave-related smoke inhalation deaths elsewhere put an end to the operation in 2004.

3.2.4 Prairie du Chien Caves

The lower Ordovician Prairie du Chien Group (OPDC) in Minnesota consists of the upper Shakopee Formation and lower Oneota Dolomite (Fig. 3.2). Both formations are primarily dolomite and limestone. There is a major, about 15-million-year long, erosional unconformity between the bottom of the overlying St. Peter Sandstone and the top of the Shakopee Formation (Fig. 3.3). The Shakopee consists of the upper Willow River Member and, to the south, a lower New Richmond Sandstone Member. A second significant, about 5-million-year long erosional unconformity separates the Shakopee from the Oneota (Fig. 3.3). The Oneota consists of the upper Hagar City Member and a thin basal Coon Valley Member. Karst processes were active during both unconformities. The St. Peter/Shakopee and the Shakopee/Oneota contacts are well-developed paleokarst surfaces.

The OPDC is the first or second bedrock from northern Washington County through most of the Twin Cities metro area, Ramsey, Hennepin, and Dakota counties and then down the west side of the Mississippi River through Goodhue, Wabasha, Winona, and Houston counties.

Washington, Ramsey, and Hennepin counties were covered by the Superior Lobe and parts of the Des Moines Lobe during the late Wisconsinan, but much of the resulting glacial drift was thin or has been removed by subsequent erosion processes and the OPDC is often at or near the land surface in these areas. In the area south of Hastings the OPDC subcrop area, while not covered by Wisconsinan glaciers, has been glaciated several times but the remaining glacial deposits are thin, patchy, and heavily eroded. The OPDC is a major, heavily utilized aquifer throughout the area and is a classic triple porosity karst aquifer (Worthington 1999). The Shakopee/Oneota paleokarst forms the OPDC High Transmissivity Zone (see Fig. 1.11 in Chap. 1) of enhanced groundwater movement across the entire area underlain by both formations (Tipping et al. 2006, Steenberg et al. 2014).

The OPDC hosts a multitude of small caves, many of which were mapped by the enthusiastic amateur "microspeleologist" Tim Stenerson of Red Wing, MN, in the early years of the new millennium. Some of them have sediments with significant nitrate concentrations, up to 35,000 ppm by weight, which were used by French fur-traders for making saltpeter (Brick 2012, 2013a, 2016a). Most of them are not parts of the current karst drainage of the OPDC and don't seem to be related to the current surface drainage. Many of the caves seem to be old or even paleo caves that have simply been intersected and/or reactivated by the current erosive incision of the land surface.

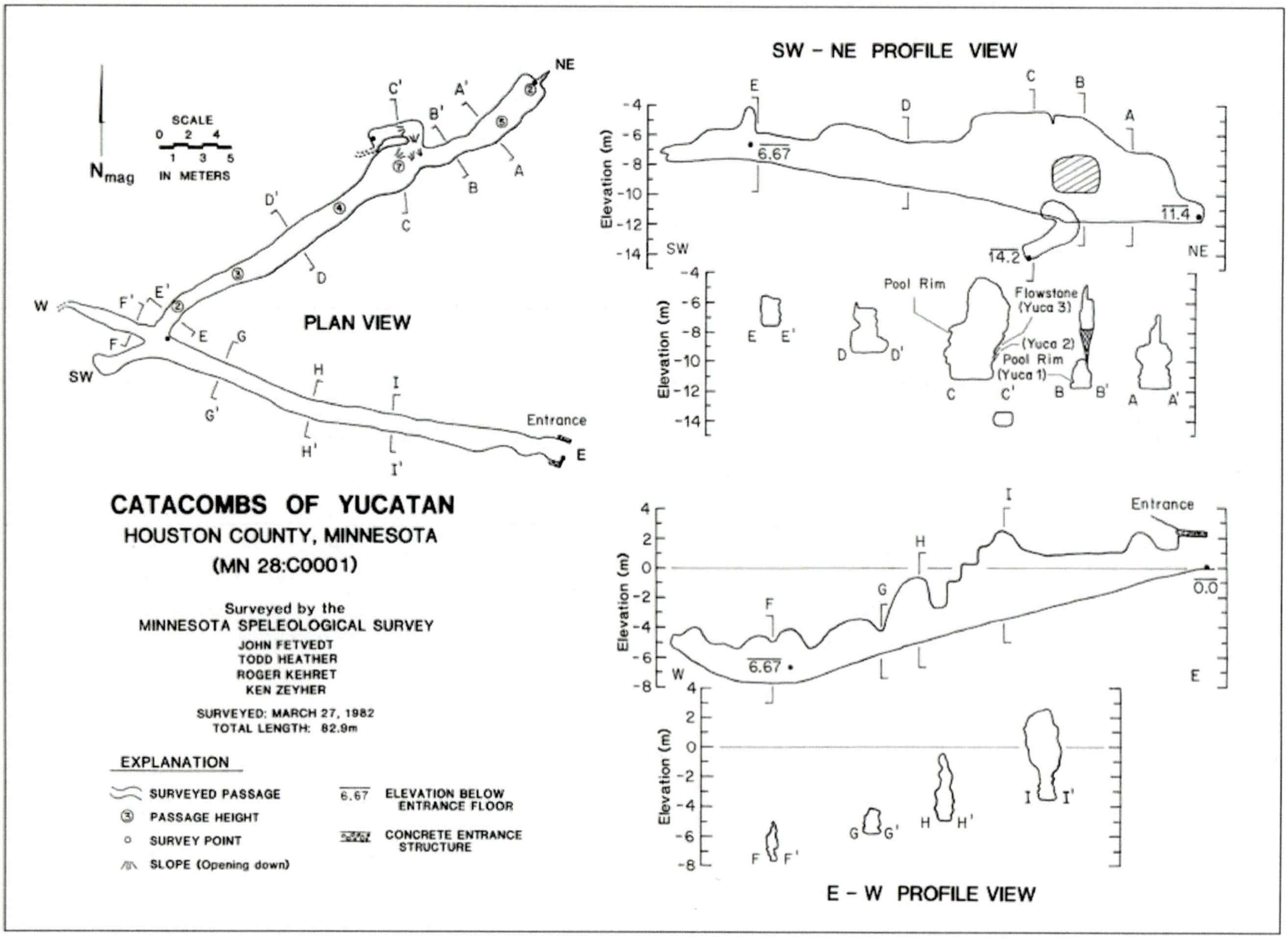

Fig. 3.44 Catacombs of Yucatan map

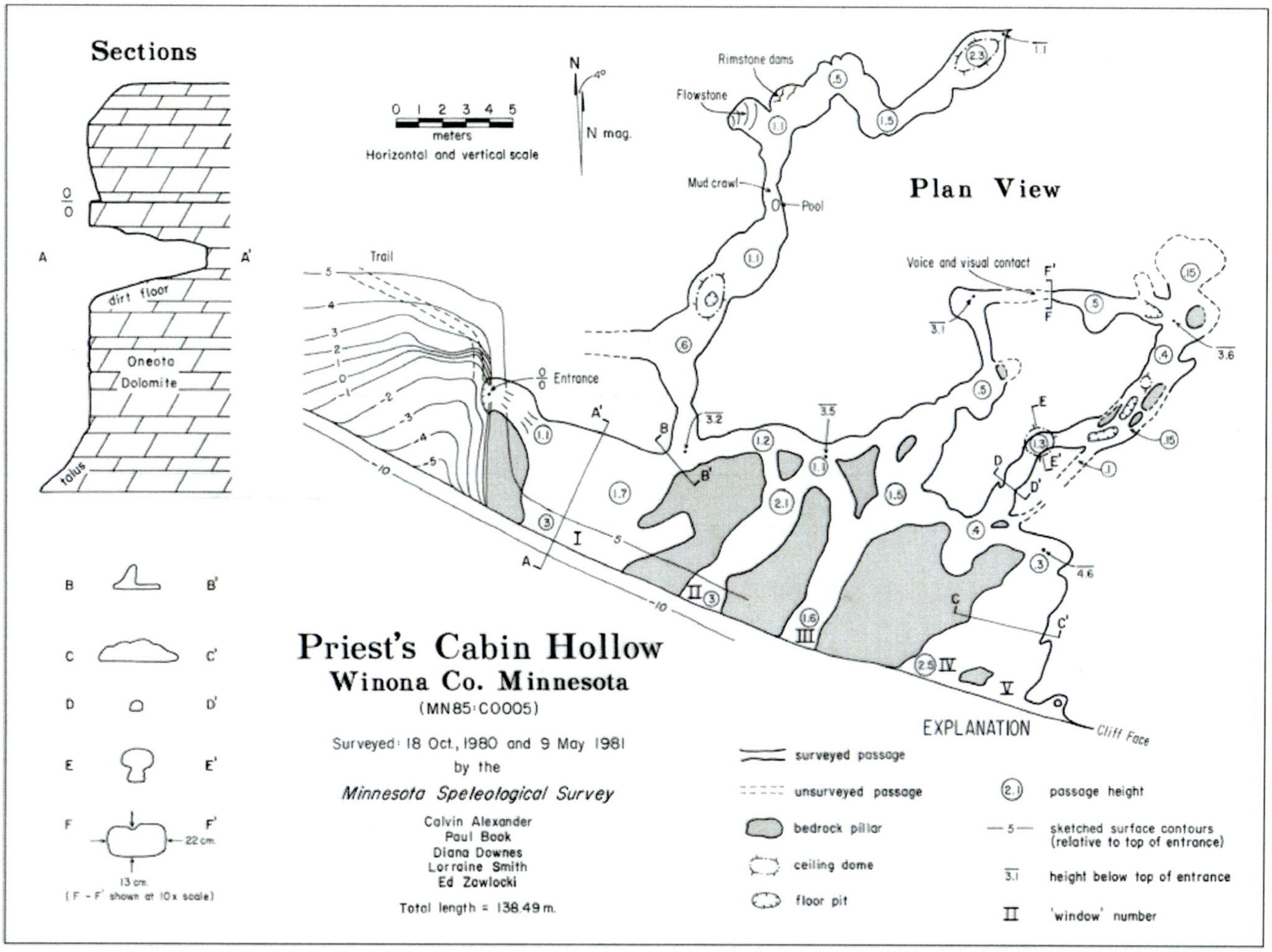

Fig. 3.45 Priest's Cabin hollow map

Starting in the south and moving north examples of OPDC caves include the following.

Catacombs of Yucatan in Houston County is an OPDC cave with an interesting history (Fig. 3.44). According to Brick (2017a, p 114–115):

> "The commercialization of this cave is a Great Depression story, a hill-top beacon in the dusty gloom. Noted as early as 1880, this hillside cave in the Yucatan Valley at the town of Black Hammer, was found to contain skeletons, assumed to be Indians, hence its name. The cave is short and simple, looking like a wishbone on a map, so it certainly did not qualify as a catacomb in terms of complexity. Its 272 feet [83 m] of passages wind through the Oneota dolomite caprock of the bluff. The cave, on private land, can be reached after an exhausting climb up the steep, sunny, hillside, picking your way carefully so as to avoid rattlesnakes.
>
> The Catacombs of Yucatan was commercialized for tours in 1934. A dance hall was constructed next to the cave, and its lights, seen from afar, were an inspiration. In 1995, composer Dr. Dan Senn reprised a bit of this magic with a "Sound & Video Installation," the remains of which were still visible years later when I visited the cave after a long absence and made inquiries about these interesting relics. Exotic musical instruments such as the Winged Pendulyre were installed by this latter-day Orpheus, where formerly only the sounds of dripping water and squeaking bats could be heard. He recorded valuable oral history on the cave, detailing the hard lives and deaths of people during this dusty era.

In 2018, the existence of the cave was successfully used to demonstrate the presence of karst features at the site and helped prevent a large hog feeding operation from being permitted on the hill atop the cave.

Carlin Caverns is an OPDC cave in Winona County. According to Spong (1980b): "In 1964 two sisters, Carmen and Linda Sens, discovered an animal den while hiking through the wooded hills east of Altura. Their curiosity piqued, Carmen and Linda returned with shovels and eventually excavated open the entrance to a crawlway cave. The girls began exploring the cave, and beyond the crawlways they discovered many chambers beautifully decorated with a variety of formations. More rooms and passages beckoned them to continue exploring. In commemoration of their discovery, the cave was named Carlin Caverns, a

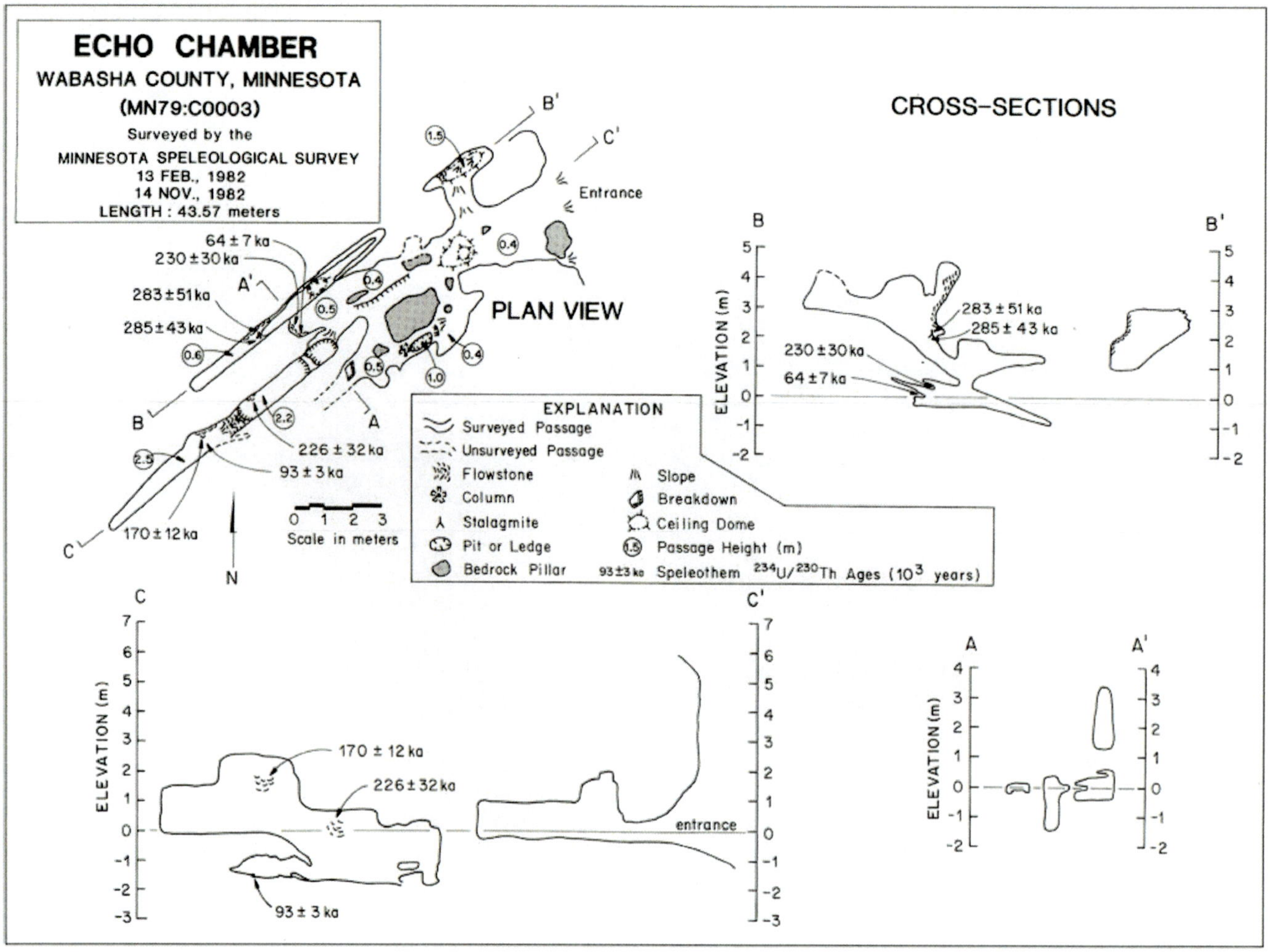

Fig. 3.46 Echo Chamber map

compound contraction of the girls' first names." In 1966 the caverns were surveyed, and more than 213 m of passages were recorded. Little constructive caving was conducted in Carlin Caverns in the 1970s "and spelunking traffic, mostly trespassers … strained cave owner relationships. As a result" by 1980 the landowner "regretfully closed the cave to all visitor, with the hope of preserving the caverns for generations to come."

Hiawatha Caverns, another former show cave in Minnesota, is an OPDC cave located near Witoka in Winona County, not to be confused with a different Hiawatha Caverns in Wabasha County (Brick 2017a, p 117). Hiawatha Caverns were discovered about 1962 and operated as a show cave during the summers of 1964, 1965, and 1966 before failing financially at the end of the summer of 1966. Soule (1974) recounted the discovery of the cave and the operation of the commercial operation for three summers. One notable novelty is that all the show cave guides were women.

After the fall of 1966, the cave was abandoned and suffered vandalism. In 2018, Martin Larsen, the president of the Minnesota Caving Club and the Minnesota Cave Preserve, began discussions with the current cave owner. In August 2019, the cave owner, Ackerman, and Larsen reached an agreement to sell the cave to the Minnesota Cave Preserve. The old dilapidated entrance was demolished and a new one installed (BCNG 2019).

Priest's Cabin Hollow is in a 20 m vertical bluff on the northeast side of the South Branch of the Whitewater River in Winona County (Fig. 3.45). The cave appears to be the remnant of a larger maze cave. That larger paleo cave was exposed and truncated as the river incised its valley into the OPDC. The cave is a dry maze cave whose truncation formed a line of five "windows" near the middle of the bluff. The cave is entered via a climb down 5 m from the land above the cave. Priest's Cabin Hollow is one example of many of the OPDC caves that are located near the tops of valleys incising the interfluves between the rivers. Many such caves are not significant parts of the current karst drainage systems. Rather they are randomly exposed/reactivated much older paleo caves.

Echo Chamber cave is a small OPDC cave in Wabasha County (Fig. 3.46). The cave exhibits strong joint control

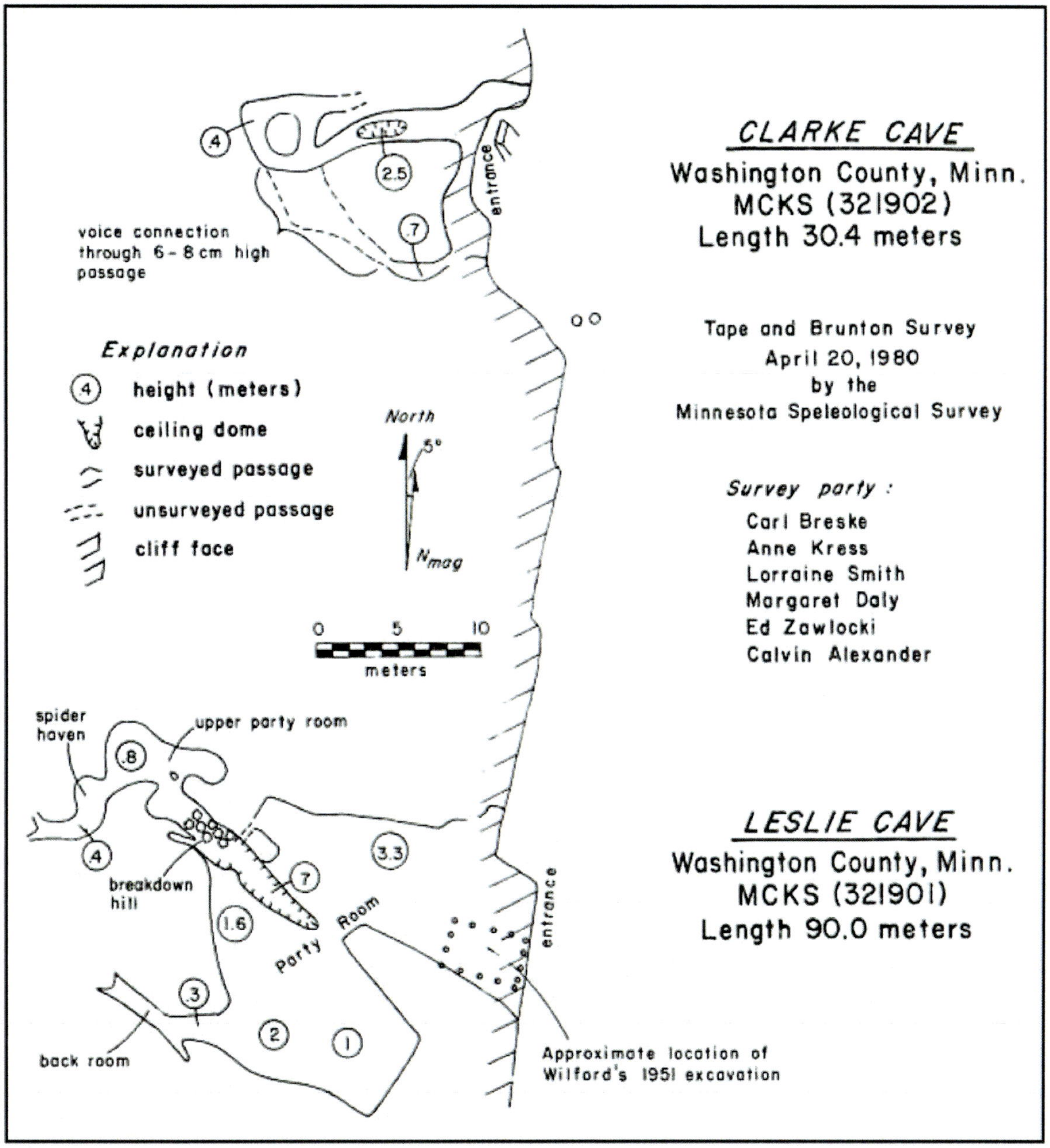

Fig. 3.47 Leslie and Clark Caves map (Alexander 1980a)

along a NE/SW joint. Echo Chamber, like many of the small OPDC caves, is located on the top edge of an incised stream valley. The cave is high and dry relative to the current stream runoff. This otherwise unremarkable little cave is of interest because Lively (1983) measured U/Th alpha counting ages on broken pieces of flowstone from several locations in the cave. Those ages, shown in Fig. 3.46, range from 64 ± 7 ka (10^3 years) back to 285 ± 42 ka. This small, inconspicuous cave is very old. Since the cave must have formed and become air filled before the flowstone can form, the cave must be older than its oldest speleothem. Echo Chamber formed well before the Wisconsinan Ice Age and provides direct evidence in support of the geomorphic field observations that indicate many of the OPDC caves predate the last glacial cycle. Visually most of the flowstone and other small speleothems in Echo Chamber were dry and dusty. No Wisconsinan or younger speleothems were found.

3.2.5 Cambrian Sandstone Caves

A series of Cambrian sandstones, siltstones, and other siliciclastic sedimentary rocks with some carbonates unconformably underlies the Oneota Dolomite. Progressing down the sequence are the Jordan Sandstone, the St. Lawrence Formation, the Tunnel City Group (formerly the Franconia

Fig. 3.48 Ice stalagmites in Robinson's Ice Cave, February 10, 1980. Photo by Ed Zawlocki

Formation), the Wonewoc Sandstone, the Eau Claire Formation, and the Mt. Simon Sandstone (Fig. 3.2). The St. Lawrence locally contains fine layers and lenses of dolomite. These formations are only exposed along the edges and bottoms of the Mississippi and St. Croix river gorges and some of their most deeply incised tributaries. Their western edges are generally buried under thick glacial deposits. The Cambrian sandstones are major aquifers throughout their geographic extent and are heavily utilized as domestic, industrial, and agricultural water supply sources.

Leslie Cave and Clarke Cave (Fig. 3.47) are developed in the bluff along the west side of the St. Croix River in Washington County. At the caves the bluff is about 70 m high and is a steep wooded slope interrupted by a 5–10 m sandstone cliff. The caves are in the base of the cliff about 20 m above river level. The sandstone is light tan to pink, friable, and is increasingly iron stained toward the top. The precise stratigraphic location is uncertain, but the caves appear to be in the Jordan Sandstone.

Leslie Cave was named after its owner, Ira Leslie, despite it being labeled "Knapps Cave" on the Scandia 7.5-min USGS topo map. Leslie Cave has been known since at least the Strong (1900) account of a visit to the cave. Allowing for a large factor of journalistic exaggeration, Strong's account fits Leslie Cave and the location he gives is consistent with Leslie Cave's location.

Leslie Cave has a large, imposing entrance which is readily visible from boats on the St. Croix River. The ceiling of the entrance alcove is roughly flat, but the soft sand floor slopes steeply toward the river. A surface ravine empties over the cliff above the entrance alcove and continues down the slope to the river. This surface drainage channel may have contributed to the formation of the entrance alcove and serves to remove material mass wasted down the floor of the entrance alcove. The shape of the entrance alcove is

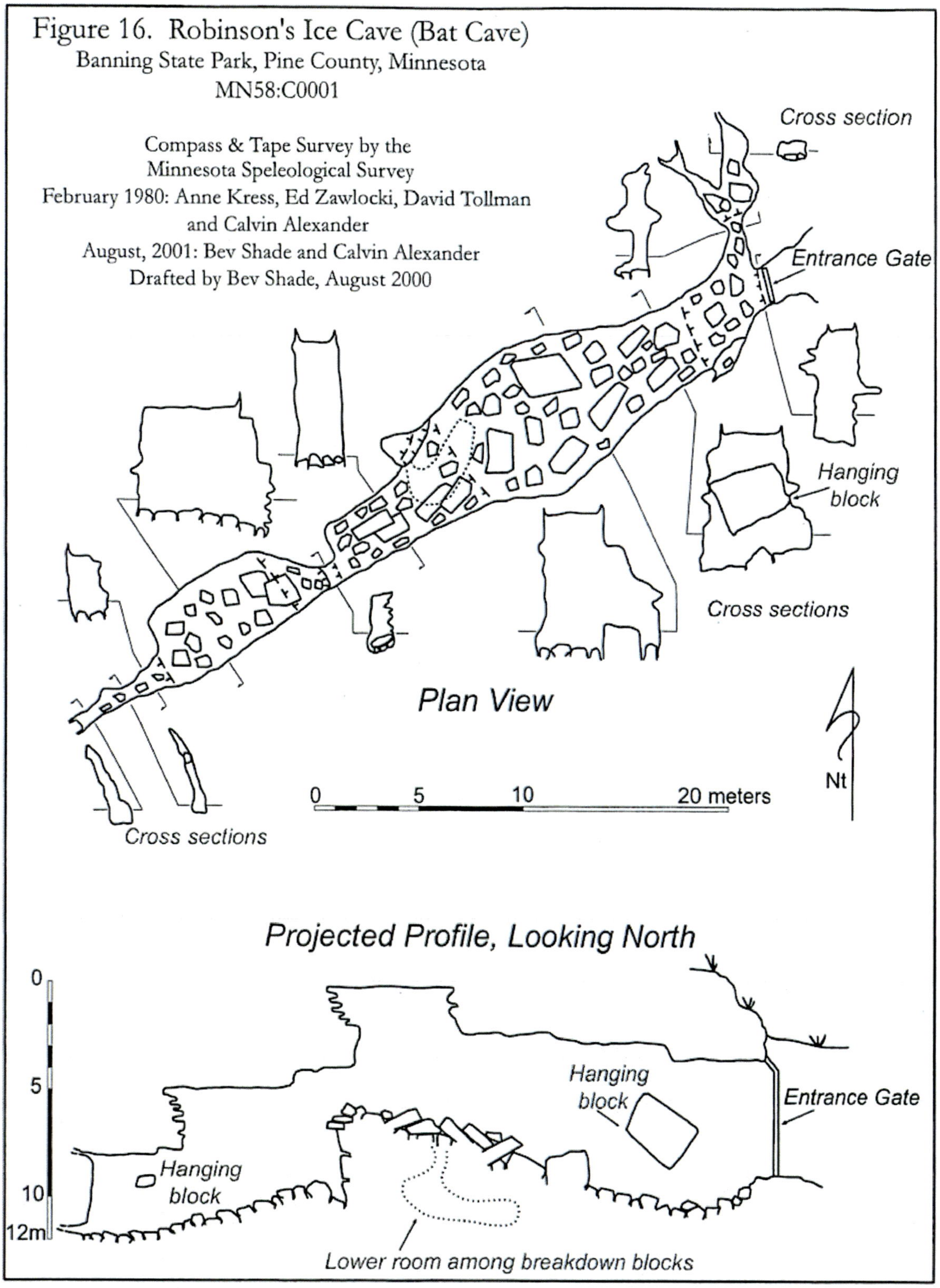

Fig. 3.49 Robinson's Ice Cave, from Shade (2002)

controlled by joint faces in the sandstone. At the top of the slope in the entrance alcove, the floor levels out in a roughly rectangular room trending northwest/southeast. The name, Party Room, illustrates the main use of the cave by local young people. A small crawlway, the Back Room, opens off the west side of the Party Room. A large, about 7 m high, breakout dome has developed in the ceiling of the Party Room. The northwest end of the Party Room is formed by the Breakdown Hill, a jumble of partially buried blocks leading up to the Upper Party Room and Spider Haven.

These upper rooms look much different from the rest of the cave. They have an amorphous, phreatic look with no obvious joint control.

In 1951, Lloyd Wilford, the indefatigable rock shelter investigator from the University of Minnesota, concluded: "The cave was used by various Indian groups as a temporary campsite, rather than a habitation site, over a long period of time. It is possible that the earliest occupants had no pottery, in which case they could be ascribed to the Archaic period. The bulk of the material may be ascribed to the Middle Woodland period, but some occupancy in the Late Woodland period is indicated" (Wilford 1951).

3.2.6 Precambrian Rock Caves

3.2.6.1 Pine County Sandstone Karst

As early as 1920, the springs of Pine County were mapped by Minnesota State fishery biologists (Brick 2019b). The landscape of north-central Pine County, in east-central Minnesota, contains over 300 sinkholes, short caves, dozens of stream sinks, and many dozens of springs. These features serve the same function as in carbonate karst terrains: sinkholes and caves focus recharge, feeding heterogeneous groundwater flow systems. These groundwater systems utilize and integrate and high permeability zones in the sandstone and discharging to springs. These landforms occur in the ~800-million-year-old Mesoproterozoic Hinckley Sandstone and thin overlying unconsolidated glacial deposits. The Hinckley Sandstone is a tan to orange, fine- to medium-grained quartz arenite that is typically about 96% quartz, 2% feldspar, and 2% felsic volcanic and metamorphic fragments and chert (Tryhorn and Ojakangas 1972). The sandstone grains are well sorted and well rounded. No carbonate grains or cements have been found in sandstone samples from the sinkhole area. No evidence has been found that calcite solution controls bedrock permeability. The sinkhole area is a sandstone karst (Shade 2002).

Robinson's Ice Cave (aka Bat Cave or Big Cave) located in Banning State Park is the largest and best known of the several caves in the Pine County sandstone karst. This description of Robinson's Ice Cave was adapted from Alexander (1980b), Nordquist (2000), and Shade (2002). Robinson's Ice Cave is in the bluff on the west side of the Kettle River immediately north of the town of Sandstone. In the winter the cave acts as a glacière, as cold, dense air flow into the lower, back cave levels. Depending on the conditions of any specific winter, very impressive arrays of ice stalagmites can develop on the cave (Fig. 3.48). Many of these ice stalagmites have complex, fluted shapes, and/or alternating horizontal bands of clear transparent and a white opaque ice. During the winters, however, strong vertical air temperature gradients exist in the cave. Despite the numerous ice formations and accumulations on the cave floor, there were no ice stalactites growing from the ceiling drips. Another minor ice formation that develops in the winter in Robinson's Ice Cave is a heavy cover of hoarfrost that forms on the ceiling and walls near the entrance. Crystals up to several centimeters in length develop in this hoar frost. The hoar frost is evidence of the circulation of relatively warmer, moister air circulation out of the upper level of the cave as the cold, dense air flows into the lower parts of the cave.

Fig. 3.50 Robinson's Ice Cave bat protective gate with signage visible within. 2018 photo by Greg Brick

Figure 3.49 is a map of Robinson's Ice Cave. The ceiling of the cave is very flat because sandstone blocks have fallen from the ceiling along horizontal bedding surfaces. The straight, vertical walls of the cave are formed by two near-parallel vertical joints in the sandstone. Near the entrance the two joints are 7 m apart and cave is 7 m wide. Along the 63 m length of the cave, the joints converge and the cave narrows to an impassable crack at the end. The cave is located near the top of the valley wall. There are several sinkholes above and west of the cave which actively drain several wetlands between the valley wall and MN Highway 23.

The floor of the cave is littered with sandstone blocks which have fallen from the ceiling.

Natural forces have removed much rock debris to create the open space. The solid floor to the cave cannot be seen.

Fig. 3.51 Kayaking through the Cave of Waves. Photo courtesy of Steve Fagin, 2014

The depth of the debris pile (and cave) is unknown, possibly as deep as the bedrock incision below the Kettle River. There is no soil covering the sandstone blocks. River flow and flood events would have deposited sediment in the cave. The lack of sediment deposits implies that the top layer of blocks fell since the Kettle River incised below this level. Further evidence of the strong vertical winter air temperature gradients is that Robinson's Ice Cave hosts a significant, year-round bat colony—hence its local alias "Bat Cave." Several hundred bats were hibernating in the cave during the February 1980 visit.

In 1988, the DNR's Nongame Wildlife Program constructed a bat gate at the entrance (Fig. 3.50). Although the gate has not kept out the most determined trespassers, it has reduced the level of winter disturbance to bats in the cave.

Fourteen smaller sandstone caves have been reported in the Pine County sandstone karst.

Eight of these caves have been mapped and are reproduced in Shade (2002). All the caves are developed in the Hinckley Sandstone. Most of these caves show a similar structural control on development that the Robinson's Ice Cave map shows. Bedrock fractures provide an initial weakness and flow path. The caves form where significant joints intersect nearly horizontal bedding that has high permeability or is mechanically weak.

3.2.6.2 North Shore Lake Superior Littoral Pseudokarst

(The information in this section is largely from Brick (2017a, p 123–124).) The North Shore of Lake Superior comprised gabbro and extrusive basalts and rhyolites of the Duluth Complex and North Shore Volcanic Group, created by magma and lava which upwelled and hardened about 1,100 million years ago during the formation of the Mid-continent Rift (Ojakagas and Matsch 1982).

The relentless pounding of the waves of Lake Superior against Minnesota's North Shore has gnawed at rock joints that are suitably oriented, carving out caves, especially in Lake County. Sometimes they are called "sea caves," but on a freshwater lake that phrase seems rather odd, so they are referred to as littoral (shoreline) caves by geologists. Strangely, no relict littoral caves from the much higher post-glacial shorelines of Lake Superior have been found. All the known caves formed at present-day lake levels.

A well-known cave along the North Shore is the **Cave of Waves** (Fig. 3.51). Formed in a rocky promontory of rhyolite forming one side of Crystal Bay in Tettegouche State Park, near the town of Silver Bay, the cave is 30 m long and elbow-shaped in map view. The cave is most often visited by kayakers but can also be reached by wading out from shore in a wetsuit on days when the lake surface is especially calm.

Thunder Caves at the mouth of the Manitou River are nearly as well known and have been depicted on postcards. Once a tourist attraction, the caves are now private. Other easily accessed shoreline caves include one near a beautiful beach of pink shingle in Iona Beach SNA, and the Two Harbors Cave, in a park in the town of that name.

In the neighboring Cook County, where the Brule River flows through Judge Magney State Park, there is a waterfall, the famous **Devil's Kettle**. Half the waterfall apparently vanishes as a waterfall into a 15 m deep pothole eroded into rhyolite bedrock. Devil's Kettle was long dubbed a mystery, even by the prestigious Smithsonian website. Anecdotal reports of the non-recovery ping pong balls and other objects poured into the "disappearing" side of Devil's Kettle added to the mystery. One guess was that there might be an unseen, secret lava tube more than 1.6 km long that conveyed this water to Lake Superior. In 2016, careful stream gauging by the DNR (above and below the pothole) demonstrated that the vanishing water is only an optical illusion. The water does not travel by a secret passage, but rather rejoins the Brule River below the water surface immediately downstream of the falls. Non-recovery of objects thrown into the pothole proves little, as they get caught up in eddies and woody debris below the waterline.

Stream gauging has also revealed that many North Shore rivers do decrease in volume as they flow downstream, a phenomenon known as "Surber's Paradox" after fisheries biologist Thaddeus Surber (1871–1949), who first described it (Surber 1922, p 5). The water sinks into rock crevices but where it re-emerges is uncertain. The hunt for the missing springs is still ongoing (Brick 2016b).

3.3 Sinkholes

Over 15,000 sinkholes are mapped in Minnesota's Karst Features Database. Over 10,000 of these are in Fillmore County. Most of the sinkholes are cover collapse sinkholes, but subsidence and other types of sinkholes exist. Individual sinkholes range in diameter from less than 0.1–90 m (Chute's Cave) and in depth from subtle, shallow depressions to shafts over 30 m deep. Compound sinkholes and blind valleys form much larger closed depressions in the landscape.

Sinkholes can be ephemeral features on the landscape—on both human and geologic timescales. New sinkholes routinely open across Minnesota's karst landscapes. Many of these new sinkholes result from natural karst processes and many of them are reactivated paleo sinkholes. However, an increasing number of these new sinkholes can be attributed to human activities. A variety of natural processes can and do fill sinkholes on geologic timescales. The several older glacial advances that covered southeastern Minnesota scraped off much of the pre-existing epikarst and filled the pre-existing sinkholes. Floods, soil erosion, and water-borne debris can plug and seal sinkholes. Loess and other wind-blown material can fill the surface depressions.

On human timescales, sinkholes are routinely filled by people. Sinkholes are real or perceived impediments and risks to a wide variety of human activities. Up until about 50 years ago many sinkholes were routinely filled with trash, household garbage, demolition debris, and the entire gamut of domestic, agricultural, and industrial waste. Through the efforts of generations of environmental workers at all levels, sinkholes are now rarely used as waste dumps. One simple, effective educational meme has been "Don't put anything in a sinkhole you don't want to drink." But sinkholes continue to be routinely filled for a wide variety of reasons.

3.3.1 Sinkhole Mapping

In the 1970s and 1980s, health and environmental officials, managers, hydrogeologists, and society began to recognize that groundwater quality was being degraded by human activities. Shallow karst aquifers are especially susceptible to anthropogenic pollution, but little information was available on where, or even what, karst aquifers actually are. Sinkholes became the visible, understandable indicator of the presence of karst processes. This was (and is) a serious oversimplification but provided a useful tool for environmental managers and regulations. Sinkholes, springs, and karst aquifers needed to be mapped and the information be made available.

The development of sinkhole maps (and karst aquifer mapping) in Minnesota is a tale of three revolutions.

3.3.1.1 Revolution 1: Paper Maps to GIS

The first revolution was transition from field observations recorded on hand-drafted maps through the co-evolution of personal computers, word and data processing software, and ultimately to GIS technology. This gradual, slow revolution began in the 1980s and continues into present as technology and software improve. This revolution gave us the ability to assemble, correlate, and present in an understandable fashion a large amount of hydrogeologic information.

Although geologists and soil scientists in the last half of the nineteenth and first half of the twentieth century were clearly aware of the presence of sinkholes and other karst features in southeastern Minnesota, very few sinkholes were shown on their maps. Farnham's (1958) Fillmore County Soil Survey was one of the first publications to show several thousand sinkholes and larger closed depressions in Fillmore County. Farnham (1958, p 2) wrote "Many small tributary streams are fed by springs, but surface water often drains into sinks and sinkholes thence to underground drainage channels. Many sinkholes occur in lines on the uplands and increase in number and size near large valleys. Underground passages connect many of them, and in places subterranean gorges can be traced for several miles [kilometers] as a succession of large sinkholes." But note that he did not use the term "karst." The sinkholes were viewed as impediments to agriculture and mapped mainly in the tilled areas. Very few sinkholes were mapped in the forested or non-tilled area.

The soil maps were published at 1:20,000 scale on 1947 photo base maps.

The 7.5-min USGS topo sheets of southeastern Minnesota began to be published in the 1960s through the 1970s. Some topo sheets in the karst areas indicated the larger sinkholes as closed depressions but, as with the soil survey maps, the coverage was incomplete in forested areas. In addition, the degree of recognition of sinkholes by the individual USGS mappers varied significantly from topo sheet to topo sheet. In addition, a one millimeter symbol on a 1:24,000 scale map corresponds to a 24 m feature. Even allowing a significant "horizontal exaggeration" only some of the largest sinkholes are shown on the topo sheets and, as with the soil survey, most of the sinkholes in wooded areas were missed.

In the early 1980s, the Minnesota Geological Survey initiated a project to map and publish 1:100,000 geologic and hydrologic atlases of Minnesota Counties. The second county mapped in the County Atlas Series was **Winona County**. The Prairie du Chien and St. Peter formations are the first bedrock under most of Winona County. The 1974 and 1976 sinkhole failures of the lagoon in the Altura Waste Water Treatment Facility (WWTF) (Alexander 1980c; Alexander and Book 1984) were fresh in the planner's minds, when the MGS Winona County Atlas was prepared. That atlas included a "Sinkholes and Sinkhole Probability" map (Dalgleish and Alexander 1984a) based on her field mapping effort (Dalgleish 1985). Janet Dalgleish talked to the residents, drove the roads, walked the fields and forests, and located 535 sinkholes, including sinkholes that had been filled. The open and closed sinkholes were given different symbols on the map. She tried to get estimates of when the sinkholes are formed. In those pre-GPS days the locations of many of the features, especially the locations of the filled sinkholes, were approximate. The sinkhole locations were plotted on paper topo sheets and then drafted onto 1:100,000 maps for publication. The Winona map was divided into five relative probability units: high, moderate to high, low to moderate, low, and no sinkhole probability.

One of Dalgleish's (1985) observations was that the sinkholes in the OPDC are concentrated near the Shakopee/Oneota contact. More recently that stratigraphic position has been recognized as the regionally mappable OPDC High Transmissivity Zone (Tipping et al. 2006). Also, that precise stratigraphic interval is where the Altura WWTF Lagoon had failed and where Lewiston's (Jannik et al. 1992) and Bellechester's (Alexander et al. 1993) WWTF lagoons would later fail due to catastrophic sinkhole collapse.

One of Dalgleish and Alexander's (1984b) notable observations was: "The rate of sinkhole formation has significantly increased in the past 50 years. Of the 535 sinkholes located, 47 have formed in the past 5 years. At the present rate of about 9 per year, all the sinkholes could have formed in the last 50–60 years. This is not the case, because many sinkholes are more than 100 years old. Although the specific causes of sinkhole collapse in Winona County are uncertain, human activities elsewhere have significantly increased the rate of sinkhole formation." Many new sinkholes are promptly filled by the land owners.

Olmsted County was the next Minnesota karst lands county for which an atlas was prepared (Alexander and Maki 1988). Olmsted County requested a Sinkhole Probability map. Geri Maki systematically searched Olmsted County for sinkholes and springs. Geri also drove the roads, talked to the residents, and walked the fields to find the sinkholes and springs. She inventoried 865 sinkholes in Olmsted County. In the northeast half of Olmsted County the OPDC formations are the first bedrock. In the southwest half the Galena Group rocks are the first bedrock. There are areas of very dense sinkhole clusters on the Galena Group formations and a sixth, higher sinkhole density category, was needed on top of the five categories used in Winona County. In these areas, sinkholes dominate the landscape and most surface runoff drains into sinkholes. I, Calvin, named this highest sinkhole density unit on the map "Karst Topography." That nomenclature turned out to be a legal tactical mistake! Lawyers representing developers argued that areas not in the Karst Topography unit areas were "not karst" and therefore not subject to protective karst regulations and/or groundwater protection regulations that Olmsted County was in the process of developing and implementing. Later Sinkholes and Sinkhole Probability maps for other counties labeled the same unit "Sinkhole Plains."

When an atlas was planned for **Fillmore County** in the early 1990s, it was evident that karst was a much more important part of the County's hydrogeology. For economic and practical reasons, the County Atlas Program was split between the MGS and the DNR. The MGS prepared and published the map plates pertaining to the geology and the DNR prepared the map plates relative to the hydrogeology of Fillmore County. In addition to the publication of a Sinkhole and Sinkhole Probability map (Witthuhn and Alexander 1995), a Springsheds map was published (Alexander et al. 1995) in Part B of the County Atlas. The latter map is discussed in Sect. 3.3. By cross-referencing the sinkholes in Farnham's (1958) Soil Survey maps with the closed depressions shown on the USGS 7.5' topo sheets, with detailed field surveys of a few selected areas, and information from many Fillmore County residents, Witthuhn and Alexander (1995) identified 6022 sinkhole locations in Fillmore County. Locations of filled sinkholes were included where they could be reliably identified. Not all the closed depressions shown on the topographic maps and the Soil Survey maps were sinkholes or other karst features. Some

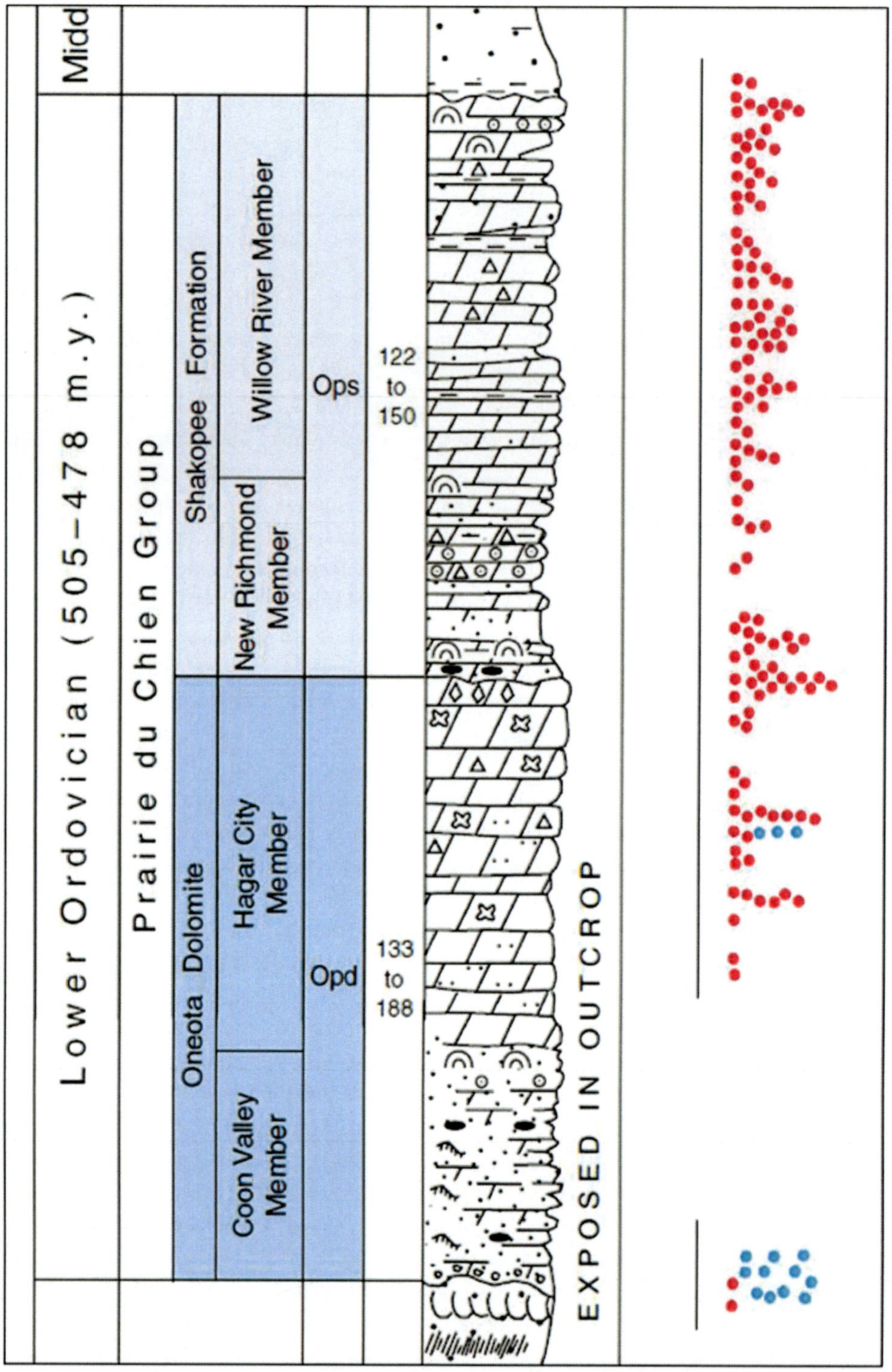

Fig. 3.52 Distribution of sinkholes (magenta dots) and springs (blue dots) in the OPDC of Wabasha County. Modified from Fig. 3.4 in Tipping et al. (2001)

closed depressions were associated with mining or other human activities. Those that could be identified were removed from the sinkhole database, but the process was not exhaustive, and some individual features mapped as sinkholes may have other origins. A comparison of sinkholes found through detailed field works and those shown in the same areas on the USGS topo maps and the Soil Survey maps indicated that about 60% of the currently active sinkholes had been mapped.

The atlas for **Wabasha County** contained a Karst Features plate (Tipping et al. 2001). The Prairie du Chien Group is the first bedrock under much of Wabasha County but is covered in some areas by more than 15 m of glacial and Cretaceous sediments. A total of 196 sinkholes were mapped in Wabasha County, all on the areas of OPDC with less than 15 m of cover.

Figure 3.52 shows the distribution of the mapped sinkholes in Wabasha County as the magenta dots on the right side of the figure. Most of the sinkholes are located between the top of the Shakopee Formation and the top half of the Oneota Dolomite. The relatively flat distribution in Wabasha County is quite different from the distribution observed by Dalgleish (1985) in Winona County which is strongly peaked at the Shakopee/Oneota contact.

When work began on the atlas for **Mower County** in the late 1990s, GIS technology had matured to the point that most of the map preparation was conducted in a GIS environment. The field work, however, was still recorded on topo maps and/or air photos.

Bedrock is only near the land surface in southeastern Mower County, in and around Le Roy, along the eastern edge of Mower County north and south of Ostrander, and in western Mower County in the Cedar River valley from north of Austin south to the Iowa border. The bedrocks are Middle Devonian carbonates of the Wapsipinicon and Cedar Valley Groups (Fig. 3.2). The broad middle of Mower County is covered by a thick blanket of pre-Wisconsinan glacial drift.

The karst-related work in Mower County included sinkhole, stream sink and sieve, and spring mapping. Five successful dye traces were conducted in and around Le Roy. Water chemistry and tritium residence time analyses were conducted on many of the springs and selected wells. Jeff Green compiled all of the information into a single Karst Hydrogeomorphic Units plate. "This plate provides a map that portrays the karst as a system with water and material moving through it. The system consists of karst hydrogeomorphic units, each with distinctive characteristics. The result is a plate that has integrated relevant geologic and hydrologic information into a comprehensive presentation that identifies located karst features and provides an interpretation of the karst flow systems" (Green et al. 2002).

Karst hydrogeomorphic units are a promising concept designed to overcome one of the fundamental weaknesses of sinkhole inventories. A sinkhole location inventory enables environmental regulations to be written in terms of arbitrary numerical setbacks from "mapped sinkholes." These setback numbers are always compromised, based on little, if any, empirical evidence or data, between groups lobbying for stronger, more restrictive regulations and groups lobbying for weaker, less restrictive regulations. Whatever each negotiated setback number turns out to be, it *a priori* will be too restrictive for some situations and not restrictive enough for others. Karst hydrogeomorphic units encourage the gentle user to think in terms of the entire local landscape and aquifer, not just the nearest sinkhole.

The sandstone karst of **Pine County** was first studied in detail by Shade (2002). Her research was incorporated into the Pine County Atlas (Shade et al. 2002a, b). Several of the closed depressions in Pine County were trenched to determine if they were indeed sinkholes collapse. Figure 3.53 is

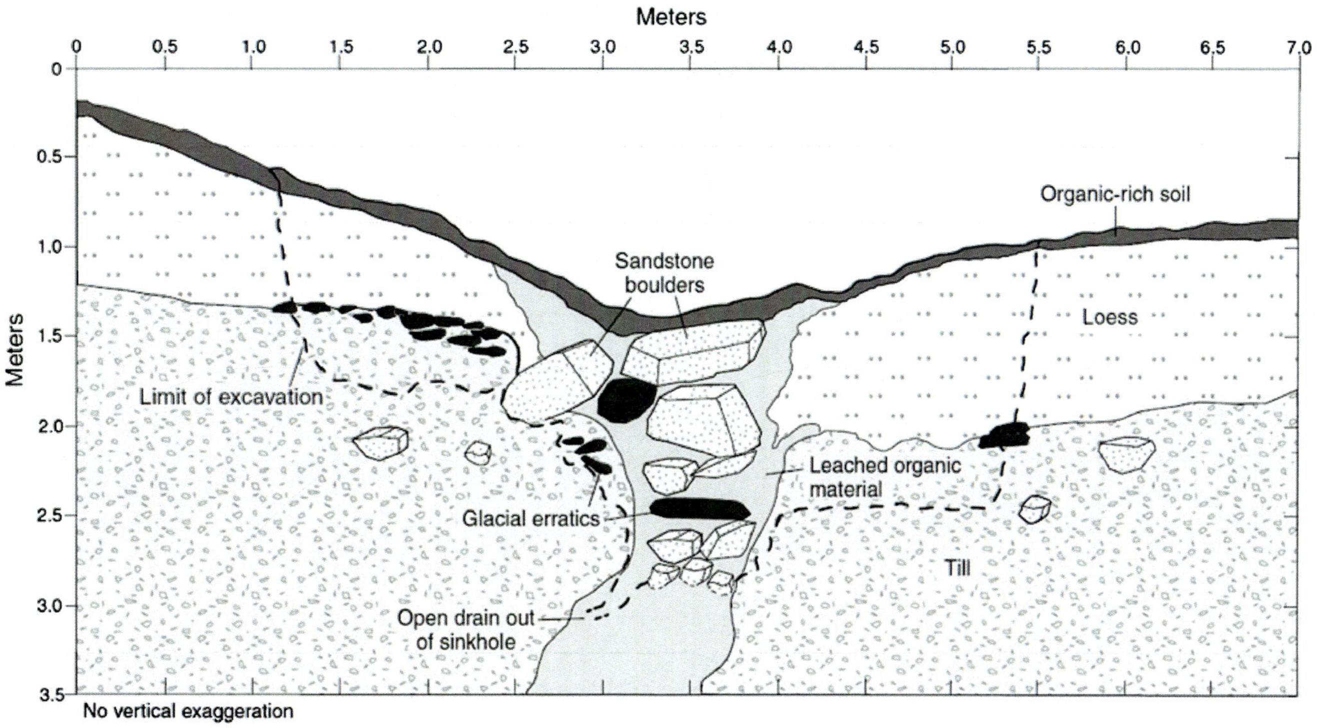

Fig. 3.53 Cross-section of sinkhole 58D222, Pine County (Shade et al. 2002b)

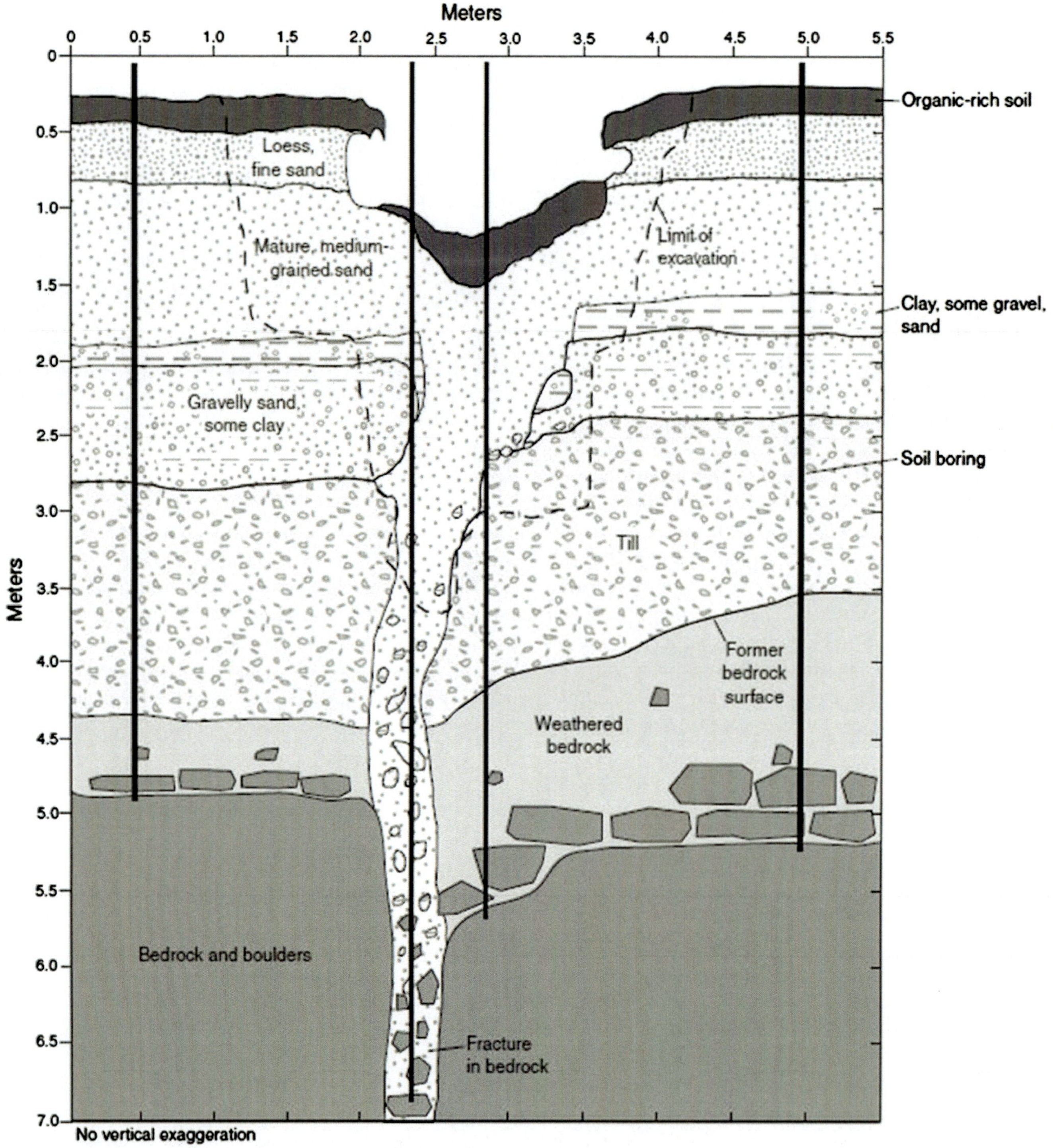

Fig. 3.54 Cross-section of sinkhole 58D127, Pine County (Shade et al. 2002b)

the cross-section revealed by the excavation of sinkhole 58D222. This sinkhole is on the east side of the Kettle River across from Banning State Park and is part of a linear array of sinkholes that includes Robinson's Ice Cave on the west side of the Kettle River. "The cross-cutting relationship of organic material within the throat of the sinkhole indicated the downward movement of surface materials. The expression of this karst feature is more pronounced in the subsurface than on the surface" (Shade et al. 2002b).

Figure 3.54 is a cross-section through sinkhole 58D127. This sinkhole was a "recent cover collapse. A 1.3 by 1.4 m topsoil plug had dropped as much as 80 cm below the flat ground surface. The grass on the displaced topsoil was still alive and the sides of the hole showed fresh soil. The topsoil

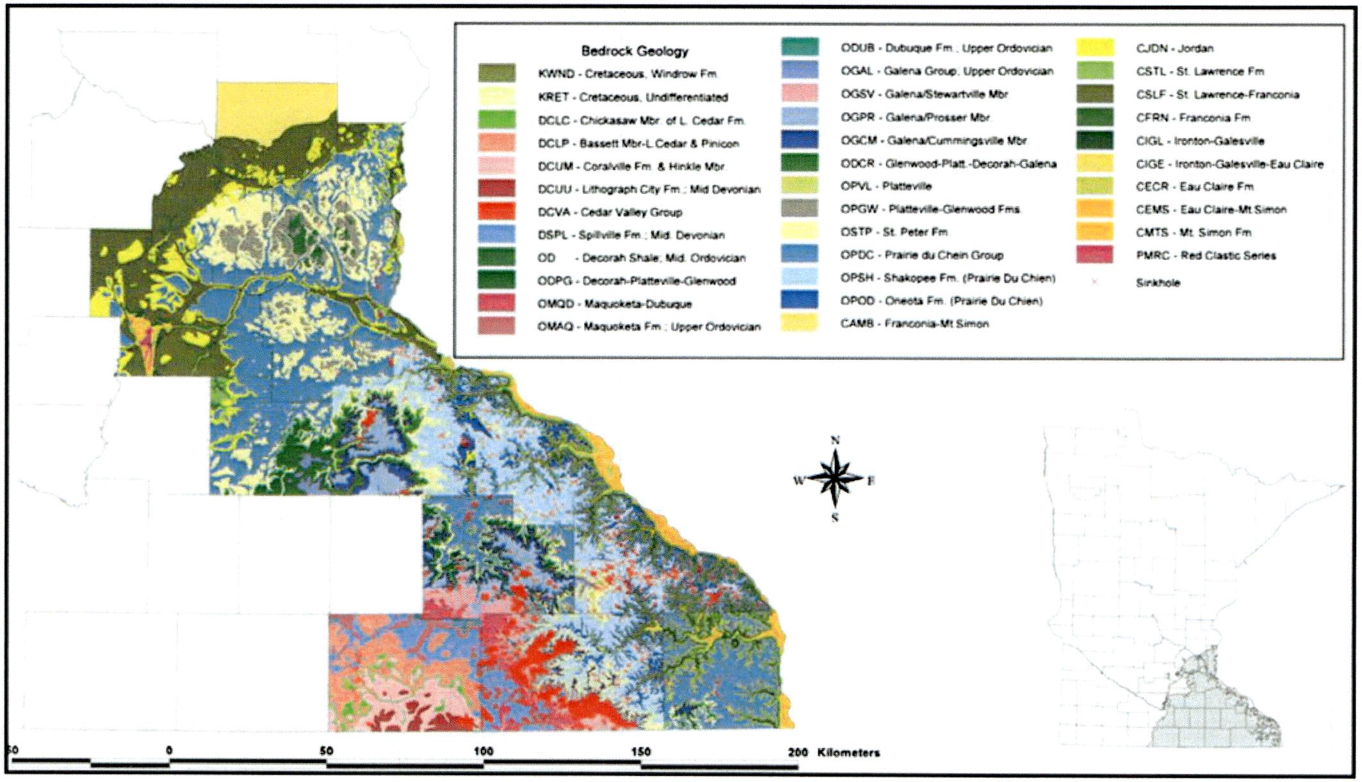

Fig. 3.55 Sinkhole distribution and bedrock geology in southeastern Minnesota (Gao et al. 2002, p 55, Fig. 3.8)

was undercut on the north and west sides, so that the actual sinkhole had a diameter of 2.0 by 1.8 m." … "The excavation revealed six matrix sedimentary strata: (1) organic rich duff, (2) mixed sand and aeolian loess, (3) mature reddish-brown sand, (4) dark reddish-brown clay, (5) immature gravelly sand, and (6) dark red till" (Shade et al. 2002b, p 61, 64). Two months after the hand trenching, four truck-mounted Giddings Soil Probe® soil borings were drilled at the site. The bedrock outside of the collapse area was about 5 m below the surface and deeper within the throat of the sinkhole. The asymmetric throat of the sinkhole extends downward at least 8 m.

Shade et al. (2002b) also described several composite features in which large, apparently glacially formed, closed depressions are internally drained by sinkholes. They also discussed numerous closed depressions caused by tree tip overs and other non-karst processes and how to distinguish such features from sinkholes.

3.3.1.2 A Minnesota Karst Features Database (KFD)

Each of the county atlases in the Minnesota karst lands had its own file of sinkholes, springs, and other karst features. Yongli Gao combined all these individual county atlas files into a single Minnesota Karst Features Database. Gao et al. (2001) wrote that "A karst feature database of southeastern Minnesota has been developed that allows sinkhole and other karst feature distributions to be displayed and analyzed across existing county boundaries in a geographic information system (GIS) environment. The central Database Management System is a relational GIS-based system interacting with three modules: GIS, statistical, and hydrogeologic modules." The KFD is described in more detail in Gao (2002). Gao illustrated the range of uses inherent in the KFD in a series of papers (Gao, 2008, Gao et al. 2002, 2005a, b, c, 2006; Gao and Alexander 2003, 2008). In a sense, the KFD is a culmination of the first revolution in sinkhole mapping in Minnesota. It is not, however, a final product. The KFD is a tool and it evolves as software and hardware keep improving and as new technologies are introduced.

Figure 3.55 is an early product of the KFD that shows the distribution of sinkholes in southeastern Minnesota. Five stratigraphic intervals host sinkholes. From the youngest to the oldest: (1) and (2) the first two of these sinkhole intervals are developed in the Devonian Cedar Valley and Wapsipinicon Groups which are the near surface bedrock only in limited areas in Minnesota in Mower and southwestern Fillmore counties. The sinkholes in the Spillville Formation of the Wapsipinicon Group appear to merge with those sinkholes in the third sinkhole interval, the Maquoketa/Galena Group. However, these sinkhole intervals extend southeast into Iowa, where the Devonian intervals separate themselves from each other and from the Maquoketa/Galena Group interval (Gao et al. 2005a, p 1087, Fig. 3.2). (3) The third interval, the Maquoketa/Galena, extends from northern Goodhue County southeast through Olmsted and Fillmore counties and south into Iowa. The Maquoketa/Galena interval contains, by far, the highest density and total number of sinkholes. (4) The fourth interval, the St. Peter/Prairie du Chien, has by far the largest area. The St. Peter/Prairie du Chien is the near surface bedrock from northern Washington County, through much of the Twin Cities Metropolitan Area and in a broad band southeast on the west side of the Mississippi. (5) The fifth sinkhole interval is in the Precambrian Hinckley Sandstone in Pine County, north of the Twin Cities.

Gao et al. (2005a) concluded: "The statistical results, along with the sinkhole density distribution, indicate that sinkholes tend to form in highly concentrated zones instead of scattered individuals. The pattern changes from clustered to random to regular as the scale of the analysis decreases from 10–100 km^2 to 5–30 to 2–10 km^2. Hypotheses that may explain this phenomenon are: (1) areas in the highly concentrated zones of sinkholes have similar geologic and topographical setting that favor sinkhole formations, and (2) existing sinkholes change the hydraulic gradient in the surrounding area and increase the soil solution and erosional processes that eventually form more new sinkholes." The KFD is a tool routinely used for research and environmental management and has facilitated the inclusion of karst information into a wide variety of human activities.

3.3.1.3 Revolution 2: GPS

The second revolution in sinkhole mapping started in 2000 when President Clinton ordered the end of the deliberate degradation of the global GPS signal. The US GPS system had been in operation for several years, but the accurate mode was only available, for national security reasons, to the armed forces. The signals available to the general public had been limited to an accuracy of about ± 100 m. Access to precision GPS allowed the location of any sinkholes, springs, or other karst features to be determined within a few meters in the field. However, it took several years for economical, small portable gps units to becine a routine part of field work. Someone still had to visit each feature and record its GPS location, but those locations can now be more accurately determined. That worked well for new mapping projects but left many of the thousands of previously mapped karst features with less accurate location information.

Goodhue County was the first of the County Atlases where GPS technology was extensively used in the karst-related field work. Sinkholes, Sinkhole Probability, and Springs and Seeps Plate (Alexander et al. 2013) were the karst contribution to the atlas. In western Goodhue County, the Galena Group is the first bedrock. The Goodhue County Atlas defined the northern limit of the Sinkhole Plains unit in

Minnesota on the east side of the Cannon River in Leon Township. (The field work missed an area of Sinkhole Plains in Sects. 13 and 14 of Warsaw Township on the west side of the Cannon River—which was found when the LiDAR DEMs became available a few years later.) In Goodhue County, the Sinkhole Plains Unit developed mainly on the Prosser Formation near the edges of bluffs overlooking incised river valleys, that is, in areas with high groundwater gradients.

In the eastern two-thirds of Goodhue County, the St Peter, OPDC, and deeper stratigraphic units are the near surface bedrock. Parts of the area have remnants of Cretaceous deposits preserved between the Paleozoic bedrock and the pre-Wisconsinan glacial sediments. Sinkholes are present on the St. Peter and OPDC subcrop areas. Sinkhole densities range up to high probability in some areas.

3.3.1.4 Revolution 3: LiDAR and High-Resolution Photos

The third revolution came to Minnesota in 2008. One-meter resolution LiDAR imaging of the Minnesota karst lands was generated and made freely available. LiDAR technology had been available for several years but had been underutilized because of its high cost. For the first time it was feasible to map sinkholes across the entire landscape in a uniform fashion. Systematic visual scanning to the shaded relief

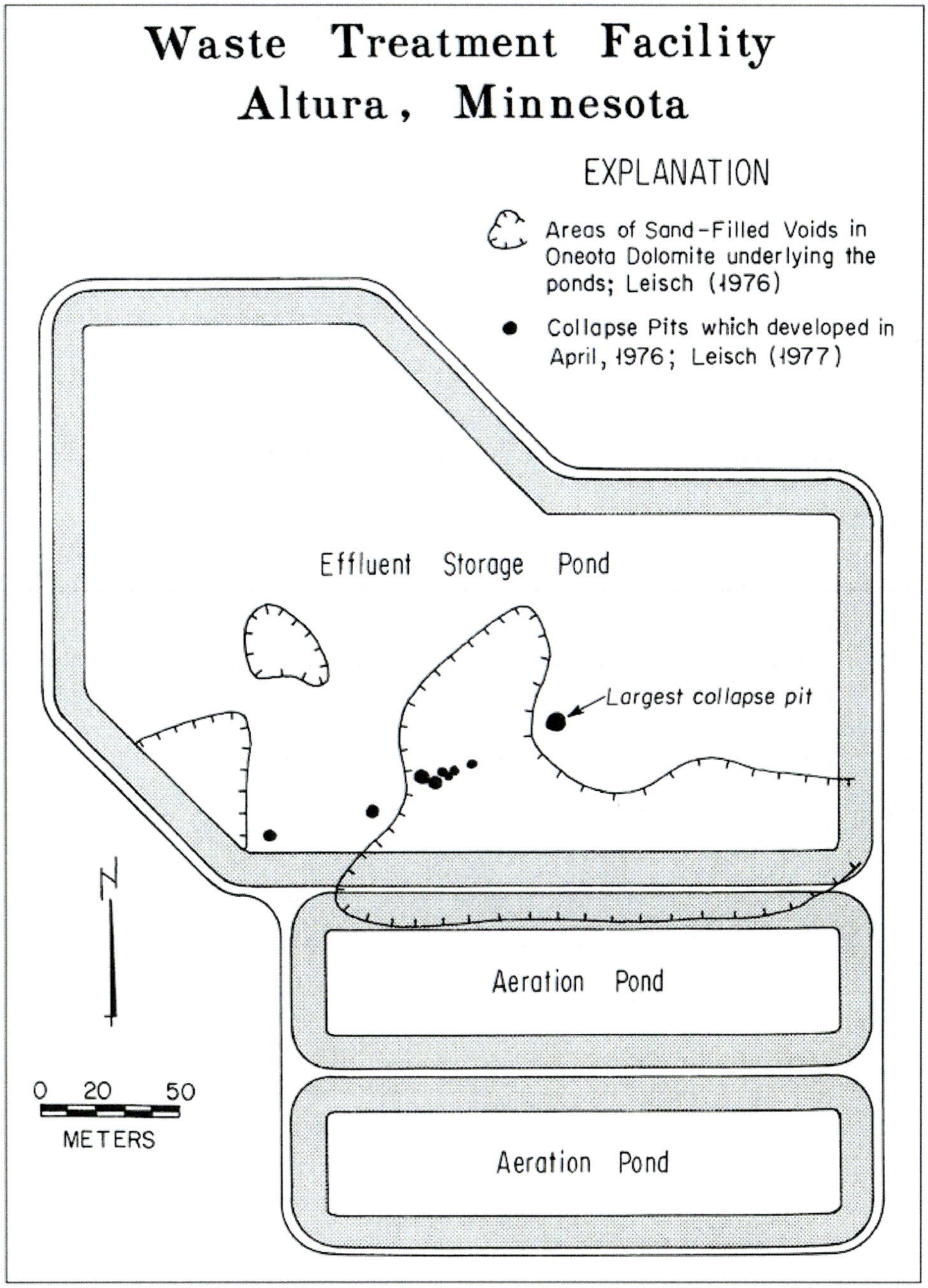

Fig. 3.56 Altura WWTF lagoon (Alexander 1980c, p 180)

digital elevation models (DEMs) revealed whole areas of sinkholes never mapped before. In addition, the locations of many of the previously mapped sinkholes were refined to an accuracy of a few meters—without a trip to each feature with a GPS unit. In the KFD it was possible to superimpose the sinkhole locations on the LiDAR DEMs and adjust the location of known sinkholes with a single "click" and to add new sinkholes just as easily.

Visual scanning of the LiDAR DEMs, although tedious, proved to be several orders of magnitude faster, more economical, and more complete than field searches that had been in the past. The minimum detectable sinkhole proved to be about 4–6 m in diameter. However, there proved to be a significant number of anthropogenic and natural closed depressions that were not sinkholes. The development of each searchers' critical skills at discriminating between sinkholes and non-sinkhole depressions required early field work to compare what was visible in the LiDAR images with what was visible in the field. Even with that training both false positives and false negatives remain an issue. In addition, there are clearly very significant sinkholes and other karst features smaller than 4–6 m in diameter. All the above notwithstanding, visual scanning of the LiDAR DEMs has doubled the number of mapped sinkholes in the KFD and, more importantly, found many sinkholes and sinkhole clusters not previously known.

The next step in analyzing LiDAR data is to automate the scanning of the DEMs for sinkholes. Filin and Baruch (2010) had suggested one approach. Mina Rahimi started the process in Minnesota with her MSc work (Rahimi and Alexander 2013). Doctor and Young (2013) reported using analogous techniques in Virginia. Wu et al. (2016) conducted an automated scanning study of the Fillmore County data set. All three of these papers obtained promising results. But all three were left with the conclusion that both false positives and false negatives remained a problem. Alexander et al. (2013) suggested that high-resolution air photos were useful in resolving some of the uncertainties in the automated LiDAR searches. Nevertheless, for any critical location, a field investigation is necessary and that may necessitate trenching, borings, and geophysical surveys to resolve the issues.

3.3.2 Catastrophic Sinkhole Collapses

Sudden catastrophe sinkhole collapses under human infrastructures are spectacular and costly phenomena. Society is unexpectedly faced with a stark choice. Either repair the sinkhole or abandon the site. Despite the much larger number of sinkholes in the Maquoketa/Galena karst, the most damaging modern catastrophic collapses in Minnesota have occurred in the St. Peter/Prairie du Chien karst areas. For example, a 90 m diameter sinkhole appeared above Chute's Cave in downtown Minneapolis in 1880 (Brick and Petersen 2004).

The Federal Water Pollution Control Act Amendments of 1972 (the Clean Water Act) focused attention and federal resources on cleaning up the United States surface waters. A major part of the Clean Water Act was the construction of sewage treatment plants for American cities and towns. Small towns with populations of a few hundred to several thousand typically do not have enough population tax base to support the ongoing operation costs of a full-scale sewage treatment plant much less than the cost of building such a facility. For many such towns the very minimal long-term operation costs made Waste Water Treatment Facility (WWTF) lagoons an attractive option—especially when 75–90% federal cost sharing was available for the construction of such facilities. Twenty-two WWTF lagoons were constructed in Minnesota karst lands often with very little, if any, design consideration of the potential risks of such lagoons in karst areas.

The first of the lagoons to fail was at Altura in Winona County (Liesch 1974, 1976, 1977; Alexander 1980c; Alexander and Book 1984) (Fig. 3.56). Sinkholes developed in the Altura WWTF lagoon twice. Two sinkholes developed in the spring of 1974 during construction.

Liesch's (1974, 1976) geophysical investigation identified sand-filled voids in the Oneota Dolomite under the lagoon. The sinkholes were filled and a bentonite clay liner was installed. The lagoon first filled to its design depth in 1976. During the planned release of the treated effluent in the storage pond in April 1976, a line of nine sinkholes developed at the bottom of the lagoon. Drilling confirmed the existence of the sand-filled voids identified in the 1974 geophysical survey. The Effluent Storage Pond was abandoned. The two smaller Aeration Ponds were not damaged and continue to be used. The wastewater is cycled through the two aeration ponds, treated in a mechanical polishing facility and then discharged to a surface valley where it normally sinks underground.

The second WWTF lagoon to suffer a sinkhole collapse was in Lewiston, also in Winona County—about 13 km south southeast of Altura (Jannik et al. 1992). In this failure a 12 m diameter by 2–4 m deep sinkhole developed at the side of one of the three ponds complex of the WWTF. The failure is estimated to have occurred on February 14, 1991 and was discovered on February 20, 1991. The sinkhole drained an estimated 29 million liters of partially treated wastewater into the groundwater. The sinkhole was filled, and a dike was erected around the collapse location, but the WWTF was ultimately abandoned and a mechanical treatment plant installed.

The Lewison WWTF failure occurred at the same exact stratigraphic interval as the Altura WWTF lagoon failure—the

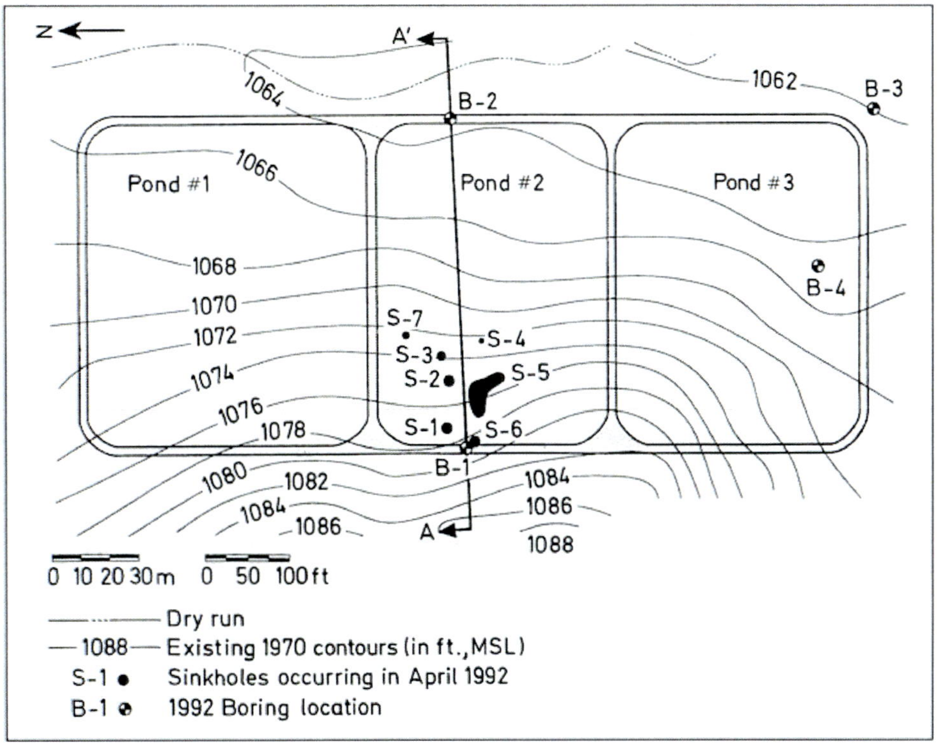

Fig. 3.57 Bellechester WWTF lagoon, plan view (Alexander et al. 1993)

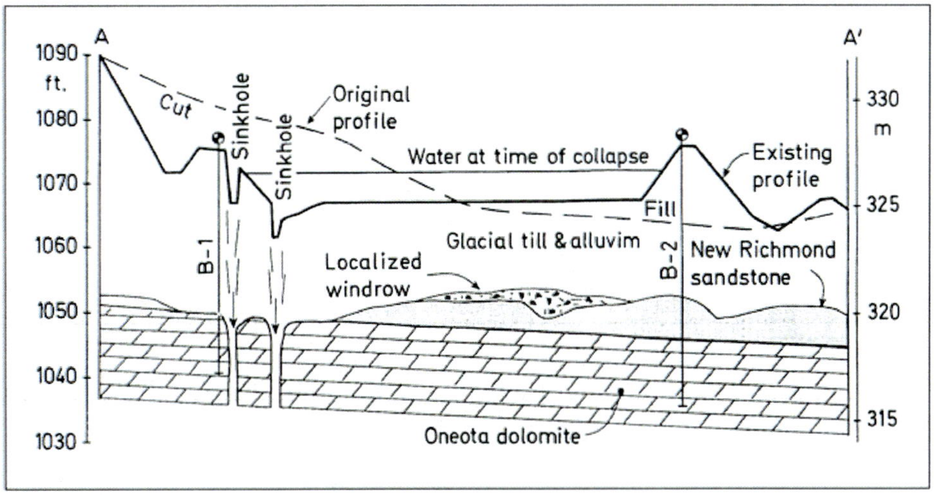

Fig. 3.58 Bellechester WWTF lagoon, cross-section (Alexander et al. 1993)

top of the New Richmond Member of the Shakopee Formation over Oneota Dolomite. This stratigraphic interval is what Tipping et al. (2006) would later recognize as the regionally extensive OPDC High Transmissivity Zone. Due to the similarity of the two WWTF lagoon collapses, the MPCA compiled an initial list of towns in southeastern Minnesota with WWTF lagoons in similar hydrogeologic environments as those of Altura and Lewiston. That initial list eventually grew that included 22 waste treatment lagoons (Alexander et al. 1993), ten of which were municipal WWTFs.

On April 28, 1992, six sinkholes were discovered to have drained cell 2 of the three-cell WWTF lagoons serving the town of Bellechester (Alexander et al. 1993). WWTF lagoons are just across the county line in Wabasha County.

Approximately 8.7 million liters of partially treated effluent and about 600 m^3 of soil drained into the subsurface. The largest of the sinkholes, S-5 (Fig. 3.57), had 2.1 m sheer walls on the west side and formed a depression 5.5–9 m wide and 9 m long with an irregular washout to the southeast. The three lagoons had been constructed in 1970. They had never been discharged to the surface drainage as per their design plan because the second pond had never filled up. The third pond was dry. After the collapses four 12 m test borings were installed, B1–B4 (Fig. 3.57).

The borings revealed a complex series of alluvial, glacial, and Cretaceous deposits overlying an irregular karst surface developed across the New Richmond and Oneota formations (Fig. 3.58). B-4 intersected a filled paleo sinkhole.

The WWTF was repaired by backfilling all the sinkholes and the entire west half of Pond 2. Membrane liners were then installed in Ponds 1 and 3 and at the reconstructed east half of the Pond 2. The entire system was put back into service and has operated thus far successfully. The ponds now fill up and the treated wastewater is periodically discharged.

In Minnesota, five municipal WWTF lagoons were built where the OPDC was the first bedrock: at Altura (1974), Bellechester (1971), Lewiston (1973), Utica (1973), and Zumbro Falls (1988). The Zumbro Falls lagoons were built with PVC membrane liners. Three of the four WWTF lagoons built on the OPDC without membrane liners in the early 1970s have catastrophically failed. That is a failure rate of 75%. Each of the WWTFs was built with majority funding from various governmental sources. The governmental programs that funded the construction of the facilities were no longer available when the lagoons failed.

In addition to the three failed WWTF lagoons, two large flood water retention structures built on the OPDC have failed due to sinkhole collapse, the Watson-Warrington structure in Goodhue County (Fay and Alexander 1994) and the Crooked Creek S3 structure in Houston County (Dogwiler and Weisbrod 2010). Both structures are at the same stratigraphic interval as the three WWTF failures and neither of them were lined.

Major collapses have also occurred in the water retention structures in the St. Peter Sandstone. The Prairie Crossing stormwater retention pond on the northern edge of Rochester in Olmsted County has experienced two sinkhole collapse episodes as described by Broberg (2015): "Following a heavy rain in the fall of 2006 nine St. Peter sinkholes opened and drained the ponded water into the subsurface. The sinkholes were filled with over 2500 cubic yards [1900 m^3] of fill. Three sinkholes collapsed again during a snow melt in the winter of 2008." Broberg (2015) also documents extensive open cavities encountered at the St. Peter/Shakopee contact at several construction sites in the City of Rochester.

In early October 2005, a newly constructed stormwater infiltration basin in the Dancing Waters Subdivision, Woodbury, Washington County, filled for the first time and 12 sinkholes were opened in the sides and at the bottom of the structure (Barr and Alexander 2009). Subsequent investigation of the newly created sinkholes found that the basin had been excavated through the surficial glacio-fluvial sediments and approximately 10 m into the top of the St. Peter Sandstone. The basin floor was about 20 m above the contact with the underlying Shakopee Formation.

In 1989, a sinkhole 12.2 m in diameter by 11 m deep opened in the backyard of a home in Mahtomedi, Washington County (Barr and Alexander 2009). The collapse "swallowed" a full-grown tree. The visible, vertical walls of the pit were in glacial deposits. The first bedrock under the site is the St. Peter Sandstone.

No one has systematically tried to count the number of farm ponds that have developed sinkholes, but there are dozens if not hundreds of such small-scale collapses in ponds in Minnesota karst lands. It has long been recognized that sinkholes are induced by impounding water on

Fig. 3.59 Irrigation induced new sinkhole 19D44. Photo provided by the Motz family

the surface (Aley et al. 1972) and Minnesota is no exception.

Pumping water out of the ground is a second, long-recognized mechanism that induces sinkholes (Jennings 1966). Center pivot spray irrigation of agricultural fields is a growing practice in Minnesota and moving into the Minnesota karst lands particularly in Dakota County. On the north edge of the glacial Cannon River valley north of the town of Cannon Falls, two new irrigation wells were drilled about 130 m apart. The first bedrock in this area is the St. Peter Sandstone. The first well was drilled in April 2012, was 105 m deep, pumped from the OPDC aquifer, and irrigated about 57 ha via two center pivots. The second well was drilled in July 2015, was 130 m deep, cased and grouted into the Jordan Aquifer, and irrigated about 15 ha. By March 2016, a cluster of five new sinkholes (19D44, Fig. 3.59, 19D45, 19D47, 19D48, 19D49) had developed immediately south of the two irrigation wells (Alexander 2016). The five sinkholes were filled. In 2019, a sixth new sinkhole (19D53) was developed in the cluster and was in turn promptly filled.

3.4 Springs

Sinkholes are the most visible "poster children" of karst landscapes, and catastrophic human-induced sinkholes have cost the society millions of dollars to remediate. However, the largest societal cost of karst is from karst phenomena's impact on groundwater supplies. As noted in the introduction of this chapter, about 75% of Minnesota's groundwater resources are in the Paleozoic aquifers comprising Minnesota's karst lands, and over half of Minnesota's population's drinking water is pumped from those aquifers.

Springs are a fundamental part of the karst hydrogeologic cycle and integrate the water flow and water quality of the entire springshed feeding each spring. Springs are a relevant part of Minnesota karst. The general observation that precipitation drains rapidly underground in karst landscapes holds true for Minnesota. Minnesota karst lands host essentially none of Minnesota's 10,000 lakes. The only perennial water sources are the baseflow rivers, the Mississippi and the Vermillion, Cannon, Zumbro, and Root Rivers, and springs. Cities, towns, and villages grew along the rivers and around the largest springs. Springs were the major water supplies of many municipalities in the twentieth century. Individual farmsteads were built immediately adjacent to springs. Spring houses were built to protect the spring and to use the water supply for the farmsteads. The largest springs were sometimes shown on maps and USGS topo sheets but there were no systematic attempts to map them.

In much of the active karst regions (Fig. 3.1), the first, near surface aquifer is already contaminated above the relevant drinking water standards. In urban and suburban areas, the contaminants of concern are often halogenated organic compounds, heavy metals, road salt, and other industrial chemicals. For example, per- and polyfluoroalkyl substances (PFAS) are fluorine containing organic chemicals whose

Fig. 3.60 Coldwater Spring, Mississippi National River and Recreation Area, Minneapolis. 2015 photo by Hisanao Kasahara

chemical and physical properties make many of them extremely persistent and mobile in the environment. PFAS were used since the 1940s in a wide range of consumer and industrial applications. PFAS were manufactured, tested, and landfilled at several sites in Washington County. Their use was largely phased out in the US between 2008 and 2015. Much of the groundwater in the near surface aquifers in the southern half of Washington County now contains PFAS at or above health advisory levels. See Yingling (2019) for an excellent summary of the rapid movement of the contaminants through the groundwater aquifers.

In agricultural areas, the contaminants of concern are typically nitrate-nitrogen, coliform bacteria, herbicides, pesticides, and viruses. In much of the Minnesota Active Karst region the near surface bedrock aquifer has been contaminated above the relevant drinking water standards and new deeper wells, cased and grouted are required by Health Department regulations. In some cases, the second or third deeper aquifers are starting to show evidence of surface contaminants. Increased pumping from the deeper aquifers increases the rate of recharge of the deeper aquifers from shallower groundwater and surface water sources.

3.4.1 Coldwater Spring

An ongoing frustration in the assessment of most water quality issues is the absence of pre-modern water quality data. However, a significant exception is Coldwater Spring in Minneapolis. An early settlement temperature record for Coldwater spring exists: French explorer Joseph Nicollet measured the temperature of Coldwater Spring and found that it was 7.8 °C in July 1836 and 7.5 °C in the winter of 1837 (Nicollet 1845). An 1880 chloride record exists: In the early 1820s, US troops lived at Coldwater Spring during the construction of Fort Snelling which overlooks the confluence of the Mississippi and Minnesota rivers. Camp Coldwater Spring was later the home to settlers in the 1830s, including a mix of nationalities such as Dakota, Ojibwe, French, and English (White 2000). Coldwater Spring was the water supply for Fort Snelling from the 1840s to 1920s. Army Captain Maguire (1880) reported the chloride level in Coldwater Spring to be 0.26 grains per gallon (4.5 mg per liter in modern units). These two reports provide a baseline against which the current temperature and chloride contents of Coldwater Spring can be evaluated.

Coldwater Spring (27A13) emerges from the Platteville Limestone near the top of the west side of the Mississippi River gorge (Fig. 3.60). The Minneapolis-St. Paul International Airport is about one km away as are many hectares of paved parking lots associated with an adjacent major Veteran's Hospital. Major four lane highways are a few tens of meters from the spring. The spring has been declared a sacred site by federally recognized Native American tribes and is currently managed by the National Park Service as part of the Mississippi National River and Recreation Area (Coldwater Spring 2020). Their website has a wealth of information about the history of Coldwater Spring.

Kasahara (2014, 2016; Kasahara et al. 2015) monitored the water chemistry of Coldwater Spring to document the human impacts on the spring's water quality. Temperature, dissolved oxygen, conductivity, pH, and anions were monitored weekly, and cations and alkalinity were monitored monthly at Coldwater Spring from February 15, 2013 to January 18, 2015. The basic chemistry of Coldwater Spring should be the calcium magnesium bicarbonate water typical of carbonate springs. However, on an equivalent basis, its water currently contains almost as much sodium as calcium + magnesium and more chloride than bicarbonate. The chloride concentrations in Coldwater Spring were about 100 times the levels that Maguire (1880) reported, and were increasing. The chloride values increased from about 320 ppm from March 2013 to about 410 ppm in December 2014. In April, May, and June of 2013 and 2014, the chloride rose about an additional 100 ppm in annual, three-month-long pulses. This major anthropogenic chloride component has a chloride to bromide ratio of 2,500 ± 300, well within the range of chloride to bromide ratios of road salt (1,000:10,000). Road salt is applied to two major multi-lane highways close to the spring and is used extensively in this heavily urbanized area throughout the winter.

The annual temperature of Coldwater Spring fluctuated smoothly between 10.7 and 13.1 °C. The spring water is coldest in May and June and warmest in October and November (see Sect. 3.4.4). The temperature of the spring's discharge is significantly higher than Nicollet's (1845) values of 7.8 °C in the summer of 1836 and 7.5 °C in the winter of 1837. The warmer temperature indicates an anthropogenic source of heat within the springshed or spring recharge area. This corroborates the detection of an anthropogenic thermal anomaly in groundwater below downtown Minneapolis first described by Brick (2006, 2009a, p 199, 2014).

3.4.2 Minnesota Spring Inventory

Minnesota's greatest spring hunter was Thaddeus Surber (1871–1949), an aquatic biologist who mapped the springs of the Root River in SE Minnesota, of Pine County, and the North Shore of Lake Superior. The latter-day equivalent is the DNR's Minnesota Spring Inventory (fully funded during 2014–2018) which, building on the existing KFD, mapped several thousand more springs, allowing the state to be divided into seven crenoregions (Brick 2019a). A protocol was developed for GPS-enabled tablet-based mapping (Brick 2015, 2017c, 2018b). Most of the other springs

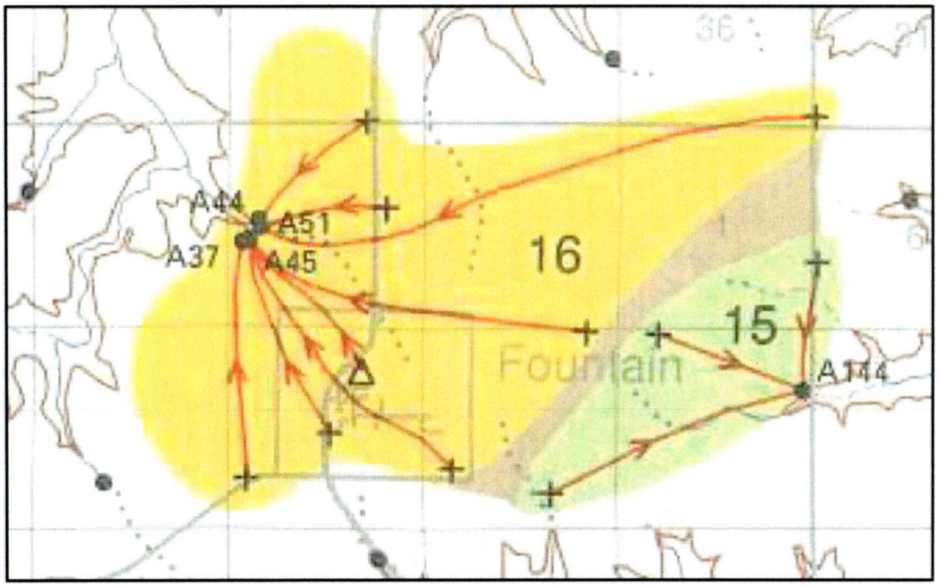

Fig. 3.61 Springsheds near the town of Fountain. Clipped from Alexander et al. (1995). Black dots are springs. Black "+" are dye input sinkholes. Red arrows are dye trace vectors

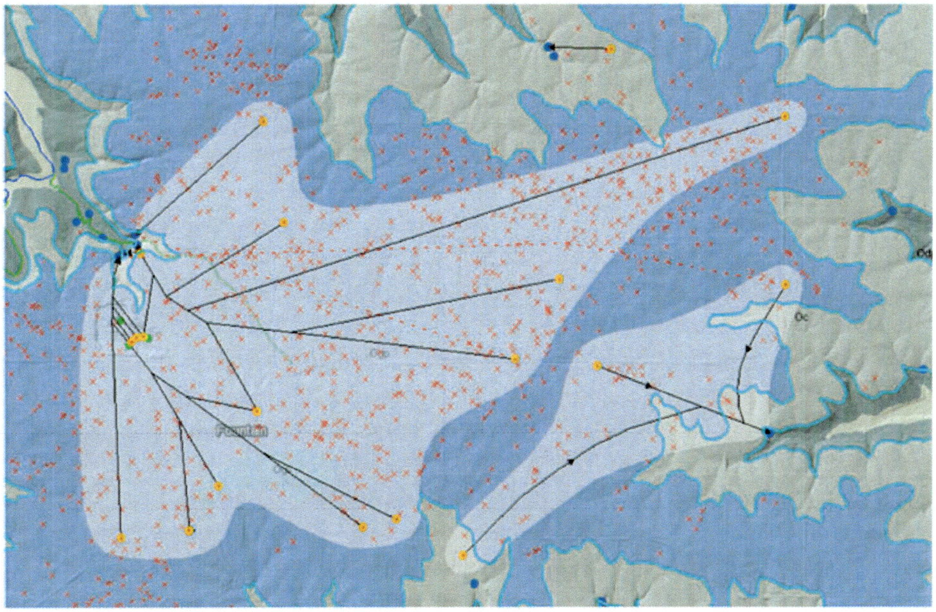

Fig. 3.62 Springsheds near the town of Fountain. The black arrows are diagrammatic dye trace vectors. Orange dots are sinkhole input points. Red **X**s are sinkholes. Blue dots are springs. The base is the bedrock geology colors draped over a shaded relief LiDAR DEM

outside of southeastern Minnesota are glacio-fluvial in origin. Several historic mineral water spas were rediscovered during this project (e.g., Brick 2018a).

Karst researchers and cavers started recording spring locations because springs were a significant, accessible part of karst hydrogeology—and springs proved to be the entrances to several large caves and may provide initial access to yet others. The KFD initially included information on spring locations and assigned unique identifying numbers to springs in Minnesota karst lands. Spring location mapping in Minnesota has an evolving history analogous to that described earlier in this chapter for sinkhole mapping. As the technology advanced from paper topo maps, to GPS to LiDAR, new springs were found, and the locations of the known springs became increasingly accurate. A characteristic LiDAR signature of springs was identified in glaciated landscapes (Brick 2020a). Springs or spring runs in inaccessible locations can often be seen in the LiDAR or high-resolution photographs. For example, LiDAR was used to discover a previously unrecognized, large-scale feature, the 100-km long Glacial Lake Lind spring line (Brick 2017b). Steenberg and Runkel (2018) list the stratigraphic locations of 163 springs in southeastern Minnesota.

The MSI is accessible at MSI (2020). Much of the initial search for new spring locations in the rest of Minnesota focused on springs on state-owned lands, mainly for ease of access reasons. Even in the karst lands only small fractions of the existing springs have been located to-date.

3.4.3 Springsheds

Dye tracing is one established method of mapping springsheds, the areas where surface recharge flows to a specific spring. Dye tracing can establish not only that a specific sinkhole or stream sink feeds a specific spring or well but can also document the underground travel time between the input and the spring. Documented dye tracing in Minnesota began in Fillmore County in 1940 when a typhoid fever outbreak prompted Kingston (1943) to use fluorescein to confirm connections between a partially treated sewage outfall into a sinkhole and a farm well in Harmony, MN. Kingston describes two additional traces in Fillmore County and one near Rochester in Olmsted County. Giammona (1973) reported the results of several traces using Rhodamine WT and Turner Model 111 Fluorometer in central Fillmore County between Wykoff and Mystery Cave.

In the late 1970s, Ron Spong began a program of dye traces in southeastern Minnesota. In the early 1980s, Eric Mohring began to conduct dye traces collecting frequent water samples and quantitative dye analyses and that produced breakthrough curves (Mohring 1983; Mohring and Alexander 1986). These and many additional dye traces by DNR and University of Minnesota staff and students were summarized as Plate 9, Springsheds of the Fillmore County Atlas (Alexander et al. 1995). Figure 3.61 is the small section of the Springsheds Map around the town of Fountain. Two colored springsheds are shown. The Fountain Springshed (16) drains west to the Fountain Spring complex (A37, A44, A45, and A51) while the Mahoney Springshed (15) drains east to A144.

Dye tracing continues to be an important tool in various parts of Minnesota karst lands and is increasingly used in other hydrogeologic environments. Individual traces have been conducted for a range of reasons and supported variously by local, county, regional, state, and federal funding. About 400 dye traces have been conducted in Minnesota. These traces have been incorporated into an online Minnesota Groundwater Tracing Database (MGTD 2020; Green et al. 2018). As an example, Fig. 3.62 shows a section clipped from the MGTD that covers roughly the same area around Fountain, as shown in Fig. 3.61. Figure 3.62 includes additional tracing done around Fountain since the Fillmore County Springsheds map was published 25 years ago. Figure 3.62 is based on a much more nuanced interpretation of the hydrogeology of the Fountain area and illustrates the utility of GIS techniques to combine different data sets.

3.4.4 Springs as Karst Hydrology Information Sources

One of the longstanding challenges of karst hydrogeology has been the attempts to use some measurable quantities at the accessible springs to deduce the properties of the inaccessible subsurface pathways between where precipitation water sinks and where it emerges—and other information about the entire system. Many karst springs are notoriously flashy, exhibiting large, rapid changes in flow. Analog depth gauges have been able to measure continuous records of water depth and thereby flow for over a century. Decades of effort have gone into studying the behavior of flow variations as karst springs respond to recharge pulses from precipitation or snow melt. The advent, in the past few decades, of economical, robust data loggers with sensors capable of monitoring not only water depth but also temperature, conductivity, pH, and a variety of other chemical parameters has opened exciting new ways to address the long-standing problem. Researchers are increasingly using data logger technology (for both applied and scientific reasons) to obtain continuous records of a variety of chemical and physical properties of karst springs.

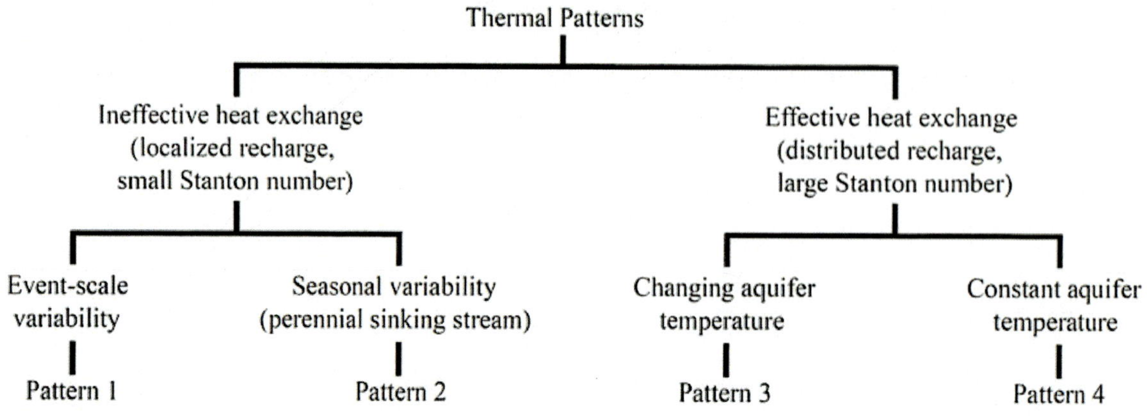

Fig. 3.63 Spring Thermal Patterns from Luhmann (2011, p 37, Fig. 2.4)

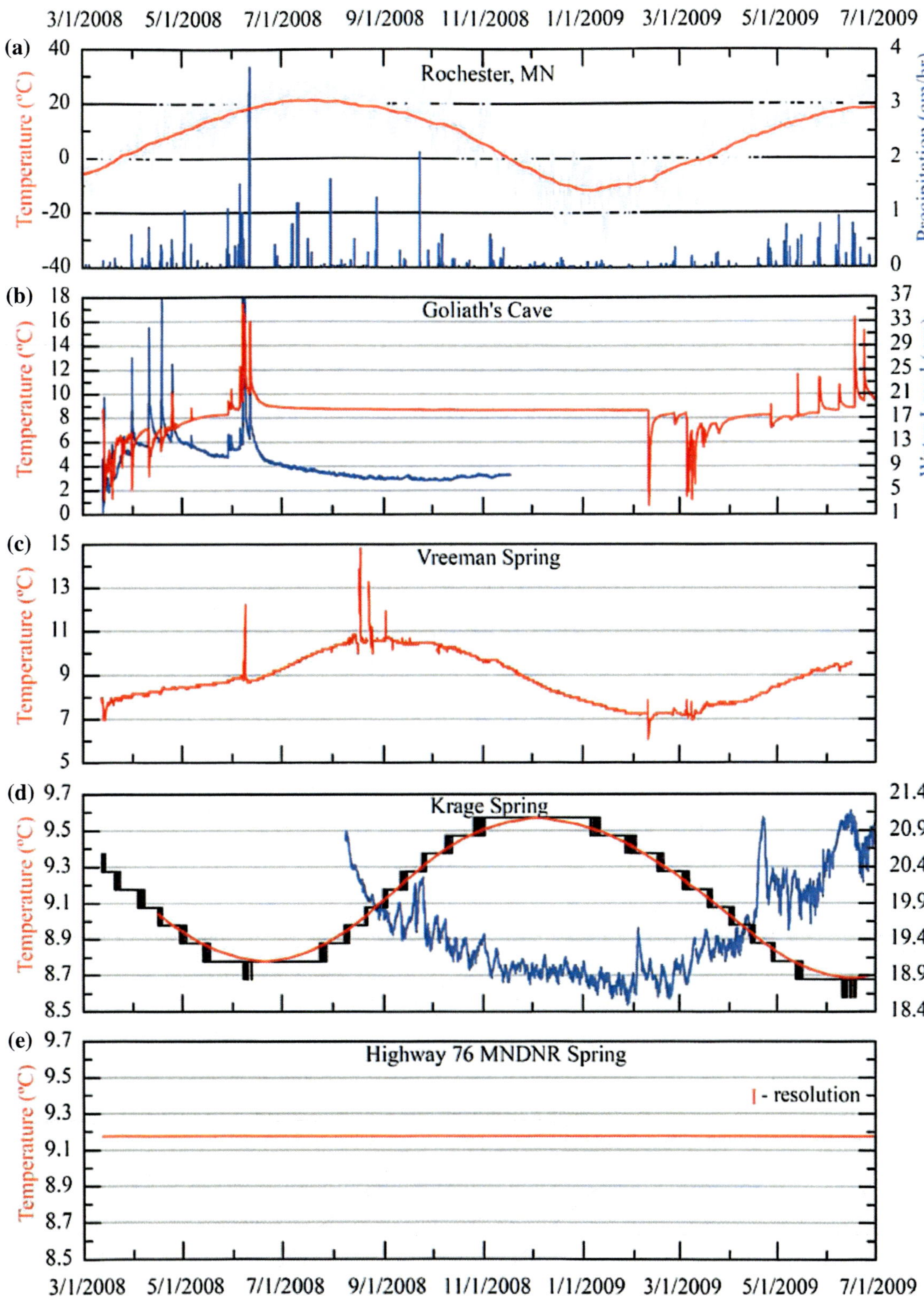

Fig. 3.64 **A** Surface temperature and precipitation at Rochester, MN **B** temperature and water level of a stream running through Goliath's Cave at David's Entrance, **C** temperature of Vreeman Spring, **D** temperature and water level of Krage Spring, and **E** temperature of Highway 76 MNDNR Spring. Note the changes in the temperature scales. Hourly surface temperatures **A** were smoothed with a two-month running average. Temperature data in **D** smoothed with a ten-week running average (Luhmann 2011, pp 28–29, Fig. 2.3)

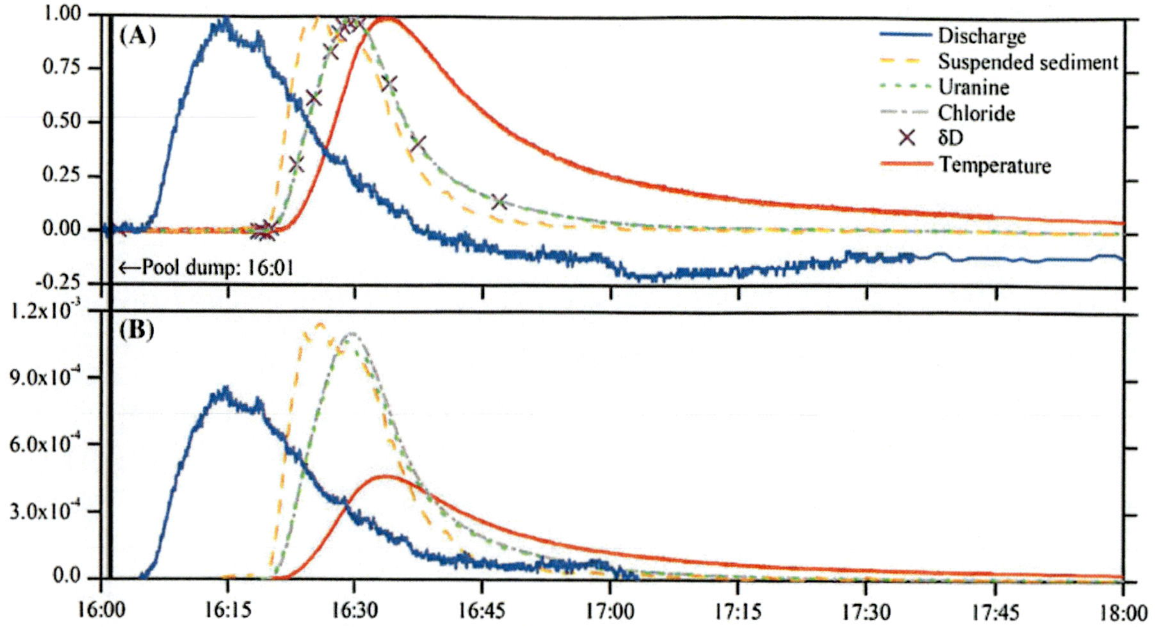

Fig. 3.65 Breakthrough curves from the August 10, 2010 Freiheit Spring multiple simultaneous tracer test. In **A** the peaks are normalized to 1, in **B** the areas under the break through curves are normalized to 1 (Luhmann 2011, p 53, Fig. 3.3)

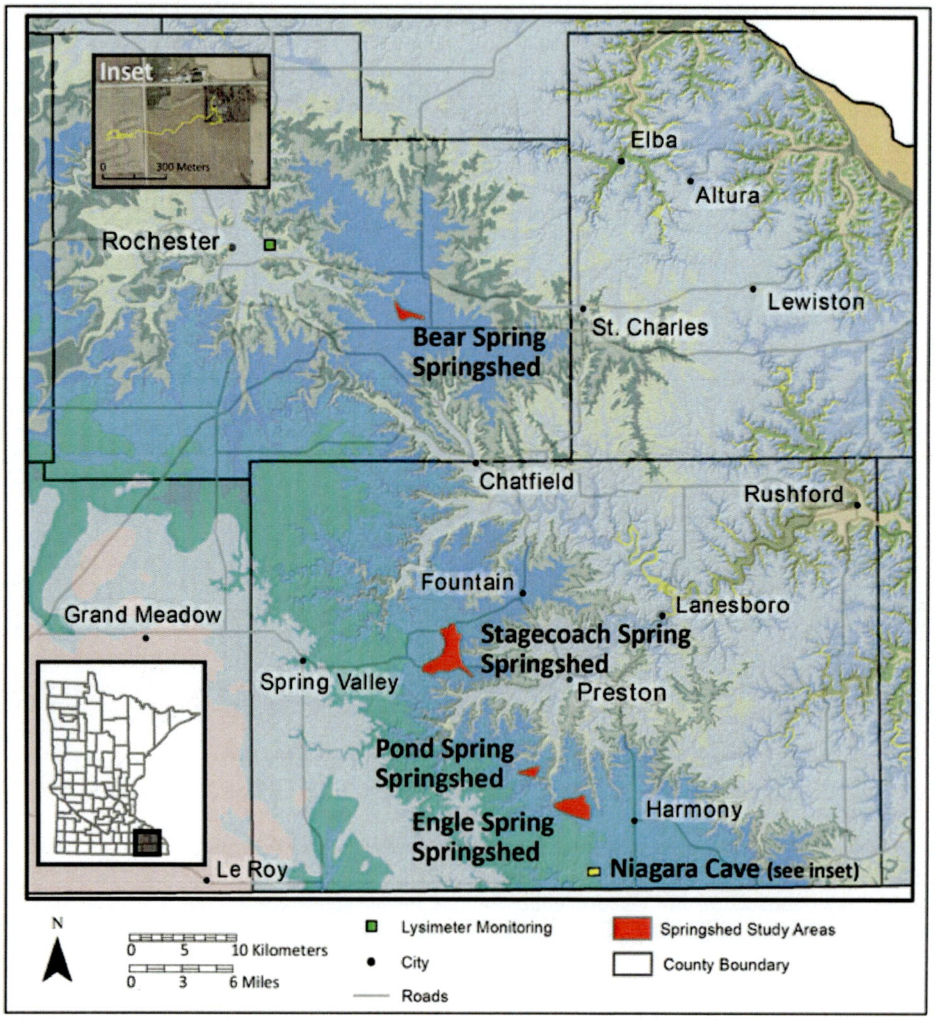

Fig. 3.66 Proposed Galena karst sentinel springs and springsheds in Fillmore and Olmsted counties. Fig. 3.1 of (Barry et al. 2020)

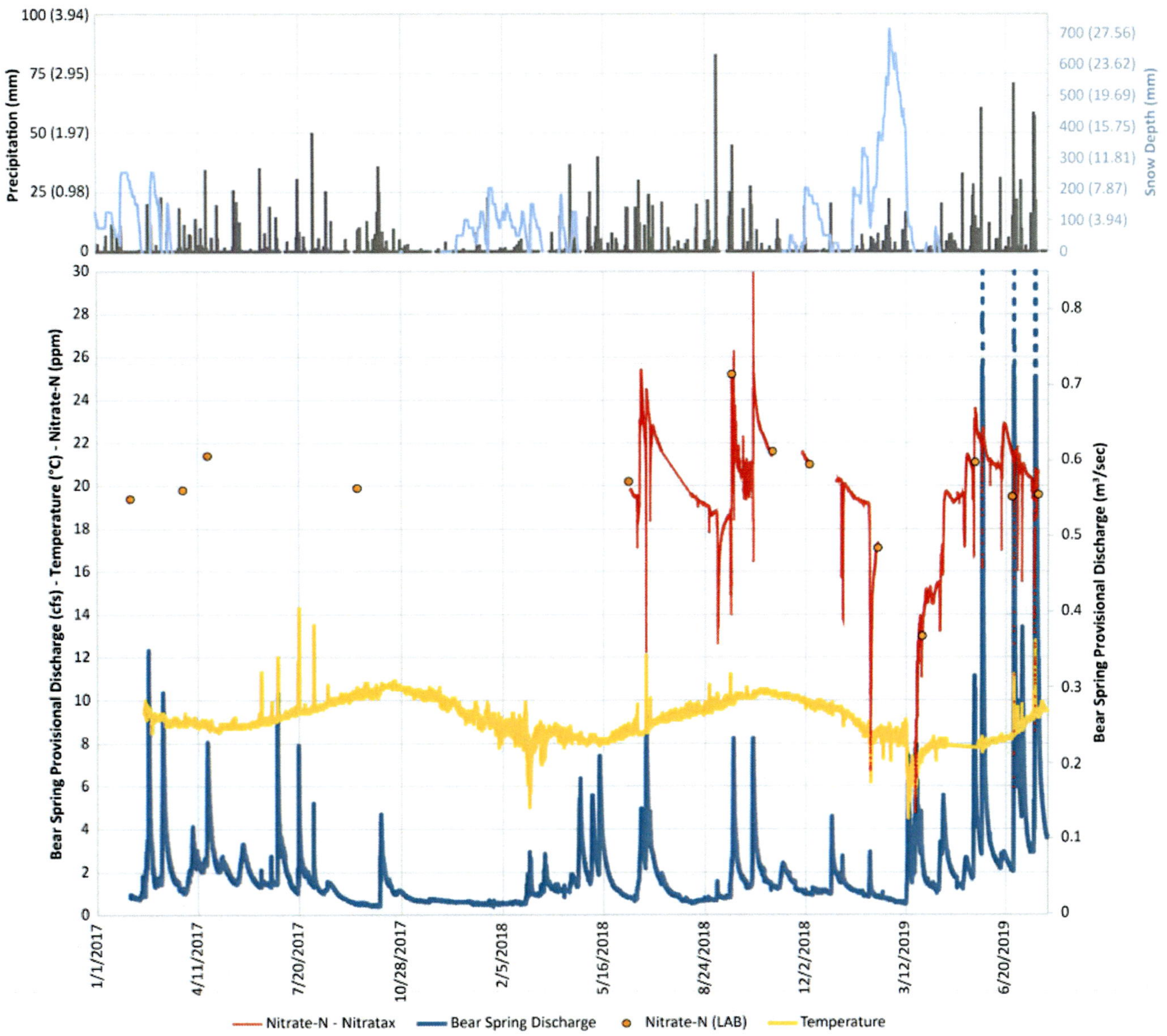

Fig. 3.67 Precipitation, nitrate-nitrogen, temperature, and discharge at the Bear Spring Sentinel Springshed monitoring station. Modified from Fig. 3.4 of Barry et al. (2020) [1 cfs = 28.3 L/s]

Shuster and White's (1972) classic work used a two-year, biweekly field campaign which sampled flow, conductivity and temperature on karst springs in Pennsylvania. They observed two types of karst spring behaviors: (1) springs with constant properties and (2) variable springs whose properties varied in phase with the seasonal air variations.

Luhmann (2011) and Luhmann et al. (2011) used digital data loggers to monitor the temperature and water level at 25 springs and cave stream in southeastern Minnesota from April 2008 to June 2009. Luhmann's data typically have hourly to sub hourly time resolution. Luhmann found four thermal patterns (Figs. 3.63 and 3.64).

Two of the patterns correspond with Shuster and White's patterns. There are springs with constant temperatures (Fig. 3.63, Pattern 4, Fig. 3.64e) corresponding with Shuster and White's constant temperature springs. There are springs with temperatures that vary on an annual time scale that it is in phase with the air temperatures (Fig. 3.63, Pattern 2, Fig. 3.64c) corresponding to Shuster and White's variable springs.

In addition, however, Luhmann found two additional temperature patterns. Luhmann identified a third set of springs with smooth annual temperature cycles whose temperatures lag the air temperature changes by up to six months (Fig. 3.63, Pattern 3, Fig. 3.64d). This type of spring was not observed by Shuster and White (1972) and may not be present in their field area. In Minnesota these springs occur along the eastern edge of Houston, Winona, and

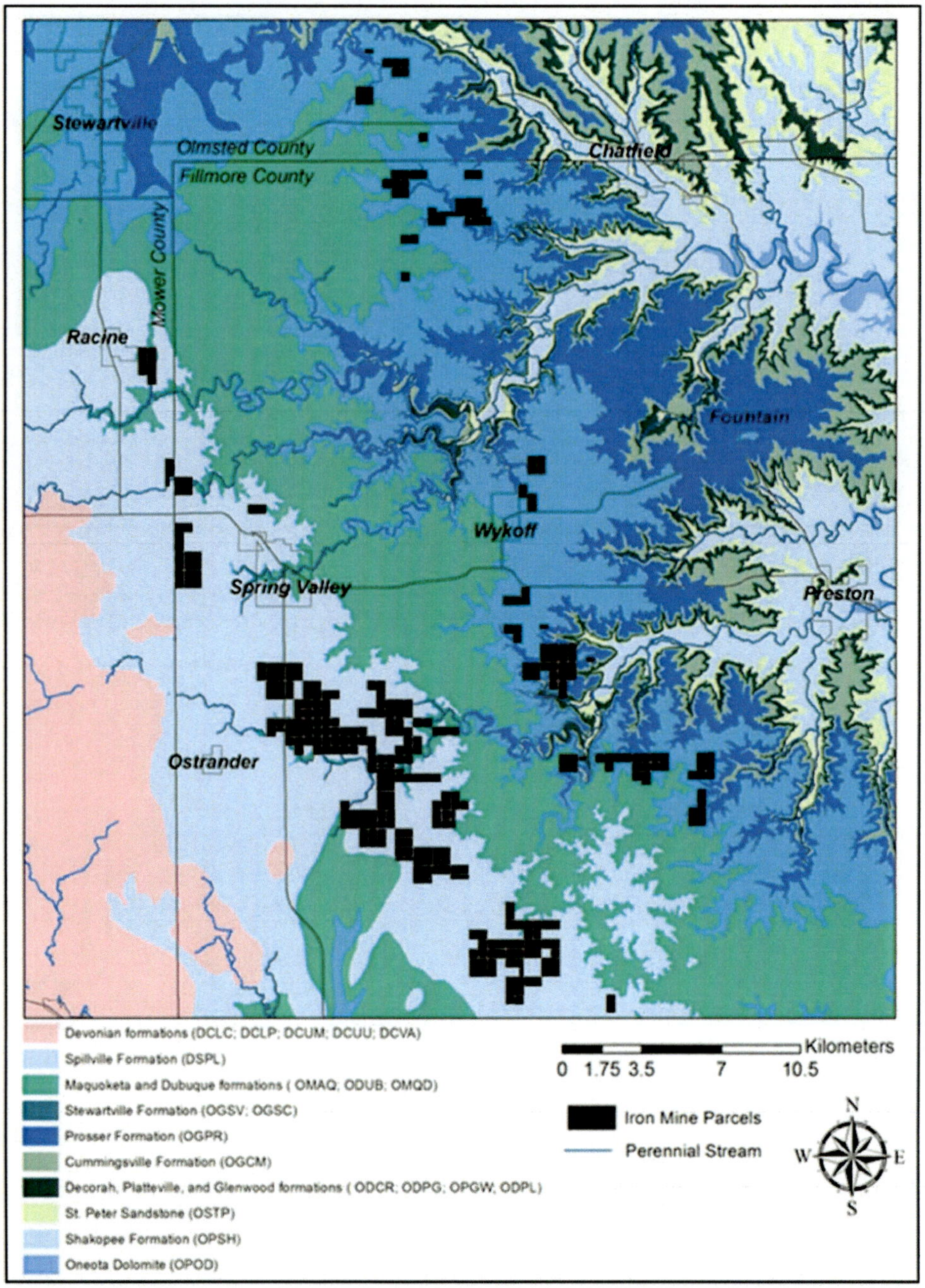

Fig. 3.68 Location of iron ore deposits in southeastern Minnesota (Alexander and Wheeler 2015; Fig. 3.1)

Wabasha counties where surface erosion has cut down into the Cambrian St. Lawrence and Lone Rock formations. These springs have warmer temperature flows in the winter than they do in the summer. Several of these springs feed important trout stream fisheries and support more robust trout populations than do streams fed by other temperature patterns.

The fourth temperature pattern Luhmann (2011) found, his Pattern 1, exhibits rapid, minutes to a few day, large temperature changes in prompt responses to recharge events (Fig. 3.64b). These events are much shorter than Shuster and White's two-week sampling interval and could not have been observed in their data set. Recharge events in the summer produce positive temperature spikes, while in the winter (often snow melt events) they produce negative temperature spikes. Figure 3.25 is a three-year-long record of the Goliath Cave stream temperature, stage, conductivity, and precipitation. Both the warm and cold recharge events decrease the conductivity of the water. The conductivity decreases are due to prompt recharge of low conductivity precipitation from surface runoff.

Pattern 1, the recharge event pattern, can be superimposed on any of the other three patterns. The presence of Pattern 1 events documents short, rapid flow pathways between the surface and that spring. Figure 3.63 shows the relationships between the four patterns. Luhmann's Patterns

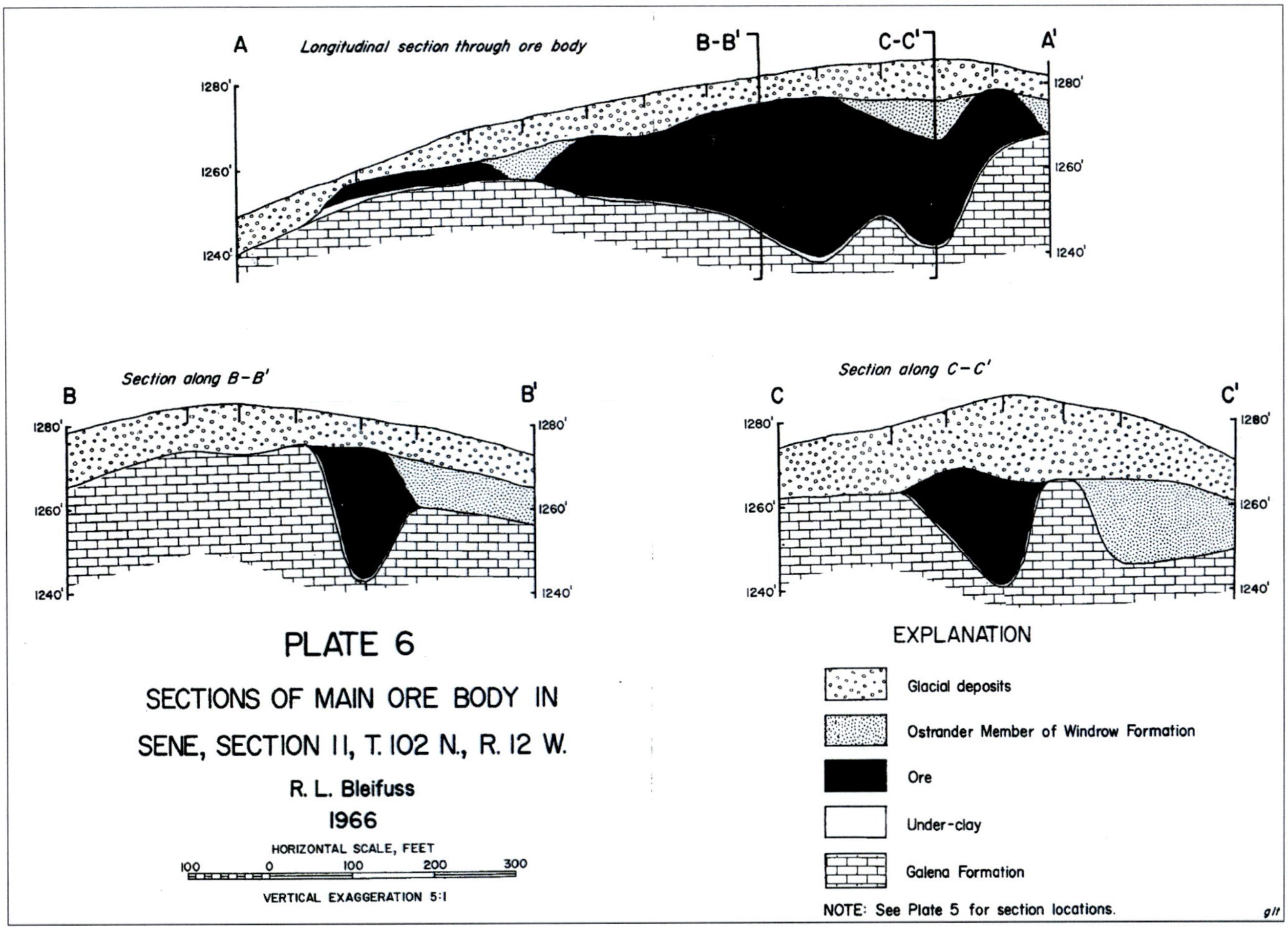

Fig. 3.69 Cross-sections of iron ore bodies in Fillmore County (Bleifuss 1966, Plate 6)

1 and 2 occur when localized, direct recharge does not have time to equilibrate completely with the thermal mass of the aquifer. Patterns 3 and 4 occur when the recharge effectively equilibrates with the thermal mass of the aquifer.

The next step was to use a combination of conservative and non-conservative tracers in a single trace and model the differences in the breakthrough curves to constrain the flow path geometry. A simple, short field site was selected (Luhmann 2011; Luhmann et al. 2012). Freiheit Spring (23A41) is a perennial spring that emerges from the base of a small bluff. On the top of the bluff sinkhole 23D2631 was an open drain that was 143 m horizontally and 29 m vertically from Freiheit Spring. A temporary weir was installed in the spring run below Freiheit Spring and instrumented with data loggers monitoring water level, temperature, and conductivity. An inflatable swimming pool was installed next to the sinkhole, filled with 13,000 L of water, the water was heated, sodium chloride, uranine, and deuterium (heavy water) were added to the swimming pool. Once mixed, the pool was then poured into the sinkhole. The sinkhole initially ponded but had drained completely by the time the pool was empty. Figure 3.65 summarizes the resulting breakthrough curves. Direct water samples were collected at the spring, initially by hand and then later using ISCO auto samplers. The water samples were subsequently analyzed for dye, sediment, and deuterium (see Luhmann 2011, p 45–48, for details).

The conduit from below the sinkhole to the spring was full and the flow of the spring responded quickly, peaked, and was declining before any of the other tracers arrived at the spring. The second "tracer" to arrive and peak was unanticipated. The pouring of the water into the sinkhole mobilized sediment from the sides and throat of the sinkhole which acted as tracer. The nominally conservative tracers, the uranine, salt, and deuterium, produced identical breakthrough curves. The thermal pulse, as anticipated, was a lagged, damped version of the other tracers. When modeled these differences resulted in an estimate that the horizontal conduit from below the sinkhole to the spring water-filled bedding plane parting 3.5 cm high and 10 m wide—which is consistent with the observed spring geometry (Luhmann 2011; Luhmann et al. 2013). Luhmann et al. (2014) present models for calculating damping of thermal pulses in karst conduits.

The combination of tracer testing defined springsheds, high-resolution LiDAR and air photo monitoring of surface

Fig. 3.70 Gushing Orange Spring (58A2), Pine County. Photo by Calvin Alexander

activity, data logger technology, and the availability of sensors to monitor nitrates means that monitoring spring water quality can be used to evaluate the impact of different agricultural best management practices (BMPs) on shallow groundwater quality in Minnesota karst. The cumulative impact of various human activities on the springshed can be monitored at the springs. And, since the springs provide the base flow to the local stream, they can be used to evaluate the impact on the water quality in those streams and rivers. The Minnesota Department of Natural Resources, in combination with other state, regional, county, and local governmental agencies and local farmers, is developing highly monitored **Sentinel Springs** and springsheds in the Galena Karst of Fillmore and Olmsted counties (Fig. 3.66). The hope is that these springsheds can become long-term monitoring sites to demonstrate the effect of various BMPs, and other changes that affect the hydrogeology of the Sentinel Springsheds.

Figure 3.67 shows the variability of the flow, temperature, and nitrate-nitrogen levels in Bear Spring from February 2017 to July 2019. The yellow curve is the temperature fluctuations of Bear Spring. Bear Spring is a Type 1 and Type 3 spring in Luhmann's (2011) classification. The spring's temperature lags about 3–4 months behind the average air temperature with superimposed recharge events causing short-term positive temperature spikes in the warm seasons and negative temperature spikes in cold seasons. The flow is very flashy with base flows below 1 cubic feet per second (cfs) and recharge flood events of over 25 cfs.

Nitrate-nitrogen levels were monitored in grab samples, the black rimmed red dots in Fig. 3.66, and then starting in June 2018 with a continuous nitrate-nitrogen sensor. The grab samples would suggest nitrate-nitrogen concentrations around 20 ppm with fluctuations between 13 and 25 ppm. The nitrate-nitrogen data logger data are in excellent agreement with the grab samples where the two methods overlap. But the continuous nitrate data reveal positive jumps of up to 30 ppm followed by rapid declines with superimposed negative, short-term dilution events associated with recharge events. "Nutrient response at Bear Spring, where both increases and decreases in nitrate occurs differs from nutrient response scenarios described in Iowa (Schilling et al. 2019) and in Ireland (Huebsch et al. 2014). At Bear Spring, the initial dilution response followed by increasing concentration found in 2019 monitoring responds as a superimposed combination of models 1 and 2 described by Schilling. This superimposed behavior is evident in the nearly instantaneous dilution followed by increasing concentration from mobilization of nitrate in the soil overburden and within aquifer storage" (Barry et al. 2020).

Continuous records of flow, temperature, and water chemistry are just beginning to reveal how rapidly changes can occur in shallow karst aquifers. These types of records promise to continue to yield fundamental new insights into karst hydrogeology.

Fig. 3.71 A complex iron oxyhydrate drapery, flow stone, rimstone dams, and gours deposit in the Soudan Mine. The photo is about 4 m high and 3 m wide. Photo by Calvin Alexander. Alexander et al (2007) suggested that these small rimstone dams may the analogs for much larger features visible on the surface of Mars

3.5 Iron Deposits and Karst

3.5.1 Fillmore County Hypogene Iron Ore Deposits

From 1942 to 1968 there was an active iron ore mining industry in western Fillmore, eastern Mower, and southern Olmsted counties of Minnesota. This mining district was 400 km south of, and 1–2 billion years younger than, the iron mines of northern Minnesota. The high-grade iron ore was mostly goethite and hematite and occurred as near-surface, relatively small pods which unconformably filled paleokarst depressions in the Devonian Spillville and Ordovician Stewartville formations. The deposits were often adjacent to or cementing discontinuous bodies of the nominally Cretaceous Ostrander Gravels. The ore bodies were covered with a few meters of Pleistocene glacial drift and loess and Holocene sediments. Figure 3.68 shows the locations of the deposits and Fig. 3.69 shows three cross-sections through one of the deposits.

The presence of iron ore deposits in southern Minnesota has been known since Winchell and Uphams's (1884) report. The source of the iron has long been controversial. The iron ores are conventionally mapped as the Iron Hill Member of the Windrow Formation (Andrews 1958). Andrews (1958, p 597) concluded "It seems probable that the Iron Hill member was deposited as a result of reaction of iron-charged waters with carbonate bedrock." Rodney Bleifuss in his Ph.D. thesis (Bleifuss 1966) and subsequent publication (Bleifuss 1972, p 498) argued "the ores are Tertiary in age and that they were developed from the oxidation of a primary marine siderite faces of the Cedar Valley Formation."

Alexander and Wheeler (2015) proposed that "available field and textural evidence is consistent with a hypogenic origin of these iron deposits. Before the current Mississippi River drainage system developed regional groundwater, flow systems could have emerged through the karst conduits in the Paleozoic carbonates. The waters in the deeply buried aquifers underlying this area currently are anoxic and enriched in dissolved ferrous iron and would have been more so before the entrenchment of the Mississippi River reorganized the regional ground water flow system. When that water emerged into the atmosphere the ferrous iron would have quickly been oxidized by a combination of biotic and abiotic processes producing the ferric oxide ores at the spring orifices."

To our knowledge none of the numerous springs issuing from the Spillville and Stewartville formations in the iron ore district of western Fillmore County are currently depositing iron oxides. However, about 45 km west, in western Mower County near Austin, the Cedar River has

Fig. 3.72 30 cm high iron oxyhydrate stalagmite in the Soudan Mine. Photo by Calvin Alexander

eroded the thick glacial sediments of central Mower County. The first bedrock there is the Spillville Formation. There are springs in that area which are currently depositing iron oxides (Green et al. 2002).

The sandstone karst of north-central Minnesota (Shade 2002; Shade et al. 2015) has many springs and seeps that are currently depositing significant amounts of iron oxyhydroxides (Alexander and Wheeler 2015). Figure 3.70 is a photograph of Gushing Orange Spring (58A2) in the sandstone karst of Pine County. This spring and several other springs in the area are actively depositing iron oxyhydroxides when their dissolved ferrous iron reacts with the atmosphere's oxygen. The leaf covered mound at the botto of the photo is a meter-thick layer of iron oxides. Based on the flow and chemistry, this spring is depositing about 3 metric tons of iron oxides per year. The entire Fillmore County iron ore deposits could have been easily produced in the available time by similar springs.

3.5.2 The Soudan Mine Siderothems

The Soudan Mine, Minnesota's first iron ore mine, opened in 1882 and operated until 1962. Unlike most of the large open pit iron mines in northern Minnesota, the Soudan Mine was a deep underground mine after 1892. The ore is a 2.722 billion-year-old banded iron formation (Jahn and Murthy

Fig. 3.73 Black soda straw stalactites growing from ceiling drips in the Soudan Mine. Photo by Calvin Alexander

1975). Mine Level No. 27, at 713.5 m below the surface, was being developed when the mine closed in 1962. In 1963, the mining company donated the mine, intact infrastructure, and the surrounding 486 ha to the state of Minnesota to create what is now the Lake Vermillion-Soudan Underground Mine State Park (LVSUMSP 2020).

At the lower levels concentrated calcium chloride brines seep into the mine workings from exploration drill holes into the surrounding rocks. The brines have elevated concentrations of soluble ferrous iron, are strongly reduced, and up to twice as salty as sea water. When that seepage reaches the mine adits it reacts with the oxygen in the mine air to produce various forms of ferrihydrite and other ferric iron precipitates. Those reactions are strongly catalyzed by a rich, very primitive microbiology (Edwards et al. 2006). Those insoluble compounds precipitate to form a wide range of speleothem morphologies.

Figure 3.71 shows a complex secondary deposit in the east side of Level 27 in the Soudan Mine. A horizontal drill hole penetrates the adit wall about 10 cm below the top of the drapery and feeds a small pool at the top of the drapery. Water dripping from the drapery has formed a flowstone mound which feeds two rimstone pools. The overflow from the rimstone pools forms a complex of gours. The test hole was drilled in 1962. The entire, bright orange structure formed in a few decades.

Figure 3.72 is a stalagmite about 25 cm tall in the Soudan Mine. The stalagmite's form strongly mimics a classic "fried egg" stalagmite except the colors are different. The center is gray, and the outer portion is yellow to brown. The stalagmite has formed on a jean jacket that was left in the adit in 1964.

Figure 3.73 shows several soda straw stalactites growing among abandoned electrical utility conduits on the ceiling of an adit in the Soudan Mine. The soda straws have the same diameter as calcite soda straws but are made of ferrihydrite. As with the other secondary deposits, these stalactites grew in a few decades.

These ferrihydrite "siderothems" (and the ice stalagmites in Robinson's Ice Cave) demonstrate that the morphology of classic calcite speleothems is not a specific function of the carbonate solution/precipitation equilibrium. But the chemical reactions are analogous. Both the carbonate and iron precipitation reactions are driven by gases moving between the local atmosphere and the drip water. But the direction of movement is reversed. In the carbonate reaction carbon dioxide diffuses from the water into the air. In the iron precipitation reaction oxygen diffuses from the atmosphere into the water. In the formation of the ice stalagmites heat is being transferred from the warmer water to the colder cave air.

Acknowledgments Karst hydrogeology research and cave exploration in Minnesota has grown through the cumulative efforts of too many cavers, students, teachers, scientists, engineers, public servants, interested private citizens, and landowners to list. Grateful thanks and appreciation to all. A major source of financial support was provided by the State of Minnesota via the Legislative Commission on Minnesota Resources (LCMR) and its successor the Minnesota Clean Water Land and Legacy Amendment and the Minnesota Environment and Natural Resources Trust Fund as recommended by the Legislative-Citizen Commission on Minnesota Resources (LCCMR). Other significant financial support has come from the various counties and other private and public organizations.

References

Ackerman J (2019) Bat river—Bat counts, unpublished report, 3 p

Adams RS Jr, Grow S, Alexander EC Jr (1985) The presence of "apparent" atrazine in groundwater of southeastern Minnesota. A Report on Field Research, University of Minnesota Agricultural Experimental Station Miscellaneous Publication 2 (revised)-1985, p 160–162

Alexander EC Jr (1980a) Leslie Cave and Clarke Cave, Washington County. In: Alexander EC Jr (ed) An introduction to Caves of Minnesota, Iowa, and Wisconsin, Guidebook for the 1980 NSS Conv. Minn Geol Surv, St. Paul, p 79–83

Alexander EC Jr (1980b) Robinson's Ice Cave, Pine County. In: Alexander EC Jr (ed) An introduction to Caves of Minnesota, Iowa, and Wisconsin, Guidebook for the 1980 NSS Conv. Minn Geol Surv, St. Paul, p 84–86

Alexander EC Jr (1980c) Stop #7. Altura Waste-Treatment Facility. In: Alexander EC Jr (ed) An introduction to Caves of Minnesota, Iowa, and Wisconsin, Guidebook for the 1980 NSS Conv Minn Geol Surv, St. Paul, p 178–183

Alexander EC Jr (2016) Irrigation pumping associated sinkhole development in Dakota County, MN. (Paper 190–192), Geol Soc Amer, Abstracts with Programs 48(7). https://doi.org/10.1130/abs/2016am-283413

Alexander EC Jr, Alexander SC, Barr KDL, Luhmann AL, Anger CT (2015) Goliath's Cave, Minnesota: epigenic modification and extension of preexisting hypogenic conduits. In: Doctor DH, Land L, Stephenson BJ (eds) National Cave & Karst Research Institute symposium 5, Sinkholes and the Engineering and Environmental Impacts of Karst. In: Proceedings of the 14th multidisciplinary conference, Rochester, MN, NCKRI, Carlsbad, NM, p 201–209

Alexander EC Jr, Alexander SC, Hanson AE, Pinta SM, Boston PJ, Gralnick JA (2007) The Soudan Mine, Minnesota: a hydrochemical analog for Rimstone Dams on the Martian Surface (expanded abstract). Lunar Plant Sci XXXVII:1758

Alexander EC Jr, Berner, DJ, Gao Y, Green JA (2003) Sinkholes, Sinkhole probability, and springs and seeps, geologic Atlas of Goodhue County, Minnesota, County Atlas Series, Atlas C-12, Part B, Plate 10

Alexander EC Jr, Book PR (1984) Altura Minnesota lagoon collapses. In: Beck B (ed) Sinkholes: their geology, engineering and environmental impact, AA Balkema, Rotterdam, Netherlands, p 311–331

Alexander EC Jr, Broberg JS, Kehren AR, Graziani MM, Turri WL (1993) Bellechester Minnesota Lagoon collapses. In: Beck BF (ed) Applied karst geology proceedings of 4th multidisciplinary conference on Sinkholes and the engineering and environmental impacts of Karst, Panama City Beach, FL, AA Balkema, Rotterdam, Netherlands, p 63–72

Alexander EC Jr, Green JA, Alexander SC, Spong RC (1995) Springsheds. Fillmore County Atlas, Atlas C-8, Part B, Minn Dept Nat Res, St. Paul, Plate 9

Alexander EC Jr, Jameson RA (1994) The Waters of Mystery Cave, Forestville State Park, Minnesota Mystery Cave Resource Evaluation (Groundwater) Management Report, 9 p. file:///C:/Users/Owner/AppData/Local/Temp/vol1.pdf

Alexander EC Jr, Maki GL (1988) Sinkholes and Sinkhole Probability. Olmsted County Geologic Atlas, Atlas C-3, Minn Geol Surv, St. Paul, Plate 7

Alexander EC Jr, Wheeler BJ (2015) A proposed hypogenic origin of iron ore deposits in Southeast Minnesota Karst. In: Doctor DH, Land L, Stephenson BJ (eds) Proceedings of the 14th Sinkhole conference, National Cave & Karst Research Institute Symposium 5, Sinkholes and the Engineering and Environmental Impacts of Karst, Proceedings of the 14th Multidisciplinary conference, Rochester, MN, NCKRI, Carlsbad, NM, p 167–175

Alexander SC, Rahimi M, Larsen E, Bomberger C, Greenwaldt B, Alexander EC Jr (2013) Combining LiDAR, aerial photography and pictometry® tools for karst features data base management. In: Lewis L, Doctor DH, Stephenson JB (eds) NCKRI Symp 2 Proc 13th multidisciplinary conf on sinkholes and the engineering and environment impacts of Karst. Carlsbad, NM, NCKRI, Carlsbad, NM, p 441–458

Aley TJ, Williams JH, Massello JW (1972) Groundwater contamination and Sinkhole collapse induced by leaky impoundments in soluble rock Terrain. Eng Geol Ser 5, Missouri Geological Survey and Water Resources, Rolla, 32 p

Andrews GW (1958) Windrow formation of upper Mississippi valley region a sedimentary and stratigraphic study. J Geol 66(6):597–624

Austin GS (1972) Paleozoic lithostratigraphy of southeastern Minnesota. In: Sims PK, Morey GB (eds) Geology of Minnesota: a centennial volume. Minn Geol Surv, St. Paul, pp 459–473

Barr KDL, Alexander EC Jr (2009) Examples of hypogenic karst collapse structures: twin Cities Metropolitan Area, Minnesota, USA. In: Stafford KW, Land L, Veni G (eds) Advances in hypogene Karst studies, NCKRI symposium I, Carlsbad, NM, National Cave and Karst Research Institute, p 51–59

Barry JD, Miller TP, Steenberg JR, Runkel TC, Kuehner KJ, Alexander EC Jr (2020) Combining high resolution spring monitoring, dye tracing, watershed analysis, and outcrop and borehole observations to characterize the Galena Karst, Southeast Minnesota, USA. Proceedings of 16th Sinkhole Conference (first edition) NCKRI Symposium 8. Carlsbad (NM): Nat'l Cave and Karst Res Inst, p 3–17.https://scholarcommons.usf.edu/sinkhole_2020/ProceedingswithProgram/Resource_monitoring_and_management/1/

BCNG, Bluff Country Newspaper Group (2019) Hiawatha Caverns purchased by caving groups. http://www.bluffcountrynews.com/content/hiawatha-caverns-purchased-caving-groups

Bleifuss RL (1966) The origin of the iron ores of Southeastern Minnesota. PhD thesis, Univ. of MN, Minneapolis, 125 p

Bleifuss RL (1972) The iron ores of Southeastern Minnesota. In: Sims PK, Morey GB (eds) Geology of Minnesota: a centennial volume. Minn Geol Surv, St. Paul MN, pp 498–505

Bretz JH (1938) Caves in the Galena formation. J Geol 46(6):828–841

Brick GA (1997a) Along the great wall: mapping the springs of the Twin Cities. Minn Ground Water Assoc Newsl 16(1):1–7

Brick GA (1997b) Classification and morphology of St. Peter Sandstone Piping Caves in Minnesota. J Cave Karst Stud 59(3):168

Brick GA (2002) Caving with Roaches: The Nicollet Mall Cave. Minnesota Speleology Monthly 34(8):91, 95

Brick GA (2003a) The University farm experimental Cave and how St. Paul became the blue cheese capital of the world. Ramsey Cty Hist 36(3):4–10

Brick GA (2003b) "Terrific Battles": pests, disease, and technological change at St. Paul's Mushroom Caves. J Spelean Hist 37(2):49–50

Brick GA (2004) "Cave Shaped Like an Inverted Bowl": the 1904 discovery of Schieks Cave. Minn Speleol Mon 37(1&2):1, 3–6

Brick GA (2006) Little Minnehaha Falls: the great subterranean spring of Minneapolis. Minn Ground Water Assoc Newsl 25(1):17–19

Brick GA (2009a) Subterranean Twin Cities. University of Minnesota Press, Minneapolis, 226 p

Brick GA (2009b) Guanophile ecosystem of an urban cave polluted with raw sewage. Proceedings of the 15th international congress of speleology, vol 3. Kerrville, TX, p 1280

Brick GA (2012) Where in Minnesota Can You Find 35,000 Ppm Nitrate? 57th Midwest ground water conference, program with abstracts, Minneapolis, MN

Brick GA (2013a) The nitrate deposits of rock crevices in the upper Mississippi Valley. Ph.D. thesis, University of Minnesota, Minneapolis, 165 p. University of Minnesota Digital Conservancy. http://hdl.handle.net/11299/157750

Brick GA (2013b) Stahlmann's Cellars: a large American Lagering Cave from the nineteenth century. Brew Hist 155:5–15

Brick GA (2014) Thermal pollution of groundwater under minneapolis. MGWA spring conference. https://www.mgwa.org/meetings/2014_spring/2014_spring_abstracts.pdf

Brick GA (2015) Legacy data in the Minnesota Spring Inventory. The 14th Sinkhole Conference Program with Abstracts, Rochester, MN, USA, p 43

Brick GA (2016a) Early French mining in Minnesota. In: Labine MA (ed) They Spoke French: French Heritage in Minnesota. French-American Heritage Foundation, Arden Hills, MN, p 65–71

Brick GA (2016b) Biological reconnaissance map rediscovered. Minn Conserv Volunt 79(464):60–63

Brick GA (2017a) Minnesota Caves: History and Lore. Arcadia/The History Press, Mount Pleasant, SC, 144 p

Brick GA (2017b) The Glacial Lake Lind spring-line of the St. Croix Valley. 2017 Research Rendezvous, St. Croix Watershed Research Station, Marine on St Croix, MN

Brick GA (2017c) Minnesota Spring Inventory Guidance Document. Minnesota Department of Natural Resources, Ecological and Water Resources, County Geologic Atlas Project, St. Paul, 32 p

Brick GA (2018a) The Sacred Heart geyser. Minn Ground Water Assoc Newsl 37(1):8

Brick GA (2018b) Spring-hunting strategy and the diversity of springs in Minnesota. Springs ecosystem symposium, Museum of Northern Arizona, Flagstaff, 12 p

Brick GA (2019a) The crenoregions of Minnesota. Minn Ground Water Assoc Newsl 38(1):11

Brick GA (2019b) The springs of Pine County, 1920. Minn Ground Water Assoc Newsl 38(2):1, 3, 15

Brick GA (2020a) The spirilliform signature: using LIDAR to map springs in glaciated landscapes. In: Geologic Mapping Forum, Minnesota Geological Survey, Minneapolis (in press)

Brick GA (2020b) Exploration and institutional history of caves of the Upper Midwest, USA. Chapter 2, this volume

Brick GA (2020c) Paradigmatic studies in Midwestern cave science. Chapter 8, this volume

Brick GA, Petersen PA (2004) Ten cent tour: the story of Chute's Cave. Hennepin Hist 63(2):4–25

Broberg JS (2015) Living with Karst in Rochester, Minnesota's Destination Medical Center, 2015 Sinkhole Conf. Field Trip Guide, p 44–46. https://digital.lib.usf.edu//content/SF/S0/05/37/49/00001/K26-03300-Field_Trip_Rochester.pdf

CGBV SNA (2020) Cherry Grove Blind Valley SNA. https://www.dnr.state.mn.us/snas/detail.html?id=sna02022

Coldwater Spring (2020) Coldwater spring, Mississippi river national river and recreation area. https://www.nps.gov/miss/planyourvisit/coldwater.htm

Costello DE, Alexander EC Jr (2006) 19 June 2006 dye trace of the cave farm Blind Valley stream sink (MN23: B0058), unpublished summer research project, 12 p +appendices. https://conservancy.umn.edu/handle/11299/185413

Dalgleish JB (1985) Sinkhole distribution in Winona County, Minnesota. MS thesis, Univ. of Minn, 95 p +maps

Dalgleish JB, Alexander EC Jr (1984a) Sinkholes and sinkhole probability, Plate 5, In: Balaban NH, Olsen BM (eds) Geologic Atlas of Winona County. Minn. Geol. Surv., St. Paul (1:100,000 scale map with text)

Dalgleish JB, Alexander EC Jr (1984b) Sinkhole distribution in Winona County, Minnesota. In: Beck B (ed) Sinkholes: their geology, Engineering and Environmental Impact, AA Balkema, Rotterdam, Netherlands, p 79–85

Dasgupta S (2008) High-resolution speleothem record of late Quaternary climate change from the upper Midwest, USA. Ph.D. thesis, Univ of Minn. 287 p

Dasgupta S, Saar MO, Edwards RL, Shen CC, Cheng H, Alexander EC Jr (2010) Three thousand years of extreme rainfall events recorded in stalagmites from Spring Valley Caverns, Minnesota. Earth Plan Sci Lett 300:46–54

Doctor DH, Alexander EC Jr (1998) Results of monitoring the Isotopic composition and hydrology of infiltrating drip Waters in Mystery Cave, Minnesota, EOS, Trans Am Geophys Union 79(17):S83–S84 (Spring Meeting)

Doctor DH, Alexander EC Jr (2005) Interpretation of water chemistry and stable isotope data from a karst aquifer according to flow regimes identified through hydrograph recession analysis. In: Kuniansky EL (ed) USGS Karst Interest Group Proc, Rapid City, SD. USGS Sci Invest Ret 2005-5160, p 82–92

Doctor and Alexander (2020) Karst Geology of the Upper Midwest. Chapter 1, this volume

Doctor DH, Alexander EC Jr, Jameson RA, Alexander SC (2015) Hydrologic and Geochemical dynamics of vadose zone recharge in a mantled karst aquifer: results of monitoring drip waters in Mystery Cave, Minnesota. In: Doctor DH, Land L, Stephenson JB (eds) Proceedings of the 14th Sinkhole Conference, Rochester, MN, NCKR Symposium 5, Sinkholes and the Engineering and Environmental Impacts of Karst, NCKRI, Carlsbad, NM, p 19–29

Doctor DH, Alexander EC Jr, Stichler W (2001) Isotopic tracers do not usurp groundwater chemistry or vice versa: combined isotopic and chemical tracers in upper mississippi valley and slovenian karst aquifers yield more information than either would alone (Abstract H41A-07), 2001 Spring Meeting Prog, Amer Geophys Union, Boston, MA, p 87

Doctor DH, Young JA (2013) An evaluation of automated GIS tools for delineating Karst Sinkholes and closed depressions from 1-m LiDAR-derived digital elevation data. In: Lewis L, Doctor DH, Stephenson JB (eds) NCKRI symposium 2 proceedings of the 13th multidisciplinary conference on Sinkholes and the Eng and Env Impacts of Karst, Carlsbad, NM, NCKRI, Carlsbad, NM, p 449–450

Dogwiler T, Weisbrod T (2010) Evaluating Sinkhole damage to a flood control structure in Southeastern Minnesota using electrical resistivity imaging. In: 2010 NC/S region GSA annual meeting, GSA abstracts with programs, vol 42, no 2, p 46. https://gsa.confex.com/gsa/2010NC/webprogram/Paper171033.html

Dorale JA (2020) Paleoclimate studies from Midwestern Caves. Chapter 11, this volume

Dornberg D (1939) Camera safari explores "Lost World" under Loop. Minneapolis Journal, p 1, 8

Downes D (1983) Editor's notes. Minn Speleol Mon 15(2):16

Edwards RA, Rodriquz-Brito B, Wegley L, Haynes M, Britbart M, Peterson DM, Saar MO, Alexander S, Alexander EC Jr, Rohwer, F (2006) Using pyrosequencing to shed light on deep mine microbial ecology. BMC Genomics 7:1–13. https://bmcgenomics.biomedcentral.com/track/pdf/10.1186/1471-2164-7-57

Farnham RS (1958) Soil survey Fillmore County Minnesota. United States Department of Agriculture, Washington, DC, 51 p + 80 11 × 17 in soil survey map sheets

Fay SJ, Alexander EC Jr (1994) Groundwater dye trace report: Watson-Warrington Group Detention Structure, Section 17, Township 112 North, Range 15 West, Featherstone Township, Goodhue County, Minnesota. Report to Goodhue County Soil and Water Conservation District, 25 p +3 appd

Feinberg JM, Lascu I, Lima EA, Weiss BP, Dorale JA, Alexander EC Jr, Edwards RL (2019) Magnetic detection of paleoflood layers in stalagmites and implications for historical land use changes. Earth Plan Sci Lett 530. https://doi.org/10.1016/j.epsl.2019.115946

Filin S, Baruch A (2010) Detection of Sinkhole hazards using airborne laser scanning data. Photogramm Eng Rem Sens 76(5):577–587

Fitzsimmons RJ (1931) Burrowing workers risk lives for public's health. Minneapolis Tribune, August 23, p 1

FMCSP (2020) Forestville/Mystery Cave State Park. https://www.dnr.state.mn.us/mystery_cave/index.html

Gao Y (2002) Karst feature distribution in Southeastern Minnesota: extending GIS-based database for spatial analysis and resource management, Ph.D. thesis, Univ. of Minn. 210 p

Gao Y (2008) Spatial operations in a GIS-based Karst feature database. Environ Geol 54(5):1017–1027

Gao Y, Alexander EC Jr (2003) A mathematical model for a map of relative Sinkhole Risk in Fillmore County, Minnesota. In: Beck BF (ed) Geotechnical special publication No. 122, Sinkholes and the Engineering and Environmental Impacts of Karst. American Society of Civil Engineering, Reston, VA, p 439–449. https://doi.org/10.1061/40698(2003)39

Gao Y, Alexander EC Jr (2008) Sinkhole hazard assessment in Minnesota using a decision tree model. Environ Geol 54(5):945–956

Gao Y, Alexander EC Jr, Barnes RJ (2005a) Karst database implementation in Minnesota: analysis of Sinkhole distribution. Environ Geol 47(8):1083–1098

Gao Y, Alexander EC Jr, Bounk M, Tipping RG (2005b) Metadata development for a multi-state karst feature database. In: Beck BF (ed) Proceedings of the tenth multidisciplinary conference on Sinkholes and the engineering and environmental impacts of Karsts, ASCE Geotechnical Special Publication 144, p 629–638

Gao Y, Alexander EC Jr, Tipping RG (2001) Application of GIS technology to study karst features of Southeastern Minnesota. In: Beck BF, Herring JG (eds) Geotechnical and environmental applications of karst geology and hydrology: proceedings of the eighth multidisciplinary conference on sinkholes and the engineering and environmental impacts of Karsts, Louisville, KY, AA Balkema, Rotterdam, Netherlands, p 83–88

Gao Y, Alexander EC Jr, Tipping RG (2002) The development of a Karst features database for Southeastern Minnesota. J Cave Karst Stud 64(1):51–57

Gao Y, Alexander EC Jr, Tipping RG (2005b) Karst database development in Minnesota: design and data assembly. Environ Geol 47(8):1072–1082

Gao Y, Tipping RG, Alexander EC Jr (2006) Applications of GIS and database technologies to manage a Karst Feature Database. J Cave Karst Stud 68(3):144–152

Giammona CP (1973) Fluorescent dye determination of groundwater movement and contamination in Permeable Rock Strata. Int J Spel 5:201–208

Green D, Brick GA (2019) Minnesota Underground, 2nd edn. HenschelHaus Publishing, Milwaukee, Wisconsin, p 191

Green JA, Alexander EC Jr, Marken WJ, Alexander SC (2002) Karst hydrogeomorphic units. Geologic Atlas of Mower County, Minnesota. Atlas C-11, Part B, Minn Dept of Nat. Res., St. Paul, Plate 10

Green JA, Tipping RG, Barry JD, Brick GA, Wheeler BJ, Rutelonis JW, Richardson BC, Alexander EC Jr (2018) Electronic access to Minnesota springs, Karst feature & groundwater tracing information. In: Sawowsky ID, Byle MJ, Land L (eds) Proceedings of the 15th multidisciplinary conference on Sinkholes and the Engineering and Environmental Impacts of Karst and the 3rd Appalacian Karst Symposium, Shepherstown, WV, NCKRI Symposium 7, Carlsbad, NM, p 167–171

Griffith CJ (2009) Opening Goliath: danger and discovery in caving. Minnesota Historical Society Press, St. Paul, p 294

Grow SR (1986) Groundwater chemical variations in a Karst Basin in Fillmore County, Minnesota. MS thesis, Univ. of Minn., p 229

Hedges J, Alexander EC Jr (1985) Karst-related features of the upper Mississippi Valley region. Stud Speleol 6:41–49

Holsinger JR (1966) A preliminary study on the effects of organic pollution of Banners Corner Cave. Virginia Intl J Spel 2(1):6

Huebsch M, Fenton O, Horan B, Hennessy D, Richards KG, Jordan P, Goldscheider N, Butscher C, Blum P (2014) Mobilization or dilution? Nitrate response of karst springs to high rainfall events. Hydrol Earth Syst Sci 18:4423–4435

Jahn B-M, Murthy VR (1975) Rb-Sr ages of the Archean rocks from the Vermilion district, northeastern Minnesota. Geochim Cosmochim Acta 39:1679–1689

Jameson RA, Alexander EC Jr (1995) The waters of Mystery Cave Forestville State Park, Minnesota Mystery Cave Resources Evaluation (Groundwater). Technical report, 414 p. https://www.leg.state.mn.us/docs/2013/other/130624/vol3.pdf

Jannik NO, Alexander EC Jr, Landherr LJ (1992). The Sinkhole Collapse of the Lewiston, MN Waste Water Treatment Facility Lagoon. In: Quinlan J, Stanley A (eds) Proceedings of 3rd conference on hydrogeology, ecology, monitoring of ground water in Karst Terranes. Nashville, TN, NGWA, Dublin, OH, p 715–724

Jennings JE (1966) Building on dolomites in the transvaal. Civil Eng S Afr: S Afr Inst Civ Eng Trans 8(2):41–62

Kasahara SM (2014) Chloride monitoring at coldwater spring. In: Snow DM, Guerrant B (co-chairs) 28th Natl Conf Undergrad Res, University of KY, Lexington. http://www.ncurproceedings.org/ojs/index.php/NCUR2014

Kasahara SM (2016) A Hydrogeological study of coldwater spring, Minneapolis, MN. Senior thesis, University of Minn, Minneapolis, 55 p. https://conservancy.umn.edu/handle/11299/182336

Kasahara SM, Alexander EC Jr, Alexander SC (2015) Human impact on water quality in coldwater spring minneapolis, MN. In: Veni G, La Moreaux J, Barr K, co-chairs. 14th Annual Sinkhole Conference, Rochester, MN, p 8–19https://scholarcommons.usf.edu/sinkhole_2015/ProceedingswithProgram/Upper_Mississippi_Valley_Karst_Aquifers/2/

Kehret R (1974) Minnesota Caves of History and Legend. Superior Printing, Chatfield, MN, p 28

KFD (2020) Minnesota Karst features database. https://gisdata.mn.gov/dataset/geos-karst-feature-inventory-ptsNote 1: Each of the mapped sinkholes, springs, stream sinks, and miscellaneous karst features in the KFD and MSI are assigned a unique ID code. In the original KFD the ID codes were ten-digit alpha numeric codes of the form "MN23A0001"—which is the original KFD code for Seven Springs (one of the major resurgent springs for Mystery Cave). "MN" is for Minnesota. "23" is the county number in an alphabetical list of Minnesota counties—Fillmore County in this case. "A" is the code for springs and seeps. "B" denotes stream sinks and sieves, "D" denotes sinkholes/dolines and "X" is used for miscellaneous point locations. The last four characters are a counting number of the features mapped in each county. When the KFD springs were moved to the MSI the IDs were changed to "Relate ID Numbers" to be consistent with other Minnesota geospatial databases. The Relate ID Numbers are ten-digit alpha numeric codes. The first two digits are the county code from an alphabetical list of Minnesota counties. The third character has the same meaning as in the KFD numbers. The KFD numbers were also converted to Relate ID numbers. The Relate ID for Seven Springs in Fillmore County is now 23A0000001. To save space and make graphics more readable, the Relate ID Numbers are often "condensed" by removing the place holding zeros. For example, the Relate ID for Seven Springs condenses to "23A1"—the first numbered spring in Fillmore County.Note 1: Each of the mapped sinkholes, springs, stream sinks, and miscellaneous karst features in the KFD and MSI are assigned a unique ID code. In the original KFD the ID codes were ten-digit alpha numeric codes of the form "MN23A0001"—which is the original KFD code for Seven Springs (one of the major resurgent springs for Mystery Cave). "MN" is for Minnesota. "23" is the county number in an alphabetical list of Minnesota counties—Fillmore County in this case. "A" is the code for springs and seeps. "B" denotes stream sinks and sieves, "D" denotes sinkholes/dolines and "X" is used for miscellaneous point locations. The last four characters are a counting number of the features mapped in each county. When the KFD springs were moved to the MSI the IDs were changed to "Relate ID Numbers" to be consistent with other Minnesota geospatial databases. The Relate ID Numbers are ten-digit alpha numeric codes. The first two digits are the county code from an alphabetical list of Minnesota counties. The third character has the same meaning as in the KFD numbers. The KFD numbers were also converted to Relate ID numbers. The Relate ID for Seven Springs in Fillmore County is now 23A0000001. To save space and make graphics more readable, the Relate ID Numbers are often "condensed" by removing the place holding zeros. For example, the Relate ID for Seven Springs condenses to "23A1"—the first numbered spring in Fillmore County

Kingston SP (1943) Contamination of water supplies in limestone formation. J Am Water Works Assoc 35(11):1450–1456

Klimchouk A (2007) Hypogene Speleogenesis: hydrogeological and morphogenetic perspective, Special Paper No. 1, National Cave and Karst Research Institute, Carlsbad, NM, 106 p

Kress A, Alexander EC Jr (1980) Farmers and Mechanics Bank Cave, Hennepin County. In: Alexander EC Jr (ed) An introduction to Caves of Minnesota, Iowa, and Wisconsin, guidebook for the 1980 NSS Conv. Minn Geol Surv, St. Paul, p 59–67 and Plate 3

Lace MJ (2020) Iowa Caves and Karst. Chapter 4, this volume

Larsen MR, Green JA, Barry JD, Kasahara SM, Wheeler BJ, Alexander EC Jr (2018) 2017 York Blind Valley dye trace Fillmore County, MN. MNDNR report, 9 p. https://conservancy.umn.edu/handle/11299/200690

Liesch BA (1974) Hydrogeologic investigation of Altura area for Altura, Minnesota, 20 p, 2 maps

Liesch BA (1976) Evaluating pollution-prone strata beneath sewage lagoon. Publ Works 107(8):70–71

Liesch BA (1977) Test drilling investigations at Altura, Minnesota waste treatment facility, 10 p

Lively RS (1983) Late Quaternary U-series speleothem growth record from southeastern Minnesota. Geology 11:259–262

Lively RS, Alexander EC Jr (1980) ^{230}Th/^{234}U aged from speleothems from Mystery Cave. Minn NSS Bull 42:34

Lively RS, Krafthefer BC (1993a) Radon concentrations, activities of radon decay products, meteorological conditions and ventilation in Mystery Cave, interpretive report. In: LCMR Mystery Cave resources evaluation, 19 p. https://www.leg.state.mn.us/docs/2015/other/155232/interpretivereport.pdf

Lively RS, Krafthefer BC (1993b) Radon concentrations, activities of radon decay products, meteorological conditions and Ventilation in Mystery Cave, Final technical report. In: LCMR Mystery Cave resources evaluation, 135 p, 10 append. https://www.leg.state.mn.us/docs/2015/other/155232/finaltechnicalrepot.pdf

Lively RS, Krafthefer BC (1995) ^{222}Rn variation in Mystery Cave. Minn Health Phys 68(4):590–594

Longnecker J (1907) In caverns of eternal night. Minneapolis Tribune, July 7.

Luhmann AJ (2011) Water temperature as a tracer in Karst aquifers. Ph. D. thesis. University of Minnesota, 164 p

Luhmann AJ, Covington MD, Alexander SC, Chai SY, Schwartz BF, Groten JT, Alexander EC Jr (2012) Comparing conservative and nonconservative tracers in Karst and using them to estimate flow path geometry. J Hydrol 448–449:201–211. https://doi.org/10.1016/j.jhydrol.2012.04.044

Luhmann AJ, Covington MD, Alexander SC, Chai SY, Schwartz BF, Groten JT, Alexander EC Jr (2013) Comparison of discharge, chloride, temperature, uranine, δD, and suspended sediment responses from a multiple tracer test in karst. Carbonates Evaporites 28:191–199. https://doi.org/10.1007/s13146-013-0127-8

Luhmann AJ, Covington MD, Myre JM, Perne M, Jones SW, Alexander EC Jr, Saar MO (2014) Thermal damping and retardation in karst conduits. Hydrol Earth Syst Sci Discuss. 18:1–54. https://doi.org/10.5194/hessd-18-1-2014, www.hydrol-earth-syst-sci-discuss.net/18/2014

Luhmann AJ, Covington MD, Peters AJ, Alexander SC, Anger CT, Green JA, Runkel AC, Alexander EC Jr (2011) Classification of thermal patterns at Karst Springs and Cave streams. Ground Water 49(3):324–334

LVSUMSP (2020) Lake Vermillion-Soudan Underground Mine State Park. https://www.dnr.state.mn.us/stateparks/park.html?id=spk00285#homepage

Maguire E (1880) Press copies of office letters sent by Captain Edward Maguire, Chief Engineer, In: The Water. Department of Dakota, p 2

MCP (2020) Minnesota Cave Preserve website. http://www.cavepreserve.com

MGTD (2020) Minnesota groundwater tracing database. https://www.dnr.state.mn.us/waters/programs/gw_section/springs/dtr-list.html

Milske JA (1982) Stratigraphy and petrology of clastic sediments in Mystery Cave, Fillmore County, MN. MS thesis, University of Minnesota. 111 p

Milske JA, Alexander EC Jr, Lively RS (1983) Clastic Sediments in Mystery Cave, Southeastern Minnesota. NSS Bull 45:55–75

Mohring EH (1983) A study of subsurface water flow in a Southeastern Minnesota Karst drainage basin. MS thesis, University of Minnesota, 99 p

Mohring EH, Alexander EC Jr (1986) Quantitative tracing of karst groundwater flow: Southeastern Minnesota, North Central, USA. In: Proceedings of the 5th international symposium on underground water tracing, Athens, Institute of Geology and Mineral Expl., Athens, Greece, p 215–227

Mossler JH (2008) Paleozoic stratigraphic nomenclature for Minnesota. Minn Geol Surv RI 65, 76 p, 1 plate

MSI (2020) Minnesota Spring Inventory. https://www.dnr.state.mn.us/waters/groundwater_section/springs/msi.html

MSS (1980) Mystery Cave, plan view, Fillmore County, Minnesota. In: Alexander EC Jr (ed) An introduction to Cave of Minnesota, Iowa, and Wisconsin, guidebook for the 1980 NSS Conv. Minn Geol Surv, St. Paul, Plate 1

Niagara Cave (2020). https://niagaracave.com/index.html

Nicollet JN (1845) Report intended to illustrate a map of the hydrographical basin of the Upper Mississippi River 1845, Document of the US House of Representatives, 28th Congress, 2nd Session, no. 52. Blair and Rives, Washington, DC. https://doi.org/10.5962/bhl.title.30763

Nordquist GE (2000) Winter use of subterranean cavities by bats in and near Minnesota. M.Sc thesis, Univ of Minn, 144 p

Ojakangas RW, Matsch CL (1982) Minnesota's Geology. University of Minnesota Press, Minneapolis, 268 p

Oothoudt JW (1980) Petrified Indian Cave: the Legend and Archaeology. In: Alexander EC Jr (ed) An introduction to Caves of Minnesota, Iowa, and Wisconsin, guidebook for the 1980 NSS Conv. Minn Geol Surv, St. Paul, p 11–29

Palmer AN (1975) The Origin of Maze Caves. NSS Bull 37(3):56–76

Palmer AN (2007) Debate on Origin of Mystery Cave: Alexander Klimchouk versus Art + Peg Palmer. Unpublished document dated July 2007, 8 p

Palmer AN (2011) Distinction between Epigenic and Hypogenic Maze Caves. Geomorphology 134(1–2):9–22. https://doi.org/10.1016/j.geomorph.2011.03.014

Palmer AN, Palmer MV (1993a) Geology and Origin of Mystery Cave, Forestville State Park, Minnesota. Technical report, LCMR Mystery Cave resources evaluation, 137 p. https://www.leg.state.mn.us/docs/2015/other/155233/technicalreport.pdf

Palmer AN, Palmer MV (1993b) Geology and Origin of Mystery Cave, Forestville State Park, Minnesota. Interpretive Report. In: LCMR Mystery Cave resources evaluation, 92 p. https://www.leg.state.mn.us/docs/2015/other/155233/interpretivereport.pdf

Rahimi M, Alexander EC Jr (2013) Locating Sinkholes in LiDAR coverage of a Glacio-fluval Karst, Winona County, MN. In: Lewis L, Doctor DH, Stephenson JB (eds) NCKRI Symposium 2 Proceedings of 13th multidisciplinary Conference on Sinkholes and the Engineering and Environmental Impacts of Karst, Carlsbad, NM, NCKRI, Carlsbad, NM, p 469–480

Runkel T (2007) Southeastern Minnesota paleozoic hydrostratigraphy: fractures in aquifers, aquitards, and aquitardifers. Luncheon Presentation, Minnesota Section, AIPG. https://aipgmn.org/meetinginfo.php?id=27&ts=1368164530

Saylor N (1980) Copeman's Cave, two Skull Cave, Windcourse Cave. In: Alexander EC Jr (ed) An introduction to Caves of Minnesota, Iowa, and Wisconsin, guidebook for the 1980 NSS Conv. Minn Geol Surv, St. Paul, p 9–11

Schilling KE, Jones CS, Clark RJ, Libra RD, Liang X, Zhang Y (2019) Contrasting NO_3–N concentration patterns at two karst springs in Iowa (USA): insights on aquifer nitrogen storage and delivery. Hydrogeol J. https://doi.org/10.1007/s10040-019-01935-y

Scobie P (1980) Old Mystery Cave, Old Still Cave. In: Alexander EC Jr (ed) An Introduction to Caves of Minnesota, Iowa, and Wisconsin, guidebook for the 1980 NSS Conv. Minn Geol Surv, St. Paul, p 7–9

Shade BL (2002) The genesis and hydrogeology of a Sandstone Karst in Pine County, Minnesota. MSc thesis, University of Minnesota, 171 p

Shade BL, Alexander EC Jr, Alexander SC (2015) The Sandstone Karst of Pine County, Minnesota. In: Doctor DH, Land L, Stephenson JB (eds), NCRI Symposium 5, Sinkholes and the Engineering and Environmental Impacts of Karst, Proceedings of the 14th Multidisciplinary Conference, Rochester, MN, NCKRI, Carlsbad, NM, p 157–166

Shade BL, Alexander EC Jr, Alexander SC, Martin S (2002a) Sinkhole Distribution. Pine County Geologic Atlas, Atlas C-13, Minn Geol Surv, St. Paul, top half of Plate 6

Shade BL, Alexander EC Jr, Alexander SC (2002b) Karst features in Pine County. Chapter 4, In: Boerboom TJ (ed) Contributions to the Geology of Pine County, Minnesota, Minn Geol Surv RI 60, St. Paul, p 55–72

Shuster ET, White WB (1972) Seasonal fluctuations in the chemistry of limestone springs: a possible means for characterizing carbonate aquifers. J Hydro 14:93–128

Sloan RE (1964) The Cretaceous System in Minnesota, Minn Geol Surv RI 5, St. Paul, 64 p, 2 plates

Soule GK (1974) Three Short Summers: The Story of Hiawatha Caverns. J Spelean Hist 7(3):47. http://caves.org/section/asha/issues/027.pdf

Spong RC (1980a) Crawlway Cave. In: Alexander EC Jr (ed) An Introduction to Caves of Minnesota, Iowa, and Wisconsin, guidebook for the 1980 NSS Conv. Minn Geol Survey, St. Paul, p 33–42

Spong RC (1980b) Carlin Caverns, Winona County. In: Alexander EC Jr (ed) An Introduction to Caves of Minnesota, Iowa, and Wisconsin, guidebook for the 1980 NSS Conv. Minn Geol Surv, St. Paul, 46–48 and Plate 2

Spong RC (1980c) Channel Rock Cavern, Hennepin County. In: Alexander EC Jr (ed) An Introduction to Caves of Minnesota, Iowa, and Wisconsin, guidebook for the 1980 NSS Conv. Minn Geol Surv, St. Paul, p 67–72

Spong RC (2006) Early spelean history of the caves of Jessies Grove. CHTHONIOS: Book FLE416.02, privately published manuscript, 16 p

Steenberg JR, Runkel AC (2018) Stratigraphic positions of springs in Southeast Minnesota. Minn Geol Surv OFR 18–2, 22 p. https://conservancy.umn.edu/handle/11299/198183

Steenberg JR, Tipping RG, Runkel AC (2014) Geologic controls on groundwater and surface water flow in Southeastern Minnesota and its impact on nitrate concentrations in streams. Minn Geol Surv OFR 14–02, St. Paul, 154 p. http://hdl.handle.net/11299/162612

Surber T (1922) A biological reconnaissance of streams tributary to Lake Superior, Baptism River to Devil Track River. typescript in MNDNR Library, St. Paul, 153 p

SVC (2020) Spring Valley Caverns. https://www.cavepreserve.com/svc.html

Tipping RG, Alexander EC Jr, Green JA (2001) Karst features. Wabasha County Atlas, Atlas C-14, Part A, Minn Geol Surv, St. Paul, Plate 5

Tipping RG, Runkel AC, Alexander EC Jr, Alexander SC, Green JA (2006) Evidence for hydraulic heterogeneity and anisotropy in the mostly carbonate Prairie du Chien group, Southeastern Minnesota, USA. Sed Geol, 184(3-4):305–330. https://doi.org/10.1016/j.sedgeo.2005.11.007

Tipping RG, Rantala M, Alexander EC Jr, Gao Y, Green JA (2015) History and future of the Minnesota Karst feature database. In: Doctor DH, Land L, Stephenson JB (eds) Proceedings of the 14th Sinkhole Conference, National Cave & Karst Research Institute Symposium 5, Sinkholes and the Engineering and Environmental Impacts of Karst, Proceedings of the 14th Multidisciplinary Conference, Oct 5-9, 2015, Rochester, MN, NCKRI, Carlsbad, NM, p 263–270

Tryhorn AD, Ojakangas RW (1972) Sedimentation of the Hinckley Sandstone of east-central Minnesota. In: Sims PK, Morey GB (eds) Geology of Minnesota: a centennial volume. Minn Geol Surv, St. Paul MN, p 416–424

UIRWP (2020) Upper Iowa River watershed project. http://northeastiowarcd.org/uirw/about.htm

Waltham A (2008) The Sandstone Caves of Nottingham, 3rd edn. East Midlands Geological Society, Nottingham, UK, 56 p

White BM (2000) Highway 55 and the camp coldwater settlement: an independent investigation. In: Turnstone historical research, St. Paul, MN. http://www.tc.umn.edu/~white067/

Widga C, Fulton TL, Martin LD, Shapiro B (2012) *Homotherium serum* and *Cervalces* from the Great Lakes Region, USA: geochronology, morphology and ancient DNA. Boreas 41(4):546–556

Wilford LA (1951) The Leslie Cave. Unpublished typescript of the Archaeology Laboratory. University of Minnesota

Witthuhn KM, Alexander EC Jr (1995) Sinkholes and Sinkhole Probability. County Atlas Series, Atlas C-8, Part B, Minn Dept Nat Res, St. Paul, Plate 8

Worthington SRH (1999) A comprehensive strategy for understanding flow in carbonate aquifers. In: Palmer AN, Palmer MV, Sasowsky ID (eds) Karst Modeling, Special Publication 5, The Karst Waters Institute, Charles Town, WV, p 30–37. http://www.speleogenesis.info/directory/karstbase/pdf/seka_pdf4493.pdf

Wright HE Jr (1972) Physiography of Minnesota. Geology of Minnesota: a centennial volume. Minnesota Geological Survey, St. Paul, p 561–577

Wu Q, Deng C, Chen Z (2016) Automated delineation of karst sinkholes from LiDAR-derived digital elevation modes. Geomorphology 266:1–10

Yingling G (2019) PFAS fate & transport, health risks, and responses for drinking water. ASDWA Webinar May 28, 2019, slides 12–14. https://www.asdwa.org/wp-content/uploads/2019/06/Yingling-MN-DOH-PFAS.pdf

Iowa Caves and Karst

4

Michael J. Lace, Raymond R. Anderson, and Patricia N. Kambesis

Abstract

Iowa landforms have been shaped by a range of processes, including multiple glacial incursions. Many of these diverse landforms harbor a variety of karst features, including numerous examples of caves formed by both solutional and mechanical processes acting on an array of dolostones, limestones, gypsum, and sandstones. Cave development has been noted in several geologic systems, ranging from upper Cambrian to Pennsylvanian sandstones and lower Silurian to middle Mississippian carbonates. Distributional patterns of cave and karst development in Iowa are examined as a function of distinct physiographic regions, respective lithologies, hydrogeologic settings, and structural morphologies illustrated by representative sites. Karst systems have played a significant role in shaping Iowa's unique surficial and subterranean habitats as well as archaeological, historical, and modern landscape uses. Ongoing exploration and study of Iowa caves continues to improve our collective understanding of past climate patterns and modern water quality issues as well as supporting comparative models of karst development on a regional scale. This chapter provides an overview examining principal karst-associated landscapes and controls associated with paleokarst, pseudokarst, and karst cave occurrences in Iowa.

M. J. Lace (✉)
Coastal Cave Survey, West Branch, Iowa, 52358, USA
e-mail: michael-lace@uiowa.edu

R. R. Anderson
University of Iowa, Iowa City, IA, USA
e-mail: raymond-anderson@uiowa.edu

P. N. Kambesis
Western Kentucky University, Bowling Green, KY, USA
e-mail: pat.kambesis@wku.edu

4.1 Introduction

Iowa karst studies date back to the mid-nineteenth century, primarily within early geological reconnaissance surveys of the region by David Dale Owen (1852) followed by a series of focused reports by members of the Iowa Geological Survey (Johnson 1977). Systematic documentation of Iowa caves has been ongoing for over 60 years by members of the Iowa chapter of the National Speleological Society, with over 1400 caves sites recorded to date. Maps have been produced for more than 1200 of these cave sites, supporting long-term multidisciplinary research and karst preservation efforts.

Exploration efforts have included cave diving in a variety of Iowa resurgences began in the late 1960s. A resurgence of exploratory dives in a number of active springs in the late 1980s also yielded numerous newly documented caves—some containing several kilometers of stream passage. Prominent vertical cave development in both joint-controlled mechanical crevices and hydrologically active drainage conduits have been noted, with most cave structures ranging from 5 to 50 m and averaging 10 m in total depth. Regional climate conditions and land use practices have occasionally made exploration less than ideal with frigid spring waters (sometimes as cold as 4 degrees C) and the sporadic encounter with partially decomposed animal carcasses and agricultural runoff.

Varying degrees of cave and karst development has been documented in four distinct physiographic regions in Iowa: the Iowan Surface, Des Moines Lobe, Paleozoic Plateau, and the Southern Iowa Drift Plain (Fig. 4.1) (Smith 1984). Caves have been recorded in 36 of Iowa's 99 counties, centering within the cuestas of the Paleozoic Plateau in the northeast quadrant of the State and bounded by a limited number of isolated outlying karst areas (Fig. 4.2). Though limited in the extent and scale of cave development on a subregional scale, these outlying karst areas provide insights which complement a more complete model of cave and karst mechanisms

G. A. Brick and E. C. Alexander Jr. (eds.), *Caves and Karst of the Upper Midwest, USA*,
Cave and Karst Systems of the World, https://doi.org/10.1007/978-3-030-54633-5_4

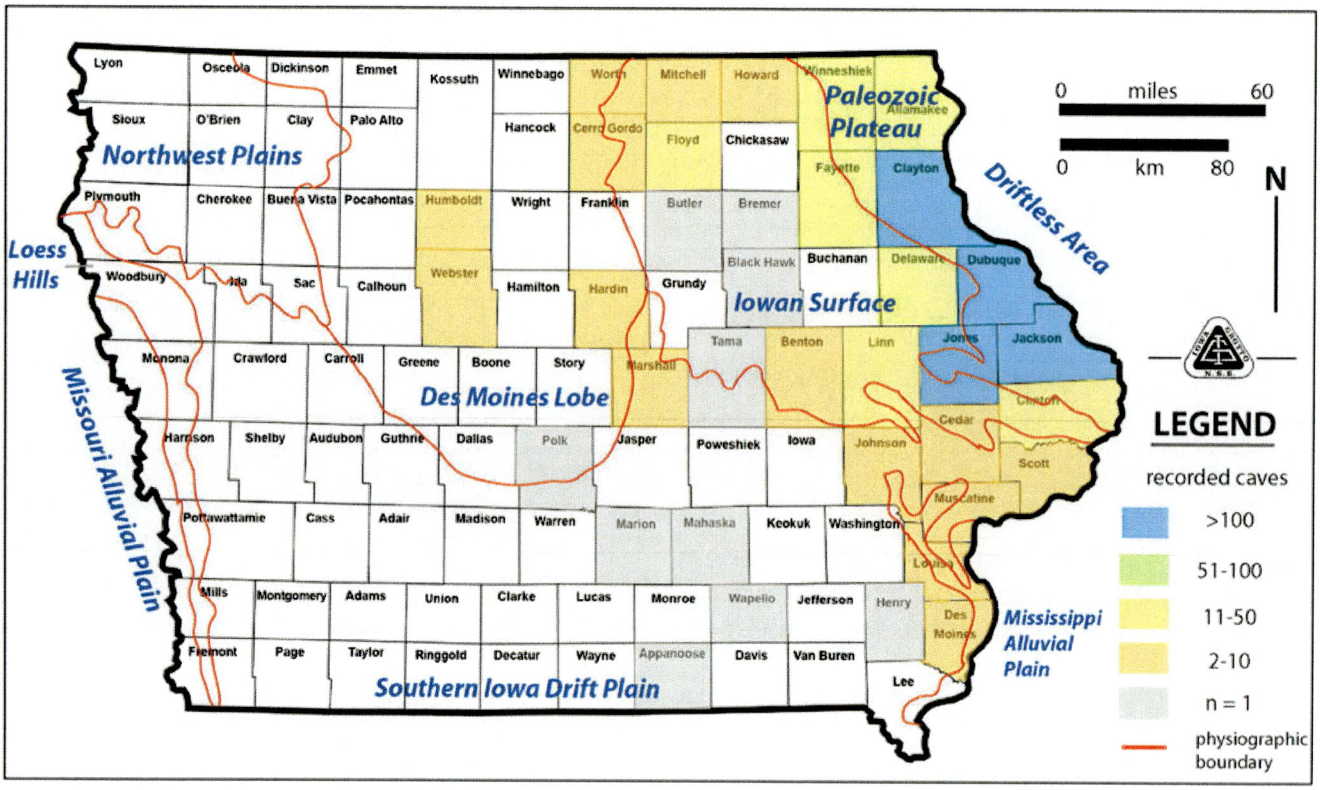

Fig. 4.1 Generalized cave density map and physiographic divisions in Iowa

in the region. Representative examples of caves formed within a wide range of bedrock lithologies and hydrologic regimes are included in the following sections to illustrate these complex patterns of karst development in Iowa.

4.2 Expressions of Iowa Karst

Iowa karst landscapes display a complex record of varying degrees of glaciation, exhibiting remnants of pre-glacial and post-glacial processes associated with episodic karst development (Hedges and Alexander 1985; Prior 1991). Cave development within these landforms has been strongly influenced by the extent of pre-Pleistocene and Pleistocene glacial mantle and the concomitant exposure of highly fractured Paleozoic bedrock systems to surface drainage and climatic factors. Many of the larger cave systems in Iowa exhibit similarities to other glaciated karst regions not only in terms of their general morphologies but their exploration challenges as well (Cooper and Mylroie 2015).

Karst and pseudokarst cave types expressed across the State include small to extensive fluvial networks, hypogene structures, shallow and deep phreatic mazes, floodwater mazes, fissures, rectilinear mazes, talus caves, glacières, pit caves, and man-made structures (Brick 2004; Kambesis and Lace 2009). Karst springs are also abundant in Iowa, comprising a critical component in water quality in eastern regions of the State, particularly within numerous watersheds of the northeast (Hruby et al. 2010).

Sinkhole networks also form a prominent component of Iowa's karst landscapes and are defined by solutional, mechanical, or anthropogenic processes or combinations of one or more of these influences (Crawford et al. 2004; Iles 1977; Launspach 2013; Palmquist 1977). They are critical components in modeling contaminant transport and effective watershed management in mantled karst areas of Iowa and similarly noted in other regions (Lindsay et al. 2010; Sasowsky 2000; Worthington 2003).

Not surprisingly, paleokarst developed within many of the same bedrock systems that supported later episodes of karst development. Paleokarst expressions have been characterized in several regions of Iowa within Ordovician, Pennsylvanian, and Devonian karst areas (Calvin 1896; Trowbridge 1966) as well as sporadic occurrences across the broader Upper Mississippi Valley Region (Hedges and Alexander 1985; Plotnick et al. 2009).

Sediment-filled paleokarst cavities are common and naturally occurring paleokarst outcrops have been recorded within sandstone-filled dolomitic voids of the Paleozoic Plateau (Anderson 2000). Humanly enterable paleokarst caves have also been reported, for example, a network of caves formed within Devonian carbonates in the Southern Iowa Drift Plain were revealed by commercial mining (McKay 1996). Dense networks of joint-controlled cavities

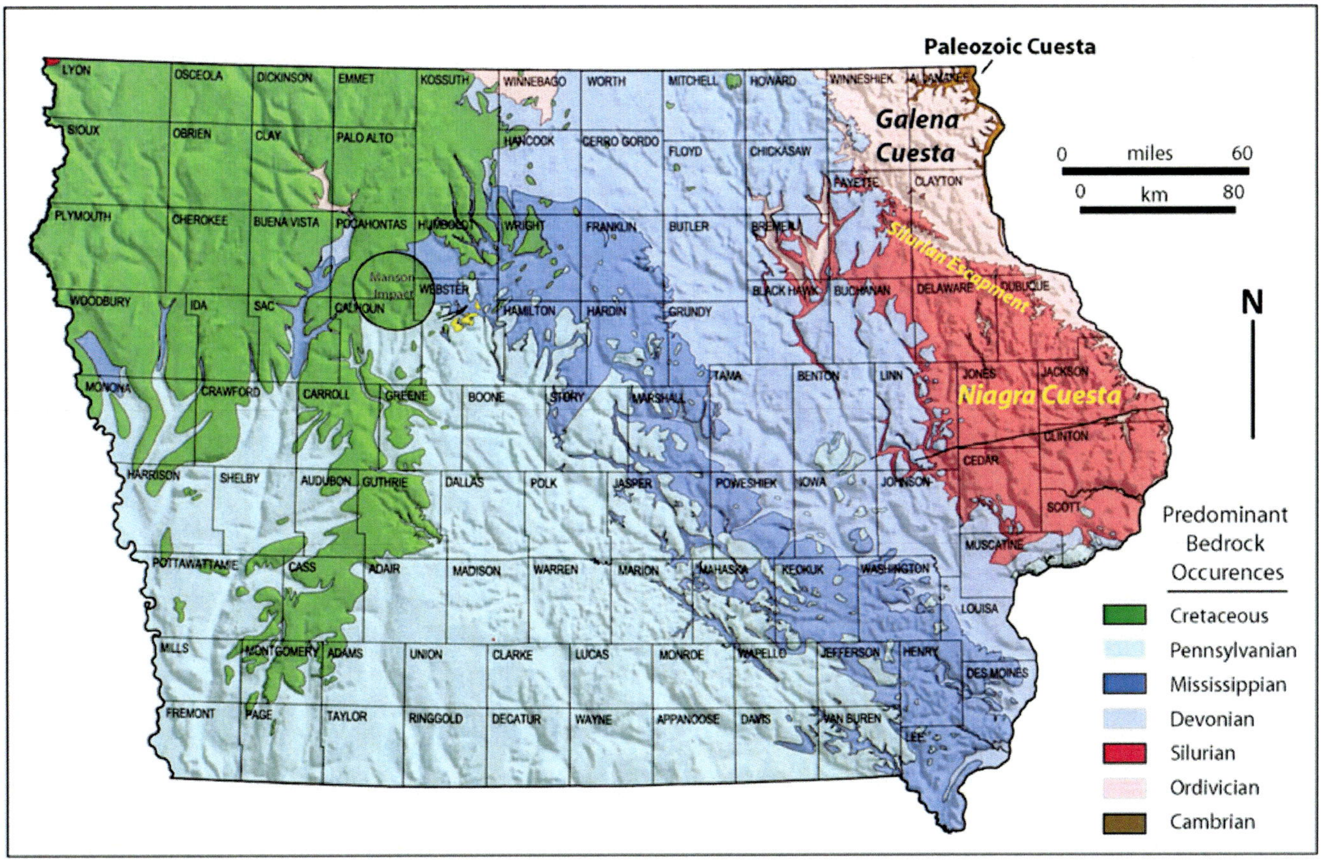

Fig. 4.2 Generalized bedrock map of Iowa with shaded relief bedrock base map prepared by the Iowa Geological Survey (Witzke 1998) and principal structures associated with karst development

in this setting display a range of morphologies generated under both vadose and phreatic conditions and are associated with unusual mineral occurrences partially consistent with those found in the Paleozoic Plateau (Garvin and Ludvigson 1993; Garvin 1995).

4.2.1 Mechanisms of Iowa Cave Development

Karst processes have modified a range of bedrock units overlain with a comparatively thin veneer Quaternary glacial sediments (Fig. 4.3). The exposure of these fractured bedrock surfaces, particularly within the Paleozoic Plateau, has contributed to initial meteoric flow associated with significant fluviokarst development (Bounk and Bettis 1984; Kambesis and Lace 2009). Numerous swallets and pit caves function as insurgence points recharging many of the larger cave systems which perennially discharge into springs of varying size. Phreatic caves have also been formed by deep-seated hydrothermal fluids in some areas. As the following sections illustrate, some of these caves consist of several kilometers of traversable cave passages.

Pseudokarst processes have shaped a significant portion of the Iowa cave inventory. Mechanical crevice development, for example, is prominent in several regions, including dense arrays of periglacial structures of Pleistocene origin (Hedges 1972; Hansel 1976). Soil piping and frost wedging have further served to modify and reveal many of these crevice structures. Talus caves also are common, particularly along the cuesta escarpments and incised valleys but are generally of limited extent.

Glacières form a fascinating subset to the Iowa cave inventory with several widely dispersed examples documented to date. A number of Iowa caves are commonly referred to as “ice caves” primarily due to their historical use as cold storage or periodic ice speleothem occurrences but only a few are considered true glacières (Fig. 4.4). Decorah Ice Cave is perhaps Iowa’s most noted example of glacière development (Fig. 4.4c) (Kovarik 1898). The cave was formed by a toreva block slide (i.e., a lithified cliff segment that slid and pivoted to form the cavity) within a highly fractured cliff face composed of the Dunleith Member of the Galena Group (Hedges and Knudson 1975). In addition to other glacières formed within the Galena limestone (Fig. 4.4a), examples have been documented as simple fissures of limited extent within Ordovician dolostones (Figs. 4.4b, d) and Desmoinesian sandstones.

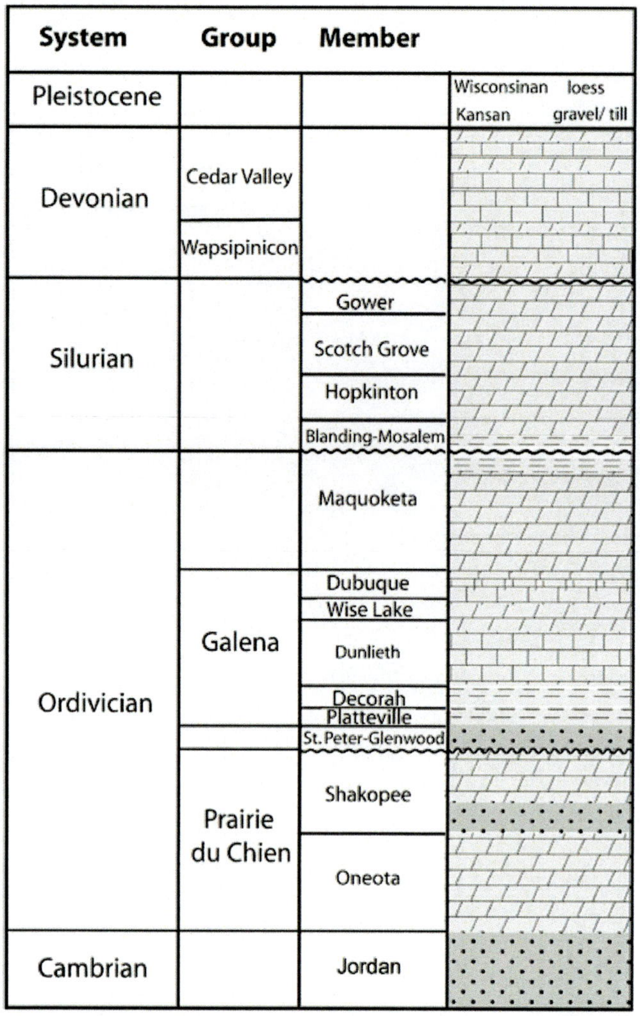

Fig. 4.3 Stratigraphic column of principal systems associated with Iowa cave development (modified from Kambesis and Lace 2009 and based on the Iowa stratigraphic column prepared by the Iowa Geological Survey 2017)

4.2.2 Principal Landforms Associated with Cave Development

4.2.2.1 Niagara Cuesta

The majority of recorded caves in Iowa occur within the Niagara cuesta. Its topography features rolling hills, sinkholes, and springs and numerous incised valleys harboring karst and pseudokarst features. Solutional caves within this area are primarily expressed as simple phreatic structures of both geological and cultural significance (Fig. 4.5a), yet more complex karst systems are occasionally encountered.

The Silurian escarpment forms the southwestern limit of the Paleozoic Plateau and the northern boundary of the Niagara Cuesta (Fig. 4.2), marking a sinuous, erosional boundary between Ordovician and Silurian systems. The escarpment features pronounced vertical relief and a significant degree of karst development (Fig. 4.5), including sinkhole networks, hydrologically active dendritic conduit caves and abandoned shallow phreatic mazes, disappearing streams and many springs associated with dolostone contact with the underlying Brainard shale (Bounk 1983a; Prior 1975). Many of the phreatic caves in this region are residual structures progressively truncated by valley head incision and in some cases partially overprinted by talus development (Fig. 4.6).

Examples of mechanical cave development are also prominently featured in this area (Howard 1963; Hansel 1976). Crevices are frequently expressed parallel to the Niagara escarpment facies as rectilinear mazes frequently formed by block slippage associated with fractures within the Silurian dolostones. Entrances associated with joint intersections are commonly revealed by a collapse of its thin overburden just below the ridge crests (Fig. 4.5d, e). Cave passage morphologies in this area display the effects of varying degrees of solutional enlargement from limited meteoric flow. Dolostones of the Hopkinton Member are host to a significant segment of karst caves within this region and the combination of comparatively high porosity within this unit coupled with joint control have generated numerous phreatic mazes and fluvial systems of limited extent (Bounk 1983b). Many of these structures may represent remnants of once larger pre-Pleistocene systems (Fig. 4.5b, c) (Hedges 1967).

Werden's Cave (Jackson County), for example, consists of nearly 300 m of interconnected chambers representing a typical shallow phreatic maze morphology common to the Niagara cuesta (Fig. 4.7). **Dancehall Cave** (Jackson County), and several other caves within Maquoketa Cave State Park, are also shallow phreatic structures formed within the Silurian-aged Hopkinton Dolomite (Bounk 1984). Formerly known as Burt's Cave and the Morehead Caves, the area was well known to visitors in the early 1900s. The main passage of Dancehall Cave extends 562 m from its southern entrance through a collapsed segment to the Natural Bridge remnant at its northern limit. The state park continues to offer casual visitors an accessible view of Iowa cave environments (Fig. 4.8).

4.2.2.2 Galena Cuesta

The Galena cuesta lies within the Paleozoic Plateau—an area characterized by steeply rolling hills and thinly mantled Ordovician carbonates (Fig. 4.3) (Hallberg et al. 1984; Levorson and Gerk 1972). Karst examples include numerous sinkholes, some representing significant material subsidence (Fig. 4.9a) and often revealing underlying phreatic conduits (Fig. 4.9c) (Bretz 1938). The area also harbors some of the most extensive fluviokarst networks in Iowa and the broader Upper Mississippi Valley region, many of which are formed within the lower Ordovician units with resurgences frequently defined by the contact with the underlying the Decorah shale (Fig. 4.9b) (Hedges 1974a; Wilder and Savage 1906). Minor cave development is also noted within dispersed surficial exposures of the Prairie du Chien dolomite (Fig. 4.9d) (Libra 2005).

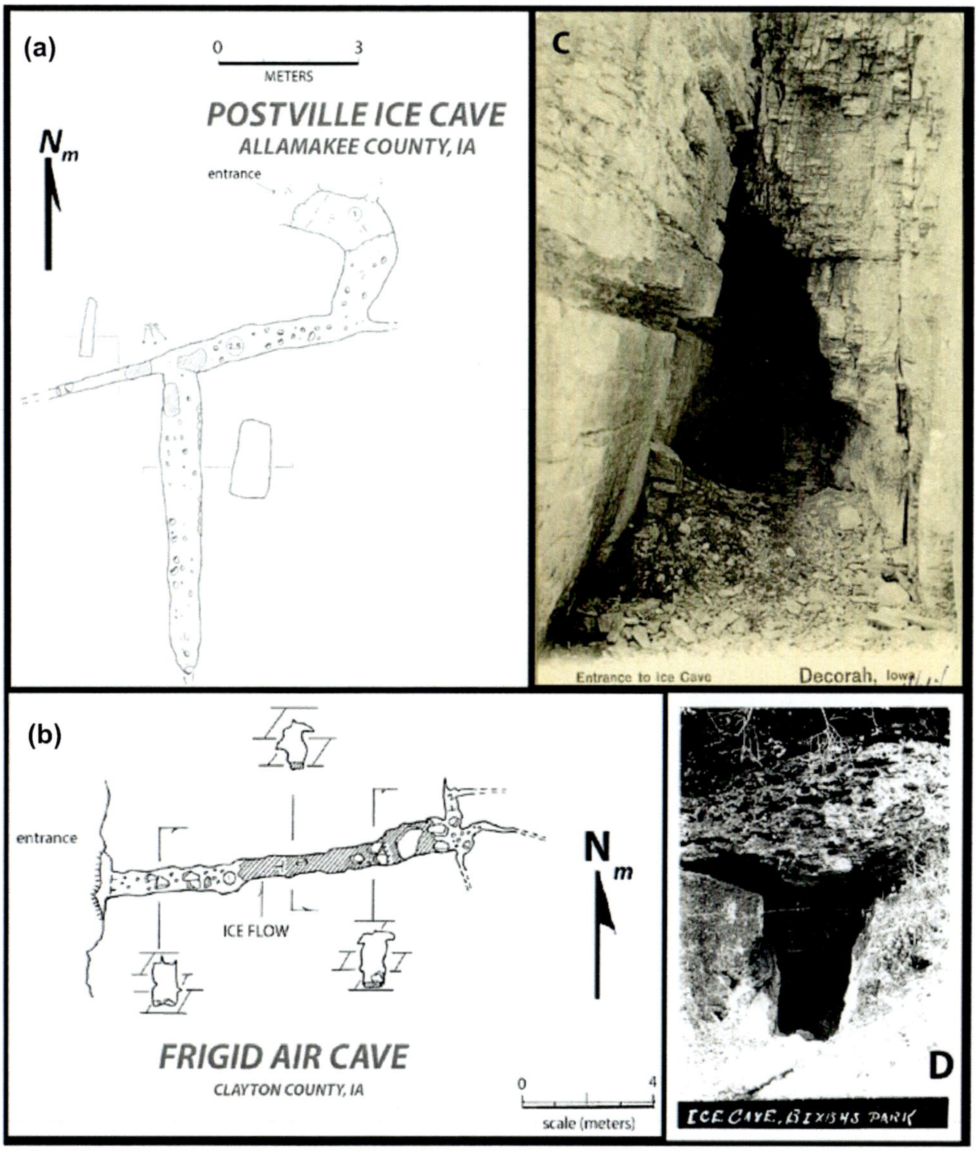

Fig. 4.4 Glacières in Iowa. **A** Postville Ice Cave (Allamakee County). **B** Frigid Air Cave (Clayton County). **C** Decorah Ice Cave (Winneshiek County). **D** Bixby Ice Cave (Delaware County) entrances (Iowa Grotto Library Archives)

4.2.2.3 Cave Development in the Paleozoic Plateau

Spanning northeastern Iowa, southwestern Wisconsin, and northwestern Illinois, the Paleozoic Plateau (also referred to by the misnomer "Driftless Area," as it features abundant evidence of glaciation) comprises one of the most diverse and dense karst areas in the region (Trowbridge 1966). The complex but partially defined speleogenesis of karst features within this district produced a wealth of mineral deposits and spectacularly decorated "spar caves" unique to this part of Iowa (Fig. 4.10). The caves in this area are anomalous in many respects, including their complex morphologies and pronounced mineral occurrences (Fig. 4.11). Joint control has strongly influenced the rectilinear maze morphology in many extensive caves in the area. The majority of the cave inventory in this area consist of single, non-branching crevices of limited extent to complex rectilinear mazes, featuring prominent east/west passage trends linked by smaller north/south passages to form complex but predictable networks (Heyl et al. 1959; Howard 1960) as illustrated in cave maps from the area (Fig. 4.12).

The prevailing model of cave development proposes that many of the caves within this Upper Mississippi Valley Lead-Zinc mining district are essentially hypogene structures formed by deep hydrothermal activity which influenced the distinctive structures of these caves (Morehouse 1968). Deep-seated fluids and tectonic effects also influenced the extensive mineral deposition and distinctive speleothem development patterns within extensive crevice systems (Ludvigson and McAdams 1980; Peck 1977). Dockal (2020), Chap. 7, this volume, explores these distinctive crevice caves and their associated mineralogy in greater detail.

Organized exploration of the caves and mines in the Dubuque area by cavers began in the 1950s, evolving into the Dubuque Underground Survey—one of Iowa's first

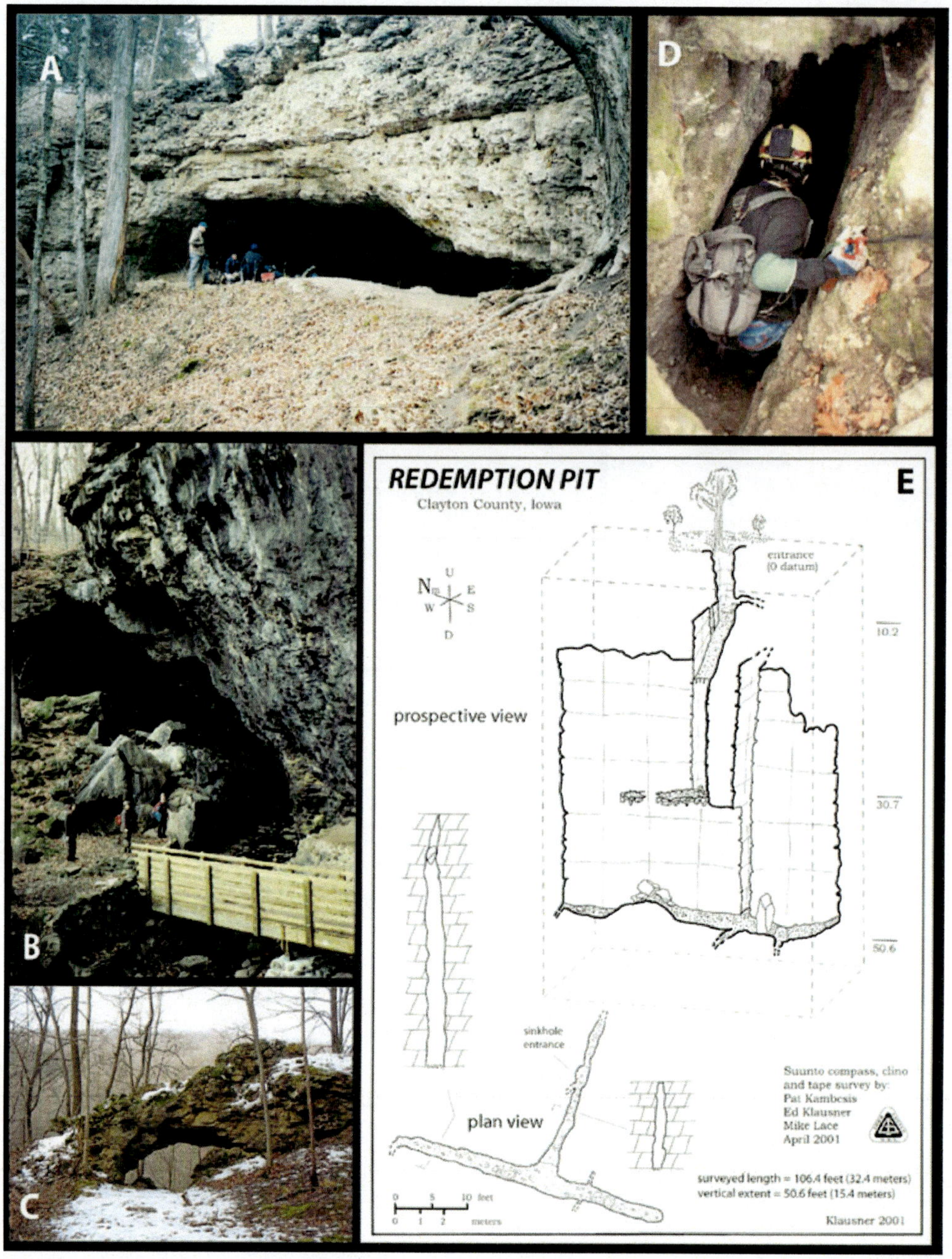

Fig. 4.5 Karst expressions in the Niagara Cuesta. **A** Typical solutional shelter. **B** Entrance to Lower Dancehall Cave (Jackson County). **C** Natural arch in Maquoketa Caves State Park (Jackson County). **D** Exploration of a mechanical crevice in the Silurian Escarpment (photo by M. Ohms). **E** Representative map of mechanical crevice (Clayton County)

organized cave exploration projects. Some of the longest caves in the State are found in this area, such as **Kemling Cave**.

"Kemling's Spar Cave" as it was originally named, was discovered at the peak of the lead rush period. Though mining records from that era are limited, extensive deposits of galena were reportedly extracted from the cave. A total of three documented shafts were used to remove the abundant deposits of lead ore. As was common at the time, skilled rockwork was used to stabilize enlarged vertical shafts by lining the walls with hand cut blocks of dolomite (a technique called "cribbing") which is still evident as one descends into the cave today via the only remaining entrance to reach a complex labyrinth. The cave harbors dense clusters of pristine and statistically rare speleothems, including the largest known occurrences of beaded helictites in the State. Local folklore relates how Lawrence Kemling was supposedly lost in the cave for two days and the extensive array of interconnected passages certainly lends credence to the story.

The Iowa City Grotto (as it was called at the time) began a survey of Kemling in 1950 (Rippey 1950), eventually producing the first map of the cave. While a useful preliminary tool, the 1950 map of Kemling Cave was drafted to mapping standards of the day and incomplete, with limited detail and many leads remaining to be surveyed. In a

Fig. 4.6 Example of valley head incision and denudation of cave structures within the Silurian escarpment (Fayette County)

contemporary revival of the Dubuque area survey project, a detailed remapping of the cave using up to date methods was initiated in 2006, compiling over 3.5 km of detailed cave survey accompanied by mineralogical and biological inventories (Fig. 4.12). Modern cave preservation efforts have included stabilization of the historic entrance, modern gate installation, and speleothem restoration.

Speleogenetic patterns in the area also include abundant examples of pit caves and mechanical crevices as well as deep phreatic maze development as a function of deep-seated waters within the same east-west joints of the Galena dolomite, as illustrated in the overall morphology of Maze Cave (Figs. 4.11a, 4.13). The cave passages extend over 830 m with depths exceeding 30 m.

Big Spring System (Clayton County) is one of the most intensively studied karst watersheds in Iowa, featuring a watershed measuring at least 267 km^2 (Fig. 4.14). A series of surface basins recharge a complex fracture flow system supporting a perennial discharge, ranging from 30 to 295 cubic feet per second (0.85–8.4 m^3/s), via its principal resurgence (Big Spring) (Heitmann 1980; Keeler 1997). A dense array of several hundred recorded sinkholes act as insurgence conduits for a significant portion of overall system discharge with several vertical pits exceeding 30 m in accessible depth. Cave Canem (Fig. 4.15a) is one such example. A 12-m deep sink leads to a 20-m shaft formed within Ordovician dolomite of the Galena Group (Fig. 4.15b). A phreatic canyon extends from the base of the entry pit to the second four-meter pitch with a sediment plug. Insurgences have been successfully dye traced over 14 km from the resurgence and the overall hydrogeology of the area indicates it could be the largest karst system in the State (Hallberg et al. 1984). In addition to supporting one of Iowa's largest fish hatcheries, the Big Spring System has served as a seminal model of karst aquifer dynamics, biodiversity, and water quality management in the State (Kennedy and Miller 1990; Kalkoff 1995).

Wonder Cave (Winneshiek County) exhibits composite effects of both phreatic and vadose development. Formed within the Dunleith Member of the Galena Group (Fig. 4.16) it is primarily expressed as a vadose canyon that extends 280 m and descends 44 m from a single, humanly enlarged entrance to a terminal domepit. Yet, sections of the cave clearly display the effects of phreatic enlargement as well (Bretz 1942). Wonder Cave routinely draws episodic meteoric flow through several side passages which eventually reached a sediment-choked drain at the base of the domepit. Unconfirmed reports indicate the cave may be hydrologically connected to nearby springs, but positive dye tracing has not yet been performed.

One of a handful of Iowa caves ever commercialized, Wonder Cave opened in 1936 and offered tours for nearly forty years, finally closing in the 1970s due to a dwindling tourism base and marketing limitations. Decaying remnants of the tour trail were removed in the early 1990s as part of a cavern-wide restoration project.

Examination of numerous vertical entrances within the sinkhole recharge areas of several Iowa drainage basins has yielded numerous caves but none have revealed accessible, extensive conduit passage to date. Exploration of these significant, hydrologically active systems has historically been accomplished via their resurgence entrances. For example, one of the most well-defined drainage basins in the State recharges the confines of its largest cave, the Coldwater Cave System.

Coldwater Cave System (Winneshiek County). Since its discovery by Iowa cave divers in 1967, over 27.8 km of surveyed stream passages have been mapped so far in this near pristine environment that was designated a National Natural Landmark in 1987 (Aley 1981). See the accompanying section by Kambesis for a detailed discussion of the hydrology and exploration of the system.

Coldwater Cave is formed within the Dunleith Member of the Ordovician Galena Group. The extensive, branching fluvial cave system exhibits morphologies and flow paths

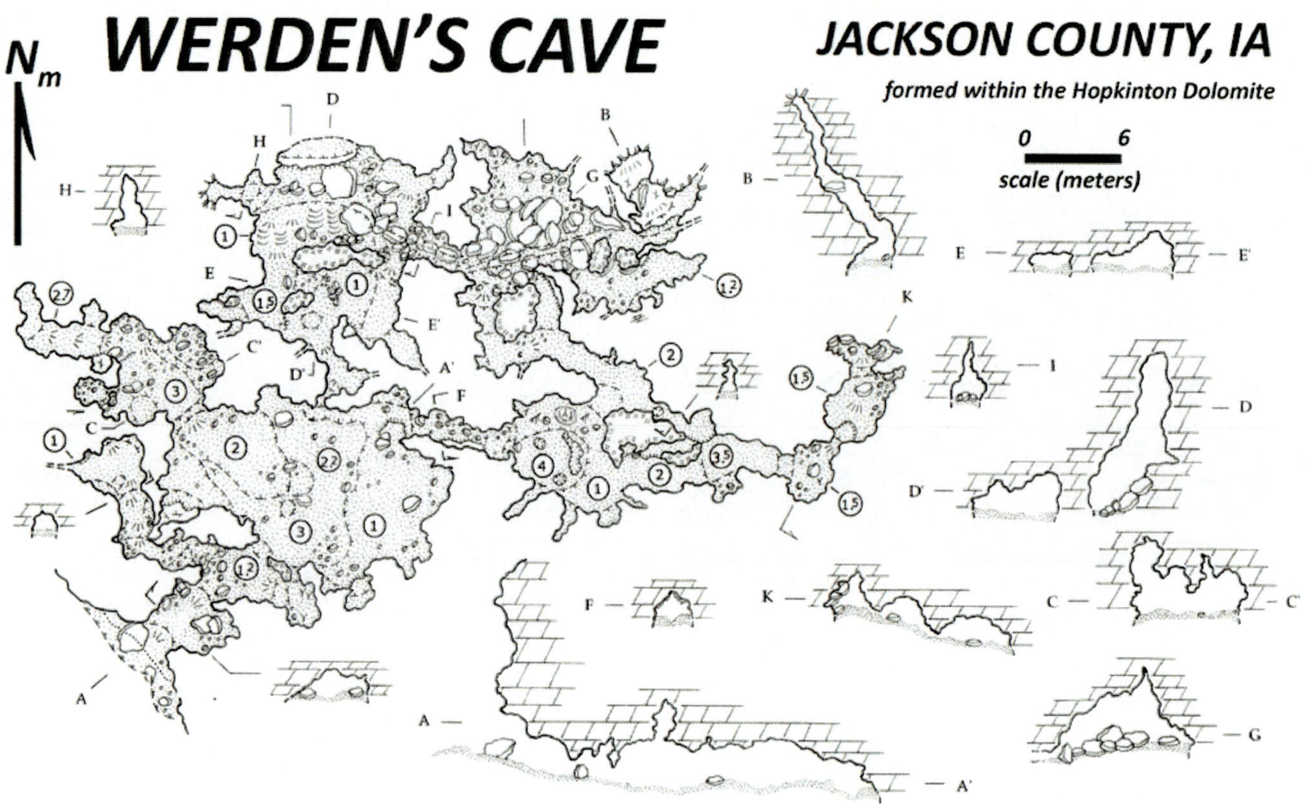

Fig. 4.7 Map of Werden's Cave (Jackson County)

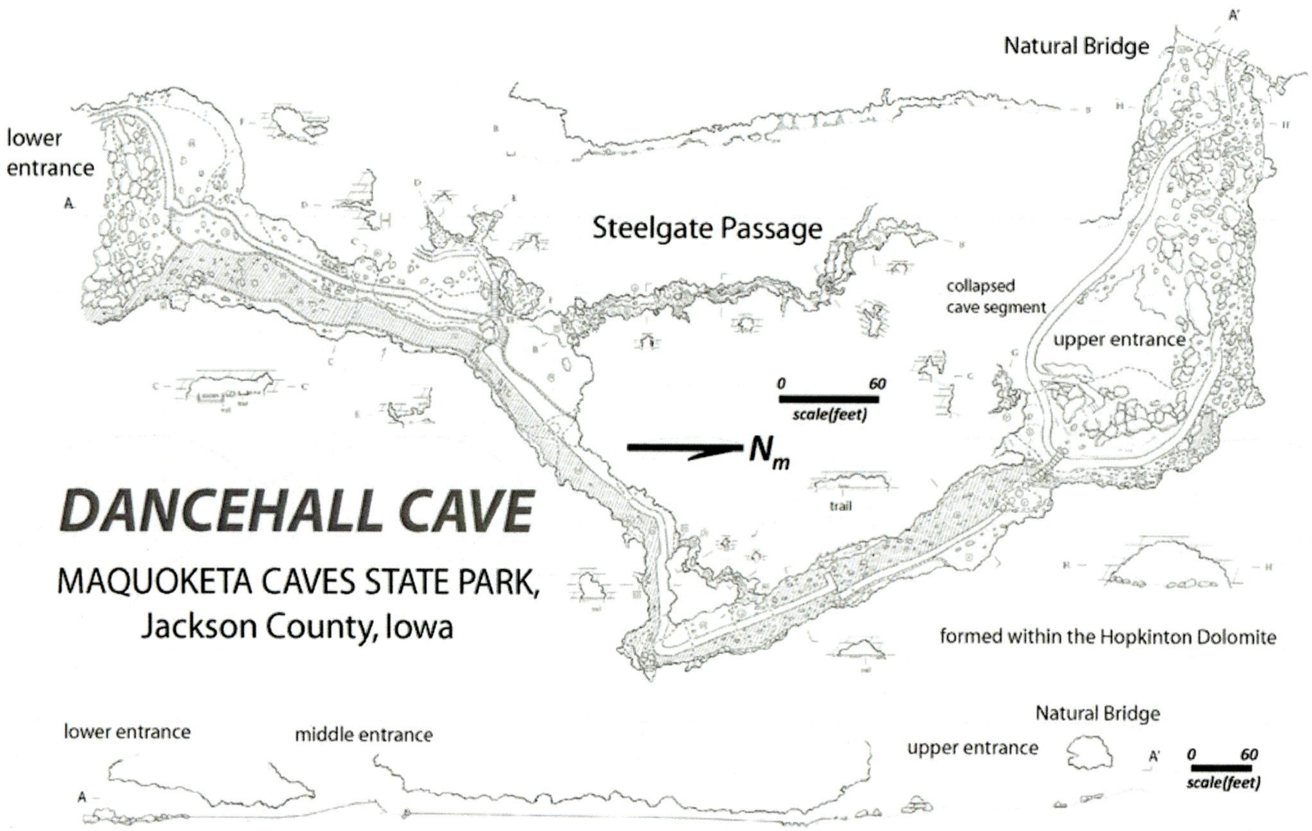

Fig. 4.8 Map of Dancehall Cave (Jackson County)

Fig. 4.9 Karst expressions in the Galena Cuesta. **A** Example of sinkhole development. **B** Biological inventory of a karst spring formed within the Dunleith Member of the Galena Group. **C** Phreatic passage development. **D** Surveying a cave formed within dolostone of the Prairie du Chien Group (Winneshiek County)

Fig. 4.10 Passage morphology and speleothem development in Kemling Cave, Dubuque County (Photo by S. Dankof)

influenced by a range of mechanisms, including meander cutoffs, stream piracy, resurgence migration, and joint control (Fig. 4.17). The Coldwater Basin defines a recharge area of 81 km^2, oriented northwest to the southeast that drains to two springs: the primary resurgence (Coldwater Spring) and a secondary resurgence (Carolan Spring). A series of parallel drainage basins are similarly oriented, yet hydrologically distinct, and flank the Coldwater System to the east and west. Collectively, these drainage basins form the southern tier of the broader Upper Iowa Watershed (McKay et al. 2006) which includes several northeastern Iowa counties as well as counties in the extreme southeast of Minnesota. The 640,000-acre (259,000 ha) watershed is characterized by over 200 km of entrenched surface streams, numerous springs, and more than 2500 sinkholes.

The Coldwater Cave System continues to offer a unique setting supporting a multidisciplinary range of research projects (Dorale et al. 1992; Harmon et al. 1979; Lewis 1981; Kambesis 2007; Shopov et al. 1994; Wheeler et al. 1988). Current studies have demonstrated that the cave microclimate and recharge patterns are far more complex than previous models indicated, revealing a hydrologically dynamic karst system with clearly identifiable segments of the watershed vulnerable to rapid contaminant transport. These studies are currently being expanded (Kambesis et al. 2013; Welch et al. 2015, 2017).

4.3 Outlying Karst Areas

The Prairie du Chien cuesta geographically defines the northeastern Iowa border and features significant outcrops of sandstone underlain with dolomites, hosting a comparatively limited number of small solutional caves of both geological and archaeological interest.

4.3.1 Caves Formed Within Iowa Sandstones

Though the majority of recorded Iowa caves are formed within Paleozoic limestones and dolostones, cave development is not limited to carbonate strata. Caves formed within sandstones are not confined to any single physiographic region in Iowa, occurring in the Southern Drift Plain, the Des Moines Lobe, and Paleozoic Plateau. Their distribution forms an outer perimeter for the overall concentration of caves in Iowa. The sandstone caves developed in Pennsylvanian, Mississippian, Ordovician, and Cambrian aged sandstones were subjected to solutional and mechanical processes. Varying cave morphologies reflect the interplay of processes acting on sandstone systems (Fig. 4.18). Floodwater maze development has been recorded within Pennsylvanian sandstone along the southernmost edge of the

Fig. 4.11 **A** Main passage of Maze Cave (Dubuque County). **B** Diagrammatic cross section of typical passage morphology in Kemling Cave (Calvin et al. 1897). **C** Speleothem linked to an adjacent limonite deposit (Kemling Cave) (Photos by S. Dankof) **D** Typical crevice altered by lead mining (Dubuque County)

Southern Drift Plain (Fig. 4.18b). Stafford's Sandstone Cave is the most extensive sandstone cave recorded in the State with an elongate central passage extending over 30 m laterally into the St Peter Formation (Fig. 4.18b). Though the observable passage is entirely within this unit, the contact cave is actually derived from initial void development within the underlying Oneota Dolomite followed by progradational collapse and exfoliation of the overlying sandstone.

Karst has also been recorded in Devonian and Jurassic gypsum in central Iowa in the form of fissures and subsidence features, as detailed in the accompanying section by Raymond Anderson.

4.3.2 Karstification of Devonian and Mississippian Carbonates

The arc of sporadically exposed Devonian carbonates extending southeast from the northern edge of the Iowan Surface to the eastern edge of the Southern Drift Plain, displays varying degrees of karst development (Figs. 4.1 and 4.2). For example, limited cave development has been noted in Floyd, Mitchell, and Cerro Gordo counties within the northern Iowan Surface (Fig. 4.19). The topography features limited vertical relief and minimal soil cover in select sections, occasionally revealing exposed surfaces of highly fractured Devonian-aged Cedar Valley limestone (Libra et al. 1984; Walters 2008). Shallow, branching phreatic caves recorded in this area are of limited extent, accessed by vertical sinkhole entrances and frequently terminating in sediment plugs or impassable narrow fissures (Fig. 4.20). Loosing streams and dense sinkhole networks are common in this area with clear influences on overall water quality and recharge patterns of a number of associated springs discharging into the Cedar River (Libra et al. 1984). As in other Iowa karst areas, the combination of a thin mantle and numerous karstic voids has presented a range of contaminant transport and landscape stability issues (Lindsey et al. 2010; Moore 1995; Munter 1980; O'Conner and Trainum 2015).

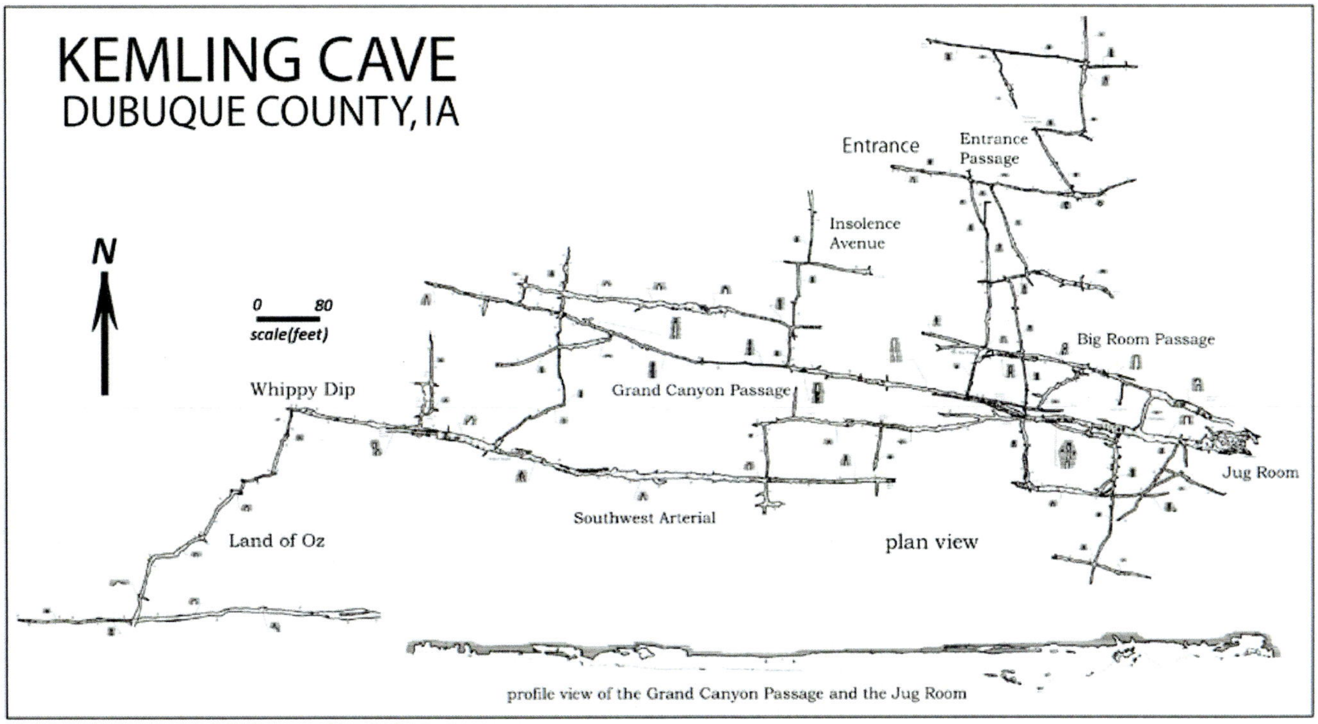

Fig. 4.12 Map of Kemling Cave (Dubuque County) (Map by E. Klausner)

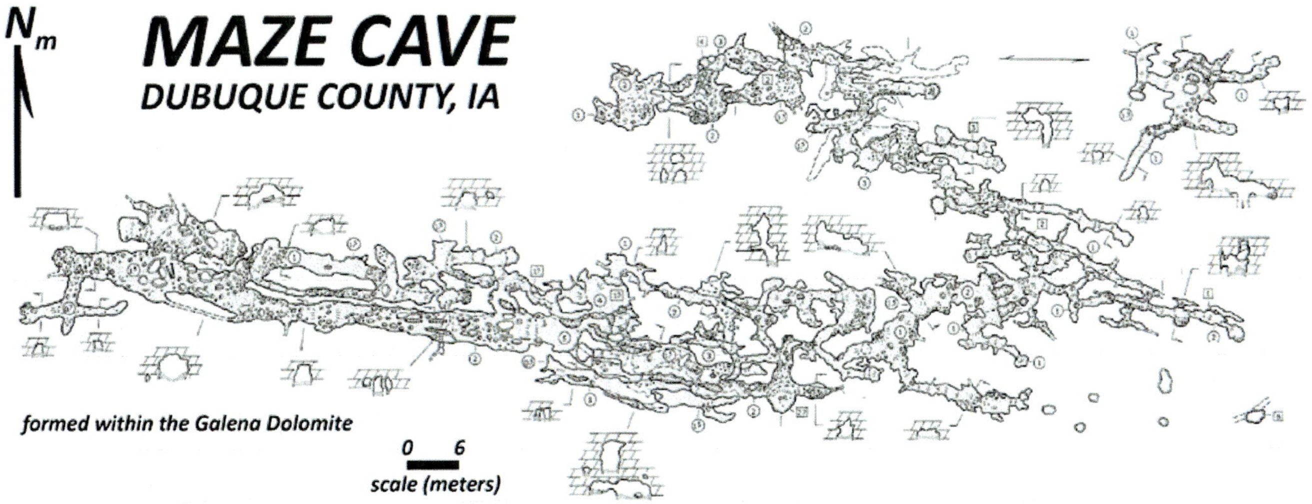

Fig. 4.13 Map of Maze Cave (Dubuque County)

Outlying occurrences of cave development are also documented within dispersed Mississippian and Devonian karst of extreme southeastern Iowa (Fig. 4.21). For example, historical records had briefly noted isolated sinkholes within the comparatively planar landscapes of Louisa County, yet karst surveys only recently documented a comparatively small karst area composed of springs and associated sinks, as well as caves revealed by streams entrenched within Devonian carbonates (Fig. 4.21a). The cliff-lined banks of Flint Creek in Des Moines County also harbor small springs and caves of limited extent, for example, Starr's Cave (Fig. 4.21b, c) but they are well-known components of complex Mississippian stratigraphy models applied to the area (Witzke and Bunker 2002).

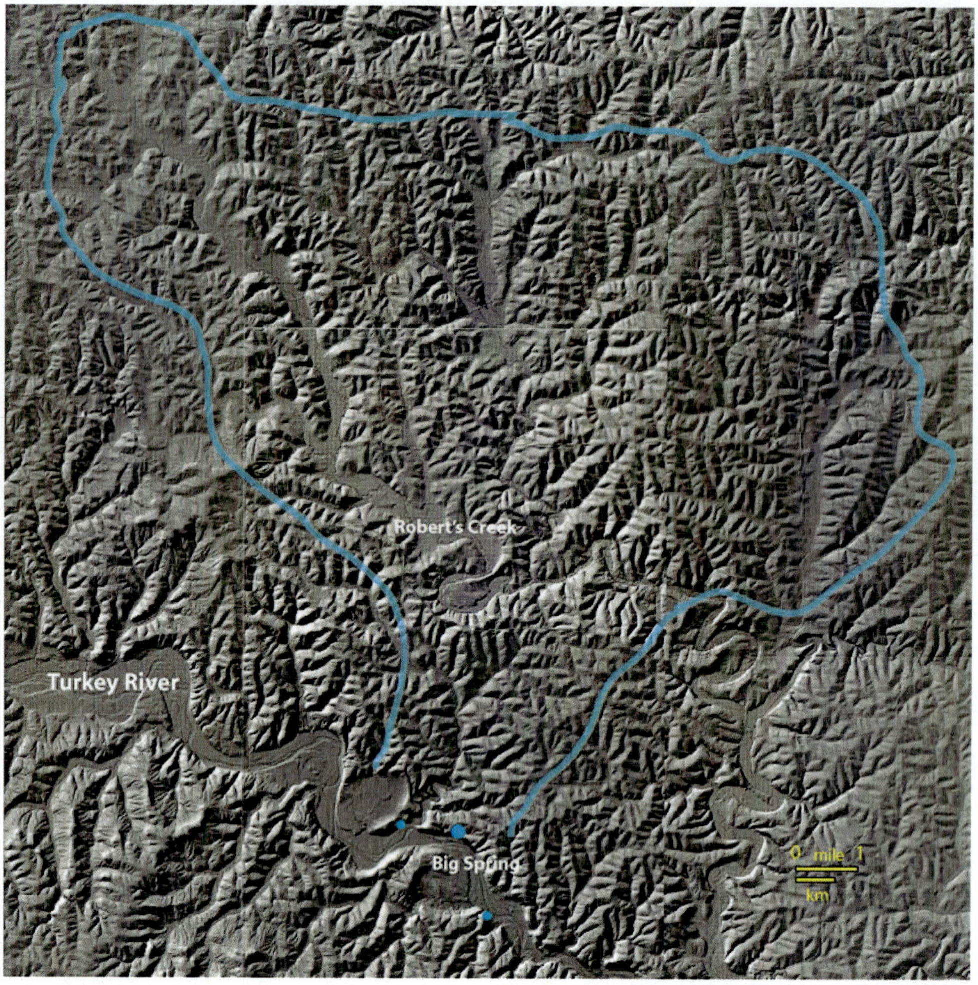

Fig. 4.14 Map of the Big Spring System (Clayton County), illustrating projected extent as per Hallberg et al. (1983) overlying 10 m resolution shaded relief LIDAR imagery (Iowa Geographic Server)

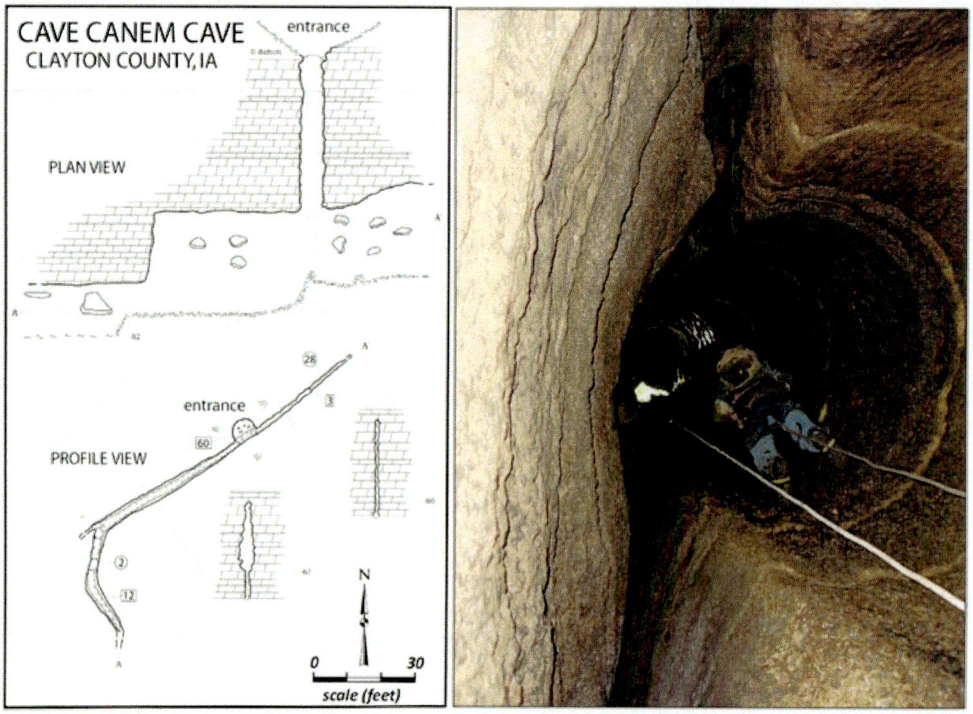

Fig. 4.15 **A** Map of Cave Canem (Clayton County) (Map by E. Klausner). **B** Entry pit morphology in Cave Canem

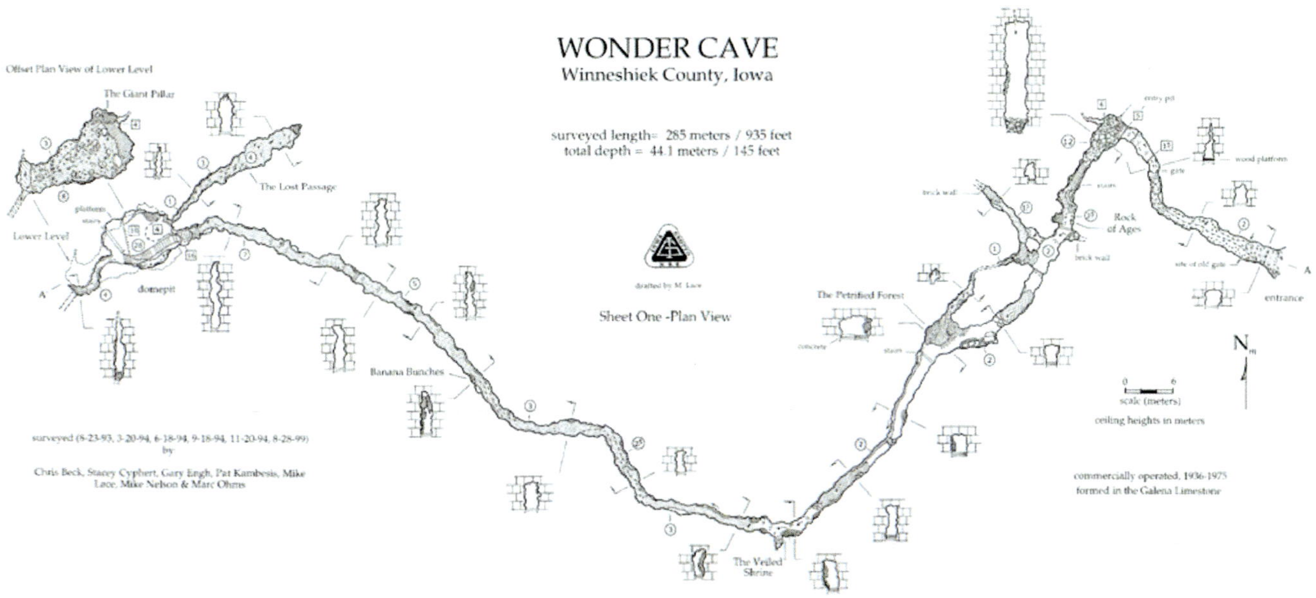

Fig. 4.16 Map of Wonder Cave (Winneshiek County)

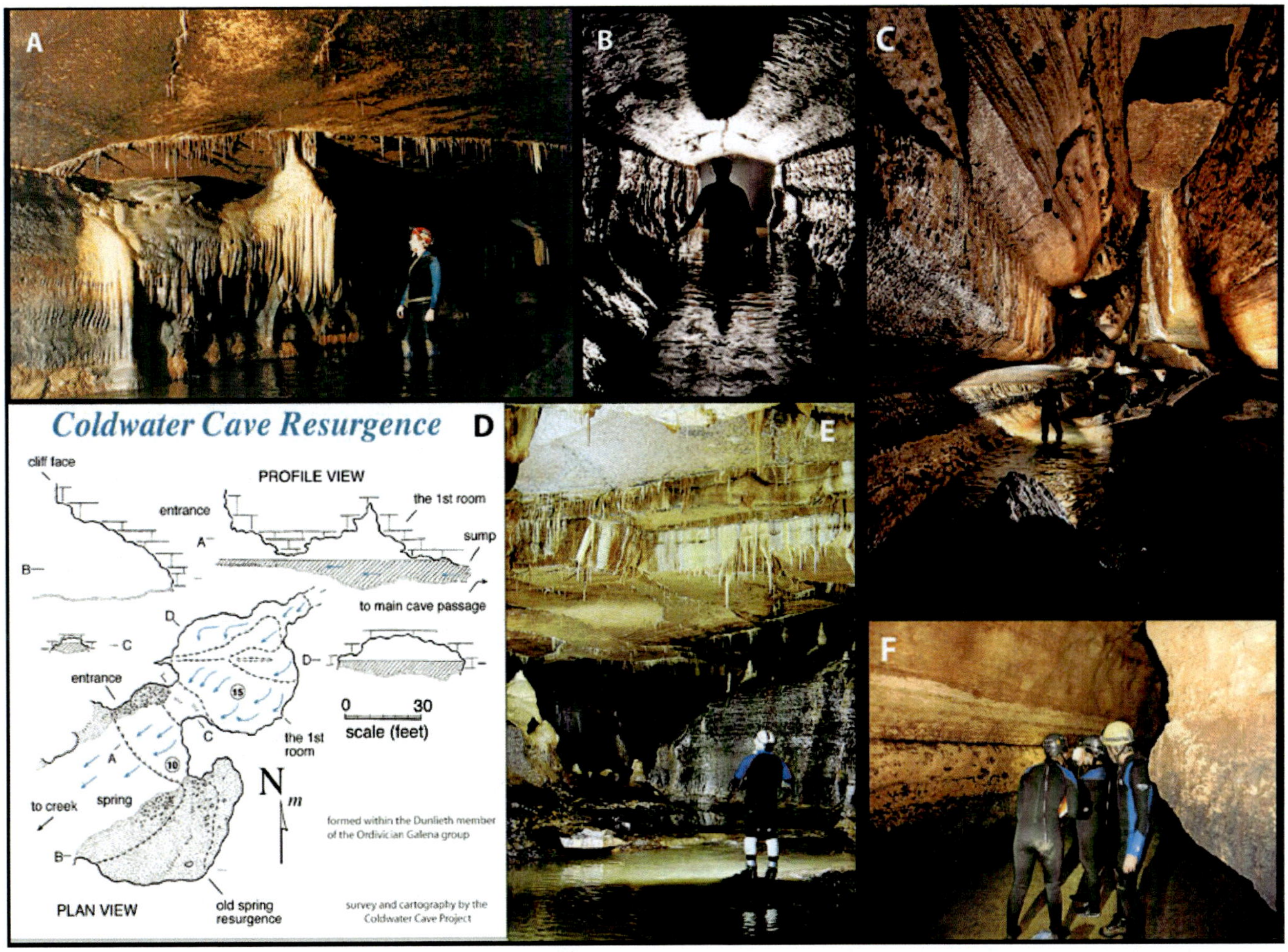

Fig. 4.17 Coldwater Cave (Winneshiek County). **A** Typical speleothem occurrence. **B** Fracture development and mainstream passage morphology. **C** Phreatic passage morphology of an infeeder passage segment. **D** Map of the primary resurgence of the cave system. **E** Breakdown area in the downstream mainstream passage (Photos by S. Dankof). **F** Geology students surveying offset beds (Photo by E. Klausner)

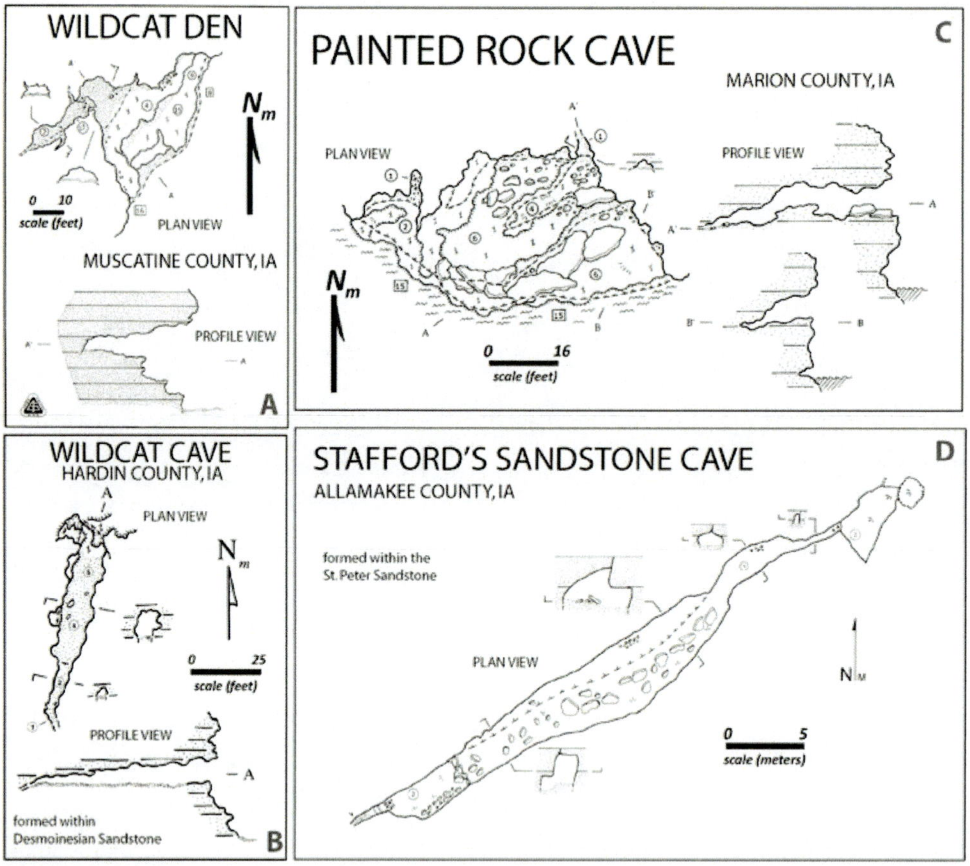

Fig. 4.18 Cave development within Iowa sandstones. **A** Wildcat Cave (Hardin County). **B** Painted Rock Cave (Marion County). **C** Stafford's Sandstone Cave (Allamakee County) (Map by M. Ohms). **D** Wildcat Den (Muscatine County)

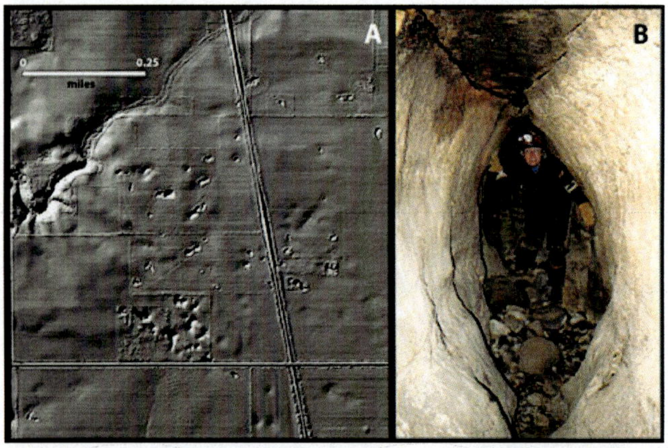

Fig. 4.19 **A** Typical sinkhole density associated with Devonian karst (Floyd County), illustrating open and closed depressions in a defined area of dense cave development (1 m resolution shaded relief LIDAR image from the Iowa Geographic Server). **B** Typical phreatic passage morphology in Floyd County (Photo by E. Klausner)

4.4 Biodiversity, Archaeology, and Changing Land Use Patterns in Iowa Karst

Iowa's karst landforms are ecologically distinct, supporting a variety of surface and cave fauna and flora and harboring detailed evidence of climate change—both past and present (Baker et al. 2001; Denniston et al. 1999; Griffith et al. 1994). Karst structures in the upper Midwest provide unique habitats for complex biosystems, such as algific slopes across the broader Driftless Area. Algific slopes are present in select Iowa karst landscapes as well (Cottrell and Strode 2005; Hallberg et al. 1984; Nekola 1999), with fragile paleorefugia harboring glacial relict species such as the Iowa Pleistocene snail (*Discus macclintocki*) (Clark et al. 2008).

Comprehensive cave bioinventories within Iowa caves have been limited but early studies revealed a range of species (Christiansen et al. 1961; Peck and Christiansen 1990; Koch and Case 1974) (see Peck et al. 2020 in this volume). Yet, systematic modern faunal surveys of small caves continue to reveal more complete profiles of macrofaunal and invertebrate distributions as a function of historic and modern landscape utilization (Dixon 2011; Gilbert 1989; Iker et al. 2010). For example, recent surveys of small cave hibernacula use patterns by a range of bat species indicate a surprisingly complex interdependency (JW Dixon, pers. comm. 2017).

4.4.1 Use and Modification of Iowa Caves and Karst

Iowa landforms, especially karst areas, harbor a rich record of prehistoric, historic, and modern land use patterns (Prior 1991). Human use and modification of karst terrains have

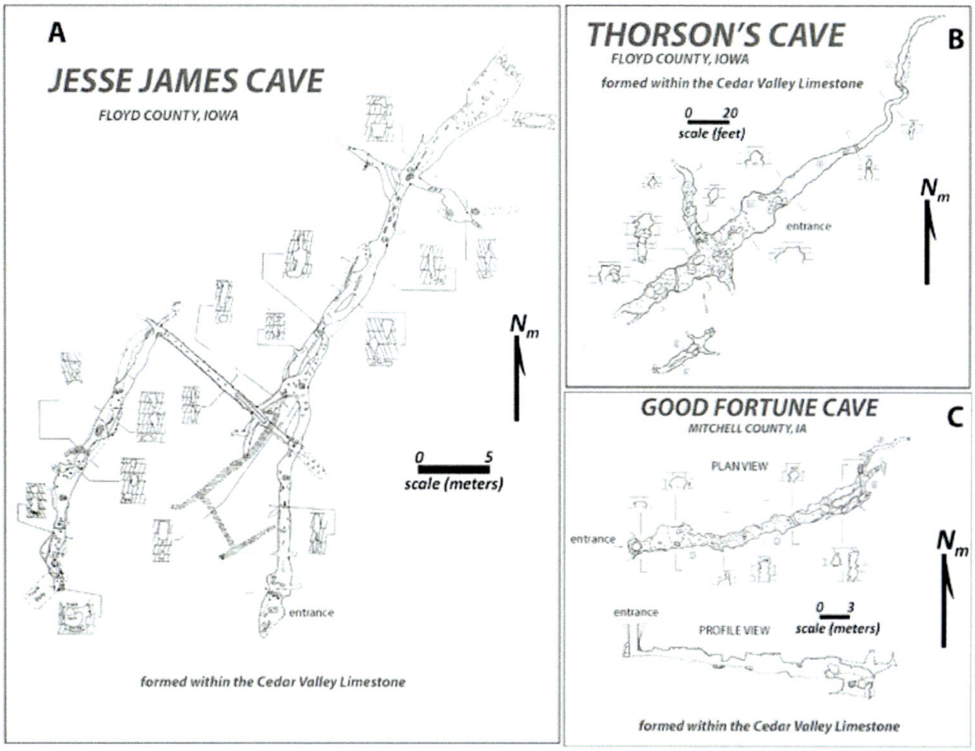

Fig. 4.20 Cave development in the Devonian carbonates. **A** Jesse James Cave (Floyd County) (map by L. Welch). **B** Thorson's Cave (Floyd County), **C** Good Fortune Cave (Mitchell County)

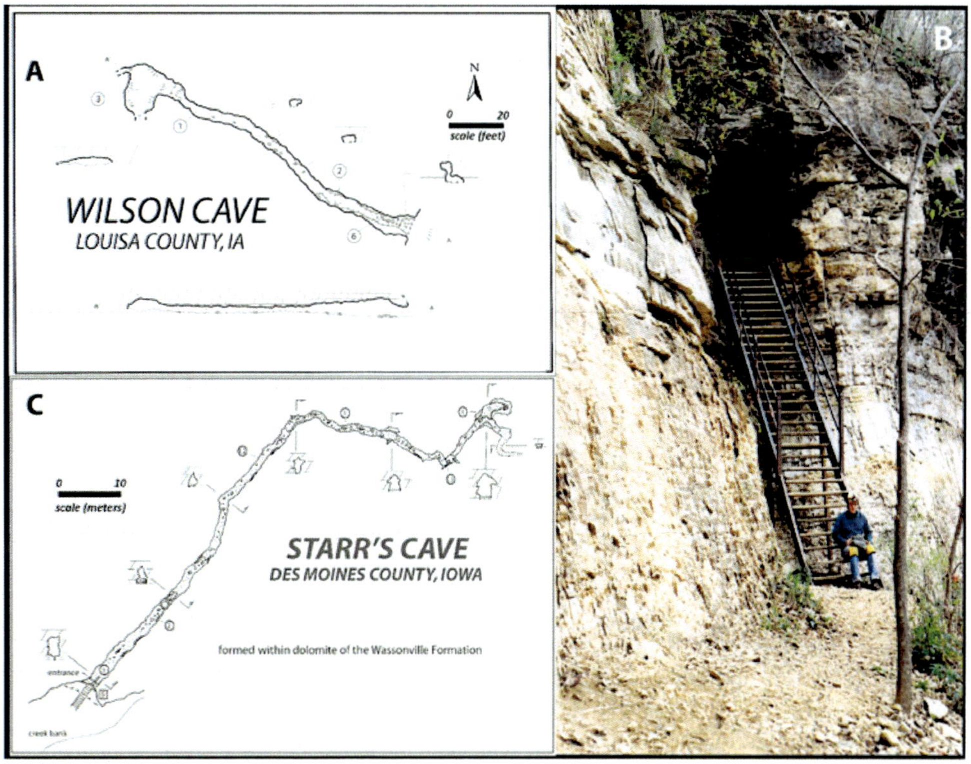

Fig. 4.21 Cave development in Southeastern Iowa. **A** Map of Wilson Cave (Louisa County) (Map by E. Klausner). **B** Entrance photo and **C** map of Starr's Cave (Des Moines County)

historically influenced many Iowa cave features, spanning Archaic, historic, and modern periods. Archaeologically significant cave uses date as far back as the Archaic period and extending through the Late Woodland period and European contact in the form of habitation, mortuary, and ritual uses by native peoples, as represented by lithics, ceramics, petroglyphs, as well as human and faunal assemblages (Stanley 2002; Thompson 2002). As noted in other karst areas, caves have offered a more durable preservation environment of cultural materials compared to dynamic surface topographies over time. Post contact historical cave utilization included nineteenth-century homestead use as shelters, improvised burial vaults, water supplies, or cold storage, as well as recreational and commercial tourism ventures (Fig. 4.22).

Water resources associated with karst areas have also been historically extracted on both small residential and large municipal scales (Steinhilber et al. 1961). For example, bedrock aquifers have long been utilized for nineteenth-century municipal water resources, such as the Jumbo Well (Benton County), an artesian water source that flooded the streets of Belle Plaines for five days before finally being capped in 1886 (Fig. 4.22a) (Mosnat 1898), or the water supply for the City of Dubuque which was derived from one of the large flooded cave systems beneath its streets (Fig. 4.22b).

Spring mills and springhouses were common adaptations of karst resurgences during the same historical period. Spring mills were erected at several northeastern Iowa springs, for example, Dunnings Spring (Winneshiek County) (Hedges 1974b). The cool and constant temperatures of springhouses were utilized by many farmsteads as natural refrigerators, in addition to domestic and animal water sources, in the initial European settlements of Iowa. The presence of karst springs dictated the locations of many pioneer farmsteads and settlements. These springhouses have been gradually abandoned as, starting in the 1930s, rural electrification provided access to mechanical refrigeration. The use of shallow karst springs for water supplies has been largely abandoned due to contamination from surface sources, but historic springhouses remain scattered throughout the Iowa karst regions.

Several karst springs in Iowa have long supported fish hatcheries, thus contributing to regional economic stability. Big Spring (Clayton County) and Siewers Spring (Winneshiek County) are notable examples (Hallberg et al. 1983; Horick and Soenksen 1989). The Coldwater Cave System also discharges into a cold stream, supporting a long-term local trout hatchery release program and a range of recreational uses along the Upper Iowa River.

Several caves were developed as private commercial tourist sites in northeastern Iowa, beginning in the late 1920s (Decorah Ice Cave) then followed by the development of caves in Winneshiek, Dubuque, and Clayton Counties in the mid to late 1930s (Hedges 1974b). While these commercial ventures varied in longevity, from a few years to a few decades, two of these caves remain operational today: Crystal Lake Cave (Dubuque County) and Spook Cave (Clayton County). Many other examples are provided by Brick (2004).

The utilization of true glaciers, which continue to hold ice even through the warmest summer months (Kovarik 1899) and modified underground cavities for cold storage for both private homestead use and commercial necessity (for example, storage for local breweries), was common in northeastern Iowa through the late 1800s and early 1900s (Fig. 4.22c). A few of these sites still retain some of the finest surviving examples of nineteenth-century barrel vaulting architecture in Iowa. The entry chambers of a few northeastern Iowa caves, featuring natural air conditioning in

Fig. 4.22 Historical cave uses in Iowa. **A** The Jumbo Well of Belle Plaines, Iowa (1886). **B** Main valve at the lower level of a flooded crevice system, used for the historic water reservoir of the City of Dubuque. **C**) Historic brewery cave, City of Cedar Rapids. **D**) Boat tour in Spook Cave (Clayton County). **E**) Historic recreation at Burt's Cave (Jackson County) ca. 1900. **F**) Early 19th century adaptation of a cave for mortuary use (Louisa County)

the muggy summer months, were also historically used as recreational sites (Fig. 4.22d), such as dance pavilions in the late nineteenth century with the last vestige of this application persisting into the mid-1970s.

Economic resources associated with Iowa's karst landscapes are commercially extracted during modern mining of silica, gypsum, and aggregate for construction as well as historical mining of limestone, coal, and mineral cave deposits, including lead, zinc, and iron ores. Perhaps the most exploited and historically well-documented mineral region of Iowa karst was within the Upper Mississippi Valley Lead and Zinc mining district which saw significant modification of existing cave features in the process of lead, zinc, and (to a lesser extent) iron ore extraction which predated Euro American settlement and became an integral part of the economic growth in the Dubuque area and on nation-wide mineral industry scale and spanned over 250 years, peaking in the mid-1800s but trailing to sporadic levels as the new century arrived (Keyes 1912; Ludvigson and Dockal 1984; Heyl et al. 1959). At that time, many fortunes were built on the extraction of lead (Galena) and zinc sulfide (Sphalerite or "Blackjack") deposits in the crevices underlying the hills of northeastern Iowa and adjacent areas (Dockal 2020; Chap. 7 in this volume). Large-scale mining ended by the early 1900s but there were sporadic exploratory mining efforts in the 1930–1940s and again in the 1950–1960s.

A total of more than 130 caves across Iowa alone have been altered by historical ore mining with a significant number noted in adjacent segments of neighboring Wisconsin and Illinois as well. While mining activity was concentrated within the dolomitic ridges of the Galena cuesta overlooking the Mississippi River, outlying karst areas were also excavated. For example, Silurian-aged dolostones across the Niagara cuesta and Ordovician dolostones of the Prairie du Chien Group along the northern edge of the Paleozoic Plateau (Garvin 1983; Heyl and West 1982).

But perhaps the most transformative long-term alterations to cave and karst areas in Iowa are rooted in changing agricultural land uses. The progressive shift from small scale to industrial-scale agriculture has been accompanied by increasing sediment load and contaminant transport while concomitantly decreasing overall water quality in vulnerable karst aquifers (Hallberg et al. 1985; Sasowsky 2000; Schilling and Helmers 2008a, b).

4.4.2 Cave and Karst Preservation in Iowa

The detailed survey and inventory of Iowa cave and karst features has yielded a more complete view of cave resources in the region and their development while providing both private landowners and public land managers with critical resource management tools. Yet, the preservation of underground wilderness areas in Iowa and the broader region faces significant challenges with clear consequences to water resource integrity as well as cultural and natural heritage management. Cave and karst resource management practices and preservation efforts have spanned both private as well as State and county-owned lands. Grassroots efforts by local cavers to actively work with private landowners and public land managers on the municipal, county, and state levels in the preservation of Iowa karst features has proven particularly successful over the past sixty years and continues to foster cooperative protection of these unique resources.

4.5 Gypsum Karst of the Jurassic Fort Dodge Formation

Raymond R. Anderson, Adjunct Professor, Department of Earth and Environmental Sciences, University of Iowa, 115 Trowbridge Hall, Iowa City, Iowa, 52242, USA. raymond-anderson@uiowa.edu.

4.5.1 Introduction

The entire Jurassic System in Iowa is restricted to rocks within a 19 square mile (50 km^2) area in central Webster County, in and around Fort Dodge (Anderson and McKay 1999). This area includes one of Iowa's most valuable natural resources, the Fort Dodge Gypsum (Fig. 4.23). The presence and potential value of this gypsum was first reported by geologist David Dale Owen who traveled up the Des Moines River in 1849. He described the gypsum

Fig. 4.23 Young girl sitting on the naturally eroded Fort Dodge gypsum surface as seen in 1917

deposits at what would be Fort Dodge as the most important bed of plaster-stone known west of the Appalachian Chain:

> …It is seen at intervals for three miles, exposed on both sides of the Des Moines, in mural faces from eighteen to twenty-five feet [7.6 m], always overlying pink shales from beneath which copious springs of excellent water issue. It has been traced in the ravines, back from the river … where it is finally lost under the deep alluvium of the vast plains that stretch away to the west. [1852:126]

Today we recognize his "pink shale" as Soldier Creek Member and the gypsum as the Gypsum Creek Member, two of the three members that comprise the Fort Dodge Formation. These are the only known survivors of the rock strata deposited during the Jurassic Period (ca. 160–140 million years ago) between central Michigan and central Nebraska.

The gypsum at Fort Dodge, like most commercial-scale gypsum deposits, had its origins in the evaporation of seawater in a restricted shallow basin. Water from the adjacent Jurassic seaway passed over a low-lying barrier into the basin, where the mineral salts became concentrated by evaporation in the hot semi-tropical sun. When the brine became sufficiently concentrated, gypsum crystals formed and settled to the floor of the basin. The beds at Fort Dodge are exceedingly pure gypsum and contain no anhydrite (a common alteration mineral contaminant.) The extent of the original depositional basin is unknown, but it was certainly much larger than the area of gypsum remaining today. As the Jurassic passed into the Cretaceous Period about 135 million years ago, North America slowly drifted northward out of the dry latitudes where the gypsum formed, into wetter, more temperate latitudes closer to the continent's present position. Cretaceous rivers flowed across Iowa, first eroding away most of the original gypsum deposit, then reburying the region, and the remaining gypsum, with river sediments. The advancing Cretaceous seaway then covered the region with marine deposits. The remaining Jurassic gypsum lay buried for tens of millions of years until a new episode of erosion uncovered it, eroded away the Cretaceous sediments, and again began to erode the gypsum bed. Beginning about 2.5 million years ago the gypsum bed was again buried, this time by glacial materials carried by Pleistocene continental ice sheets that made several advances into Iowa. After the final retreat of glacial ice, the modern Des Moines River cut through the glacial materials at Fort Dodge and once again exposed the ancient Jurassic gypsum beds.

4.5.2 Geology of the Fort Dodge Formation

The Fort Dodge Formation consists of three members, the basal Shady Oak Member, the Gypsum Creek Gypsum Member, and the upper Soldier Creek Member. The Shady Oak Member consists of shales, sandstones, and conglomerates, deposited in an arid fluvial (river) environment, probably by intermittent streams that experienced flash flooding. Shady Oak sediments are found in only a limited area near the confluence of Gypsum Creek and the Des Moines River. The Gypsum Creek Gypsum Member overlies Pennsylvanian shales and locally Shady Oak rocks. It consists of discontinuous, thin basal plastic, dark grey shale overlain by a very pure banded gypsum unit that reaches thickness up to about 30 feet (9.1 m). The Gypsum Creek Member is the most widespread of the three Fort Dodge Formation members. The upper unit is the Soldier Creek Member, a red bed sequence of sandstones, siltstones, and mudstones, mottled grey-green in some beds, and reaching a maximum known thickness of about 45 feet (13.7 m).

The three members of the Fort Dodge Formation are representative of different but related depositional environments. They are related in that they all demonstrate deposition in a very arid environment, understandable since during the Jurassic the central portions of North America lay in the horse latitudes, a region of the globe where constant high-pressure systems drive away the rain clouds. The Shady Oak Member is dominated by clasts of limestones eroded from Mississippian and Pennsylvanian rocks exposed in the region. The dominance of these carbonate clasts identify the streams that deposited them as intermittent in an arid environment. The Gypsum Creek Gypsum Member was deposited in a large terrestrial evaporite basin, recharged by seawater from the adjacent Jurassic Sundance Sea. The uppermost Soldier Creek Member was deposited in an arid, braided fluvial environment by rivers flowing to the receding Sundance Sea.

4.5.2.1 Age of the Fort Dodge Formation

The age of the Fort Dodge Formation has been a topic of much geologic speculation through the years. The unit does not contain any minerals that can be dated using radiometric techniques, and fossils are limited to terrestrial palynomorphs in the Gypsum Creek gypsum and reworked Mississippian and Pennsylvanian marine fossils in Shady Oak and Soldier Creek members. Owen (1852), who first reported the presence of gypsum in the area, speculated that the gypsum was of Pennsylvanian age. Keyes (1895) concluded that the gypsum was correlated with the Cretaceous Niobrara chalks to the west. A few years later Keyes (1915) changed his ideas and proposed a Miocene age for the gypsum. The first detailed investigations of the Fort Dodge Formation rocks was undertaken by Wilder (1903), who observed thin beds of gypsum interbedded with red shales along Soldier Creek and concluded that they were a part of the same depositional system, which he observed were similar to Permian rocks in Kansas. Moore et al. (1944) reported the identification of Pennsylvanian fusulinids from

Fort Dodge Formation rocks as evidence of a mid-Virgilian (Pennsylvanian) age for the rocks. Hale (1955) observed that the fusulinids were worn and abraded, assumed they were reworked, and cautiously concurred with a Permian age for the strata. Cross (1966), reporting on his investigation of palynomorphs recovered from the gypsum in an abstract republished by Cody et al. (1996: 21), concluded that the gypsum was mid-Mesozoic, "*probably Upper Jurassic Kimmeridgian (Morrison?) although possibly lowermost Cretaceous.*" More recently, Ludvigson (1996) used paleolatitude and paleoclimate data to argue that the Fort Dodge Formation rocks were Upper Jurassic, probably coeval with deposition of the Sundance and possibly lower Morrison formations and equivalent units in the Rocky Mountain States. Cody et al. (1996) concluded that the best age assignment for the gypsum was Jurassic, probably Kimmeridgian (~155 ma) but perhaps as old as Bajocian (~180 ma). Klug (1996) reinvestigated the palynology of the Fort Dodge Formation gypsum and concluded that the gypsum was deposited between the upper Lower Jurassic Toarcian Stage (~190 ma) and the lower Upper Jurassic Oxfordian Stage (~160 ma). Incorporation of the palynological interpretations of Klug (1996) with marine transgression/regression interpretations by Vail et al. (1977) and Brenner (1983) suggest that the deposition of the Fort Dodge Formation gypsum was associated with the maximum transgression of the Sundance Seaway (the maximum eastward advance of the Jurassic seaways) during the Oxfordian Stage (161.2 ± 4.0–155.7 ± 4.0 Ma).

4.5.3 Gypsum Creek Gypsum Member

The Gypsum Creek Gypsum Member is dominated by bright white to grey banded, remarkably pure gypsum (>96% $CaSO_42H_2O$). It is distinctly bedded, the layers ranging from 2 to 10 inches (5 to 25 cm) in thickness and separated by a thin clay band along which it is easily parted (Fig. 4.24). The upper surface of the gypsum has undergone varying amounts of dissolution. This dissolution is especially pronounced at locations where the rocks of the overlying Soldier Creek Member are missing and the gypsum is directly in contact with Pleistocene glacial till. Although the maximum thickness of the gypsum is about 30 feet (10 m), considerable variation exists because of topographic relief on the evaporite basin floor and later erosion and dissolution which locally reduced thickness and totally removed most of the original deposit. The Gypsum Creek gypsum displays a variety of other structures and features, including ripple marks (Fig. 4.25), minor folds, recrystallization features such as nodules, satin spar (a silky fibrous form of gypsum), and most prominently solution-enlarged joints (fractures).

Fig. 4.24 Horizontal banding in Fort Dodge formation gypsum

Probably the most striking features of the Gypsum Creek gypsum itself are the distinctive bands or laminations consisting of white lamina alternating with thinner dark ones. From a distance, the bands give the impression of continuous laminations, but a closer study shows that individual bands extend for a few meters or less before blending with others and disappearing. Lamellae are typically irregular, and many are contorted to various degrees. Dark banding is siltier and predominately blue-grey (but yellow, red, or brown colors also occur). These bands are probably the product of seasonal changes in gypsum deposition, modified by diagenetic changes when the gypsum was buried by younger rocks.

4.5.3.1 Gypsum Mining Near Fort Dodge

The gypsum deposit near Fort Dodge has been a valuable resource since the area was settled. When the first settlers moved into the Fort Dodge area they found exposures of gypsum along the Des Moines River and several tributary creeks. This stone, which proved to be so pure that it could easily be worked with normal woodworking tools, made ideal building stone, vulnerable only to running water. So these exposures were initially quarried by hand, with horses

Fig. 4.25 Ripple Marks on the surface of gypsum at U.S. Gypsum North Welles Quarry

Fig. 4.26 Geology students examine solution-enlarged fractures in the Gypsum Creek gypsum at the U.S. Gypsum North Welles Quarry. Overlying glacial till and a meter of Soldier Creek Member mudstone was removed and the joints cleaned out prior to mining (2010)

and mules used to move heavier stones. Gypsum stone was used to construct walls, sidewalks, foundations, and entire buildings. But the real economic value of the gypsum deposits in the area began to be realized after the railroads reached Fort Dodge in 1869, opening new markets for gypsum products. One of the strangest "products" was the Cardiff Giant, a human replica carved from gypsum that was exhibited as genuine in the late nineteenth century (Brick 2004, p 126).

In 1872 the Fort Dodge Plaster Mill was constructed to mine, grind, and prepare gypsum for stucco and other products. It was followed by other mills (13 in total have operated in the Fort Dodge area) driving a rapid expansion in the mining of gypsum to supply these mills. As the surface exposures of gypsum began to be depleted the quarrying advanced into the bluffs, where overburden thickness increased until it became more economic to extract the gypsum from the subsurface. Underground mining of gypsum at Fort Dodge began in 1895 and continued until about 1950 (Natte 2008). Wilder (1918) noted that where mined underground, gypsum ranged from about 25–30 feet (7.6–9.1 m) in thickness and was overlain by up to 60 feet (18.3 m) of overburden, including an average of about 11 feet (3.4 m) of Soldier Creek Member rocks capped by Pleistocene glacial drift. About 10–15 feet (3–5 m) of gypsum was left as roof rock in the mines and about 6 feet (1.8 m) in the floor, with mine room heights ranging from 15 feet (4.5 m) to as little as 8 feet (2.5 m) (Crawford et al. 2004). Hundreds of miles of tunnels were blasted in about a dozen area mines, removing about 20,000 tons of gypsum per acre (44,833 ton per ha). After about 1950, when the volume of gypsum required by the newer, modern mills exceeded the capacity of the mines, more efficient surface mining procedures were adopted. The gypsum resources nearest the mills were often the most economic to produce, so many of the undermined areas were stripped to remove the remaining gypsum. This mining exposed some of the old mine rooms (Fig. 4.28). Gypsum mining continues today with four surface mines currently in operation.

4.5.4 Karst Features in the Fort Dodge Formation Gypsum

The gypsum beds in the Gypsum Creek Member of the Fort Dodge Formation display features characteristic of karst development. Fractures in the gypsum have been enlarged by the movement of groundwater and sinkholes develop above areas where the gypsum was mined underground, especially where the joints intersect.

4.5.4.1 Solution-Enlarged Joints

The most striking feature of the upper surface of the gypsum, where it is exposed almost exclusively in quarries, is the presence of a pattern of deep channels (or slips as they are

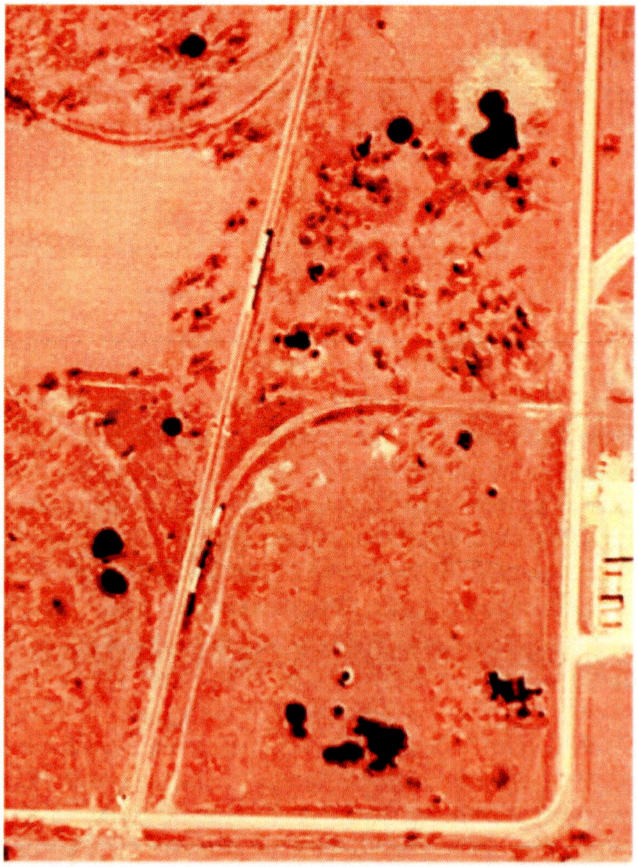

Fig. 4.27 Color infrared aerial photo showing sinkholes created by collapse into underground mines near Fort Dodge. The red color identifies healthy vegetation, the lighter lines are roads and railroads, and the dark areas are water-filled sinkholes

Fig. 4.28 Collapsed and partially collapsed mine drifts in the Gypsum Creek gypsum encountered during surface mining in the U.S. Gypsum Old Mill Meadows Mine (2010)

called by miners) extending downward into the gypsum (Fig. 4.26). These slips result from dissolution of the gypsum by groundwater moving along earlier fractures. Channels often >1 m wide may extend deep into the gypsum bed. According to miners, the channels are particularly well-developed at locations lacking a cover of Soldier Creek Member rocks, where the gypsum is directly in contact with Pleistocene glacial till. The slips are filled with these tills, either Pre-Illinoian (2.5 million-500,000-year-old) or Wisconsinan (40,000–10,500 years ago), although in some locations they may be filled by red Soldier Creek Member, sediments that have apparently slumped into the fractures. The orientation of the joints in the gypsum was studied by Hayes (1986) at three Fort Dodge gypsum quarries. Although the gypsum channels generally occur in intersecting parallel sets, he noted that the orientation of the enlarged fractures was essentially random when all three quarries are compared. He attributed the fractures and their orientation to the underlying bedrock topography and differential compaction of underlying sedimentary rocks by the weight of overlying ice during the multiple Pleistocene glacier advances.

4.5.4.2 Sinkholes

Sinkhole structures can be observed (Fig. 4.27) at several locations in the Fort Dodge area that are underlain by Gypsum Creek Member gypsum. These sinkholes are not totally natural features but are the product of groundwater moving through early gypsum mine drifts that were excavated by miners in the first half of the 1900s. The natural movement of groundwater through these gypsum tunnels dissolves wall rock, widening the drifts until they are no longer able to support the roof, and the tunnel collapses opening a sinkhole at the land surface. Locations where two natural joints intersect in the drift roof are especially susceptible to collapse. Groundwater movement along the joints weakened the gypsum, facilitating collapse. Once roof collapse begins it spreads outward creating solution chimneys which can stope through the overlying materials (Crawford et al. 2004), ultimately reaching the surface if the chimneys grow wide enough. Figure 4.29 shows photographs of such stoping as in cross section in the U.S. Gypsum Old Mill Meadows Quarry, and Fig. 4.30 is an aerial view of the U.S. Gypsum South Welles Quarry (after surficial materials had been removed to expose gypsum) clearly showing the slips and several chimneys. Subsequent surface mining activity has exposed and destroyed many miles of drifts, but many still remain, and sinkhole development is a recurring problem (Fig. 4.28).

4.6 Coldwater Cave System, Winneshiek County

Patricia N. Kambesis, Center for Human GeoEnvironmental Studies, Western Kentucky University, 1906 College Heights Boulevard, Bowling Green, Kentucky, 42101, USA. pat.kambesis@wku.edu.

Fig. 4.29 Dissolution causes piping in gypsum as viewed in the U.S. Gypsum Old Mill Meadows Quarry (from Crawford et al. 2004)

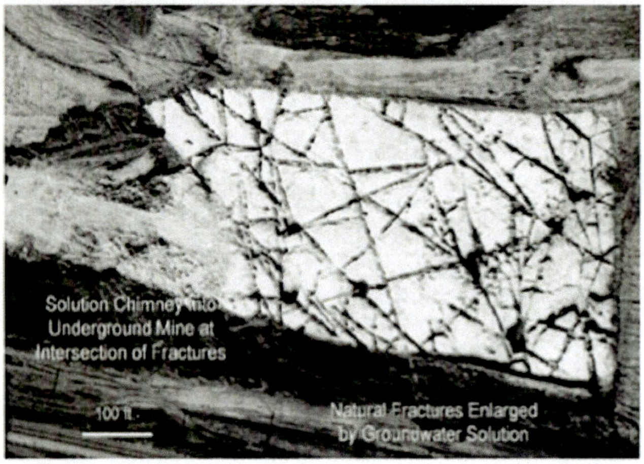

Fig. 4.30 Aerial photo of exposed gypsum showing dissolution enlargement of joints and piping at joint intersections in the U.S. Gypsum South Mills Quarry

4.6.1 Introduction

The Galena cuesta of the Paleozoic Plateau of northeastern Iowa contains some of region's most spectacular karst features and landscapes. The Coldwater Cave System is a prime example of the active hydrogeological systems that drain the cuesta. The cave system is the main drain of the Coldwater Cave groundwater basin, located within the Corn Belt region of the Upper Midwest in northeastern Winneshiek County, Iowa and southeastern Fillmore County, Minnesota, USA (Fig. 4.31). Coldwater Cave is the longest and one of the state's most significant caves. In 1987 it was designated as a National Natural Landmark by the U.S. Department of the Interior, a status accorded to geologic and ecologic features considered to be of national significance.

Fig. 4.31 Location map for Coldwater Cave area

The Coldwater Cave area is located within the "Driftless Area" of the Upper Midwest, (Ruhe 1969). Although not covered by Wisconsinan ice the area has been glaciated several times during the Pleistocene Epoch. The area is a stepped plateau with a dip to the southwest of less than one degree (Huppert et al. 1988). As the gradient of the Upper Iowa River increases on its route to the Mississippi River to the east, it erodes into the carbonate bedrock. Valleys are deeply incised, and massively bedded dolostones from successive cap rock layers which reflect the stepped plateau morphology (Huppert et al. 1988). The plateau displays a well-developed fluviokarst topography which has formed on middle to lower Paleozoic carbonates.

The Upper Iowa River is the major base-level drainage of the region and flows 60 km east to its confluence with the Mississippi River. As the Upper Iowa River approaches the Mississippi River, its gradient becomes steeper and it down-cuts into gently rolling uplands that are mantled with glacial till and loess. Agricultural land use predominates in the upland areas and also on the valley floors which are flat and alluviated. The steep slopes in between the uplands and valleys are forested.

The landscape above Coldwater Cave is characterized by deeply incised stream valleys, steep-sided bluffs, and a mantle of Quaternary-age glacial sediments. The valleys have cut into the relatively flat-lying massively bedded limestones and dolostones that form a successive caprock layer which reflects the stepped morphology of the Galena cuesta.

4.6.2 Geology

Northeastern Iowa and southeastern Minnesota are underlain by nearly flat-lying sedimentary rocks of Lower and Middle Paleozoic age. The strata were deposited in seas that covered the Hollandale Embayment, a shallow depression located between the Wisconsin Dome to the northeast and the Transcontinental Arch to the northwest (Austin 1972). The local bedrock consists of Cambrian, Ordovician, and Devonian sandstones, shales, limestones, and dolostones that were deposited in a series of transgressive and regressive cycles (Fig. 4.32). The lowermost units are predominantly sandstones with shale and carbonate beds. These strata grade upward into carbonate sequences containing subordinate sandstones and shales. The uppermost sedimentary sequences are composed entirely of carbonates (Sims and Morey 1972). Locally, the Ordovician-aged Galena Group directly underlies the land surface and in descending order consists of the Dubuque, Wiselake, Dunleith, and the basal unit of the Decorah Formation. The rock units form the Galena aquifer that is one of the major agricultural and domestic water sources for most of the region (Hallberg et al. 1983) and to a lesser degree still serves as a water source for some of the residents of the area (Friest 2003). The upper unit of the Decorah Formation is composed of calcareous shale which serves as an aquaclude that retards shallow groundwater from entering the deeper carbonate aquifers of the region (Fig. 4.33).

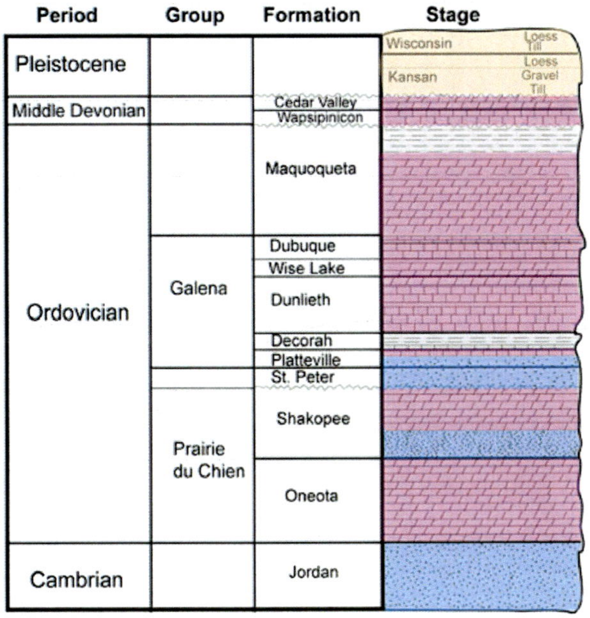

Fig. 4.32 Stratigraphy of the Coldwater area

4.6.3 Soils

Soils, derived mainly from Wisconsin-aged loess and glacial till of Kansan age, mantle the topography of the Coldwater Cave area (Ruhe 1969). Soil thickness varies from as much as 22 m in the northern part of the area to less than two meters on steeper slopes in the south.

There are two major soil types that cover most of the region (46%): Fayette soil and Downs series. The remainder (44%) of the area is made up of 14 different soil types, but no type accounts for more than 10%. These units consist of very deep, well-drained soils formed in loess and till and are found on convex crests, interfluves and side slopes, uplands, and treads and risers on high stream terraces. Slopes in Fayette soils range from 0 to 60% and those in the Down series have slope ranges from 0 to 25% (Kiel 2004). Both soil types are well-drained and surface runoff potential is negligible to high depending on the slope. In general, soil slopes range from gentle to steep with thin soils occurring on the hillsides (Kiel 2004). Valley bottoms contain thicker soils composed of alluvium and colluvium (Soil Conservation Service 1968). The native vegetation of Fayette soils is deciduous trees, mainly oak and hickory. The native vegetation of Downs soils is big bluestem, little bluestem, switchgrass, other grasses of the tallgrass prairie and widely spaced oak and hickory trees (NRCS 2004).

4.6.4 Fluviokarst

Coldwater Cave (Fig. 4.34) is located within a well-developed fluviokarst landscape that has formed as the result of the evolving interaction between fluvial and karst processes that are driven by climate. Formations of the Galena Group have a strongly developed system of joints and fissures that provide the impetus for karst development in the region (Hallberg et al. 1983). Sinkholes, swallets, stream sieves, caves, and springs characterize the fluviokarst landscape.

The surface water, which is now groundwater, solutionally enlarges joints, fissures, and bedding planes ultimately forming conduits that transport the groundwater through the aquifer to springs that discharge it to the surface at the contact between the Dunleith limestone and Decorah shale of the Galena Group. The groundwater discharged from the springs forms spring runs which flow into the Upper Iowa River.

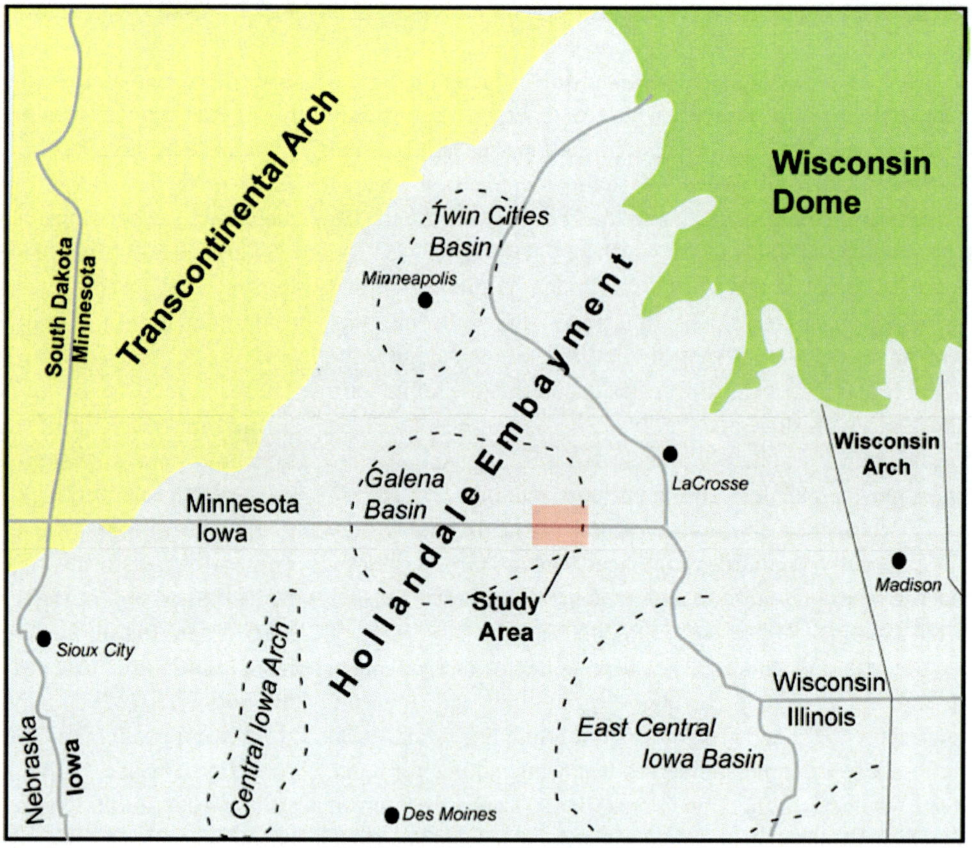

Fig. 4.33 Regional geology of the study area

4.6.5 Epikarst

The epikarst is the upper boundary of a karst landscape. This zone stores and directs percolating recharge waters to the underlying karst aquifers. Epikarst permeability decreases with depth below the surface. Epikarst is an important element of karst drainage basins (Williams 1983) and based on local observations and preliminary studies, plays a significant role in the Coldwater Cave area. Discharge versus precipitation documented in an 11-year study by the Iowa Department of Natural Resources (Bounk 1987) within the Coldwater Cave noted lag times between precipitation and changes in the underground stream level. Cave explorers reported that during heavy rain, it can take up to seven hours for a storm event to affect water levels within the cave system (Nelson 2002). These indicate that the downward flow of drainage is restricted during storm events and results in storage of water within the epikarst zone.

Epikarst flow is concentrated along a few major fissures. These are preferentially enlarged from the base of the epikarst zone downward and can form "hidden" vertical shafts (Klimchouk 1995, 1997). The shafts function as headwaters that drain the epikarst zone above into the conduit drainage system below. Epikarst dome features are common in the Coldwater Cave System and serve as points of recharge into the karst aquifer.

4.6.5.1 Hydrogeology

The Coldwater Cave region is part of the greater Upper Iowa River watershed, which is situated on a highly karstified landscape that is drained by the surface creeks and by underground conduit flow. During precipitation or freeze-thaw events, sinkholes in the study area also contribute recharge to the groundwater basin (Fig. 4.35). The limestone bedrock underlying the watersheds forms an extensive carbonate aquifer that is used for agricultural purposes as well as being a drinking water source for some residents in the area.

A series of dye tracing studies (Wheeler et al. 1988; Kambesis 2007) revealed that there are four karst groundwater basins in the Coldwater Cave region: the Coldwater Cave, Serendipity, East Pine Creek, and Silver Creek Basins (Fig. 4.36). The northern limit of all of four groundwater basins roughly corresponds with the surface topographic divide with the Root River drainage which is located near Harmony, Minnesota. The southern limit of all of the basins is the Upper Iowa River.

4.6.6 Karst Groundwater Basins

The Serendipity, East Pine Creek, and Silver Creek groundwater basins are distinctly separate from the

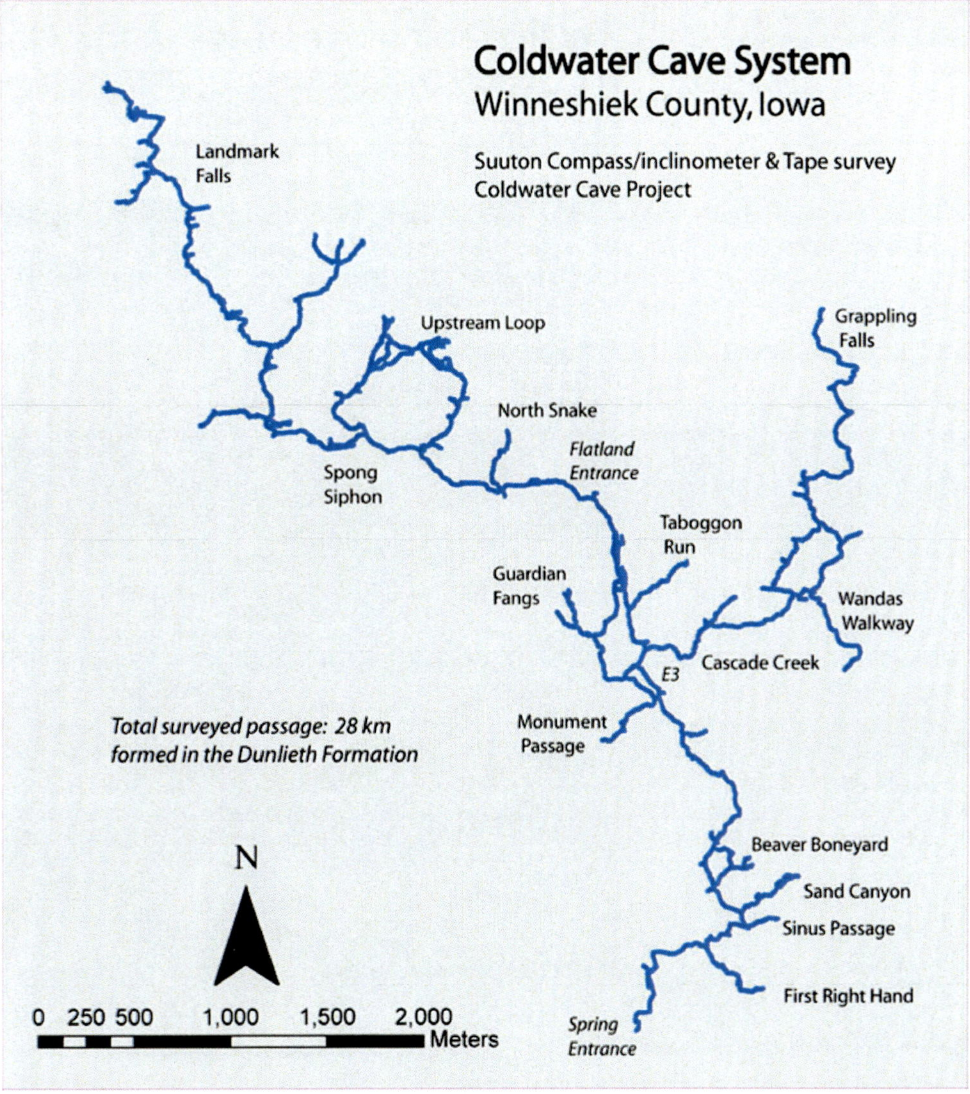

Fig. 4.34 Map of Coldwater Cave

Coldwater Cave groundwater basin. Tracer results during base-level flow and after precipitation events demonstrated that the underground flow routes can shift between adjacent groundwater basins or can change within a groundwater basin (Kambesis 2007).

The Coldwater Cave groundwater basin (Fig. 4.37) has an area of 80 km^2 and is the largest of the groundwater basins in the area. During base flow conditions, the upper, spring-fed reaches of Deer Creek, Coldwater Creek, and Pine Creek Proper sink (Fig. 4.38) into the subsurface and recharge the Coldwater Cave groundwater basin. The capacity of the three streams' various sinking reaches to accept water is exceeded during high flow periods. During such high flow periods part of the surface runoff remains on the surface and flows to the Upper Iowa River.

The subsurface route of the surface creeks is the Coldwater Cave System which consists of a series of conduits many of which are humanly enterable. The conduit series is composed of a master conduit (the main stream passage of the cave), a smaller parallel conduit (Wanda's Walkway—Fig. 4.34), and a series of minor conduits (infeeding side passages) which drain into the master conduit. During base flow conditions the main conduit and auxiliary infeeders of the Coldwater Cave System resurge at Coldwater Spring. The parallel conduit, Wanda's Walkway, resurges at Carolan Spring.

During high flow conditions, the Wanda's Walkway conduit overflows to the west across a subterranean drainage divide into the main stream passage via the infeeding conduits. Overflow also crosses a divide in the Monument Passage but the spring this overflow discharges to is currently unknown. Coldwater Spring discharges water from the main conduit of Coldwater Cave and Wanda's Walkway conduit (Carolan Spring).

The results of the three quantitative traces conducted in 1986 during high flow (Wheeler et al. 1988) and a quantitative trace done in 2003 (Kambesis 2007) during base flow, illustrate that the velocity of water from the surface injection

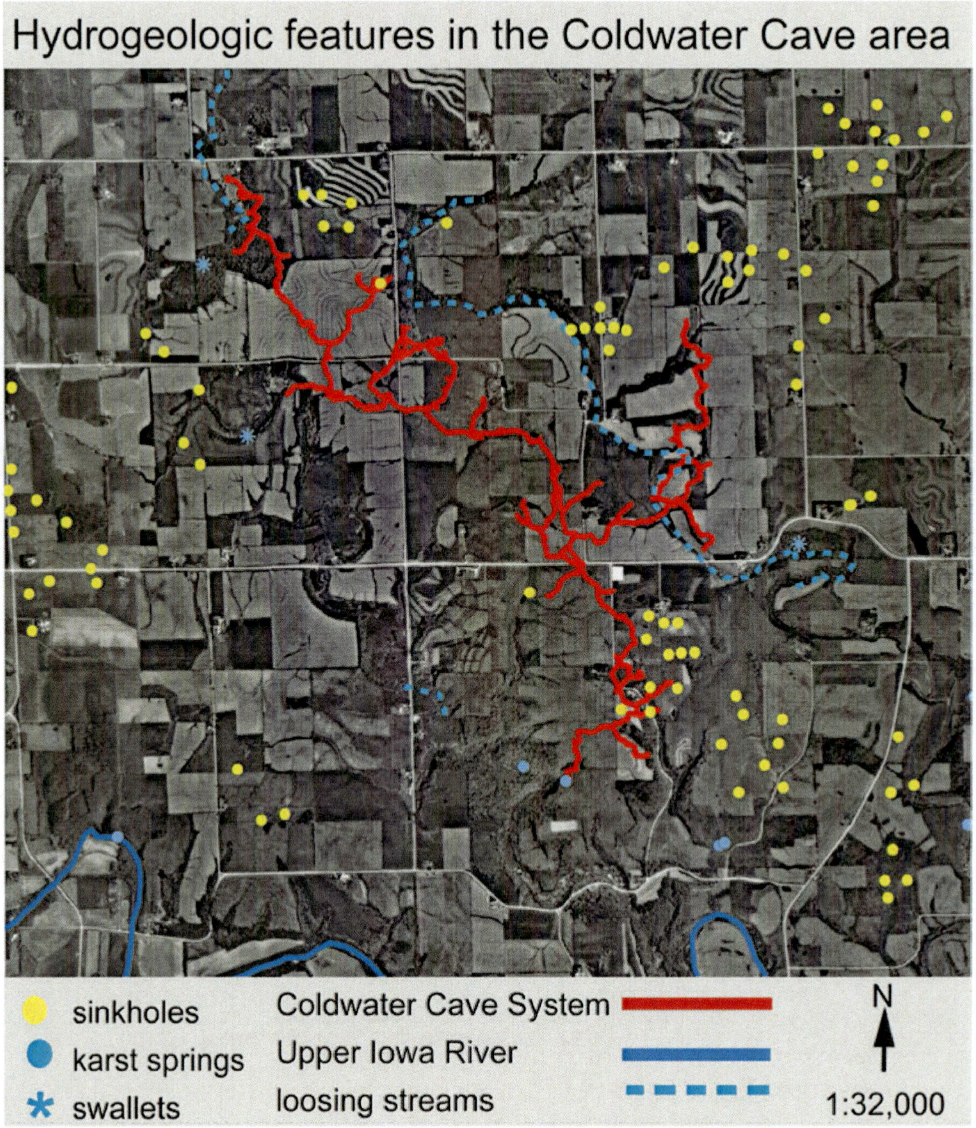

Fig. 4.35 Hydrogeologic features in the Coldwater Cave area

point to the detection site ranged from 600 m/h during high flow to 150 m/h during very low base flow.

The Serendipity groundwater basin drains 33 km^2 and is the westernmost groundwater basin identified in this study (Fig. 4.37). The basin is recharged by spring-fed Elliot Creek (Fig. 4.38) during base flow conditions. Dye trace results indicate that groundwater moves through the basin via conduit flow. Discharge is from Serendipity Spring during base flow and from Serendipity Spring and Coldwater Cave Spring during high flow conditions. The Serendipity groundwater basin abuts the Niagara Cave groundwater basin to the west.

The East Pine Creek groundwater basin (Fig. 4.36) covers an area of 39 km^2 and is located due east of the Coldwater Cave groundwater basin. It is recharged by a series of diffuse flow springs in its northern reaches and discharges from the Hoppin Spring series and Marlow Spring. More dye traces during base and high flow conditions will be necessary to better delineate the boundaries of this drainage basin.

The Silver Creek groundwater basin (Fig. 4.36) is the easternmost one, with an area of at least 28 km^2. It is drained via conduit flow by Rimstone River Spring. The cave system associated with the spring contains over a kilometer of humanly enterable conduit. It is possible that the cave system is recharged through an unidentified flow route that originates in the northern section of the East Pine Creek drainage or it could also be part of the Casey Spring Creek surface watershed.

4.6.7 Recharge of the Coldwater Cave Groundwater Basin

The Coldwater Cave groundwater basin is mantled with Quaternary glacial sediments. In the study area, the

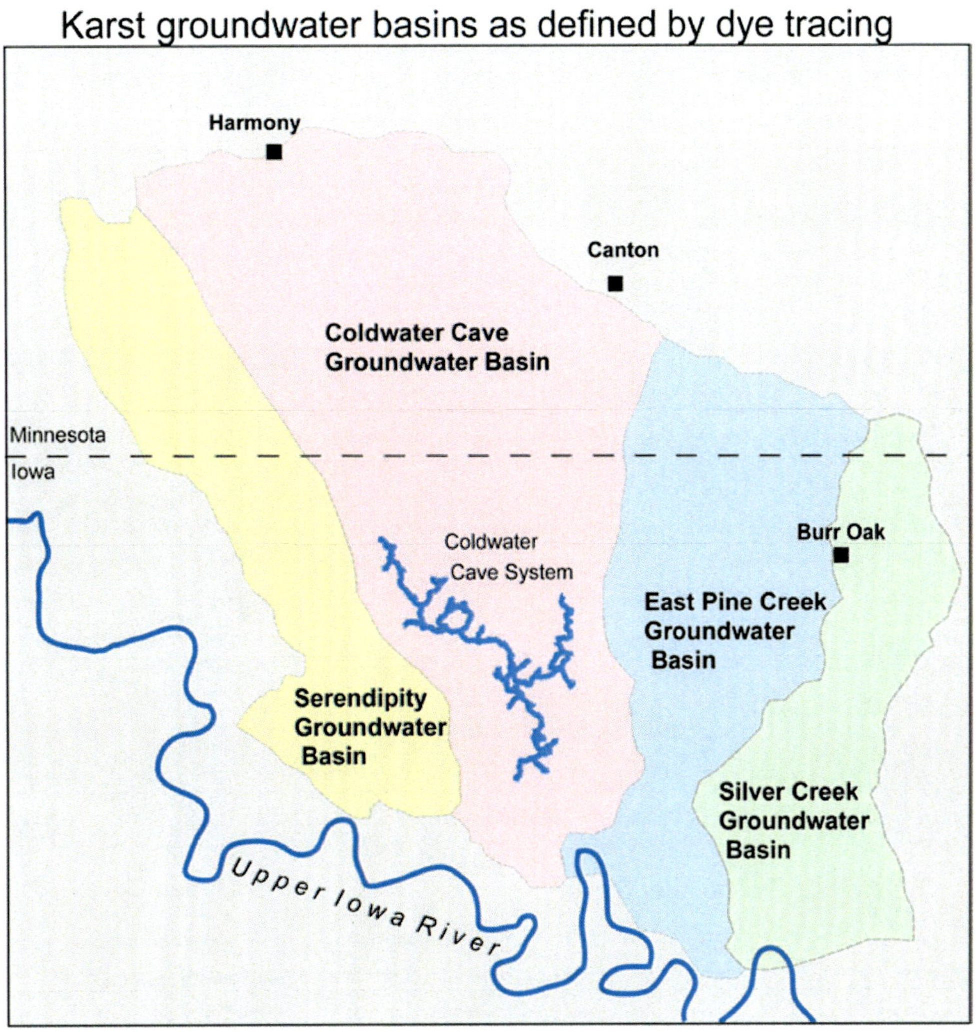

Fig. 4.36 Karst groundwater basins as defined by dye tracing

sediments range in thickness from 15 to 22 m in the northern reaches of the groundwater basin and thin to as little as two meters or less to the southwest (Figs. 4.39 and 4.47). The mode of recharge within the basin is a function of the thickness of the Quaternary sediment that mantles it.

In the northern part of the basin where Quaternary sediment thicknesses are greater than 15 m (Fig. 4.47), the basin displays both diffuse allogenic and concentrated allogenic recharge (Fig. 4.40). An example of the former is the series of diffuse flow springs and wetlands which drain a glacial till/loess perched aquifer and serve to recharge the surface streams. The latter is the surface streams that flow across the mantled bedrock and sink when the sediment cover thins.

Where the Quaternary sediments thin to eight meters or less, the allogenic streams sink into the carbonate bedrock and autogenic recharge, both concentrated and diffuse, becomes the dominant mode of recharge. Concentrated autogenic recharge occurs as precipitation that falls on the basin flows into bedrock fractures, swallets (Fig. 4.41a) and sinkholes (Fig. 4.41b) and into stream sieves, coalescing in the epikarst. The concentrated autogenic recharge has solutionally enlarged many of the fissures in the epikarst, forming vertical shafts (Fig. 4.42). Because the shafts were discovered and explored from within the cave system, they will be referred to from here on as epikarst domes. These features are common in the Coldwater Cave system where 102 epikarst domes have been documented (Fig. 4.43). Epikarst domes ranged in floor diameter from one to 10 m and some reach heights of 20 m or more. The domes function as headwaters that drain the epikarst above into the conduit system below. There are three perennial waterfall domes located within the cave system that actively drain the epikarst. Most of the other domes also contribute recharge into the cave system though the amount is a function of climatic conditions. All of the epikarst domes display solutional features such as flutes and rills which indicate that the waters forming them are undersaturated with respect to carbonate.

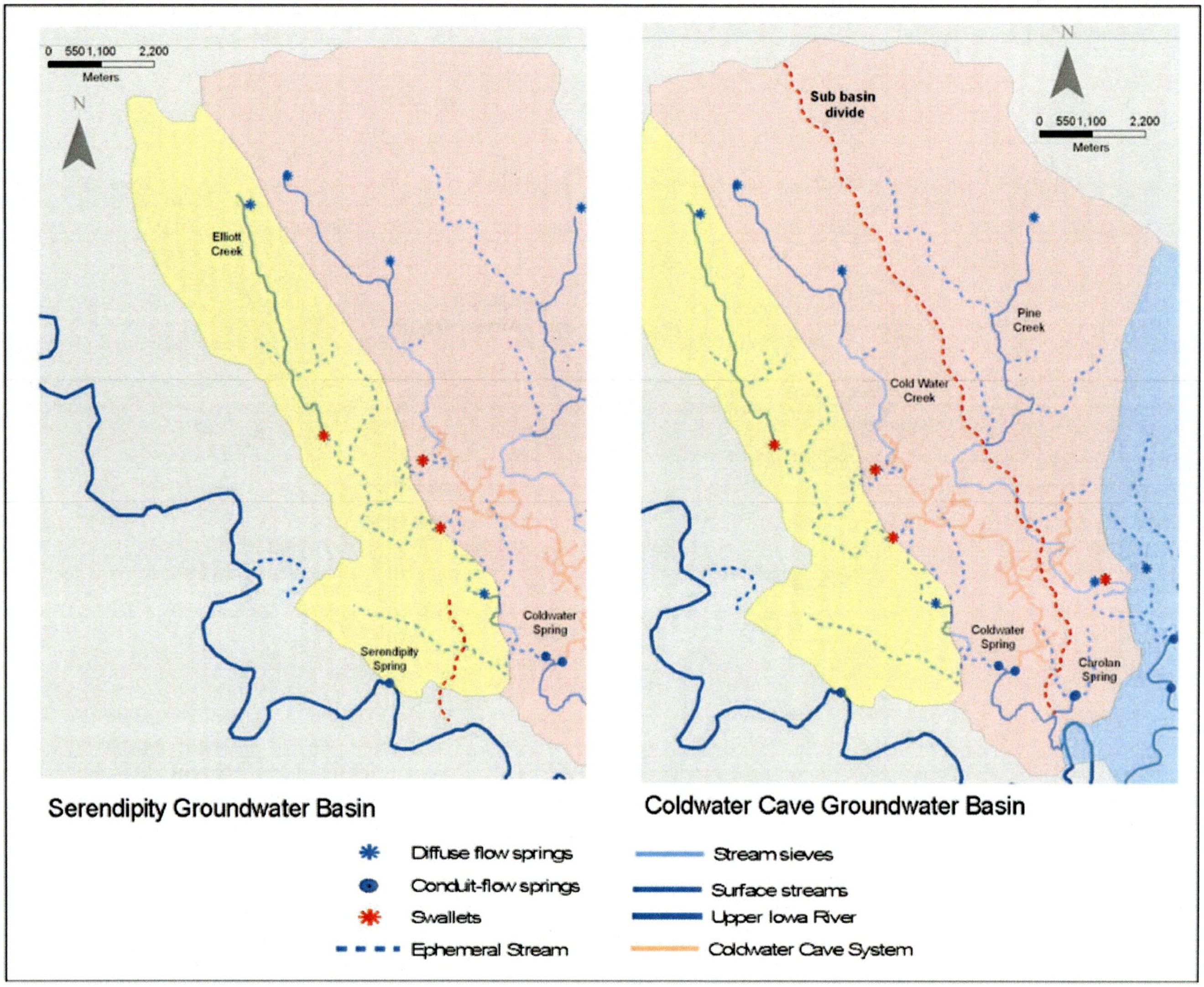

Fig. 4.37 Coldwater Cave and Serendipity groundwater basins

Diffuse autogenic recharge occurs when precipitation falls on the basin and slowly percolates through the soil and small openings, making its way through the epikarst and eventually to the conduit system below. This type of recharge has a longer residence time in the epikarst and as a consequence becomes supersaturated with respect to carbonate. The results of this type of flow are evident in many of the solutionally enlarged epikarst domes which also contain speleothems and whose walls are coated with flowstone. Speleothem development is also evident on the ceilings, walls, and on the floor of sections of the main stream passage and in many of the infeeders (Fig. 4.44).

4.6.8 Spring Discharge of Coldwater Cave Groundwater Basin

The base-level stratigraphic unit of the Coldwater Cave aquifer is the Decorah shale which serves as an aquitard. The underground streams flow at or near the contact of the shale with the Wiselake Formation until valley erosion truncates the cave passage and water discharges to the surface as springs. The surface discharge is carried in spring runs that join the Upper Iowa River.

The Coldwater Cave groundwater basin is drained by Coldwater Spring (Fig. 4.45a) and Carolan Spring (Fig. 4.45b) which are perennial. Base flow discharge at

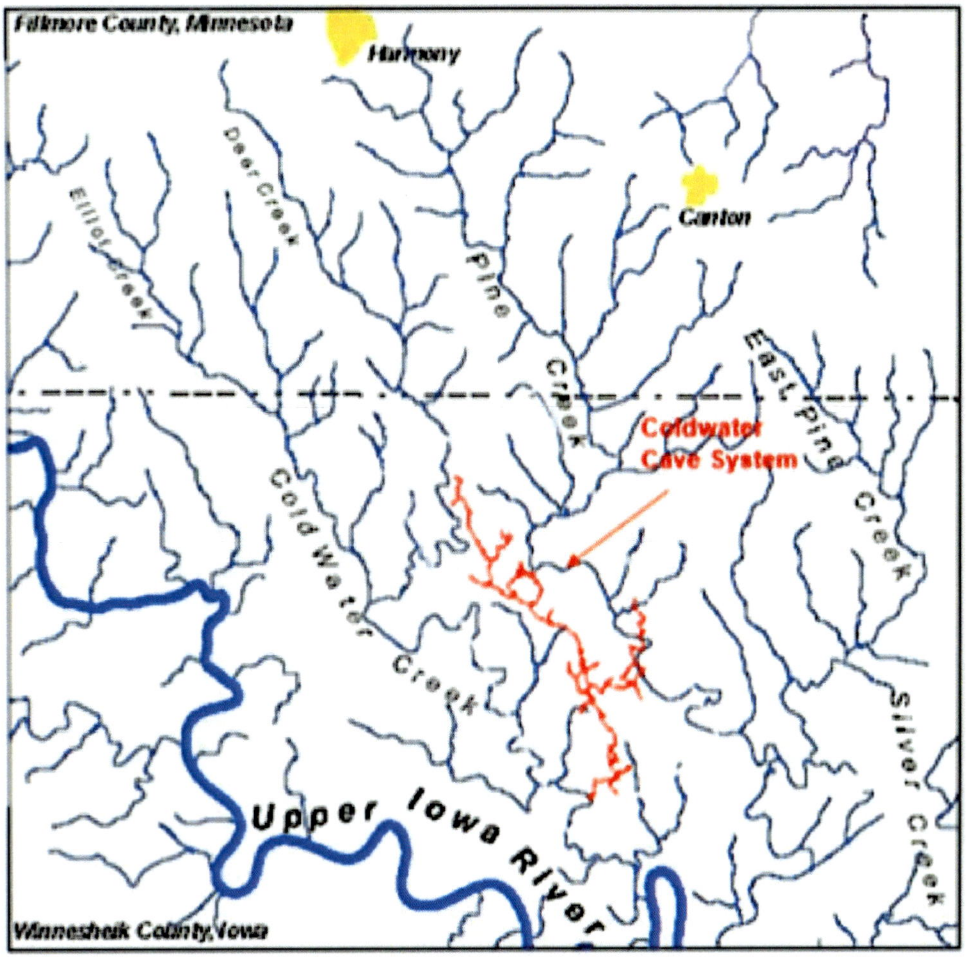

Fig. 4.38 Streams of the Coldwater Cave area

Coldwater Spring is 549 L/s and at Carolan Spring is 150 L/s. A rise pool located 400 m northwest of Coldwater Spring serves as overflow drainage. During high flow conditions an overflow spring begins to flow from a small outlet 50 m north of Carolan Spring. Two paleosprings suggest that flow routes of the groundwater basin have changed over time.

Though the stream resurging from a karst spring would be classified, by the Strahler stream order method (Strahler 1952), as a first order stream, the fact that underground streams can have many tributaries means that it is not unusual for karst resurgence springs to be a second order or higher stream. Coldwater Spring is at least a third order stream and Carolan Spring is at least a second order stream.

4.6.9 Relationship of Surface and Subsurface Meteorology

Cave air and water temperatures within a fluviokarst cave system are affected by airflow from open surface entrances, and recharge. Cave airflow in this setting is a function of air exchange with the surface environment driven by barometric fluctuations. Variations in cave stream temperature from recharge can cause significant fluctuations in cave air temperature. Coldwater Cave has two spring entrances that are water-filled and two air-tight man-made shafts, i.e., there are no open-air entrances to the cave. Initial studies of water temperature of the cave spring indicated an average reading of 8.9 °C and consequently cave temperatures in the deep

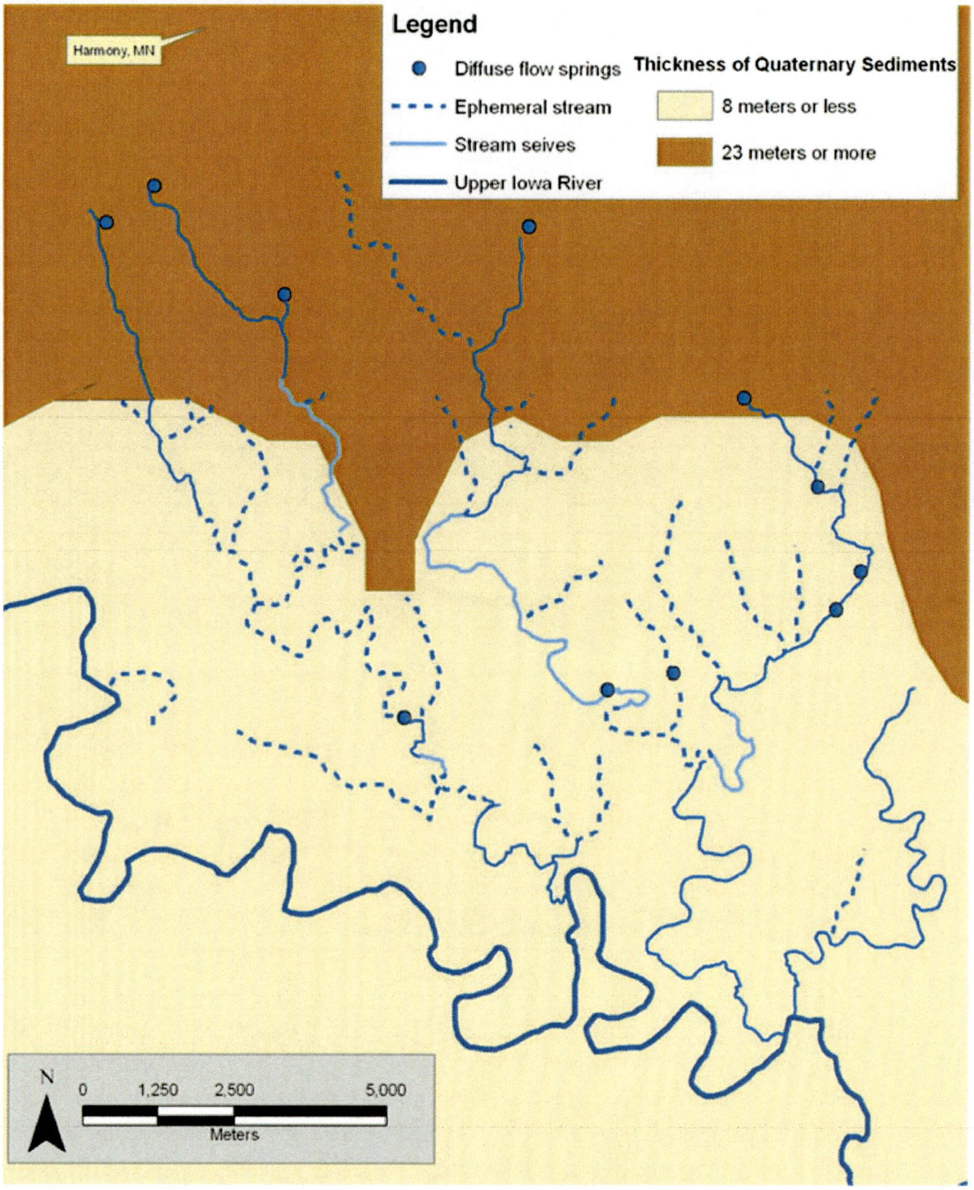

Fig. 4.39 Thickness of glacial deposits in the Coldwater Cave area

cave environment were assumed to be stable throughout the year (Koch and Case 1974).

A cave meteorology study conducted at Coldwater Cave revealed fluctuations of water temperature to be as much as 9° above and seven degrees below the mean annual temperature of the spring water and five degrees above and three degrees below average air temperature (Kambesis et al. 2013). Significant ranges in subsurface water and air temperatures were documented over an eight-year period (Fig. 4.46).

The study found that cave air and water and surface temperatures were highest from April to October, lowest from November to January, and fluctuate above and below freezing in February and March. The period between March and October has the most precipitation which is reflected in the spiky nature of the cave stream temperature graphs during that time span. Even the North Snake Passage, which shows the least effect from surface temperature changes, displays a little noise in the March through October time period. Though the period between November and March did not receive storm event precipitation, the cave temperature graphs showed major fluctuations during this time period because of the diurnal freeze-thaw affect. There are significant temperature variations between different parts of the cave system. The cave air and streams that showed significant temperature ranges had headwaters that were

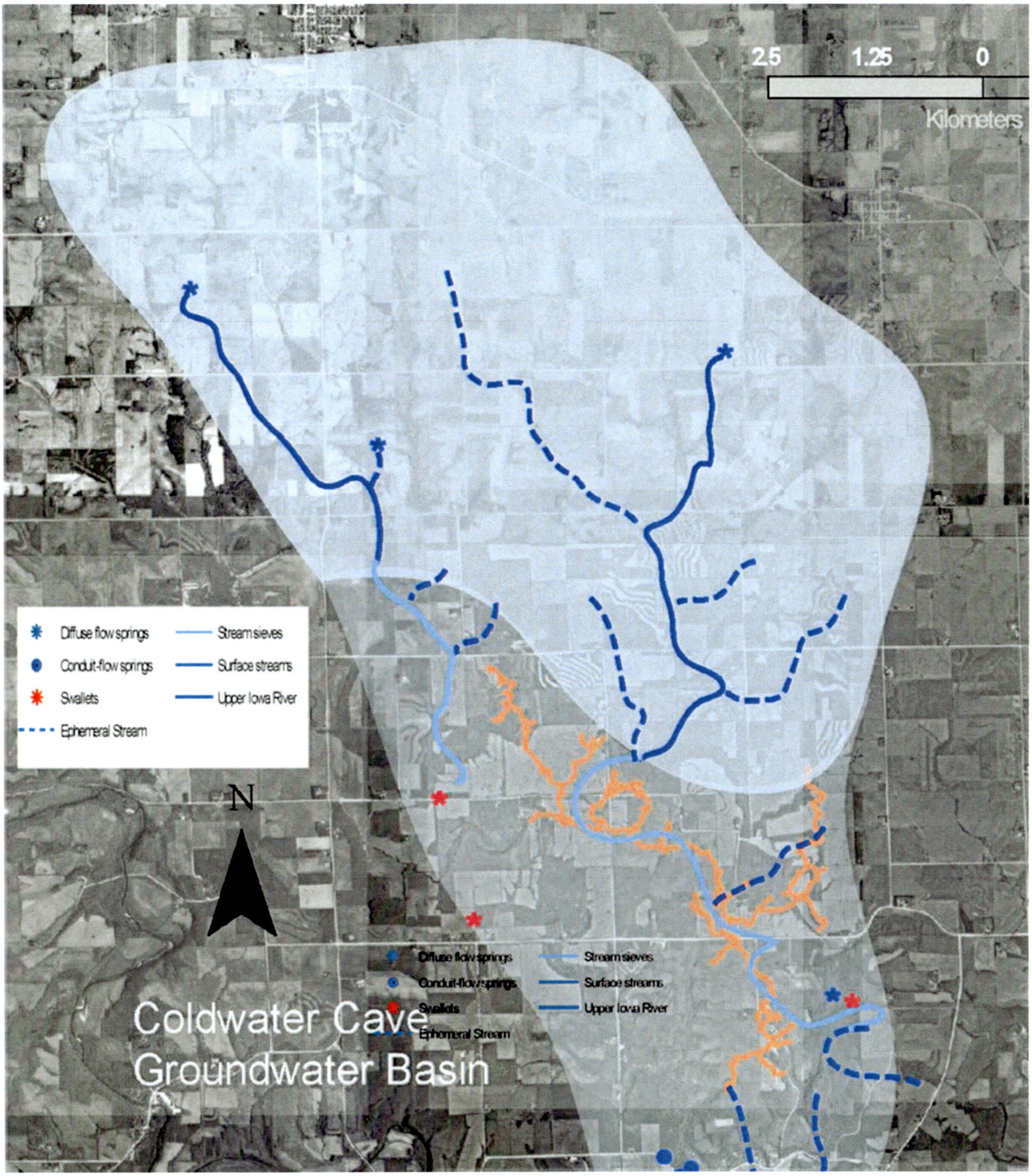

Fig. 4.40 Allogenic (light grey) and autogenic (dark grey) recharge in the Coldwater Cave groundwater basin

Fig. 4.41 **A** Stream swallet (Photo by John Lovaas). **B** Sinkhole taking water during recharge event (Photo by M. Lace)

Fig. 4.42 Epikarstic dome in Coldwater Cave (Photo by Scott Dankof)

located directly under or within 200 m of surface streams. Surface waters enter and move through the system at high velocities such that they do not equilibrate with the ambient underground temperature. As a consequence, the springs that discharge the cave stream water also display significant ranges in temperature (Fig. 4.46).

4.6.10 Cave Description

Coldwater Cave (Fig. 4.35) is developed in the Ordovician-age Dunleith Formation of the Galena Group and displays a curvilinear branchwork pattern in its layout which reflects many points of surface recharge. The curvilinear character is indicative of bedding plane control on passage development however joint control is evident in many of the side passages.

Coldwater Cave consists of seven km of main stream passage, nearly two km of parallel passages, and another 18 km of infeeders. The cave is developed within a subtle carbonate ridge bounded by surface drainages; some of the side passages cross under these drainages. The Dubuque Formation serves as the ridge cap rock under which the cave system has formed. Coldwater Spring is the only known natural entrance to the system and issues from the base of a 30-m-tall bluff located within the Cold Water Creek Conservation Area. Access to the historic entrance requires SCUBA and the underwater entrance is currently gated. There are also two additional active springs and two paleo springs all of which are not humanly enterable. Primary access to the cave is through a 29-m shaft (Flatland entrance)

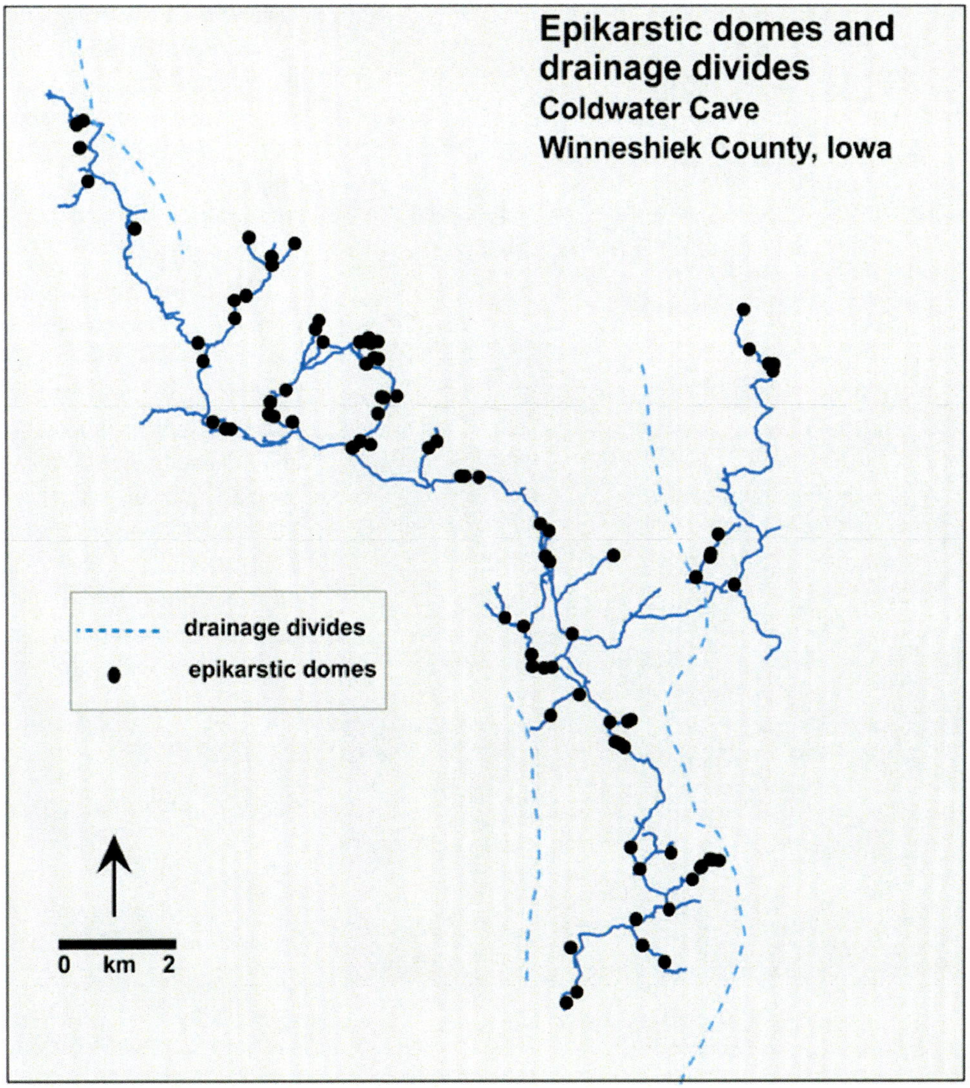

Fig. 4.43 Epikarstic domes and drainage divides in Coldwater Cave

that was drilled by the State of Iowa for researcher access in the early 1970s. A second privately owned shaft entrance was drilled in 2003 and is located approximately two kilometers downstream from the Flatland entrance.

Three of the epikarst domes contain perennial waterfalls that recharge the system throughout the year. Thunder dome feeds the Upstream Loop in the northeastern section of the cave. Landmark Falls provides recharge into the northern area of the cave system. Grappling Falls Dome recharges the southeastern section of the cave (Fig. 4.34).

The epikarst domes recharge the cave system during base flow conditions, in storm events, and during snowmelt. When base flow is at its lowest, the domes contribute diminished recharge into the main conduit system. As recharge levels increase, more waterfalls become active.

There are four drainage divides within the cave that have been identified by dye tracing and direct observation (Fig. 4.43). A drainage divide located in the northernmost reaches of the cave system beyond the upstream sumps, bifurcates approximately 100 m from the main conduit. One arm of the cave stream flows south/southwest into passages that are too small for human traverse, and the other arm flows to the east/southeast into the main conduit.

Another bifurcation in flow was observed in several of the side passages in the Cascade Creek area (Fig. 4.43). Streamflow bifurcates in these areas with one branch flowing south in a tributary that parallels the main conduit and the second flows southwest into the main conduit. A third bifurcation is stage-driven and is located in Sand Canyon in the extreme downstream section of the cave. Sand Canyon

Fig. 4.44 Speleothem development in main stream passage, Coldwater Cave (photo by Scott Dankof)

usually drains west into the main stream passage but during storm events or freeze-thaw events when water levels rise, the stream bifurcates and flows west into the main stream passage and to the east away from the main conduit. The east and southeast flowing stream bifurcations are related to the conduits on the east side of the cave which drain to Carolan Spring. The west and southwest flowing streams all drain to Coldwater Spring.

The fourth divide is located in the Monument Passage (Fig. 4.43). This passage is one of only two passages that have significant development due west of the main stream conduit in the downstream portion of the cave. Perennial epikarst flow is minimal and moves away from the main conduit and to the south/southwest during base flow. In flood events the main stream passage can back flood into the Monument Passage. The back-flooded water will flow into a small conduit that trends south-southwest. The destination of the water flow is currently unknown. The Monument Passage may be part of an older conduit complex that includes the Cascade-Well pipe area passages that have been intersected by the currently active conduit.

4.6.11 Exploration History

Coldwater Cave was discovered in 1967 by cave explorers who were investigating the cave potential of the many springs in the area. Their successful dive of Coldwater Spring resulted in one of the most significant speleological discoveries of the UMV. They explored and did a compass and pace survey of over five kilometers of sizeable stream

Fig. 4.45 A Coldwater Cave spring discharge of the Coldwater Cave groundwater basin (Photo by P. Kambesis). B Carolan Spring discharge of the Coldwater Cave groundwater basin (Photo by P. Kambesis)

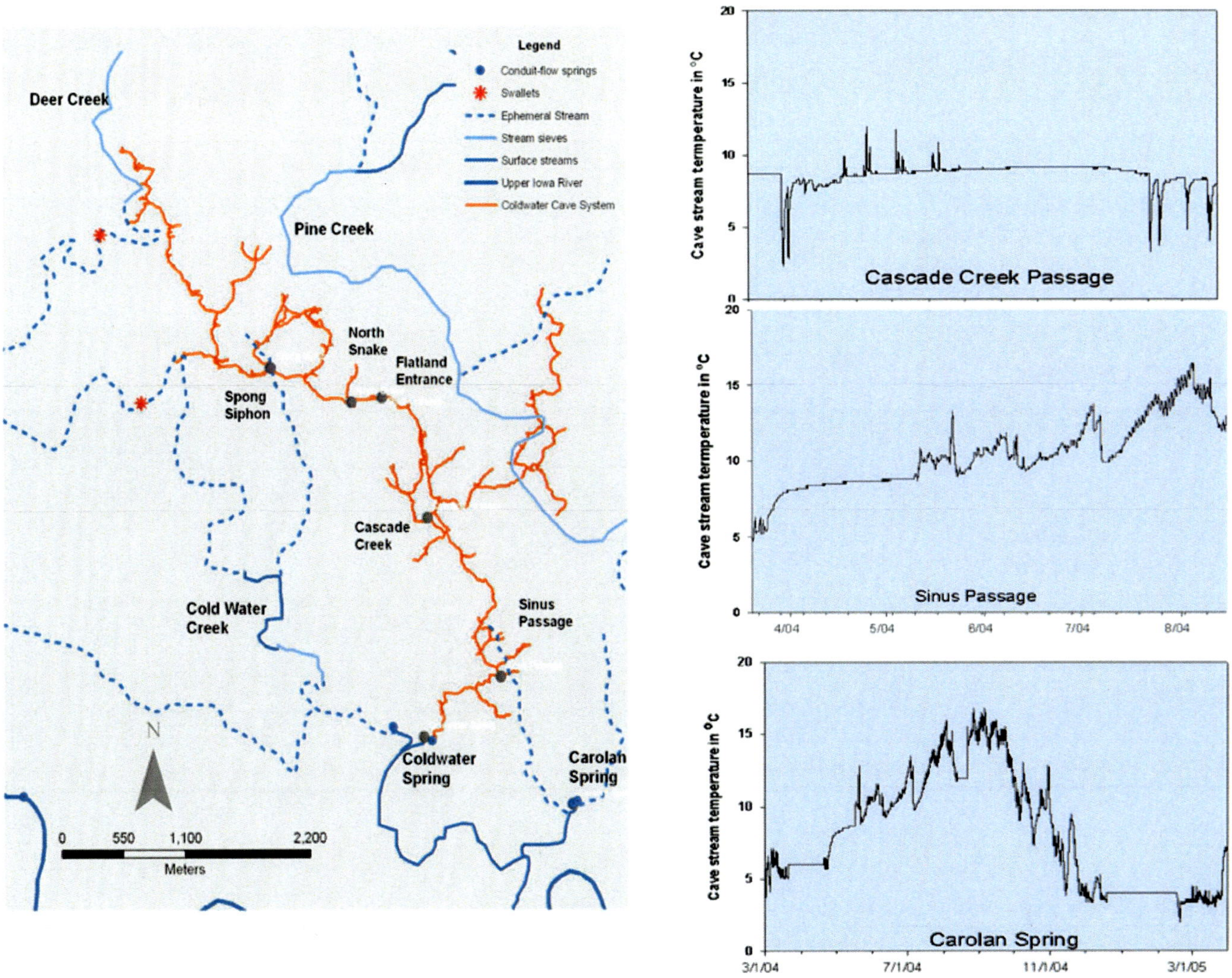

Fig. 4.46 Variations of in cave temperatures—Coldwater Cave

passage (Fig. 4.35). See Kambesis and Bain (1988) for more history of the early exploration of Coldwater Cave. After their initial explorations and survey, the cavers brought their discovery to the State of Iowa with the hope that the state would take measures to preserve the cave.

The State of Iowa did conduct a two-year study of the cave in order to determine if development was feasible for tourism (Koch and Case 1974). In order to facilitate the studies, they drilled a 29-m entrance shaft, installed a wooden platform at the base of the shaft, and constructed a metal building over the top. The building currently serves as a field house and research station.

The results of the State's study indicated that developing the cave would not be cost-effective considering the cave's remote location from tourist traffic. In 1974, the state lease on the cave expired and the landowners took over the responsibility of overseeing the cave. Their willingness and desire to open the cave for exploration, survey and study instigated the establishment of the Coldwater Cave Project.

Cave exploration and mapping have continued from 1976 to the present. Coldwater Project cavers continue to extend the limits of known cave. Mapping trips to many small, miserable side passages slowly increased total survey footage. Digging projects were initiated in downstream infeeders. As

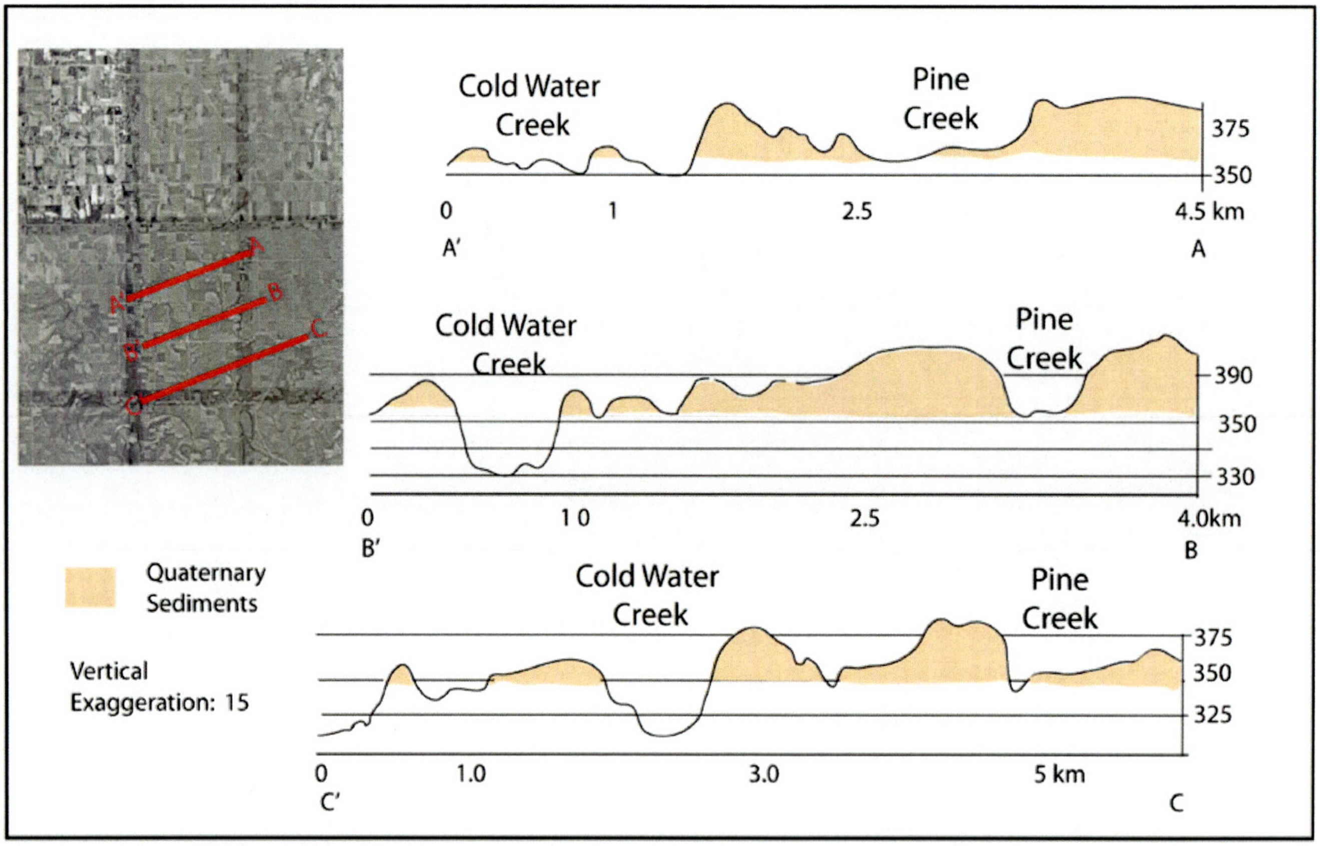

Fig. 4.47 Sediment profiles of the Coldwater Cave area

explorations proceeded, many epikarst domes were documented, some with promising climbing leads. However, attempts at reaching an upper level via any of these leads were unsuccessful, not in terms of the climbing attempts, but because the cave did not "cooperate," i.e., there is no continuous and extensive upper level in Coldwater Cave.

In the early 1980s, Coldwater Project cavers succeeded in breaching a drainage divide in the Cascade Creek Passage and getting into Wanda's Walkway, a stream trend that paralleled Coldwater's main passage. A major infeeder, Grappling Falls, was climbed and passage explored and mapped to a breakdown terminus. Dye tracing confirmed that Carolan Spring drained Wanda's Walkway.

Multiple attempts had been made at cracking the upstream sumps with SCUBA and, in mid-winter, without. During very dry years, the latter was successful and survey teams were able to get beyond low-air passages into going stream canyons. Many kilometers of cave has been explored and mapped beyond the upstream sumps through the years. Exploration and survey continues in the Coldwater Cave System and the cave's surveyed length currently stands at 28 km.

Acknowledgments Significant contributions to this chapter were derived from the digital archives of the Coldwater Cave Project and the Iowa Grotto Library (National Speleological Society). The authors also wish to extend thanks to public land managers and private cave owners who have long supported cave research and preservation in Iowa.

References

Aley T (1981) Evaluation of Coldwater Cave, Winneshiek County, Iowa for designation as a National Natural Landmark. Ozark Underground Laboratory Report prepared for the Rocky Mountain Region, National Park Service, 41p

Anderson RR (ed) (2000) Natural history of Pike's Peak State Park. In: Geological Society of Iowa Guidebook, no 70, pp 103–105

Anderson RR, McKay RM (eds) (1999) The geology of the Jurassic Fort Dodge Formation, Webster County, Iowa. In: Geological Society of Iowa Guidebook, vol 67, 89 p

Austin GS (1972) Paleozoic lithostratigraphy of southeastern Minnesota. In: Sims, PK and Morey, GB (eds) Geology of Minnesota: a centennial volume. Minnesota Geological Survey, p 459–473

Baker RG, Bettis EA, Denniston RF, Gonzalez LA (2001) Plant remains, alluvial chronology, and cave speleothem isotopes indicate abrupt Holocene climatic change at 6 ka in midwestern USA. Glob Planet Change 28(1):285–291

Bounk MJ (1983a) Karstification of the Silurian Escarpment in Fayette County, Northeastern Iowa. GSI Guidebook 40:31p

Bounk MJ (1983b) Some factors influencing phreatic cave development in the Silurian strata of Iowa. Proc Iowa Acad Sci 90(1):19–25

Bounk MJ (1984) Development and evolution of the Maquoketa Cavern system. In: Anderson R (ed) The natural history of Maquoketa Caves State Park. Geological Society of Iowa Guidebook, Jackson County, Iowa, pp 23–29

Bounk MJ (1987) Stream level record from Cold Water Cave, Winneshiek County, Iowa, and its relationship to rainfall and stream flow volume. In: Proceedings of the geological society of America, North-central Section, Annual Meeting, p 25–27

Bounk MJ, Bettis EA III (1984) Karst development in northeastern Iowa. Proc Iowa Acad Sci 91(1):12–15

Brenner RL (1983) Late Jurassic tectonic setting and paleogeography of Western Interior North America. In: Reynolds RM, Dolly ED (eds) Mesozoic Paleogeography of the West-Central United States. Rocky Mountain Section, Society of Economic Paleontologists and Mineralogists, Denver, Colorado, pp 119–132

Bretz JH (1938) Caves in the Galena formation. J Geol 46(6):828–841

Bretz JH (1942) Vadose and phreatic features of limestone caverns. J Geol 50(6):675–811

Brick GA (2004) Iowa Underground: A Guide to the State's Subterranean Treasures. Trails Books, Black Earth, WI, 223 p

Calvin S (1896) Apparent anomalies of stratification of Postville well. American Geologist, XVII, pp 195–203

Calvin S, Leonard AG, Bain HF (1897) Geology of the Dubuque S. Quadrangle IA-IL. Iowa Geological Survey, vol. VI: Report on Lead-Zinc District and Artesian Wells (Fig. 2, p 30)

Christiansen K, Wilson M, Tecklin J (1961) Preliminary study of the microarthropod ecology of Hunter's Cave. Bull Natl Speleolo Soc 23(2):63–70

Clark WR, Henry CJ, Dettman CL (2008) Demographic processes influencing population viability of the Iowa Pleistocene snail (*Discus macclintocki*). Am Midland Natl 160(1):129–139

Cody RD, Anderson RR, McKay RM (1996) Geology of the Fort Dodge formation (Upper Jurassic) Webster County, Iowa. In: Cody RD, Anderson RR, and McKay RM (eds), Geology of the Fort Dodge Formation, Iowa Geological Survey Bureau Guidebook Series, no 19, p 3–24

Cooper MP, Mylroie JE (eds) (2015) Glaciation and speleogenesis: interpretations from the Northeastern United States. Springer, Dordecht, p 142p

Cottrell TR, Strode KS (2005) Summer soil temperatures on algific talus slopes in Northeast Iowa. J Iowa Acad Sci 112(1–2):13–16

Crawford J, Blazic J, Benzie M, Price K (2004) Subsidence management at New Mineral City Mine, Fort Dodge. Report presented to Interstate Technical Group on Abandoned Underground Mines 5th Biennial Workshop, Tucson, AZ

Cross AT (1966) Palynologic evidence of mid-Mesozoic age of Fort Dodge (Iowa) gypsum. In: Geological Society of America, Annual Meeting, Abstract, Special Paper 101, p 46

Denniston RF, Gonzalez LA, Semken HA Jr (1999) Integrating stalagmite, vertebrate and pollen sequences to investigate Holocene climate change in the Southern Midwestern United States. Quatern Res 52:381–387

Dixon JW (2011) The role of small caves as bat hibernacula in Iowa. J Cave and Karst Stud 73(1):21–27

Dockal JA (2020) Lead-Zinc Crevices of Iowa, Illinois, and Wisconsin. Chapter 7, this volume

Dorale JA, Gonzalez LA, Reagan MK, Pickett DA, Murrell MT, Baker RG (1992) A high resolution record of Holocene climate change in speleothem calcite from Cold Water Cave, Northeast Iowa. Science 258:1626–1630

Friest L (2003) Personal communication

Garvin PL (1983) Sulfide mineralization at Mineral Creek Mines, Allamakee County, Iowa. Proc Iowa Acad Sci 90(2), Article 4

Garvin PL (1995) Paleokarst and associated mineralization at the Linwood Mine, Scott County Iowa. J Iowa Acad Sci 102(1–2):1–21

Garvin PL, Ludvigson GA (1993) Epigenetic sulfide mineralization associated with Pennsylvanian paleokarst in eastern Iowa, USA. Chem Geol 105(4):271–290

Gilbert CC (1989) Fence Cave local fauna: changing historic land use patterns interpreted via faunal distributions in Northeast Iowa. Dissertation, University of Iowa

Griffith GE, Omernik JM, Wilton TF, Pierson SM (1994) Ecoregions and subregions of Iowa: a framework for water quality assessment and management. J Iowa Acad Sci 101(1):5–13

Hale WE (1955) Geology and groundwater resources of Webster County, Iowa. Iowa Geological Survey Water Supply Bulletin No. 4, 257 p

Hallberg GR, Bettis EA III, Prior JC (1984) Geologic overview of the paleozoic plateau region of Northeastern Iowa. Proc Iowa Acad Sci 91(1):5–11

Hallberg GR, Hoyer BE (1982) Sinkholes, hydrogeology, and ground-water quality in northeast Iowa, vol 82, no 3, Iowa Geological Survey, Iowa City

Hallberg GR, Hoyer BE, Bettis EA III, Libra RD (1983) Hydrogeology, water quality and land management in the Big Spring Basin. Clayton County, Iowa, Iowa Geological Survey, Iowa City, IA, pp 118–120

Hallberg GR, Libra RD, Hoyer BE (1985) Nonpoint source contamination of ground water in karst carbonate aquifers. Perspectives on Nonpoint Source Pollution Proceedings, St. Louis, MO, EPA 440/5-85-100, Washington, D.C., Office of Water Regulations and Standards, U.S. Environmental Protection Agency, p 109–114

Hansel AK (1976) Mechanically induced karst development along the Silurian Escarpment in northeastern Iowa. Doctoral dissertation, University of Northern Iowa

Harmon RS, Schwarcz HP, Ford DC, Koch DL (1979) An Isotopic Paleotemperature Record for Later Wisconsinan Time in Northeast Iowa. Geology 7:430–433

Hayes DT (1986) Origin of fracture patterns and insoluble minerals in the Fort Dodge gypsum, Webster County, Iowa: MS thesis. Iowa State University, Ames, p 76

Hedges J (1967) Geology of Duttons Cave, Iowa: Bull. Nat. Speleo. Soc 29(3):73–90

Hedges J (1972) Expanded joints and other periglacial phenomena along the Niagara Escarpment. Builtyn Peryglac 21:87–126

Hedges J (1974a) Areal geology of the Upper Mississippi Valley Cave Region. National Speleological Society 1974 Convention Guidebook, Decorah, Iowa: National Speleological Society and Iowa Grotto, p 27–33

Hedges J (1974b) Commercial Caves. National Speleological Society 1974 Convention Guidebook, Decorah, Iowa: National Speleological Society and Iowa Grotto, p 93–118

Hedges J, Alexander EC Jr (1985) Karst-related features of the Upper Mississippi Valley Region. Stud Speleol VI:41–49

Hedges J, Knudson G (1975) The Ice Cave at Decorah, Iowa. Annal Iowa 43:113–131

Heitmann N (1980) Water source of Big Spring Trout Hatchery, Clayton County, Iowa. Proc Iowa Acad Sci 87(4):143–147

Heyl AV Jr, Agnew AF, Lyons EJ, Behre CH Jr (1959) The geology of the Upper Mississippi Valley Zinc-Lead District. U.S Geological Survey Professional Paper 309, 310 p

Heyl AV, West WS (1982) Outlying mineral occurrences related to the Upper Mississippi Valley mineral district, Wisconsin, Iowa, Illinois, and Minnesota. Econ Geol 77(8):1803–1817

Horick PJ, Soenksen PJ (1989) Water resources of northeast Iowa. Iowa Department of Natural Resources Water Atlas, No. 8

Howard AD (1960) Geology and origin of the crevice caves of the Iowa, Illinois, and Wisconsin lead-zinc district. J. Yale Speleol. Soc 2(4):61–95

Howard AD (1963) The development of karst features. Bull Natl Spel Soc 25:45–65

Hruby CE, Wolter CF, McKay RM, Bounk MJ (2010) Northeast Iowa watershed and karst Map. Open file map OFM-7-10. Iowa Geological and Water Survey

Huppert G, Wheeler B, Alexander EC Jr, Adams R (1988) Agricultural Land Use and Groundwater Quality in the Cold Water Cave Groundwater Basin, Upper Iowa River Karst Region, USA. Part I. Resource Management in Limestone Landscapes: International Perspectives. In: Proceedings of the international geographical union study group man's impact on Karst, Sydney, Australia, pp 235–247

Iker BC, Kambesis PN, Oehrle SA, Groves C, Barton HA (2010) Microbial atrazine breakdown in a karst groundwater system and its effect on ecosystem energetics. J Environ Qual 39(2):509–518

Iles DL (1977) Variations in the geometry and density of dolines, Allamakee County, Iowa. Dissertation, Iowa State University of Science and Technology

Johnson ME (1977) Early geological explorations of the Silurian system in Iowa. Proc Iowa Acad Sci 84(4):150–156

Kalkoff S (1995) Relation between stream-water quality and geohydrology during base-flow conditions, Roberts Creek watershed, Clayton County, Iowa. Water Resour Bull 31(4):593–604

Kambesis PN (2007) Agricultural contaminant source and transport in a karst groundwater basin. MSc thesis, Dept. of Geography and Geology, Western Kentucky University, Bowling Green, KY. 156 p

Kambesis PN (2009) Coldwater Cave System, Winneshiek County, Iowa. In: Palmer AN, Palmer MV (eds) Caves and Karst of the USA. National Speleological Society, Huntsville, AL p, pp 154–155

Kambesis PN, Bain B (1988) Coldwater Cave 20 years of exploration and still going. NSS News 46(6):141–147. http://caves.org/project/coldwater/cwp1b.html

Kambesis PN, Lace MJ (2009) Glaciated central lowlands: Iowa. In: Palmer AN, Palmer MV (eds) Caves and Karst of the USA. National Speleological Society, Huntsville, pp 151–154

Kambesis PN, Lovaas J, Lace MJ (2013) High resolution temperature sampling of cave climate variation as a function of allogenic recharge, Coldwater Cave, Iowa, USA. In: Proceedings of the 16th international congress of speleology, Cave Climate and Paleoclimate Record, pp 413–416

Keeler RR (1997) Modeling groundwater flow in a fractured-karst aquifer in the Big Springs Basin, Iowa. Doctoral dissertation, University of Iowa

Kennedy JO, Miller JG III (1990) A survey of the benthic macroinvertebrates of the Big Spring basin, Iowa. J Iowa Acad Sci 97 (2):46–54

Keyes CR (1895) Gypsum deposits in Iowa. Iowa Geolo Surv Annu Report 3:257–304

Keyes CR (1912) Spanish Mines: an episode in primitive American lead-mining. Annals of Iowa 10:539–546

Keyes CR (1915) Miocene age of Fort Dodge gypsum. Eng Min J 25:587–598

Kiel A (2004) Upper Iowa river GIS analysis. http://www.northeastiowarcd.org/uirw/publications.htm

Klimchouk AB (1995) Karst morphogenesis in the epikarstic zone: Caves and Karst. Science 21:45–50

Klimchouk AB (1997) The nature and principal characteristics of epikarst. In: Proceedings of the 12th international congress of speleology, La Chaux-de-Fonds, vol 1, p 306

Klug C (1996) Investigation of the palynomorphs of the Fort Dodge Gypsum, North-Central Iowa. unpublished open file report, Iowa Geological Survey Bureau, 5 p

Koch DL, Case JC (1974) Report on Coldwater Cave: a Summary of Research Results with Inclusion of Information Related to Potential Development of a New Recreation Facility by the State of Iowa. Iowa Geological Survey, 80 p

Kovarik AF (1898) The Decorah Ice Cave and its explanation. Sci Am Suppl (1195):19158–19159

Kovarik AF (1899) Ice Cave observations. Decorah Public Opinion

Launspach J (2013) Automated sinkhole extraction and morphological analysis in northeast Iowa using high-resolution LiDAR data. MA thesis. University of Northern Iowa, 104 p

Levorson CO, Gerk AJ (1972) A preliminary stratigraphic study of the Galena Group of Winneshiek County, Iowa. Proc Iowa Acad Sci 79

Lewis WC (1981) The "Breathing" of Coldwater Cave. In: Proceedings of the 8th international congress of speleology, Bowling Green, Kentucky, pp 89–90

Libra RD (ed) (2005) Living in Karst. Iowa Geological Survey Guidebook no. 25. Iowa Department of Natural Resources, Iowa City. 50 p

Libra RD, Ressmeyer GG, Bunker BJ, Hoyer BE (1984) Groundwater quality and hydrogeology of Devonian-Carbonate aquifers in Floyd and Mitchell Counties, Iowa, vol 84, no 2. Iowa Geological Survey

Lindsey BD, Katz BG, Berndt MP, Ardis AF, Skach KA (2010) Relations between sinkhole density and anthropogenic contaminants in selected carbonate aquifers in the eastern United States. Environ Earth Sci 60(5):1073–1090

Ludvigson GA (1996) Paleoclimate Controls on Detrital Modes of Jurassic and Cretaceous Sandstones. Geological Society of America, Cratonic Margin, Western Interior Basin (abstract)

Ludvigson GA, Dockal JA (1984) Lead and zinc mining in the Dubuque Area. Iowa Geology, no 9, Iowa Department of Natural Resources

Ludvigson GA, Mcadams MP (1980) New evidence of early Ordovician tectonism in the Upper Mississippi Valley. Iowa Geolo Surv Tech Inf Ser 9:27p

McKay R (1996) Paleokarsts in Iowa. Iowa Geol 21:20–23

McKay R, Tassier-Surine S, Liu P, Young J (2006) Geologic mapping of the Upper Iowa River Watershed

Moore F (1995) Floyd County groundwater protection project and sinkhole cleanout. Geology and hydrogeology of Floyd-Mitchell counties, north-central Iowa. Geological Society of Iowa Guidebook (62):75

Moore RC, Wanless H, Weller J, Williams JS, Read C, Bell W, Ashley G, Cheney M, Cline I, Condra G (1944) Correlations of Pennsylvanian formations of North America. Geol Soc Am Bull 55 (6):657–706

Morehouse DF (1968) Cave development via the sulfuric acid reaction. Natl Speleol Soc Bull 30:1–10

Mosnat HR (1898) Artesian wells of the Belle Plaine area. Iowa Geol Survey Ann Rept 9:521–562

Munter JA (1980) Evaluation of the extent of hazardous waste contamination in the Charles City area. Iowa Geol Surv Contract Report, 64 p

Natte RB (2008) Images of America: Fort Dodge 1850–1970. Arcadia Publishing, Chicago, p 127

Natural Resources Conservation Service. https://www.nrcs.usda.gov/wps/portal/nrcs/main/soils/survey/

Nekola JC (1999) Paleorefugia and Neorefugia: the influence on colonization history on community pattern and process. Ecology 80 (8):2459–2473

Nelson M (2002) Personal communication

O'Connor K, Trainum M (2015) Monitoring the threat of sinkhole formation under a portion of U.S. 18 in Cerro Gordo County, Iowa using TDR measurements. In: Proceedings of the 14th Sinkhole conference, pp 569–578

Owen DD (1852) Report of a geological survey of Wisconsin, Iowa, and Minnesota; and incidentally a portion of Nebraska Territory. Lippincott Grambo & Co., Philadelphia, p 683

Palmquist RC (1977) Distribution and density of dolines in areas of mantled karst. Hydrologic Problems in Karst Regions

Peck SB (1977) A growth rate for cave gypsum needles. Natl Speleol Soci Bull 39:104–105

Peck SB, Christiansen K (1990) Evolution and zoogeography of the invertebrate cave faunas of the Driftless Area of the Upper Mississippi River Valley of Iowa, Minnesota, Wisconsin, and Illinois, USA. Can J Zool 68(1):73–88

Peck SB et al (2020) This volume

Plotnick RE, Kenig F, Scott AC, Glasspool IJ, Eble CF, Lang WJ (2009) Pennsylvanian paleokarst and cave fills from northern Illinois, USA: a window into late Carboniferous environments and landscapes. Palaios 24(10):627–637

Prior JC (ed) (1975) Karst topography along the Silurian Escarpment in southern Clayton County. Geological Society of Iowa Guidebook, no 27, 13 p

Prior JC (1991) Landforms of Iowa. University of Iowa Press, Iowa City, p 153p

Rippey J (1950) Iowa City Grotto maps Kemling Cave. Natl Speleol Soc News 3:1–2

Ruhe RV (1969) Quaternary Landscapes in Iowa. Iowa State University Press, Ames, p 255

Sasowsky ID (2000) Groundwater flow and contaminant transport in carbonate aquifers, 193 p

Schilling KE, Helmers M (2008a) Effects of subsurface drainage tiles on streamflow in Iowa agricultural watersheds: exploratory hydrograph analysis. Hydrol Process 22(23):4497–4506

Schilling KE, Helmers M (2008b) Tile drainage as karst: conduit flow and diffuse flow in a tile-drained watershed. J Hydrol 349:291–301

Shopov YY, Ford DC, Schwarcz HP (1994) Luminescent microbanding in speleothems: high-resolution chronology and paleoclimate. Geology 22(5):407–410

Sims PK, Morey GB (1972) Geology of Minnesota: a centennial volume. In: Minnesota geological survey. University of Minnesota Digital Conservancy. http://hdl.handle.net/11299/57318

Smith DD (1984) Iowa's Driftless area: symposium. Proc Iowa Acad Sci 91(1), Article 3

Soil Conservation Service (1968) Soil survey: Winneshiek County. U. S. Department of Agriculture, Washington D.C, Iowa, p 225

Stanley LA (2002) A century of rock art research. J Archaeol Soc Iowa 49:65–85

Steinhilber WL, Van Eck OJ, Feulner AJ (1961) Geology and ground-water resources of Clayton County, Iowa. Iowa Department of Natural Resources. Geol Surv Bur Water-Supply Bull (7), 142

Strahler AN (1952) Dynamic basis of geomorphology. Geol Soc Am Bull 63:923–938

Thomson JB (2002) Prehistoric archaeology in Northeast Iowa. J Archaeol Soc Iowa 49:27–46

Trowbridge AC (1966) Glacial drift in the "Driftless Area" of Northeast Iowa. Iowa Geological Survey Report of Investigations, vol 2

Vail PR, Michum RM Jr, Thompson S III (1977) Seismic stratigraphy and global changes of sea level, part 4: global cycles of relative changes of sea level. In: Payton CE (ed) Seismic stratigraphy—Applications to hydrocarbon exploration. AAPG Mem vol 26, pp 83–97

Walters JC (2008) The Floyd karst area. In: Carbonate platform facies and faunas of the middle and upper devonian Cedar Valley group and Lime Creek Formation, Northern Iowa, p 11

Welch LE, Paul BE, Jones MD (2015) Measurement of radon levels in caves: logistical hurdles and solutions. In: Proceedings of the American association of radon scientists and technologists (AARST) international radon symposium, pp 1–17

Welch LE, Paul BE, Jones MD (2017) Use of electret ionization chambers to measure radon in Caves. In: Proceedings of the American association of radon scientists and technologists (AARST) international radon symposium, pp 1–18

Wheeler BJ, Alexander EC Jr, Adams RS, Huppert GN (1988) Agricultural land use and groundwater quality of the Coldwater Cave groundwater Basin, Upper Iowa River Karst Region, USA. In: Gilleson D, Smith DI (eds) Resource management in limestone landscapes: international perspectives. In: Proceedings of the international geographical union study group man's impact on Karst, Sydney, Australia, pp 249–260

Wilder FA (1903) The age and origin of the gypsum of central Iowa. Iowa Geological Survey XXVIII, Annual report 1917, pp 723–748

Wilder FA (1918) Gypsum: its occurrence, origin, technology, and uses. In: Kay GF, Lees JH (eds) Iowa geological survey XXVIII, annual report 1917, 1918. Des Moines, State of Iowa, pp 47–558

Wilder FA, Savage TE (1906) Iowa geological survey annual report, vol. XVI, 673 p

Williams PW (1983) Role of the subcutaneous zone in karst hydrology. J Hydrol 61(1–3):45–67

Witzke BJ (1998) Bedrock geologic map of northeast Iowa. Iowa DNR Geological Survey Bureau. Iowa Department of Natural Resources Iowa

Witzke BJ, Bunker BJ (2002) Bedrock geology in the Burlington area, southeast Iowa. Pleistocene, Devonian, and Mississippian Stratigraphy of the Burlington, Iowa Area Iowa Department of Natural Resources. In: Geological Survey Bureau Guidebook Series, vol 23, pp 23–48

Worthington S (2003) Comprehensive strategy for understanding flow in carbonate aquifers. In: Palmer AN, Palmer MV, Sasowsky ID (eds) Karst modeling. Special Publication #5, Karst Waters Institute, Charlestown, WV, pp 30–37

5 Illinois Caves and Karst

Samuel V. Panno, Donald E. Luman, and Joseph A. Devera

Abstract

The Illinois Basin occupies much of Illinois and contains Paleozoic sedimentary rocks consisting of Cambrian through Pennsylvanian strata. Most of the formations are dominated by limestone and dolomite. The carbonate rocks range from Ordovician through Mississippian in age, with lesser amounts occupying Cambrian and Pennsylvanian strata. Most of these carbonate rocks are exposed in the northern, western, and southern margins of Illinois and adjacent states. All bedrock has been extensively fractured by tectonic forces, and were exposed beneath a thin cover of unconsolidated sediment, recharge has dissolved the carbonate rocks along fracture and bedding planes, creating caves, solution-enlarged crevices, and cover-collapse sinkholes. Illinois has five major karst areas. The northernmost karst region is the Driftless Area in northwestern Illinois, where dolomites of the Ordovician Galena and Silurian formations are exposed in outcrops or near the surface and form cover-collapse sinkholes, short network-type caves, sinuous conduits, and small karst springs. Some of the caves and conduits were created and mineralized by Mississippian-type ore-forming solutions during the Permian Period, which were mined for lead and zinc sulfides from the 1700s to the mid-1900s. The North-Central karst region is similarly dominated by dolomites of the Ordovician Galena Group that form small sinkholes and short network-type caves. Bedrock in both areas has abundant fractures and solution-enlarged crevices exposed in road cuts and quarries. Situated in western Illinois, the Lincoln Hills karst region contains rocks that range from the Silurian to Mississippian Periods that form cover-collapse sinkholes and caves. Most of the caves are network-type caves that are relatively short owing to infilling of fine-grained sediment. Located in southwestern Illinois, the Salem Plateau karst area is composed of the St. Louis and Ste. Genevieve Limestone formations, forming the largest cover-collapse sinkholes, the largest springs, and the longest caves in the state, most of which are branchwork type. The Shawnee Hills karst area in southern Illinois has abundant sinkholes and network-type caves in a variety of formations that range from the Silurian through Mississippian Periods. This area is home to Cave-In-Rock, a well-known historic remnant cave.

5.1 Introduction

Multiple glacial events during the Pleistocene Epoch (2.63 million years (Ma) to 11,800 years before present (yr BP); Curry et al. 2010) buried much of the preexisting rugged bedrock topography of Illinois with glacial sediments up to 152 m in thickness (Piskin and Bergstrom 1975). However, in several areas of the state along the margins of the Illinois Basin, the original topographic relief has been retained because glacial drift is absent. These areas include the Driftless Area of northwestern Illinois, the western margin of Illinois, and far southern Illinois. In areas where Paleozoic carbonate rock constitutes bedrock and glacial till is thin or absent, five areas of karst terrain constitute about 10% of the land surface of Illinois (Fig. 5.1). Native Americans colonized the Americas, including Illinois, near the end of the Wisconsin Glacial Epoch, as early as 17,000 yr BP (Berkson 2009). Most of Illinois was covered with extensive prairie vegetation prior to European settlement in the late 1600's. Today, row crop agriculture dominates most of the state and very little of the prairie vegetation remains.

The karst features within Illinois typically include solution-enlarged crevices, cover-collapse sinkholes, caves and conduits, large springs, and lost and disappearing

S. V. Panno (✉) · D. E. Luman · J. A. Devera
Illinois State Geological Survey, 615 E. Peabody Drive, Champaign, IL 61820, USA
e-mail: s-panno@illinois.edu

G. A. Brick and E. C. Alexander Jr. (eds.), *Caves and Karst of the Upper Midwest, USA*, Cave and Karst Systems of the World, https://doi.org/10.1007/978-3-030-54633-5_5

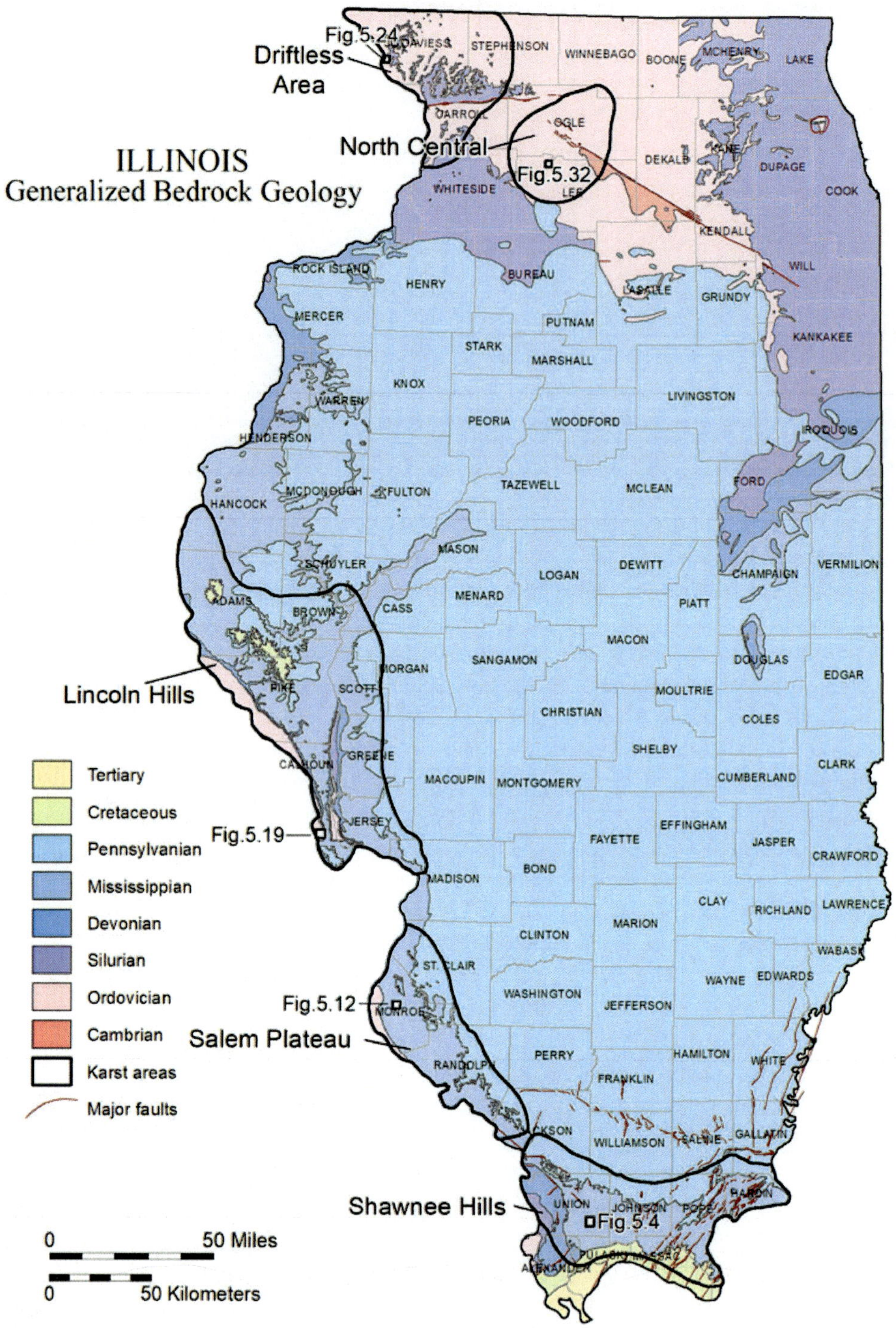

Fig. 5.1 Generalized bedrock map of Illinois showing the location of the five major karst areas of the state based on surface and subsurface karst features, modified from Kolata (2005). Copyright © 2005 University of Illinois Board of Trustees. Used with permission of the Illinois State Geological Survey. Karst areas modified from Weibel and Panno (1997). Copyright © 1997 University of Illinois Board of Trustees. Used with permission of the Illinois State Geological Survey

streams. The Paleozoic rocks within Illinois are well fractured, and solution-enlarged crevices are abundant throughout the state. Fractures and crevices in carbonate rock are common in road cuts, caves, and quarry walls and floors (Bretz and Harris 1961; Panno et al. 1997). These features are reflected in aligned cover-collapse sinkholes and recently discovered crop line phenomena within the Driftless Area of Illinois (Panno and Luman 2012a; Panno et al. 2015).

Cover-collapse sinkholes, caves, and large springs are often considered the major indicators of karst terrain. However, James F. Quinlan (1947–1995), formerly of Quinlan and Associates, Inc. and an expert in karst geology and hydrogeology, stated that if bedrock within a region is composed of carbonate rock, we should assume the area is karst and is underlain by a karst aquifer unless proven otherwise. In line with that assumption, the presence of carbonate bedrock, sinkholes, and caves became the basis for preparing a statewide map of karst terrain published by the Illinois State Geological Survey (Weibel and Panno 1997) which identified five karst areas of the state (Fig. 5.1) based on the presence of carbonate bedrock, the drift thickness, outcrops of creviced carbonate bedrock, and the presence of cover-collapse sinkholes, caves, and karst springs.

Bretz and Harris (1961) were first to describe the caves of Illinois. Oliver and Graham (1988) later estimated the number of caves within Illinois to be at least 480. These caves included both branchwork and network types, most of which are located in the five major karst areas of Illinois. The approximate locations of the caves of Illinois may be found in Weibel and Panno (1997), Panno et al. (1997), but precise locations are no longer made available for security reasons. Caves have been described as repositories for climatic and geologic history, as shown in their cross-sectional and plan-view geometry, remnant alluvial sediments, megafauna fossils, and speleothems extending back tens of thousands to millions of years (Dorale et al. 1998).

5.1.1 Geology

The geology and geologic history of Illinois are summarized in *Geology of Illinois*, edited by Kolata and Nimz (2010). The Illinois Basin, which constitutes much of the area of the state, is an intracratonic basin containing a sequence of Paleozoic sedimentary (carbonate, shale, and sandstone) rocks that extend from the surface along the basin margins in northern, western, and southern Illinois to a depth of up to 9.1 km near the center of the basin in southern Illinois. Glacial deposits up to 100 m or more in thickness extend throughout most of Illinois, creating large areas of low relief or nearly flat terrain ideal for the intensive agricultural uses; these account for more than 75% of the state's land area (Luman et al. 2004). Areas that are driftless or that possess only a thin veneer of glacial materials are located at or near the basin margins. They typically exhibit karst features in bedrock outcrops, road cuts, and quarries and may contain sinkholes and karst aquifers. Geologic structures such as the La Salle Anticlinorium (Fig. 5.1) can bring karstified carbonate bedrock ranging in age from Ordovician through Pennsylvanian to the surface or near surface. Additionally, paleokarst features are occasionally intersected with drill holes and excavations (e.g., Bretz and Harris 1961; Plotnick et al. 2014).

A wide variety of formations in the state are susceptible to karstification because the Illinois Basin is dominated by carbonate rocks (Fig. 5.2) and continental-scale tectonism

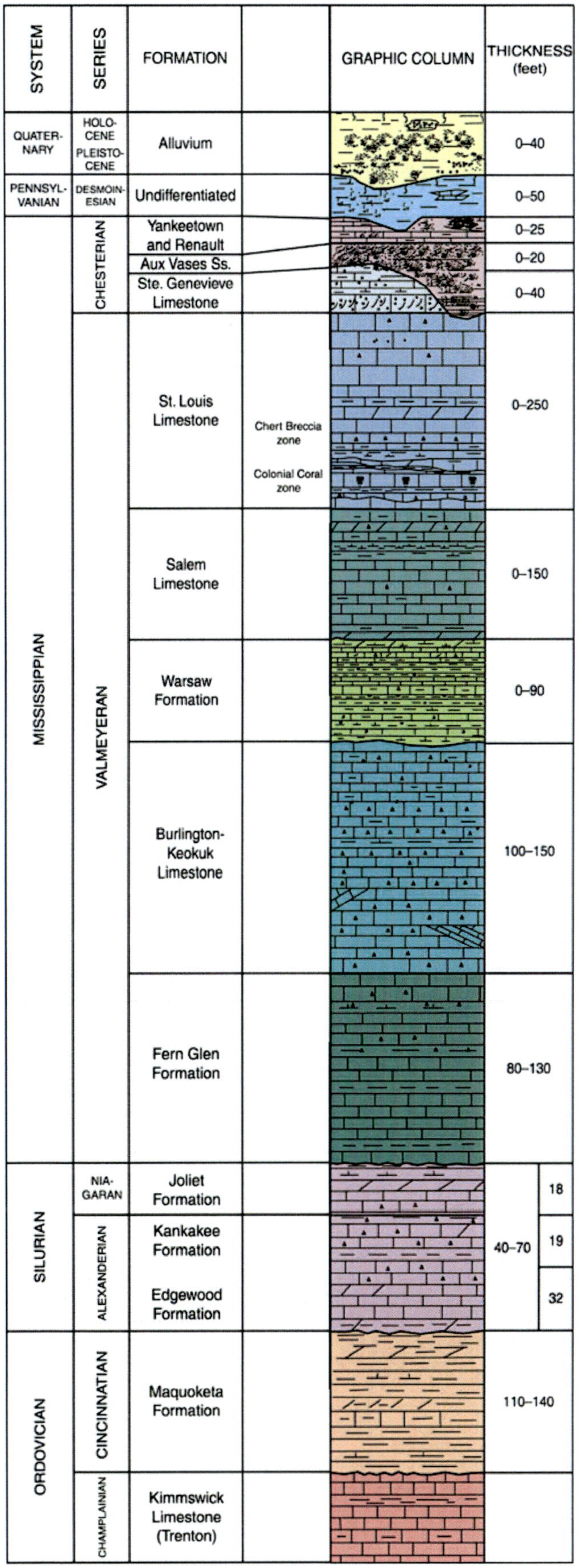

Fig. 5.2 Stratigraphic column for Paleozoic bedrock and glacial drift ▶ in Illinois, modified from Devera (2000). Copyright © 2000 University of Illinois Board of Trustees. Used with permission of the Illinois State Geological Survey

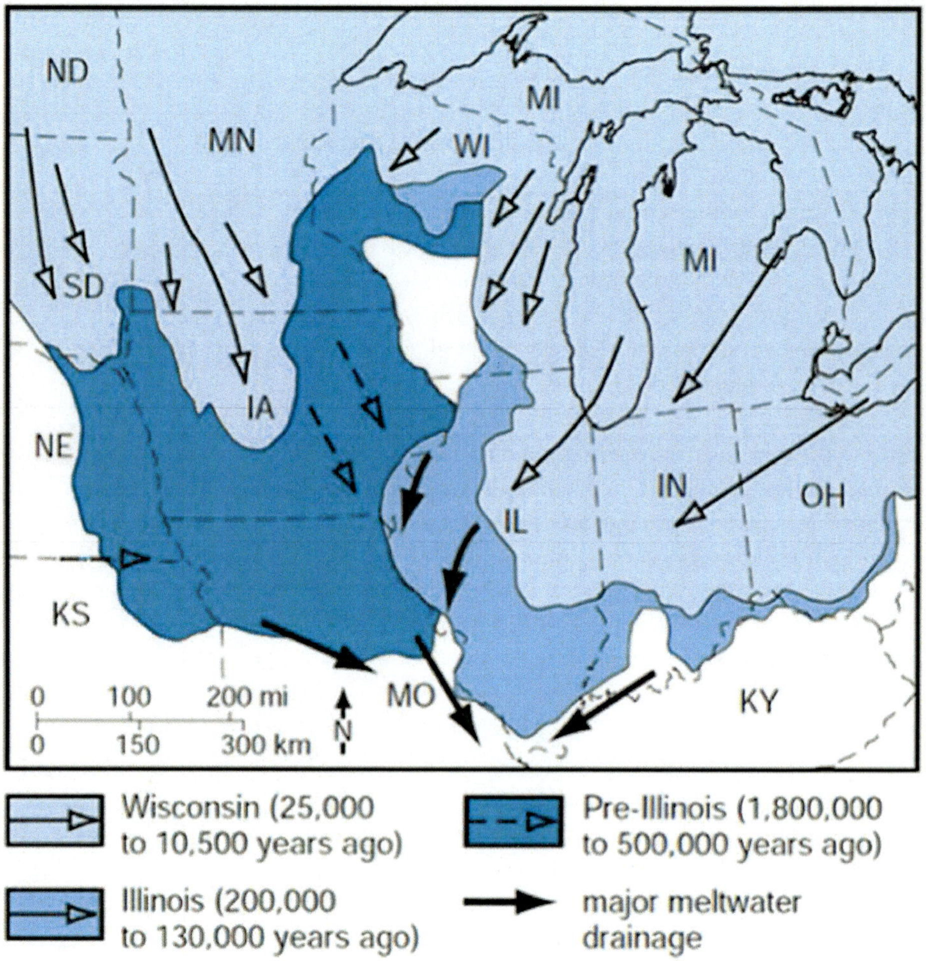

Fig. 5.3 Glaciation within North America in the vicinity of Illinois showing the extent of pre-Illinois, Illinois, and Wisconsin glaciers, from Killey (2007). Copyright © 2007 Illinois State Board of Trustees. Used with permission of the Illinois State Geological Survey

has folded, faulted, and fractured these rocks. Ordovician (485–444 Ma BP) bedrock of the Galena Group is dolomitic in northern Illinois, whereas its time-equivalent Kimmswick Subgroup is dominated by pure limestone in southern Illinois (Willman et al. 1975). Both of these strata display karst features throughout the state. Silurian (444–419 Ma BP) bedrock contains abundant karst features in the northwestern Driftless Area of northeastern and far western Illinois, where they are exposed in quarries and outcrops. Silurian rocks are predominantly composed of dolomite in the northern part of the state, whereas in the Shawnee Hills karst area, the rocks are dominated by red marl and limestone. Carbonate rocks of the Devonian Period (419–359 Ma BP) are rarely exposed in Illinois, but exposures do occur in the Lincoln Hills and the western margin of the Shawnee Hills karst areas. The Grand Tower Limestone of middle Devonian age contains small remnant caves at the type section at Grand Tower in Jackson County, in the Shawnee Hills karst area. Units of the Mississippian strata exhibit the most karst features in the state and are the most varied relative to other Paleozoic carbonate rocks. Mississippian (360–325 Ma BP) strata susceptible to karstification include the limestones of the Fort Payne and Paoli Formations, and the Ullin, Salem, St. Louis, Ste. Genevieve, Renault, Haney, Glen Dean, Menard, and Kinkaid Limestones. Limestones of the St. Louis, Ste. Genevieve, Kinkaid, and Ullin Limestones are the most susceptible to karstification and are exposed or near the surface in the Lincoln Hills, Salem Plateau, and Shawnee Hills karst areas. The Kinkaid Limestone is a limestone–shale unit that has its greatest thickness in southern Illinois, is resistant to erosion, and hosts abundant sinkholes and caves. The Ullin Limestone outcrops in Hardin and Pope Counties in southeastern Illinois, is porous and relatively soft, and dissolves readily along fractures, resulting in numerous sinkholes.

Pleistocene glaciation had a profound effect on the formation and evolution of caves and karst within Illinois. The effects are varied, depending on the location within the state and whether glaciers covered those areas or remained unglaciated or driftless (Fig. 5.3). The proximity of glaciers to areas with carbonate bedrock in the Upper Mississippi Valley (UMV) has resulted in the recording of flood events in cave sediments and stalagmites in Illinois, as well as in neighboring states (e.g., Panno et al. 2004a, 2016).

5.1.2 Historical and Recent Cave and Karst Investigations

Bretz (1938), Bretz and Harris (1961) conducted some of the earliest cave and karst studies in Illinois. The latter publication was a compilation titled *Caves of Illinois* that describes the locations and character of more than 60 caves in Illinois. When discussing the caves of Illinois, Bretz and Harris stated, "As a cave state, Illinois is not outstanding" (Bretz and Harris 1961, p 15). This is relative to neighboring "cave states," including Missouri, Indiana, and Kentucky, where show caves are common and attract visitors from all over the world. Only a single Illinois cave has ever been successfully commercialized—Illinois Caverns in Monroe County, within the Salem Plateau of southwestern Illinois (Panno et al. 2004b).

Research in southwestern and northwestern Illinois included water quality studies of springs (Panno et al. 2001; Hackley et al. 2007), caves, and wells (Panno et al. 1997; Hackley et al. 2007); geomorphology studies of sinkholes (Panno et al. 1997; Panno and Luman 2018); speleogenesis studies (Panno et al. 2004a, 2016); earthquake studies (Panno et al. 2009); and remote sensing studies involving the crop line phenomena (Panno et al. 2015).

Illinois caves are broadly categorized as two types, branchwork and network; both types can be found throughout the state. The longest caves, the greatest sinkhole density, and the largest sinkholes in the state are found in the Salem Plateau and Shawnee Hills karst areas (Fig. 5.1). The Salem Plateau karst area, known informally as the "sinkhole plain," possesses sinkhole densities up to 90 per km^2, with sinkhole diameters up to 300 m and depths up to 20 m (Panno and Luman 2012a, b). Abundant shorter caves and similarly large cover-collapse sinkholes at lower densities are present in the Shawnee Hills and Lincoln Hills karst areas. Smaller and lower density cover-collapse sinkholes (relative to all other karst areas of Illinois) are present overlying Ordovician bedrock of the Driftless Area and the North-Central karst area. The smaller nature of the sinkholes (in both diameter and depth) overlying these formations is primarily due to the narrower crevices and thinner overburden. Cover-collapse sinkholes overlying Silurian dolomite in the Driftless Area are typically 30 m or more in diameter. Here, thicker unconsolidated sediments with underlying bedrock contain crevices one meter in width that were in part formed by the movement of bedrock blocks on underlying Maquoketa Shale Group (Panno and Luman 2016). Along the margins of the Illinois Basin, caves in Illinois are commonly filled or partially filled with clay-rich, fine-grained sediment, as discussed in detail by Bretz and Harris (1961).

5.2 Carbonate Bedrock and Karst Areas

5.2.1 Shawnee Hills

Paleozoic carbonate bedrock in the Shawnee Hills karst area is either exposed or near the surface beyond the southern extent of Pleistocene glaciation. At these locations, a wide variety of carbonate rocks are exposed, originating from the Mississippian through the Silurian Periods. The St. Louis, Ste. Genevieve, Kinkaid (part of the Chesterian Series), and Ullin Limestones of Mississippian age are the geologic formations most susceptible to karstification in this area (Fig. 5.1). Karst features include abundant cover-collapse sinkholes (Fig. 5.4), intermittent stream beds (Fig. 5.5a, b), and caves (Fig. 5.6a, b).

Lidar elevation data (Fig. 5.4) show cover-collapse sinkholes near the town of Dongola in Union County, which have developed in unconsolidated sediments overlying Mississippian Ste. Genevieve Limestone (northeastern side of the stream course) and St. Louis Limestone (southwestern side of the stream course). The sinkholes overlying the Ste. Genevieve Limestone range in depth from 3 to 15 m (white rectangle), in part because erosion has breached the cave roofs of large cave systems in multiple places, and the caves are in the process of collapse. The sinkholes overlying the St. Louis Limestone are related to the collapse of overlying soils into bedrock crevices where the water table is well below the soil–bedrock interface. These circular sinkholes range in depth from 2 to 3 m (white circle) and are similar in size and shape to those of the Salem Plateau karst area.

Cover-collapse sinkholes were formed in the Dongola area in 1993 after a new municipal well was tested. The well intersected the Mississippian-age Ullin Limestone, and the potentiometric surface of the underlying karst aquifer dropped below the soil–rock interface as groundwater was pumped from the karstified limestone. This resulted in the formation of cover-collapse sinkholes at multiple locations within and adjacent to the playground of an elementary school. Three sinkholes 4 m in diameter formed in the sediment, one of which swallowed a 15-m-high light pole. The formation of these sinkholes forced the permanent shutdown and abandonment of the well (Panno et al. 1994).

Caves in this region are mostly joint controlled. The best known and longest caves are located in the eastern part in the Shawnee Hills karst area. These include Cave-In-Rock, Equality Cave, Shelterville Cave, and Frieze Cave. Cave-In-Rock, one of the most famous caves in Illinois, has the appearance of a large remnant cave system (Fig. 5.6a, b). However, in its present state, the cave is relatively short (30 m) and extends from the entrance to a floor of clay-rich

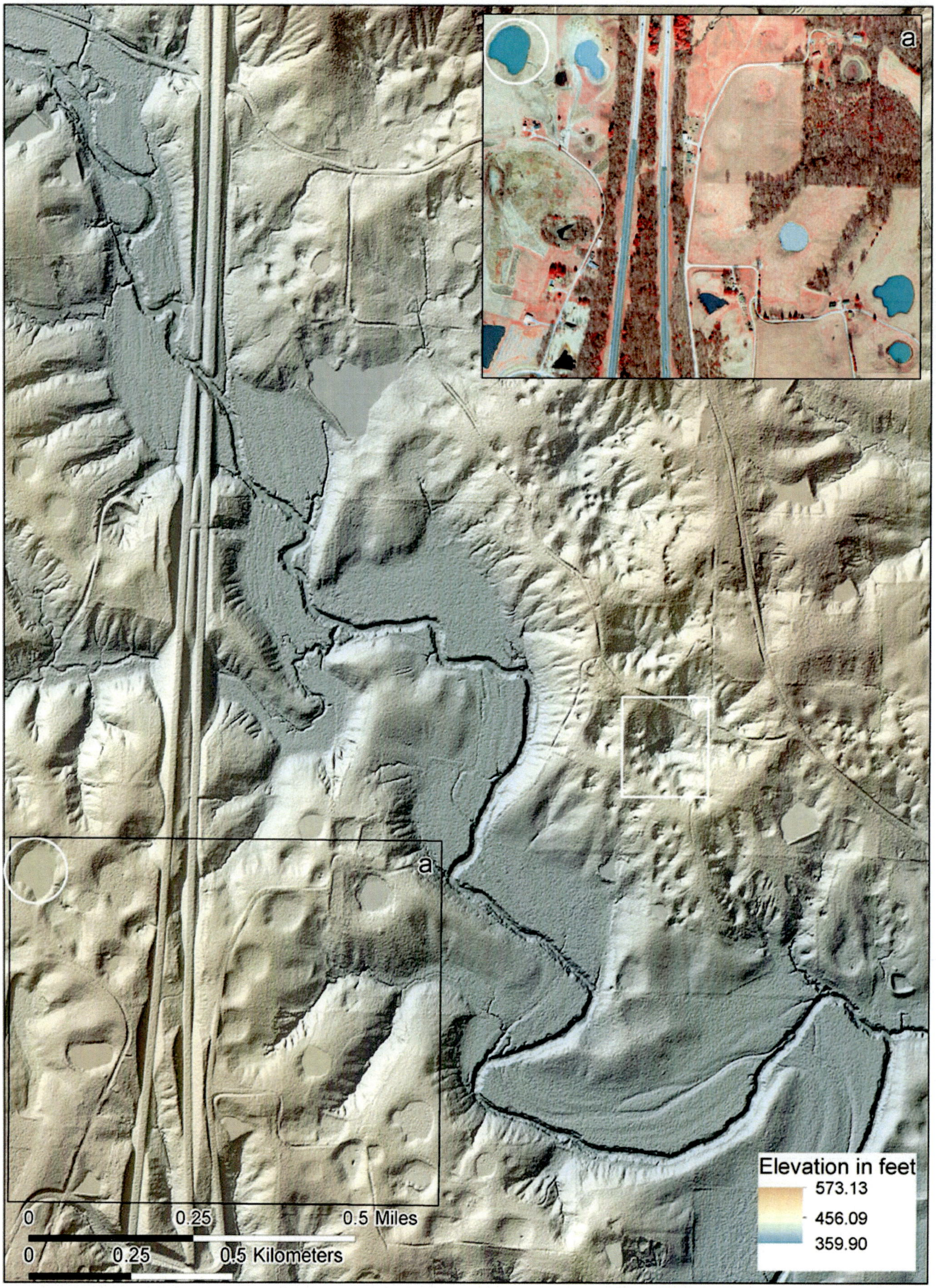
a
a
0
0.25
0.5 Miles
0
0.25
0.5 Kilometers
Elevation in feet
573.13
456.09
359.90

◄ **Fig. 5.4** Shaded relief 2011 lidar showing cover-collapse sinkhole development within the Shawnee Hills karst area. Inset **a** at upper right is a color near-infrared digital orthoimage showing relative amounts of turbidity in the ponded sinkholes. A lighter color indicates higher turbidity and a darker color indicates lower turbidity. Data provided by the Illinois Height Modernization Program (clearinghouse.isgs.illinois.edu) supported by the Illinois Department of Transportation

sediment at the rear of the cave. The cave has a large 17-m cross-sectional area and is situated on a bluff of St. Louis Limestone facing the Ohio River. The keyhole shape near circular cross section of the cave suggests it originally formed as a phreatic conduit, with later vadose modifications of a sinuous nature to the base of the conduit. The origin and extent of the cave to the north and south are only conjecture; however, other karst features are present in the area, including sinkholes and dome pits that extend to the north and south for many kilometers. The largest sinkhole in Illinois, referred to as Big Sink, is located less than 3.5 km north of Cave-In-Rock. The presence of large caves in the area would not be surprising, given the relationships among proximity to large caves, sinkhole size, and morphology (Panno et al. 2013). Conversely, their only relationship may be related to the karstified bedrock they share. Panno and Bourcier (1990) suggested that caves and karst features near the southern margin of the Illinois Basin could have been initiated, or at least accelerated, by brines discharging from the basin during periods of glacial melting that mixed with freshwater (mixing corrosion) within carbonate aquifers.

The Kinkaid Limestone is host to many caves in the Shawnee Hills karst area. Equality Cave, Shelterville Cave, Frieze Cave, and Rich's Cave are examples of relatively long network caves 100 m or more in length located in southeastern Illinois (Saline, Hardin, Pope, and Union counties, respectively) within the Kinkaid Limestone (Fig. 5.7a, b). Passages are joint controlled (network type),

Fig. 5.5 **a** Stream beds in the western Shawnee Hills karst area with exposed hoodoos or spires of Ste. Genevieve Limestone. One can hear water running beneath the stream bed during dry periods. **b** Another exposure of creviced Ste. Genevieve Limestone reveals anastomoses (sinuous, dilution-related conduits) along a bedding plane beneath the stream channel. Copyright © University of Illinois Board of Trustees. Photographs by J.A. Devera; used with permission of the Illinois State Geological Survey

Fig. 5.6 a The entrance of Cave-In-Rock along the Ohio River in southern Illinois and b the interior of the cave. Copyright © University of Illinois Board of Trustees. Photographs by S. V. Panno; used with permission of the Illinois State Geological Survey

Fig. 5.7 a Equality Cave and b Rich's Cave are network caves within the Kinkaid Limestone in the Shawnee Hills karst area. The entrance to Rich's Cave is about 2 m high. Copyright © University of Illinois Board of Trustees. Equality Cave photograph by S.J. Taylor; used with permission of the Illinois Natural History Survey. Rich's Cave photograph by J.A. Devera; used with permission of the Illinois State Geological Survey

with two perpendicular joint sets trending northwest and northeast (Fig. 5.8). The cave floors are typically filled with fine-grained sediment (laminated silty clay) that can completely fill passages. Caves in the central and western parts of the Shawnee Hills are also network-type caves found predominantly in the Kinkaid Limestone (Bretz and Harris 1961). Caves in the western part of southern Illinois northeast of Dongola are an exception to this. In this area, a series of relatively large and complex collapsed cave passages formed within the Ste. Genevieve Limestone (Fig. 5.4).

5.2.2 Salem Plateau

The Salem Plateau (Leighton et al. 1948), situated along the western margin of the Illinois Basin, includes one of the state's five major karst areas and is also referred to as the Illinois sinkhole plain (Fig. 5.1). Karst features are abundant in this region because of the relatively thin unconsolidated sediments overlying karstified Mississippian-age carbonate rock. The sinkhole plain has the greatest density of cover-collapse sinkholes (Fig. 5.9a, b), the longest caves, and

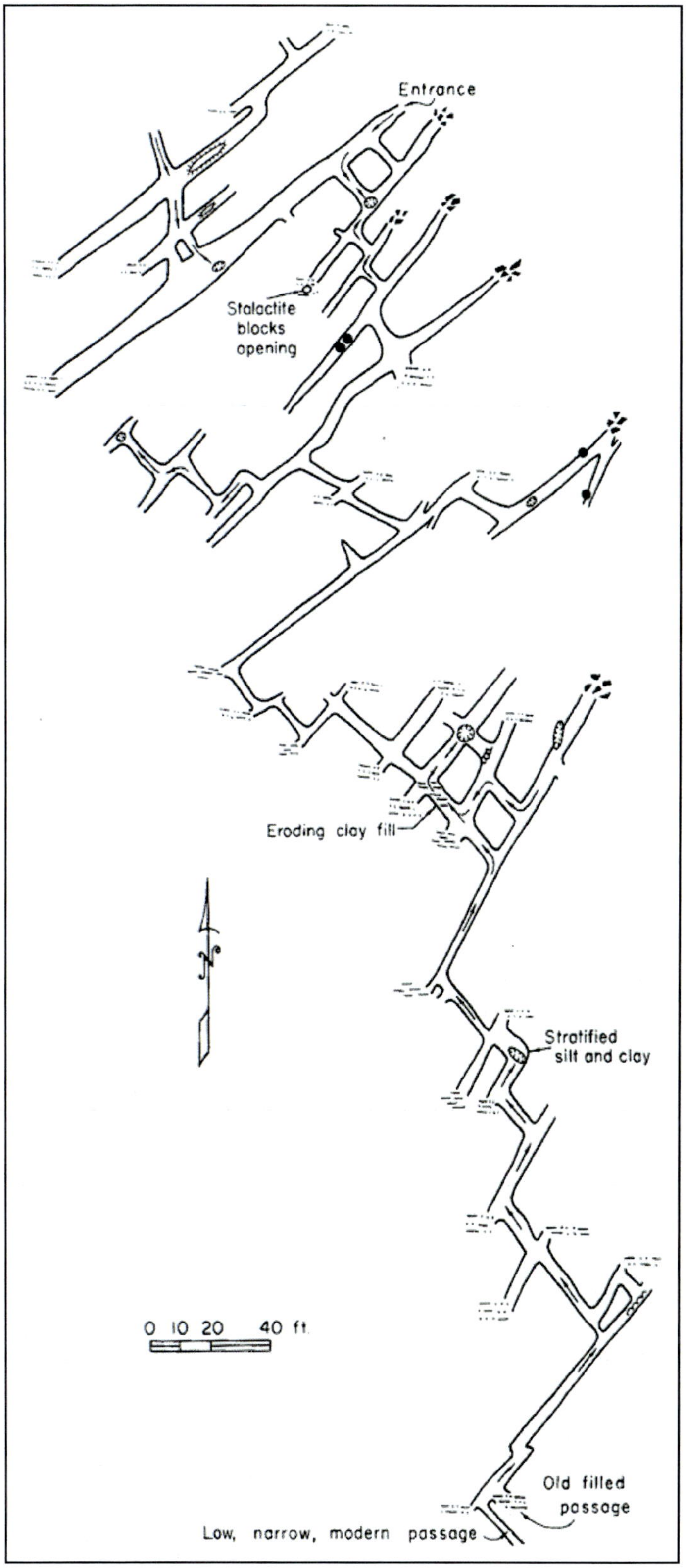

Fig. 5.8 Map of Frieze Cave in the Shawnee Hills area showing a network of joint-controlled cave passages that are typical of caves in this area, from Bretz and Harris (1961). Copyright © 1961 University of Illinois Board of Trustees. Used with permission of the Illinois State Geological Survey

the largest springs in the state (Panno et al. 1997; Panno and Luman 2012a, b). Karstified carbonate bedrock is overlain by Illinois glacial till, residuum, and Wisconsin loess that typically ranges from 0 to 10 m in thickness. The two most karstified formations in this region are the St. Louis and Ste. Genevieve Limestones, and both are composed of exceptionally pure marine calcium carbonate (Panno et al. 1997).

Panno and Luman (2012a, b) inventoried a total of 8,890 cover-collapse sinkholes from direct interpretation of historical and recent aerial orthophotography of Monroe

Fig. 5.9 **a** Cover-collapse sinkholes are ubiquitous in the Illinois sinkhole plain and occur predominantly in glacial till and loess overlying the St. Louis Limestone. **b** Sinkholes in this area are typically circular to elliptical in plan view and are either dry, with a cluster of trees in the center, or ponded, from Panno et al. (2004b). Copyright © 2004 University of Illinois Board of Trustees. Used with permission of the Illinois State Geological Survey. Photographs by S. V. Panno

Fig. 5.10 **a** Excavation of a sinkhole in the Salem Plateau revealed a solution-enlarged crevice up to 1 m wide that led to a 45 × 50 cm cave containing flowing water. **b** These crevices are similar to those found on road cuts along Route 3 near Columbia, Monroe County. Copyright © University of Illinois Board of Trustees. Photographs by S. V. Panno; used with permission of the Illinois State Geological Survey

County, the most karstified area within the Illinois sinkhole plain; the calculated maximum density of sinkhole features was 125 per km^2 (48 mi^2). Sinkholes form along solution-enlarged crevices with east–west and north–south orientations (Fig. 5.10a, b), as seen in road cuts and quarries. Excavated sinkholes and naturally exposed bedrock surfaces at the bottom of sinkholes reveal crevice widths ranging from 0.6 to greater than 1.0 m; many of the wider crevices lead to small conduits and caves with flowing water (Fig. 5.10a). Sinkholes range from several meters to more than 100 m in diameter and are typically circular to elliptical in plan view (Panno and Luman 2012a, b); sinkhole densities can be as high as 95 per km^2 (Angel et al. 2004). Some sinkholes form ponds, whereas others host dense clusters of trees (Fig. 5.9a, b).

Sinkhole morphology in the Salem Plateau karst area can be affected by their proximity to cave passages because sinkholes leading directly to caves have the ability to drain runoff faster than those draining to crevices. Caves can accommodate much more runoff than crevices; therefore, sinkholes draining directly into caves tend to be erosive features (Fig. 5.11a, b), whereas sinkholes drained by crevices tend to be depositional environments (Panno et al. 2013). In addition, sinkhole morphology changes with the evolution of sinkholes within groundwater basins. This evolutionary process is countered to the currently accepted model of simple sinkholes coalescing into compound sinkholes. Specifically, stream valley-shaped sinkholes in Stage 1 initially form at the leading edge of a groundwater basin. In Stage 2, an amoeboid-shaped compound sinkhole

Fig. 5.11 Cover-collapse sinkholes overlying or in the vicinity of cave passages are erosive in nature because of the ability of the caves to receive large volumes of runoff without back flooding. **a** Erosion gullies typically lead to these sinkholes and **b** relatively large open cavities in bedrock serve as entry points for runoff. Copyright © University of Illinois Board of Trustees. Photographs by S. V. Panno; used with permission of the Illinois State Geological Survey

develops, with multiple drains to bedrock. Stage 3 follows erosion of the compound sinkholes to form simple circular and elliptical sinkholes closest to the discharge point of the basin (Panno and Luman 2018; Fig. 5.12).

The thickness of the unconsolidated sediment in the region is about 10 m, and sinkhole depths range from 1 to 15 m. Panno and Luman (2018) conservatively estimated that more than 10 million m^3 of sediments will be missing from a 26.5 km^2 groundwater basin in the sinkhole plain because of sinkhole formation and continued erosion (Fig. 5.12). The bivariate modeling of sinkhole complexity and circularity illustrates and supports the three-stage sinkhole evolution model. Sinkhole initiation ensues at the distal end of streams draining the landscape along a groundwater divide (Stage 1), evolves to compound sinkhole formation (Stage 2), and then progresses to the end state of simple sinkholes (Stage 3; Panno and Luman 2018).

Numerous caves in this region result from slightly acidic rainwater, snowmelt, and soil water entering crevices leading to bedding plane partings. Water flowing along bedding plane partings eventually leads to the formation of conduits 10 cm in diameter, and some of these conduits become cave passages with stream channels. Three of the longest caves in the state, Fogelpole Cave (15 km), Illinois Caverns (10 km), and Pautler Cave (9.5 km; Unklesbay and Vineyard 1992; Panno et al. 2004b) are located in this area, and like most of the caves in the sinkhole plain, are the branchwork type, which are sinuous in plan view. These caves have active streams flowing through them (Fig. 5.13a–c), discharging to resurgence springs (Fig. 5.14). Passages within these two caves can be 10 m or more wide and up to 20 m high. The caves in this area are typically well decorated with abundant speleothems and flowstone (Fig. 5.13a, b). Speleothems in Illinois Caverns can be more than 4 m tall. Illinois Caverns is one of the most well-known caves in Illinois and is part of the Illinois Caverns State Natural Area, a 49 ha nature preserve in southwestern Monroe County. Historically, Illinois Caverns was a popular attraction during the 1904 World's Fair in St. Louis, Missouri, and became a commercial venture that ended several years after the fair. The cave was again commercialized when electric lights were installed in 1947, but because of the remote location of the cave and the small number of visitors, the enterprise soon ended (Panno et al. 2004b). Illinois Caverns became the preferred cave for students and visitors during the 1960s through almost the first decade of 2000. Unfortunately, because of a combination of state budgetary cuts to Illinois State Parks and State Natural Areas, and the subsequent incidence of white-nose syndrome in bats, at the date of this writing, the cave and associated natural area are closed to visitors.

The Salem Plateau is home to the largest springs in Illinois, having discharge rates as great as 900 L/s during periods of high flow (Hackley et al. 2007). Discharge is typically from bedding planes, bedding plane caves, or both, to surface streams (Fig. 5.14a, b). Because of the open nature of mature karst in the sinkhole plain, discharge from springs can be heavily laden with fine sediment from soil erosion caused by surface runoff associated with construction and farming activities draining directly into sinkholes.

5.2.2.1 Paleoclimate

Climatic inferences have been interpreted from alluvial flood deposits and speleothems within two branchwork caves of southwestern Illinois. Panno et al. (2004a) examined fine-grained, laminated lacustrine sediment within Illinois Caverns and Fogelpole Cave located in the Illinois sinkhole plain, and they acquired radiocarbon dates and $\delta^{13}C$ values from the organic carbon fraction of the sediments. The silty

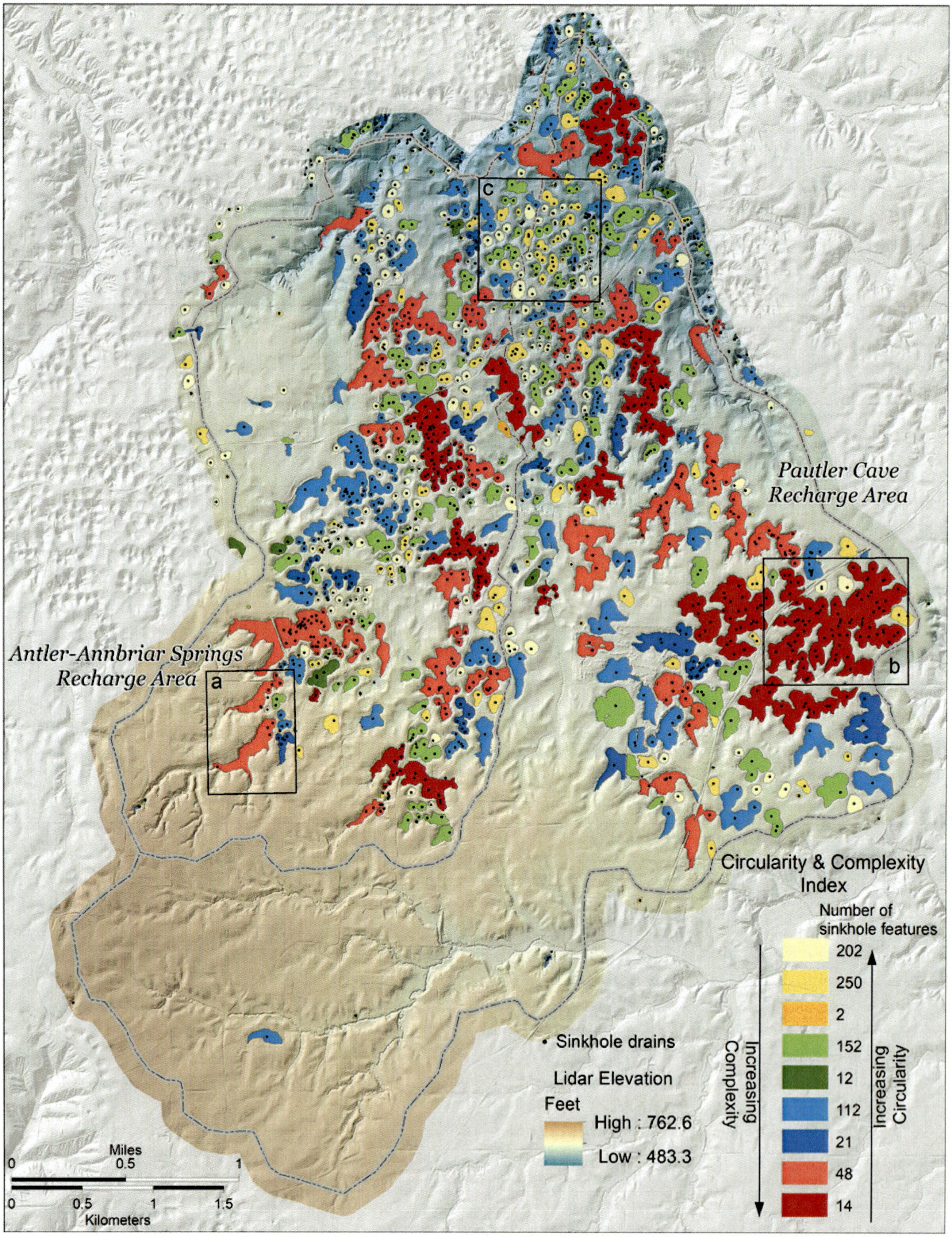

Fig. 5.12 Shaded relief lidar of a 26.5 km^2 area of the sinkhole plain within the Salem Plateau in Monroe County, showing 610 cover-collapse sinkholes developed in the Mississippian-age St. Louis Limestone formation. Each colored area depicts a sinkhole feature at least 0.3 m in depth; each sinkhole feature is characterized on a bivariate distribution of sinkhole complexity and circularity. Groundwater flow is toward the north. Data provided by the Illinois Height Modernization Program (clearinghouse.isgs.illinois.edu) supported by the Illinois Department of Transportation. Reprinted from Panno and Luman (2018). Copyright © 2018, with permission from Elsevier

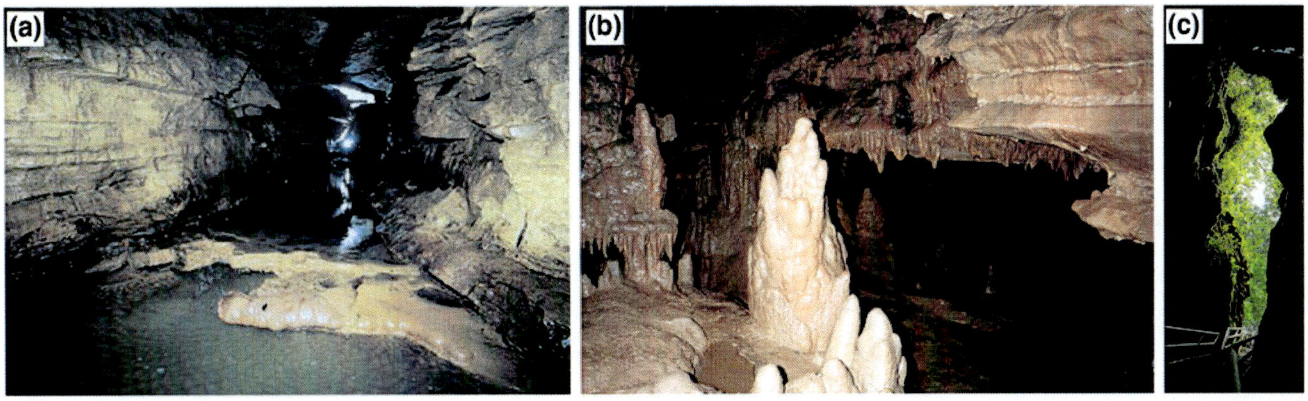

Fig. 5.13 Caves within the sinkhole plain, such as Illinois Caverns, are sinuous in nature, containing **a** flowing streams and **b** abundant speleothems, the largest in this photograph is 1 m high. **c** Entrances are often solution-enlarged crevices, this crevice opening is 15 m. From Panno et al. (2004b). Copyright © 2004 University of Illinois Board of Trustees. Cave stream photograph by Joel Dexter. Speleothem and cave entrance photographs by S. V. Panno; used with permission of the Illinois State Geological Survey

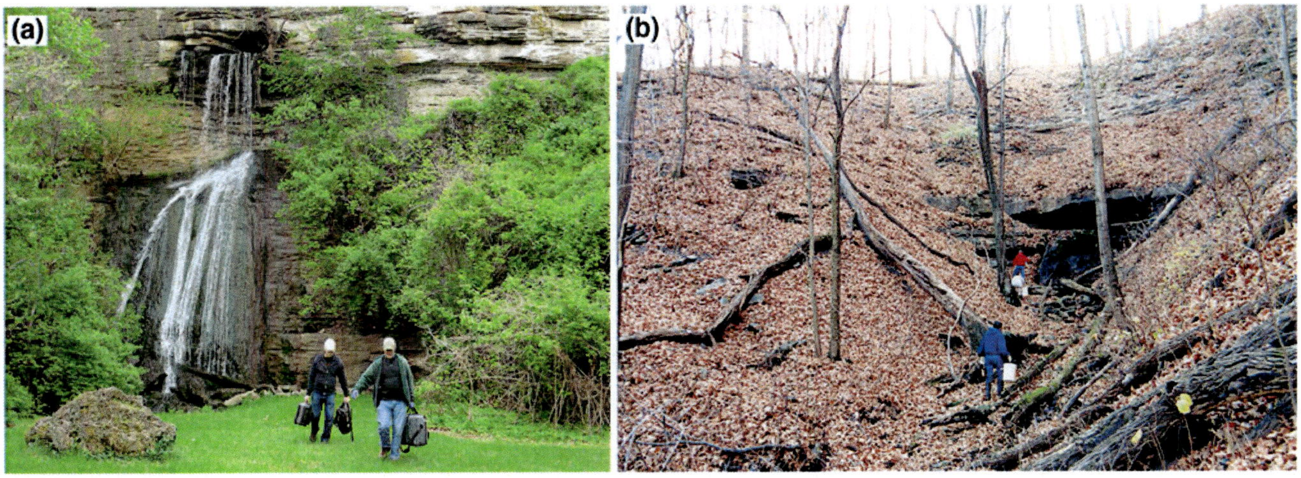

Fig. 5.14 **a** Falling Springs and **b** Auctioneer Cave are springs that discharge from small caves along bluffs of Mississippian-age St. Louis Limestone. Both have well-formed tufa layers. A remnant of the tufa formed at Terry Spring in the same region was dated at 10,840 ^{14}C yr BP (Webb et al. 1996) near the end of the Wisconsin. The tufa growing beneath Falling Springs is recent (Panno et al. 2011). Copyright © University of Illinois Board of Trustees. Falling Springs photograph by S.J. Taylor; used with permission of the Illinois Natural History Survey. Auctioneer Cave photograph by S. V. Panno; used with permission of the Illinois State Geological Survey

sediment was deposited in subterranean slackwater lakes in the caves between 42,500 and 31,200 cal yr BP. During this dry period, the sediment almost completely filled the caves in this region, based on the dominance of the drought-tolerant C_4 plants that thrived during this period (Fig. 5.15a, b). Zhou et al. (2005) dated a large stalagmite from Fogelpole Cave that contained abundant fine-grained materials and which had stopped growing at 31,500 cal yr BP. Panno et al. (2004a) suggested this was a period of accelerated erosion on the surface. The mineralogy of the cave sediments indicated that the materials were derived from the entire glacial succession and not from the upper soil horizon, suggesting this was a period of rapid sinkhole formation, the effects of which are visible today (Fig. 5.15a, b). Most of the sediment fill was flushed from the caves sometime after 28,800 cal yr BP as conditions became wetter. Bones of Pleistocene megafauna, including the ribs of a ground sloth and teeth of an American Mastodon, have been found in these caves; their remains likely washed into sinkholes overlying passages within the caves. Relatively thin laminated flood deposits dated at 11,430 cal yr BP and 5100–4100 cal yr BP recorded the youngest flood events. The 5100–4100 cal yr BP deposits relate to a Holocene dry period (e.g., see Curry et al. 2010) that extended from 6300 to 4000 cal yr BP. Present-day deposits, often visible following precipitation events, probably are the result of accelerated erosion caused by construction and intensive farming activities in the vicinity of sinkholes in the area. In addition, Panno et al. (2012) determined, based on the age of the fine-grained sediments and speleothems from Illinois

Fig. 5.15 **a**, **b** A sediment-filled passage in Fogelpole Cave. The sediment dates back to at least 42,475 cal yr BP. Panno et al. (2004a) estimated that the cave was initiated at about 125,000 cal yr BP. Copyright © University of Illinois Board of Trustees. Photographs by S. V. Panno; used with permission of the Illinois State Geological Survey

Caverns and Fogelpole Cave, that at least the two longest caves of the sinkhole plain were initiated at about 125,000 yr BP, near the end of the Illinois glacial episode. Panno et al. (2016) also dated speleothems and sediment collected from Donnehue's Cave in south-central Indiana. Sediment ages were as old as 247,000 yr BP, and stalagmite ages ranged from greater than 400,000 yr to the present, suggesting that Donnehue's Cave is much older than the longest caves in southwestern Illinois. See Dorale (2020), Chap. 11 in this volume, for an extended discussion of paleoclimate in the UMV.

5.2.2.2 Earthquakes

Since the late 1920s, researchers have recognized the possibility of using caves and cave deposits as indicators of earthquake activity in seismic-prone areas of the world (Becker 1929). Subsequent work had continued to focus on damage within caves until Postpischl et al. (1991) correlated the ages of irregularities in the central growth axes of speleothems with known seismic events. Today, seismic studies in caves are based on these parameters (e.g., Gilli 2005; Kagan et al. 2005), as well as cave sediment deformation, rock falls, fault displacement (see Becker et al. 2006; Sebela 2008), and the initiation and reinitiation of stalagmite growth (Panno et al, 2009). In the United States, Tinsley et al. (2011, 2015), Panno et al. (2009, 2016) have explored the occurrence of seismic events in the UMV by using speleothems from caves in close proximity to the New Madrid Seismic Zone (located just south of Illinois near the intersection of Missouri, Arkansas, Kentucky, and Tennessee) and the Wabash Valley Seismic Zone (located near the state line between southeastern Illinois and southwestern Indiana; Fig. 5.16a, b). Panno et al. (2009) speculated that the disruption of stalagmite growth probably involved dilation or contraction of the hydraulic pathways of speleothem-forming waters seeping into the cave from above and precipitation or sedimentation within their pathways. The periodicity of the occurrence of these stalagmites is consistent with the 300- to 500-yr periodicity identified by Kelson et al. (1996), Tuttle et al. (1999). However, no research has been conducted on the role of hydraulic pathways.

5.2.3 Lincoln Hills

Bedrock in the Lincoln Hills karst area includes carbonate rock of the Silurian through Mississippian periods (Fig. 5.1). Cover-collapse sinkholes and caves are common in this region (Fig. 5.17); the cover-collapse sinkholes form in thick loess deposits overlying creviced limestone. In the eastern part of the Lincoln Hills region near Alton, Illinois, in Madison County, karst features are found within the Mississippian-age St. Louis and Ste. Genevieve Limestones. In the western part, karst features are found within the Kimmswick and St. Louis Limestones (Rubey 1952). Cover-collapse sinkholes in this area are similar to those of the Salem Plateau karst area: circular to elliptical bowl-shaped depressions containing trees, or shallow circular to elliptical ponds ringed by trees. Farther north in this region, cover-collapse sinkholes coincide with Mississippian-age Burlington–Keokuk Limestone bedrock on either side of the Illinois River and along the east side of the Mississippi River.

Caves in this region are described by Bretz and Harris (1961) as being relatively short, often partially collapsed, and containing abundant fine-grained sediment. These features limit the extent of the caves and often make them difficult to explore. Bretz and Harris suggested that the caves, which typically are found along the bluffs of the

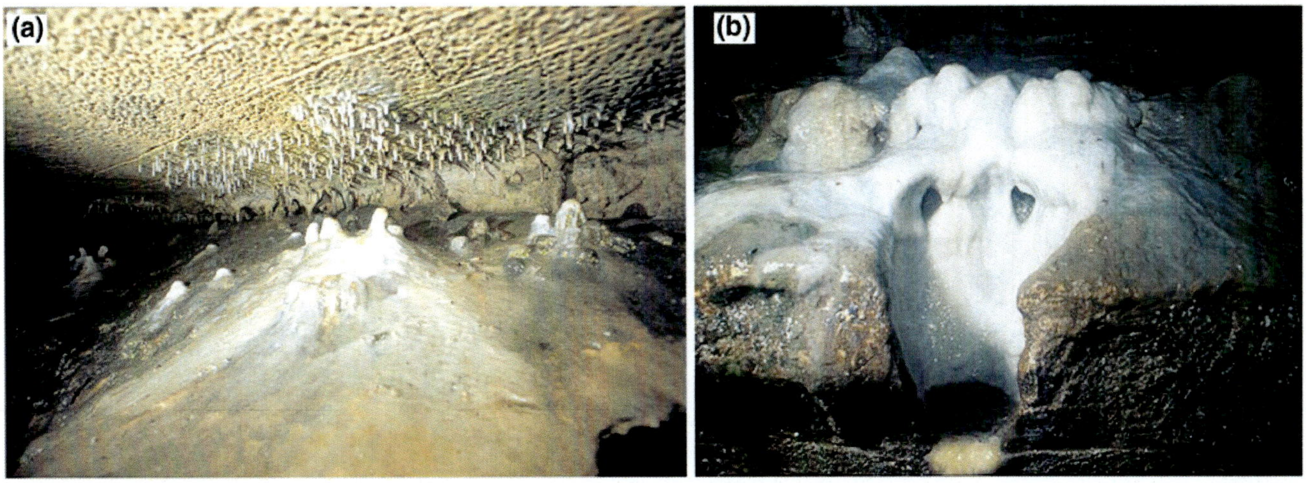

Fig. 5.16 White stalagmites about 5 cm in height in **a** Illinois Caverns and **b** Fogelpole Cave located in Monroe County, were initiated after the earthquakes of 1811–1812 in the New Madrid Seismic Zone. Copyright © University of Illinois Board of Trustees. Photographs by S. V. Panno; used with permission of the Illinois State Geological Survey

Fig. 5.17 Cover-collapse sinkhole in the western part of the Lincoln Hills karst area overlying Mississippian-age Burlington–Keokuk Limestone. Copyright © University of Illinois Board of Trustees. Photograph by S. V. Panno; used with permission of the Illinois State Geological Survey

Mississippi and Illinois Rivers, are erosional remnants partially filled or floored with fine-grained sediment emplaced by flooding some time during the Pleistocene Epoch (Fig. 5.18a, b).

The morphology of cover-collapse sinkholes in the Lincoln Hills area is similar to that of the Shawnee Hills (Fig. 5.4) and the Salem Plateau (Fig. 5.9) karst areas in that sinkholes within agricultural lands are relatively shallow and smooth in appearance. Sinkholes in densely wooded areas tend to be deeper, have steeper slopes, and appear more rugged. The differences are caused when sinkholes within the agricultural areas have been repeatedly infilled and their margins smoothed as a result of decades of cultivation, whereas the deeper, wooded sinkholes have never been modified by farming activities. The wooded cover-collapse sinkholes range from 12 to 15 m deep and 65 to 140 m wide, whereas the shallower sinkhole features situated within intensive agricultural areas range from 3 to 5 m deep and 110 to 130 m wide (white circle and white rectangle, respectively, in Fig. 5.19). Figure 5.19 shows lidar elevation data of cover-collapse sinkholes in the southern part of the Lincoln Hills karst area that has developed in Mississippian-age Limestone in a manner consistent with karst features present elsewhere in the Salem Plateau karst area. A recent aerial photograph of the same geographic area (Fig. 5.19, inset) shows erosion channels visible within these deeper sinkholes, indicating still-active erosion and expansion. Barely visible are the shallower sinkholes in the upper right-hand corner of the inset, which are in intensive agricultural land use.

5.2.4 Driftless Area

The bedrock geology of the Driftless Area of northwestern Illinois is composed of Middle Ordovician dolomites of the Galena and Platteville Groups and the Maquoketa Shale Group, as well as Silurian dolomite. The Galena Group and

Fig. 5.18 **a** The entrance to Burton Cave located in Adams County, in the Lincoln Hills karst area and **b** a large column within the cave. Copyright © University of Illinois Board of Trustees. Photographs by S.J. Taylor; used with permission of the Illinois Natural History Survey

the Maquoketa Shale Group compose the underlying bedrock for the lowland areas devoted to agricultural activities and Silurian dolomite accounts for the bedrock highland areas that are distinctive to this karst area (Fig. 5.21). The Ordovician-age Galena Group and Platteville Group host the karst aquifers that provide water resources to rural areas throughout northwestern Illinois. Remote sensing interpretation and analyses of 2012 multitemporal aerial imagery acquired during an extreme drought aerial of the Driftless Area in Jo Daviess County, revealed more than 18,000 crop lines that closely mimicked shallow groundwater-filled bedrock fractures and crevices of the Galena Group (Panno et al. 2015; Fig. 5.22a). These crevices can be seen in detail in road cuts and quarries in the area (Fig. 5.22b, c). Fracture and crevice orientations are predominantly east–west and north–south in this area, and these features compose the underlying bedrock karst aquifer (Panno et al. 2015).

Cover-collapse sinkholes are found in soils overlying both Silurian- and Ordovician-age dolomites. Sinkholes that developed in soils overlying the Galena Group are typically small, usually about 1 m in diameter and slightly less than 1 m deep because of the relatively thin soils in the area. Farm operators often infill these sinkholes with soil material as a way to expand the area of arable land. However, the Maquoketa Shale Group can behave like soil and collapse into crevices within the Galena Dolomite. This is particularly noticeable in the vicinity of low-lying areas where the water table can fall below the top of the Galena Dolomite (Fig. 5.23a, b). In addition, sinkhole springs can form in shale of the Maquoketa Group and drain to nearby streams. Crevice widths within the Galena Dolomite range from several centimeters to 1 m, as observed in quarries and road cuts. Crevices within the Galena Group tend to narrow with depth but can retain their widths to depths of 20 m or more, based on exposures in quarries and road cuts. Crevices within the Silurian dolomite extend the full thickness of the formation and tend to widen as dolomite blocks move on the Maquoketa Shale. Crevices in both formations are often partially filled with fine-grained sediment, particularly nearest the surface (Fig. 5.23a, b).

Cover-collapse sinkholes in soils overlying Silurian dolomite are on the order of 20–30 m in diameter and 2–7 m deep (Panno et al. 2014). These sinkholes typically aligned along east–west orientations, which are consistent with the dominant crevice orientation within the Silurian dolomite (Panno et al. 2015; Fig. 5.24). Crevice widths in this formation typically remain constant with depth and terminate in the Maquoketa Shale Group at the Silurian–Ordovician boundary. Crevices within the Silurian dolomite can widen as large blocks slide along underlying shale (Fig. 5.25a, b) and create large sinkholes and open crevices along the bluff faces (Panno et al. 2016; Fig. 5.26a, b).

The crevice caves that dominate in northwestern Illinois are relatively short (<100 m) and occur in the strata of both the Galena Group and the Silurian dolomites (Fig. 5.25a, b), although phreatic caves are found in both formations (Fig. 5.26c). The caves explored by Webb et al. (1993) in Jo Daviess County and adjoining Carroll County are described as dry crevice caves with no speleothems. One of the longest

Fig. 5.19 Shaded relief lidar showing cover-collapse sinkholes in the southern part of the Lincoln Hills karst area developed in fine-grained sediment overlying Ordovician-age limestone. At the lower right of inset **a** is an aerial photograph showing deeper sinkholes in densely wooded areas (white circle). Barely visible are the shallower sinkholes, which are in intensive agricultural land use (white rectangle). Cover-collapse sinkholes in and around Alton in Madison County, have formed within Ste. Genevieve Limestone and average 12–15 m deep. Ravines visible within some of the sinkholes (Fig. 5.20) indicate that these sinkholes are still actively being eroded, making such features a potential health and safety hazard. Some of the sinkholes that occupy an urban setting (black outline on Fig. 5.20a) have been stabilized, and at least one has been modified into a park (Fig. 5.20b). Data provided by the Illinois Height Modernization Program (clearinghouse.isgs.illinois.edu) supported by the Illinois Department of Transportation

caves in the Driftless Area is Bat Cave within the Mississippi Palisades State Park. This cave extends 58 m and is 26 m deep (Webb et al. 1993). Many of the caves in the Galena Group of the Driftless Area are filled with clay and dolomitic sand; these caves were discovered during mining operations in the area and were excavated in the search for base metal ores (Bretz and Harris 1961). Some crevice caves in the Galena Group were formed during an episode of ore mineralization during the Early Permian Period (Brannon et al. 1992). These cavities or crevices have been observed in mine workings within the Galena and Platteville Groups throughout the Driftless Area.

Other crevices have been filled with ore minerals (primarily galena- and sphalerite-filled gash-vein deposits) that constitute the Upper Mississippi Valley zinc–lead district in the Driftless Area of northwestern Illinois, northeastern Iowa, and southwestern Wisconsin (Fig. 5.27). For example, the California Diggings located just southwest of the town of Galena, Illinois, was described as having "bonanzas of sulfide-encrusted cave walls and ceilings" (Bretz and Harris 1961, p 85) in the Galena Group. The galena and sphalerite were initially deposited on the cave walls. At the time of mining, most of the ore had fallen off the walls and was found in reddish clay near the bottom of the cave, possibly dislodged at some point by flooding after deposition of the ore minerals (Bretz and Harris 1961). Bretz (1938) published a summary of caves in the Galena Group within the Driftless Area and found that caves in the area had formed along predominantly east–west-trending joints. See Dockal (2020), Chap. 7 in this volume, for an extended discussion of lead–zinc mineralizations in the UMV.

Both Silurian dolomite and Galena Dolomite have meter-wide crevice caves and phreatic tubes (Fig. 5.26c and Fig. 5.28). Some of the crevice caves in the Silurian

Fig. 5.20 **a** Shaded relief lidar showing the southern Lincoln Hills karst area and sinkholes within a high-density residential area of Alton in Madison County. Data provided by the Illinois Height Modernization Program (clearinghouse.isgs.illinois.edu) supported by the Illinois Department of Transportation. **b** One of the sinkholes (black outline in panel a) has been modified into Riverview Park. The sinkhole has been modified to a rectangular shape and is 8 m deep. Copyright © University of Illinois Board of Trustees. Photograph by S. V. Panno; used with permission of the Illinois State Geological Survey

Fig. 5.21 View from the Silurian dolomite highlands of Illinois' Driftless Area, northwestern Illinois. The agriculturally dominated lowlands are underlain by Galena Dolomite. Copyright © University of Illinois Board of Trustees. Photograph by S. V. Panno; used with permission of the Illinois State Geological Survey

dolomite are located near talus slopes and, in a complex sinkhole–cave–talus slope relationship, form a unique habitat known as an *algific talus slope*, first discovered in 1981 (Post 1998; Fig. 5.29a, b). Here, winter temperatures cool bedrock surrounding the cave, water from melting snow and ice drips into the sinkhole and freezes within the cave atop Maquoketa Shale, and the ice slowly, but never completely, melts during the summer months. The sinkhole–cave–slope system leads to the discharge of cold air from crevices present in bedrock along the talus slope during the warmer summer months, creating a microclimate similar to that in Pleistocene times. Living within these microclimates are species of Pleistocene plants and, within the mossy cover, two species of snails that before the discovery of algific talus slopes were thought to have been extinct since the Pleistocene Epoch (Fig. 5.29).

Fig. 5.22 **a** Low-altitude oblique photograph of an alfalfa field exhibiting vegetated crop lines taken on July 19, 2012, in Jo Daviess County. North is to the left, field of view (width) is 140 m. **b** The crop lines reflect the bedrock fractures and crevices of the Galena Dolomite visible in the quarries of northwestern Illinois. The height of the highwall is 20 m. **c** Rose diagram of azimuth orientations of crop line fractures and crevices determined for northwestern Illinois. The strongly dominant trend is east–west; the azimuth of the subdominant trend is nearly north–south. Alfalfa field photograph and rose diagram from Panno et al. (2015). Copyright © 2015 University of Illinois Board of Trustees. Used with permission of the Illinois State Geological Survey. Quarry wall photograph from Panno et al. (2017). Copyright © 2017 University of Illinois Board of Trustees. Photograph by S. V. Panno; used with permission of the Illinois State Geological Survey

Fig. 5.23 **a** Shaded relief lidar showing cover-collapse sinkholes along the bluffs of the Apple River in Jo Daviess County, that formed in shale of the Maquoketa Group overlying dolomite of the Galena Group (1 and 2). Data provided by the Illinois Height Modernization Program (clearinghouse.isgs.illinois.edu) supported by the Illinois Department of Transportation. **b** The largest sinkhole on the south side of the river has a diameter of 20 m and a depth of 4 m. The depth from the thalweg of the river to the upland is 50 m at this location. The angular nature of the erosion gully (3) suggests fracture control. Copyright © University of Illinois Board of Trustees. Photograph by S. V. Panno; used with permission of the Illinois State Geological Survey

Research on springs in the Driftless Area has revealed that springs in the area are typically small, discharging from small crevices at a rate of only about 0.5–0.7 L/s, but can have as much as 3 L/s of baseflow (Fig. 5.30a). Springs discharge from crevices in the Galena Group and Silurian dolomites, and often along the Silurian dolomite–Maquoketa Group Shale and the Galena Dolomite–Maquoketa Group Shale contacts (Panno et al. 2019a). The discharge point of the spring commonly creates a 2- to 3-m-wide stream choked with watercress, an edible aquatic plant (*Nasturtium officinale*), before discharging to a nearby stream (Fig. 5.30b).

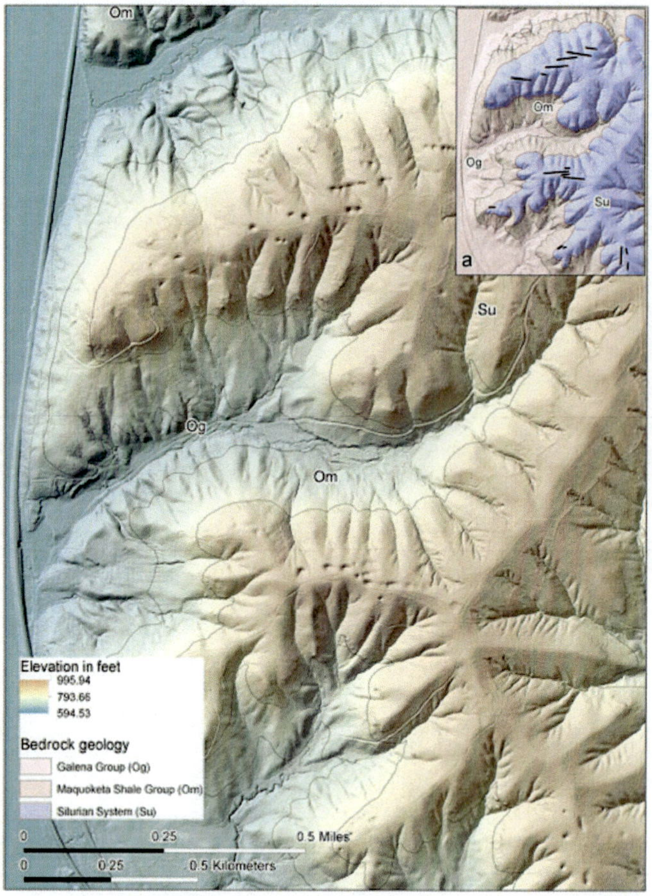

Fig. 5.24 Shaded relief lidar showing east–west-trending cover-collapse sinkholes overlying crevices within Silurian dolomite in Jo Daviess County. Data provided by the Illinois Height Modernization Program (clearinghouse.isgs.illinois.edu) supported by the Illinois Department of Transportation. Inset **a** shows the geologic formations (boundaries shown as gray lines on the main image) and the linear nature of sinkhole formation (after Panno et al. 2015)

Fig. 5.25 **a** Silurian dolomite blocks separated along joints near the top of a bedrock ridge in Jo Daviess County. Copyright © University of Illinois Board of Trustees. Photograph by S. V. Panno; used with permission of the Illinois State Geological Survey. **b** Dolomite blocks sliding downslope on a base of Maquoketa Shale. Copyright © E.L. Baranski; used with permission

Fig. 5.26 **a** Cover-collapse sinkholes range in depth from 1 to 7 m and in width from 10 to 30 m, from Panno et al. (2004b). Copyright © 2004 University of Illinois Board of Trustees. **b** Crevices within the Silurian dolomite are visible in outcrops and road cuts, from Panno et al. (2017). Copyright © 2017 University of Illinois Board of Trustees. **c** Sinuous phreatic caves are present at the base of the Silurian dolomite where the dolomite is in contact with the Maquoketa Shale. Copyright © University of Illinois Board of Trustees. Photographs by S. V. Panno; used with permission of the Illinois State Geological Survey

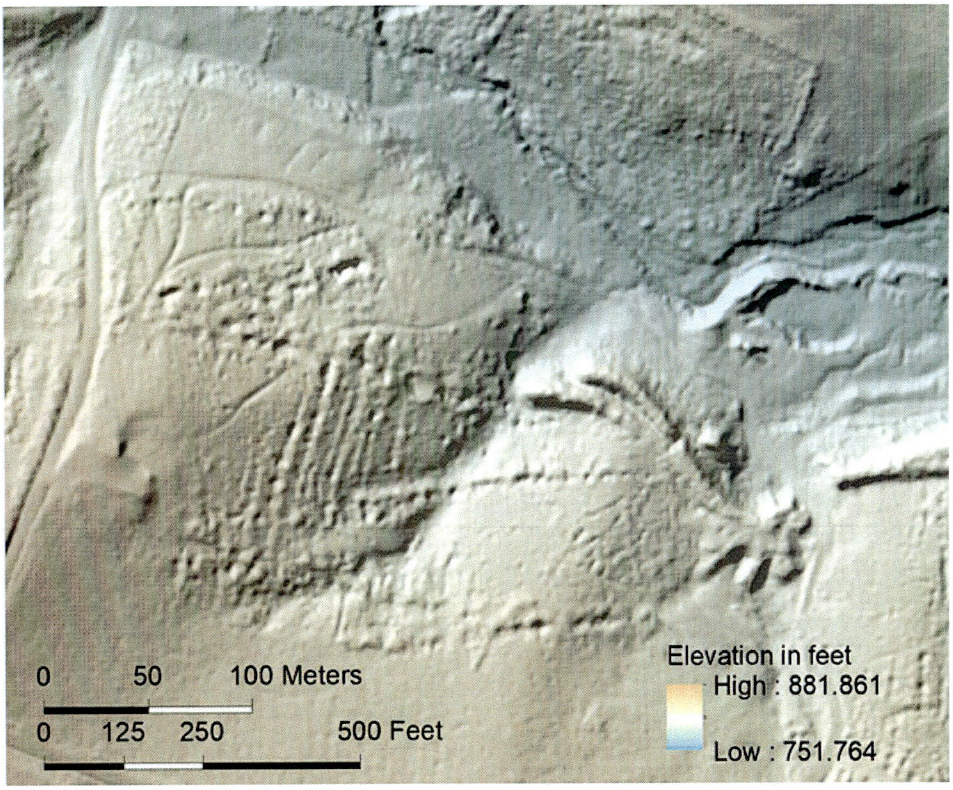

Fig. 5.27 Shaded relief lidar showing the Vinegar Hill Mine site in Jo Daviess County with numerous mining dog holes or diggings. The digging alignments are consistent with the major bedrock crevice orientations in the Driftless Area. Data provided by the Illinois Height Modernization Program (clearinghouse.isgs.illinois.edu) supported by the Illinois Department of Transportation. Modified from Panno et al. (2017). Copyright © 2017 University of Illinois Board of Trustees

5.2.5 North-Central

Located within the Rock River watershed, the North-Central karst area consists of numerous cover-collapse sinkholes overlying carbonate bedrock of the Lower Ordovician Shakopee Dolomite of the Prairie du Chien Group and the Middle Ordovician-age Platteville and Galena Groups. Exposures of these rocks along road cuts (particularly along Interstate 39 and U.S. Route 20), quarries, and outcrops along streams reveal creviced dolomite (Fig. 5.31a). Weibel and Panno (1997) have mapped sinkholes in this area (Fig. 5.1) but the few documented caves are network type

Fig. 5.28 **a** Crevice cave exposed within a quarry in Galena Group dolomite that follows a roughly east–west joint pattern, from Panno et al. (2017). Copyright © 2017 University of Illinois Board of Trustees. Used with permission of the Illinois State Geological Survey. **b** Interior of a similarly oriented cave in the Driftless Area. Copyright © University of Illinois Board of Trustees. Photographs by S. V. Panno; used with permission of the Illinois State Geological Survey

Fig. 5.29 **a** Algific talus slope showing **b** cold microclimate along the slope. Algific talus slope from Panno et al. (2016). Copyright © 2016. Used by permission of the Illinois State Geological Survey. Microclimate photograph copyright © Richard Mattas; used with permission

Fig. 5.30 **a** Ice Spring and **b** Malon Spring in Jo Daviess County. The temperature of the springs was about 10 °C in July 2016. Watercress **b** often dominates the discharge paths to nearby streams. Ice Spring photograph from Panno et al. (2017). Copyright © 2017 University of Illinois Board of Trustees. Used with permission of the Illinois State Geological Survey. Malon Spring photograph copyright © University of Illinois Board of Trustees. Photographs by S. V. Panno; used with permission of the Illinois State Geological Survey

and typically short (Fig. 5.31b). The longest cave in the North-Central karst area is Settlers Cave, located 10 km south of Rockford, Illinois. Settlers Cave is a network-type cave with 153 m of passages (Fig. 5.28b) that occurs within the Wise Lake Formation of the Galena Group (Carpenter and Ekberg 2009). As with the caves in the Driftless Area, Settlers Cave is dry and formed along preexisting fractures. Carpenter and Ekberg (2009, p 498) stated that caves in the area "tend to form on bentonite layers that act as permeability barriers to downward percolating vadose water." Passage orientations of the cave are consistent with the fracture orientations in that area described by McGarry (2000), Panno et al. (2015).

Immediately east of Dixon in Lee County, fractures and solution-enlarged crevice traces on the Galena Dolomite bedrock surface (Fig. 5.32, areas a and b) and cover-collapse sinkholes are aligned with the dominant crevice orientations. A lidar shaded relief image shows that cover-collapse sinkholes have developed in unconsolidated sediments overlying the Ordovician Galena and Platteville Groups (Op formation). Glacial till in this area is relatively thin, providing optimal geologic conditions for sinkhole formation over solution-enlarged crevices. The northern group of sinkholes (white rectangle, Fig. 5.32) ranges in width from 2 to 3 m and in depth from 10 to 20 m. The southern group of sinkholes (white rectangle, Fig. 5.32) ranges in width from 3 to 5 m and in depth from 15 to 30 m. Cover-collapse sinkholes in the northern group appear to be linear depressions in sediment that likely formed along crevices. Formerly buried crevices and fractures are visible as linear vegetated patterns on the excavated floor of the quarry (Fig. 5.32, areas a and b). Healthy green vegetation growing in the water-bearing fractures and crevices enhances their discrimination. These karst features are similar in orientation to those documented in the Driftless Area of Jo Daviess County (Panno et al. 2015) and include an additional northwest–southeast-trending set.

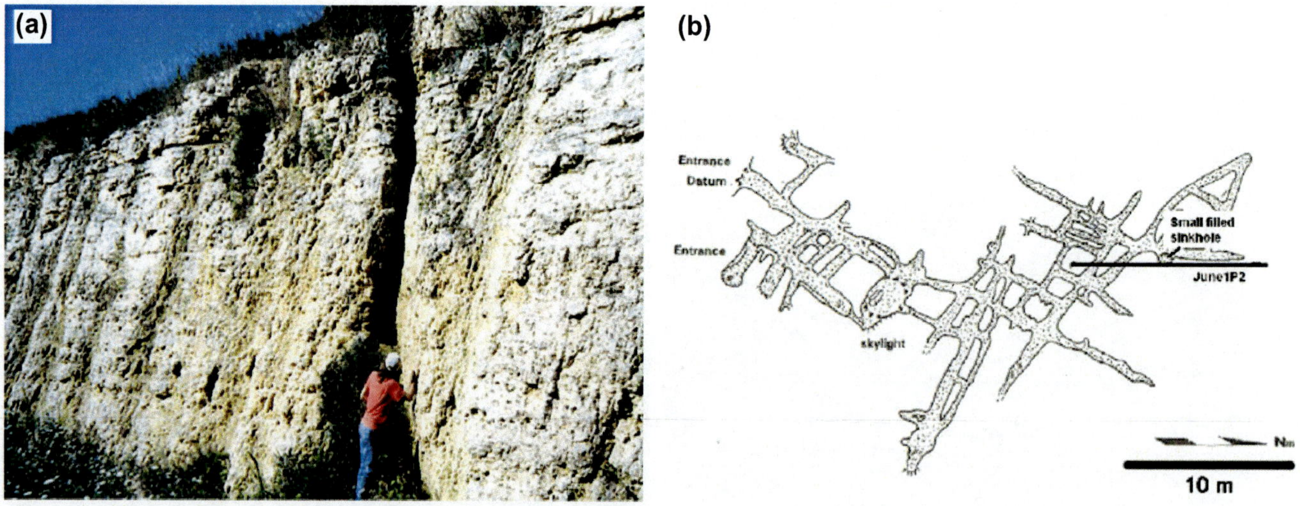

Fig. 5.31 **a** Solution-enlarged crevices are common along road cuts near Freeport in Livingston County. From Panno et al. (2017). Copyright © 2017 University of Illinois Board of Trustees. **b** Plan view of Settlers Cave reflects the orientation of fractures and crevices in the North-Central karst area. From Carpenter and Ekberg (2009); used with permission

5.2.6 Minor Occurrences of Karst

Isolated karst features are found throughout Illinois where carbonate rock is exposed at the surface or near surface by natural or man-made exposures, or both. These occurrences are indicators of paleokarst features.

5.2.6.1 The Kankakee Plain

Will County Silurian limestone exhibits karst and paleokarst features that are exposed in outcrops, quarries, and along the Illinois River in northeastern Illinois (Fig. 5.33). Bretz (1940) first described solution cavities in the Niagaran Series limestone of the Silurian System in a quarry near the town of Joliet. These features were found on limestone bedrock surfaces; they are linear joint-controlled cavities filled with Pennsylvanian stratified shale, sandstone, and large fragments of limestone. The cavities range from a few centimeters to 25 m wide and can be up to 45 m long (Bretz 1940). In photographs by Bretz, the cavities appear to be at least 10 m deep. Samuel Panno visited a closed oil refinery near Joliet that contained a number of these cavities. The cavities were aligned along fractures and crevices but did not appear to be connected by anything other than a narrow fracture. They also contained gravels that included pyrite nodules up to 5 cm in diameter.

Samuel Panno also examined several cover-collapse sinkholes in a wooded area west of the town of Manhattan near the Illinois River in 1989. These sinkholes were in close proximity to one another in relatively thin soil (3 m thick) overlying Silurian limestone; they measured 5–7 m in diameter and 2–3 m deep. Quarries to the east contained crevices that were large enough to be responsible for sinkhole formation in areas of relatively thin soil cover, less than 15 m thick. Sinkholes formed in this area because the soils are thinner near the river; they thicken away from the river to depths greater than 15 m.

Kankakee County Several small caves are located along the Kankakee River Valley. The longest of these is Horsethief Cave, which was described by Bretz and Harris (1961) as high enough to be entered without crouching, extending for about 10 m, and apparently the "lateral toe" of a naturally excavated filled solution cavity common in the area.

5.2.6.2 Bloomington Ridged Plain

LaSalle County Near the town of Utica, numerous limestone mines (the Blackball Mines) formed within the Pennsylvanian-age La Salle Limestone Member of the Bond Formation. In 1969, Samuel Panno explored a cave within this limestone near the base of a bluff adjacent to Pecumsaugan Creek; other caves may have been present but were hidden by the dense foliage. The cave was a phreatic tube, about 50–55 cm in diameter, that formed along a near-vertical fracture on a single level, with an entrance along a steep bluff. The cave extended into the bluff for about 30 m and continued beyond, where slight undulations in the cave made it difficult to traverse. Cover-collapse

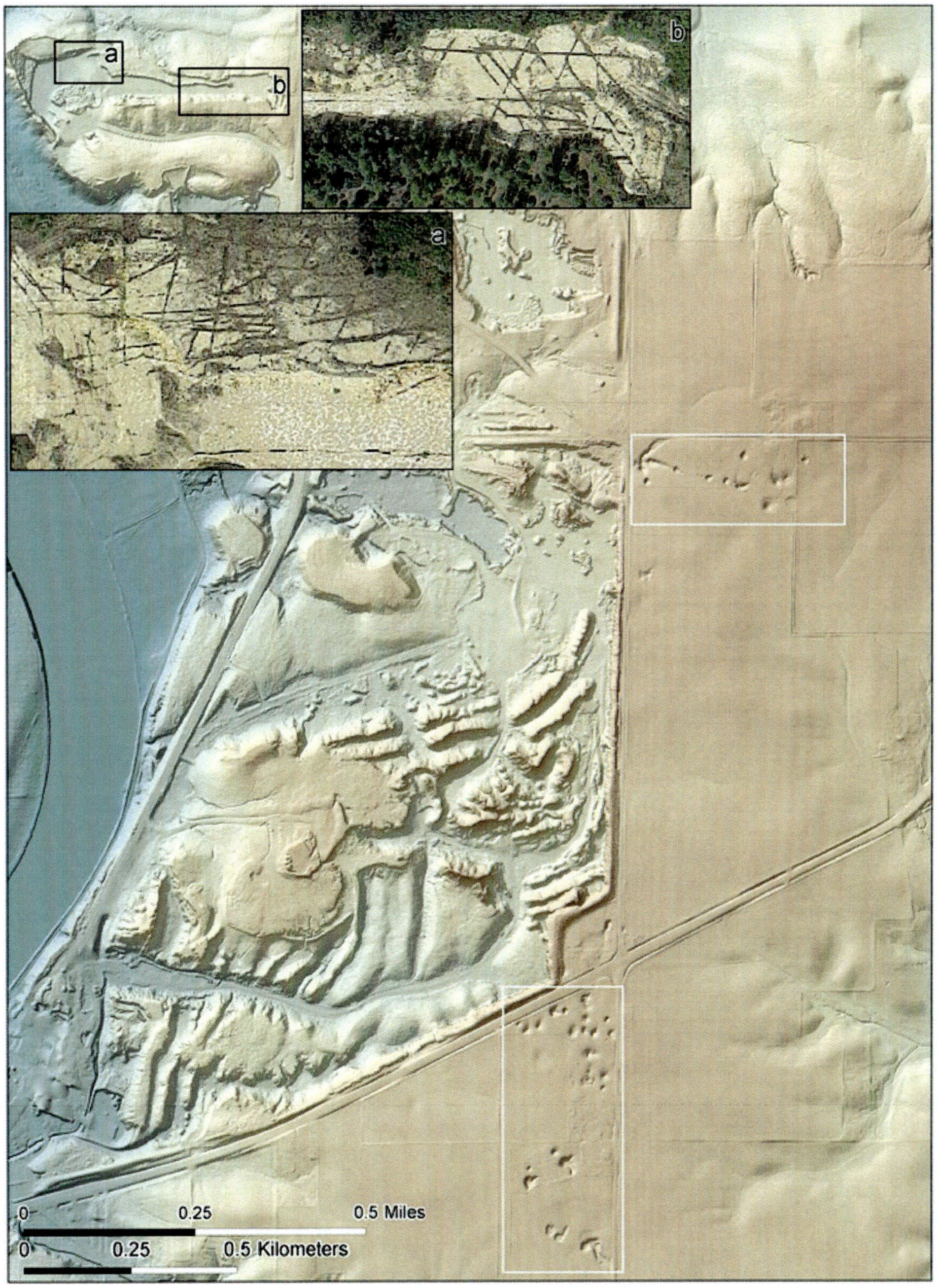

Fig. 5.32 Shaded relief lidar showing two groups of cover-collapse sinkholes overlying Ordovician-age Galena–Platteville limestone near Dixon in Lee County. This represents the North-Central karst area, where glacial till is relatively thin. Lidar data provided by the Illinois Height Modernization Program (clearinghouse.isgs.illinois.edu) supported by the Illinois Department of Transportation. The linear features within the quarry (**a**, **b**) are due to vegetation growing along crevices and fractures within the quarry floor (imagery © Google)

sinkholes occur in sediment overlying the La Salle Limestone Member and are visible on aerial photographs of the area.

As part of the Utica Hydraulic Cement Company, the limestone near Utica was mined for its high purity in the early 1900s using room and pillar methods. The La Salle Limestone Member is apparent where mine openings are present (Fig. 5.34a). This member is about 10 m thick and is exposed along the western flank of the La Salle Anticlinorium (Willman et al. 1975, p 197). After mining, the limestone was placed into kilns on site and alternately layered with coke; burning of the coke converted the limestone to quicklime. Remnants remain of the kilns in use in the late 1880s. Because the mines contain a major hibernaculum for at least four species of bats, entrances to the mines are now gated and access is now limited only to researchers. In

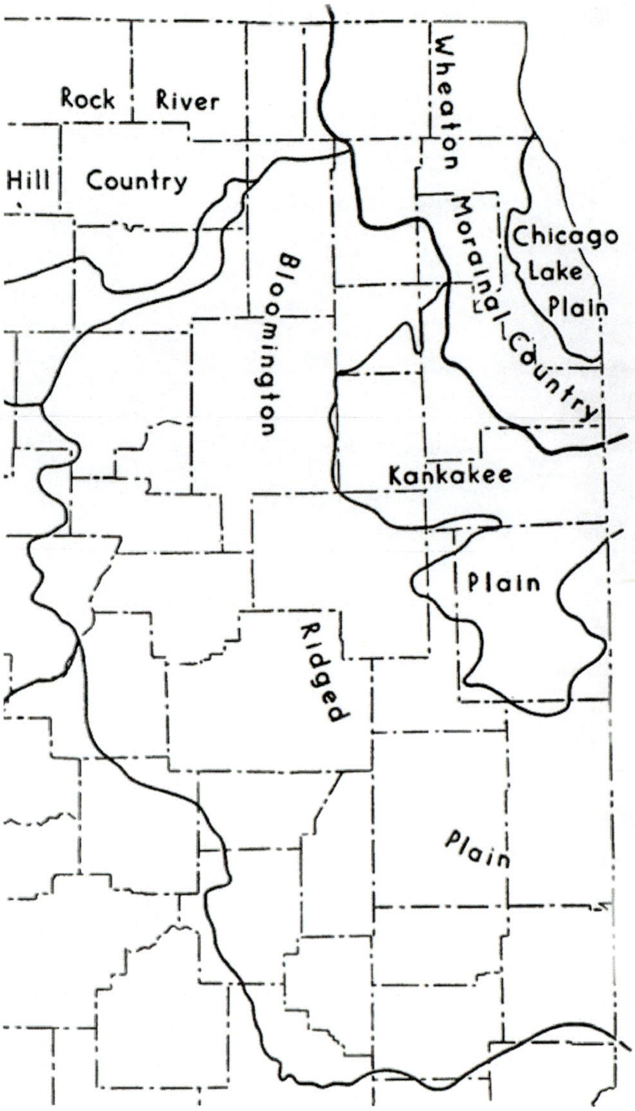

Fig. 5.33 Physiographic divisions of northeastern Illinois. Adapted from Leighton et al. (1948). Copyright © 1948 University of Illinois Board of Trustees. Used with permission of the Illinois State Geological Survey

addition, the mines contain stalactites up to 5 cm in length and often contain ice stalagmites near at least one of the entrances during winter months (Fig. 5.34b).

Starved Rock and Matthiessen State Parks are the locations of exposures of Ordovician-age St. Peter Sandstone, a poorly cemented, nearly pure marine quartz sand that is mined extensively for glassmaking and other uses. In northern Illinois, the St. Peter Sandstone also contains associated karst features, such as rare sinkholes (Carpenter and Schroeder 2015) and small caves. Characteristic of the park area is shelter-type caves that occur within canyons in the St. Peter Sandstone. Small conduits associated with fractures are also visible in canyon walls along the bedding planes (Fig. 5.35a, b).

Douglas County A relatively small cave was exposed in a borrow pit east of the town of Tuscola in Devonian-age limestone near the axis of the La Salle Anticlinorium. No sediment was found in the cave, suggesting the cave was hydrogeologically connected and still active. However, crevices in limestone in a quarry 1.5 km east of the cave site are filled with fine sediment (Panno et al. 1997). The borrow pit where the cave was found is now filled with water, forming a small pond, and the cave is no longer accessible.

5.2.7 The Effects of Karst in Illinois

Citizens living within the five major karst areas of Illinois often have problems with sinkholes and water quality similar to those of people living in other areas of the UMV, although the specific types of problems vary across the state. Bedrock crevices are narrower in the dolomites of the Driftless Area of northwestern Illinois, where soils are thinner. As a result, cover-collapse sinkholes are relatively small features that typically do not create problems for property owners. Conversely, bedrock in southern and southwestern Illinois consists of highly erodible limestone formations that exhibit wider crevices in association with thicker soils. In these locations, cover-collapse sinkholes can be relatively large and can cause property damage. For example, construction can change surface-water runoff patterns, resulting in erosion along the sinkhole margins. This can contribute to sinkhole growth or the formation of other sinkholes in close proximity to buildings and roadways, resulting in damage to these structures.

The effects of Illinois karst on groundwater quality are similar for all karst areas in Illinois. Regardless of their size, sinkholes are direct conduits to bedrock aquifers, and groundwater movement in karst aquifers is very rapid; thus, surface-borne contaminants can enter and migrate through the aquifer virtually unchecked. Contaminants found in groundwater in these areas included road salt and components of septic effluent, such as enteric bacteria, nutrients, pharmaceuticals, personal care products, animal waste, and nitrogen and potassium fertilizers (Panno et al. et al. 2001, 2006; Panno and Kelly 2004; Zhang et al. 2014; Dodgen et al. 2016). Most recently, microplastic fibers were found in karst springs and shallow wells $\leq$ 45 m deep within the Driftless Area of northwestern Illinois and in springs of the sinkhole plain of southwestern Illinois. In both areas, these fibers likely emanated from private septic systems (Panno et al. 2019b). Microfibers are an indicator of the open nature of the karst aquifers of Illinois—direct evidence that particulates can enter and migrate through these aquifers regardless of bedrock lithology and crevice size.

Fig. 5.34 **a** One of the openings to the Blackball Mines near Utica, outlined with a picturesque brick façade that highlights the limestone of interest for mining. Copyright © S. V. Panno; used with permission. **b** Bamboo-shaped ice stalagmites near one of the entrances to the Blackball Mine, ranging from 30 to 45 cm high, from Panno et al. (2002). Copyright © 2002 University of Illinois Board of Trustees. Photographs by S. V. Panno; used with permission of the Illinois State Geological Survey

Fig. 5.35 **a** Shelter caves and **b** canyons form within the St. Peter Sandstone in Matthiessen State Park. Copyright © University of Illinois Board of Trustees. Photographs by S. V. Panno; used with permission of the Illinois State Geological Survey

Acknowledgements This work was supported by the Illinois State Geological Survey (ISGS), Prairie Research Institute, University of Illinois. The authors thank Brandon Curry (ISGS) for his review of the manuscript, Michael W. Knapp for his assistance with the photographs, and especially Susan Krusemark for her efforts in editing the many details in this chapter.

References

Angel JC, Nelson DP, Panno SV (2004) A comparison of manual and a GIS-based method for determining sinkhole density: An example from Illinois' sinkhole plain. J Cave Karst Stud 66(1):9–17

Becker HK (1929) Hohle und Erdbeben. Mitteilungen Uber Hohlen Und Karstforschung 14:130–133

Becker A, Davenport CA, Eichenberger U, Gilli E, Jeannin PY, Lacave C (2006) Speleoseismology: a critical perspective. J Seismol 10:371–388

Berkson A (2009) Discover Illinois archaeology, 3rd printing. Illinois Association for Advancement of Archaeology and the Illinois Archaeological Survey, Springfield, 27p

Brannon JC, Podosek FA, McLimans RK (1992) Alleghenian age of the Upper Mississippi Valley zinc-lead deposits determined by Rb-Sr dating of sphalerite. Nature 356:509–511

Bretz JH (1938) Caves in the Galena formation. J Geol 46:337–384

Bretz JH (1940) Solution cavities in the Joliet Limestone of northeastern Illinois. J Geol 48:337–384

Bretz JH, Harris SE (1961) Caves of Illinois. Report of Investigations 215. Illinois State Geological Survey, Champaign, 87p

Carpenter PJ, Ekberg DW (2009) Identification of buried sinkholes, fractures and soil pipes using ground-penetrating radar and 2D electrical resistivity tomography. In: Anderson NL (ed) Proceedings of the 2006 highway geophysics–NDE conference, St. Louis, Missouri, Paper VI-B-4, pp 437–499

Carpenter PJ, Schroeder LM (2015) Investigation of a sinkhole in Ogle County, northwestern Illinois, using near-surface geophysical techniques. In: Proceedings of the 14th multidisciplinary conference on sinkholes and the engineering & environmental impacts of karst, Rochester, Minnesota, 5–9 October 2016, 8p

Curry BB, Wang H, Panno SV, Hackley KC (2010) Quaternary paleoclimate. In: Kolata D, Nimz CK (eds) Geology of Illinois. Illinois State Geological Survey, Champaign, pp 248–260

Devera JA (2000) Bedrock geologic map: Renault Quadrangle, Monroe County, Illinois. Illinois Geologic Quadrangle Map–Renault BG. Illinois State Geological Survey, Champaign

Dockal JA (2020) Lead-zinc crevices of Iowa, Illinois, and Wisconsin (Chap. 7 in this volume)

Dodgen L, Kelly WR, Panno SV, Taylor SJ, Wiles KN, Zhang Y, Zhang W (2016) Characterizing pharmaceutical, personal care product and hormone contamination in a karst aquifer of southwest Illinois, USA, using water quality and stream flow parameters. Sci Total Environ 578:281–289

Dorale JA (2020) Paleoclimate studies from Midwestern Caves (Chap. 11 in this volume)

Dorale JA, Edwards RL, Ito E, Gonzalez IE (1998) Climate and vegetation history of the midcontinent from 75 to 25 ka: A speleothem record from Crevice Cave, Missouri, USA. Science 282:1871–1874

Gilli E (2005) Review on the use of natural cave speleothems as palaeoseismic or neotectonics indicators. C R Geosci 337 (13):1208–1215

Hackley KC, Panno SV, Hwang HH, Kelly WR (2007) Groundwater quality of springs and wells of the sinkhole plain in southwestern Illinois: Determination of the dominant sources of nitrate. Circular 570. Illinois State Geological Survey, Champaign, 39p

Kagan E, Agnon A, Bar-Matthew M, Avalon A (2005) Dating large infrequent earthquakes by damaged cave deposits. Geology 33 (4):261–264

Kelson KI, Simpson GD, Van Arsdale RB, Harris JB, Haradan CC, Lettis WR (1996) Multiple Holocene earthquakes along the Reelfoot fault, central New Madrid seismic zone. J Geophys Res 101(B3):6151–6170

Killey MM (2007) Illinois' ice age legacy. Geoscience Education Series, 14. Illinois State Geological Survey, Champaign, 74 p

Kolata DR (2005) Bedrock geology of Illinois. Illinois Map 14, scale 1:500,000. Illinois State Geological Survey, Champaign, 2 sheets

Kolata DR, Nimz S (eds) (2010) Geology of Illinois. Illinois State Geological Survey, Champaign, 530p

Leighton MM, Ekblaw GE, Horberg L (1948) Physiographic divisions of Illinois. Report of Investigations 129. Illinois State Geological Survey, Champaign, 33p

Luman D, Tweddale T, Bahnsen B, Willis P (2004) Illinois land cover. Illinois Map 12, scale 1:500,000. Illinois State Geological Survey, Champaign

McGarry CS (2000) Regional fracturing of the Galena–Platteville aquifer in Boone and Winnebago Counties, Illinois: geometry, connectivity and tectonic significance. MS thesis. University of Illinois at Urbana-Champaign, 209p

Oliver JS, Graham RW (1988) Preliminary inventory of natural resources in select caves in Illinois. Final report. Illinois Department of Natural Resources, Illinois State Museum, Springfield, 155p

Panno SV, Bourcier WL (1990) Glaciation and saline/fresh-water mixing as a possible cause of cave formation in the eastern midcontinent region of the United States: a conceptual model. Geology 18(8):769–772

Panno SV, Kelly WR (2004) Nutrient and herbicide loading in two groundwater basins in Illinois' sinkhole plain. J Hydrol 290:229–242

Panno SV, Luman DE (2012a) Karst map of Monroe County. Illinois County Geologic Map, ICGC Monroe County-SD, 1:62,500. Illinois. Illinois State Geological Survey, Champaign

Panno SV, Luman DE (2012b) Mapping palimpsest karst features of Illinois' sinkhole plain using historical aerial photography. Carb and Evap.https://doi.org/10.1007/s13146_012_0107_4

Panno SV, Luman DE (2016) Characterization of Illinois' sinkhole plain using airborne LiDAR data and a new Arc-GIS program. North-Central Geol Soc Am Abstr Progr 48(5). https://doi.org/10.1130/abs/2016NC-275647

Panno SV, Luman DE (2018) Characterization of cover-collapse sinkhole morphology on a groundwater basin-wide scale using lidar elevation data: a new conceptual model for sinkhole evolution. Geomorphology 318:1–17

Panno SV, Weibel CP, Heigold PC, Reed PC (1994) Formation of cover-collapse sinkholes in a karst area of southern Illinois: Interpretation and identification of associated buried cavities. Environ Geol 23(3):214–220

Panno SV, Weibel CP, Li W (1997) Karst regions of Illinois. Open File Series 1997-2. Illinois State Geological Survey, Champaign, 42p

Panno SV, Hackley KC, Hwang HH, Kelly WR (2001) Determination of the sources of nitrate contamination in karst springs using isotopic and chemical indicators. Chem Geol 179(1–4):113–128

Panno SV, Greenberg SE, Weibel CP, Gillespie PK (2002) A field guide to the Illinois Caverns State Natural Area: Teacher's curriculum. GeoScience Education Series 2002. Illinois State Geological Survey, Champaign

Panno SV, Curry BB, Wang H, Hackley KC, Liu C-L, Lunstrom C, Zhou J (2004a) Climate change in southern Illinois, USA, based on the age and $\delta^{13}C$ of organic matter in cave sediments. Quat Res 61 (3):301–313

Panno SV, Greenberg SE, Weibel CP, Gillespie PK (2004b) Guide to the Illinois Caverns State Natural Area. GeoScience Education Series 19. Illinois State Geological Survey, Champaign, 106p

Panno SV, Hackley KC, Kelly WR, Hwang HH, Wilhelm F, Taylor SJ, Stiff BJ (2006) Potential effects of recurrent low oxygen conditions on the Illinois Cave amphipod. J Cave Karst Stud 68(2):55–63

Panno SV, Lundstrom C, Hackley KC, Curry BB, Fouke B, Zhang Z (2009) Major earthquakes recorded by speleothems in Midwestern U.S. caves. Bull Seismol Soc Am 99(4):2147–2154

Panno SV, Hackley KC, Kelly WR, Luman DE (2011) Illinois' sinkhole plain: classic karst terrain of the Midwestern United States. Geological field trip guidebook for the 12th multidisciplinary conference on sinkholes and the engineering & environmental impacts of karst, 10–14 January, 2011, St. Louis, Missouri, USA. Guidebook 39. Illinois State Geological Survey, Champaign, 45p

Panno SV, Curry BB, Wang H, Hackley KC, Zhang Z, Lundstrom C (2012) The effects of climate change and glaciation on speleogenesis and karstification in southwestern Illinois' sinkhole plain. Carbonates Evaporites 24:87–94

Panno SV, Kelly WR, Angel JC, Luman DE (2013) Hydrogeologic and topographic controls on evolution of karst features in Illinois' Sinkhole Plain. Carbonates Evaporites 28:13–21

Panno SV, Millhouse PG, Nyboer RW, Watson D, Kelly WR, Anderson L, Abert CC, Luman DE (2014) Guide to the geology, hydrogeology, botany, history and archaeology of the driftless area of northwestern Illinois, Jo Daviess County. Field Trip Guidebook 2014A. Illinois State Geological Survey, Champaign, 70p

Panno SV, Luman DE, Kolata DR (2015) Characterization of karst terrain using remotely sensed data in Jo Daviess County, Illinois. Circular 589, 1 map, scale 1:62,500. Illinois State Geological Survey, Champaign, 29p and digital appendix. https://isgs.illinois.edu/publications/58/appendix. Accessed 22 May 2019

Panno SV, Chierinco MI, Bauer RA, Lundstrom CC, Zhang Z, Hackley KC (2016) Earthquakes recorded in stalagmites from a cave in south-central Indiana. Bull Seismol Soc Am 106(5):2364–2375

Panno SV, Luman DE, Kelly WR, Larson TH, Taylor SJ (2017) Karst of the driftless area of Jo Daviess County, Illinois. Circular 586. Illinois State Geological Survey, Champaign, 39p

Panno SV, Kelly WR, Baranski BL (2019) Hydrogeochemical controls on aquifers of northwestern Illinois' driftless area, USA. Environ Earth Sci 78(276):1–15

Panno SV, Kelly WR, Scott J, Zheng W, Holm N, Baranski BL, Hoellein T, McNeish R (2019) Microplastic contamination in karst groundwater systems. Groundwater 57(2):189–196

Piskin K, Bergstrom RE (1975) Glacial drift in Illinois: Thickness and character. Circular 490. Illinois State Geological Survey, Champaign, 35p

Plotnick R, Kenig F, Scott A (2014) Using the voids to fill in the gaps: Caves, time and stratigraphy. Special Publication 404. The Geological Society, London, pp 233–250

Post SL (1998) The driftless area: an inventory of the region's resources. Illinois Department of Natural Resources, Critical Trends Assessment Program, Springfield, 22p

Postpischl D, Agostini S, Forti P, Quinif Y (1991) Paleoseismicity from karst sediments: the Grott del Cervo cave case study (central Italy). Tectonophysics 193:33–44

Rubey WW (1952) Geology and mineral resources of the Hardin and Brussels quadrangles (in Illinois). Professional Paper 218. US Geological Survey, Reston, Virginia, 179p

Sebela S (2008) Broken speleothems as indicators of tectonic movements. Acta Carsologica 37:51–62

Tinsley JC, Maher KA, O'Henley G (2011) Earthquake damage to delicate cave formations in Missouri's Ozarks caused by the New Madrid seismic zone. Seismol Res Lett 82(2):303

Tinsley JC, Paces JB, Panno SV (2015) Exploring paleoseismic signals from caves in the central and eastern USA. Seismol Res Lett 86 (2B):597

Tuttle MP, Collier J, Wolf LW, Lafferty RH III (1999) New evidence for a large earthquake in the New Madrid seismic zone between A. D. 1400 and 1670. Geology 27(9):771–774

Unklesbay AG, Vineyard JD (1992) Missouri geology: Three billion years of volcanoes, seas, sediments and erosion. University of Missouri Press, Columbia, 189p

Webb DW, Taylor SJ, Krejca JK (1993) The biological resources of Illinois caves and other subterranean environments. Technical note 1993, no 8. Illinois Natural History Survey, Champaign, 143p

Webb DW, Wetzel MJ, Reed PC, Phillippe LR, Young TC (1996) Karst springs in the sinkhole plain of Illinois: Their community diversity and hydrogeology. Technical report 1996, no 11. Illinois Natural History Survey, Center for Biodiversity, Champaign, 122p

Weibel CP, Panno SV (1997) Karst terrains and carbonate bedrock of Illinois. Illinois Map Series 8, scale 1:500,000. Illinois State Geological Survey, Champaign

Willman HB, Atherton E, Buschbach TC, Collinson C, Frye JC, Hopkins ME, Lineback JA, Simon JA (1975) Handbook of Illinois stratigraphy. Bulletin 95. Illinois State Geological Survey, Champaign, 261p

Zhang Y, Kelly WR, Panno SV, Liu W-T (2014) Tracing fecal pollution sources in karst waters by a combination of host-specific Bacteroidales genetic biomarker, bacterial indicators, and environmental variables. Sci Total Environ 490:1082–1090

Zhou J, Lunstrom CC, Fouke B, Panno SV, Hackley KC, Curry BB (2005) Geochemistry of speleothem records from southern Illinois: Development of ($^{234}U/^{238}U$) as a proxy for paleoprecipitation. Chem Geol 221:1–20

Wisconsin Caves and Karst

6

Michael J. Day

Abstract

Wisconsin's caves and karst landscape are more widespread than is commonly appreciated and they merit greater attention. Roughly, 20% of the state exhibits karst landscape and carbonate-cemented sandstones contribute to carbonate-dominated groundwater underlying nearly 45% of the state. The caves and karst are formed primarily in Ordovician and Silurian dolostones and in underlying and interbedded Cambrian and Ordovician sandstones. Pleistocene glaciation has obscured much of the eastern, central, and south-central karst, leaving two main areas of exposure, in the Door Peninsula and adjacent counties of east-central Wisconsin and in the southwestern Driftless Area. In eastern, Wisconsin surface karst is essentially limited to Door County, where there are extensive open joints, sinkholes, sinking streams, karren, caves, and springs, plus glaciokarstic benches and steps. The caves are of dissolutional, littoral, and mechanical origin. Rapid karstic hydrologic connections between the surface and shallow groundwater pose serious threats of contamination, necessitating appropriate land use regulation. The karst hydrology of the Silurian aquifer south of Door County continues to be underappreciated but became manifest during construction of a deep tunnel stormwater management network in Milwaukee in the 1980s and 1990s. The southwestern karst is better known and escaped extensive Pleistocene glaciation. It may have evolved more-or-less continuously since the Paleozoic, but hypogenic contributions, long-term environmental episodicity and recent Pleistocene influences may all be involved. The karst is a fluviokarst, integrated into the fluvial landscape with a wide array of dry valleys, sinkholes, caves, and springs. Dissolution remains active, but the karst hydrogeology deserves greater study, as do the springs and caves, which have probably been influenced by hydrothermal processes, hypogenic dissolution, and possibly by dissolution under saline–freshwater mixing conditions. Lead ores were mined extensively from the Platteville and Galena formations during the nineteenth century and quarrying remains a significant industry. Land use has been dominated by small-scale agriculture, but intensive agriculture and sand mining have emphasized the potential for land-use conflicts and detrimental environmental impacts. Regional karst science has yet to be incorporated adequately into planning mechanisms, but the karst is becoming a central theme for NGOs and the regional press. There are also caves and karst developed in Paleozoic sandstones in southwestern and central Wisconsin, including pavements, natural bridges, and other fragile rock formations.

6.1 Introduction

Wisconsin's karst is underappreciated, both scientifically and from land-use management perspectives. About 37,000 km^2 of the state is underlain by carbonate bedrock—nearly 22% of the land area (Mauel, pers. comm. 2016; Nicholson et al. 2007; Mudrey et al. 1982). Moreover, although karst landscape features—dry valleys, sinkholes (dolines), caves, and springs—are apparent, although not dramatically conspicuous, in both southeastern and southwestern Wisconsin, karst is effectively more widespread and relevant by virtue of the karst hydrology of subsurface carbonate-cemented rocks, including those that are overlain by geologically younger strata and/or mantled by glacial deposits. Cambrian sandstones, in particular, have soluble carbonate cement matrixes and aquifers with carbonate-dominated chemical signatures underlie some 76,000 km^2, or nearly 45% of the state

M. J. Day (✉)
Department of Geography, University of Wisconsin-Milwaukee, 410 Bolton Hall, 3210 N. Maryland Avenue, Milwaukee, WI 53211, USA
e-mail: mickday@uwm.edu

G. A. Brick and E. C. Alexander Jr. (eds.), *Caves and Karst of the Upper Midwest, USA*, Cave and Karst Systems of the World, https://doi.org/10.1007/978-3-030-54633-5_6

(Mauel, pers. comm. 2016; Kammerer 1981). Thus, the statewide relevance and impact of karst far exceed that which is suggested by the understated surface landforms.

The karst is developed in Paleozoic carbonates primarily of the Ordovician to Silurian periods, which were deposited in transgressing shallow seas on the flanks of the subsiding Michigan and Forest City basins east and west, respectively, of the anticlinal Wisconsin Arch. Conventionally, two distinct potentially karstic areas are recognized within the state —one of about 24,000 km^2 in southeastern Wisconsin and the other of some 10,000 km^2 in the southwestern corner of the state (Day 2009)—but this distinction is somewhat arbitrary and oversimplifies the hydrogeological situation. Additionally, because Pleistocene glaciation has obscured much of the west-central, southeastern, and south-central karst, karst in the southwest is better known and karst in west-central, south-central, and east-central Wisconsin is poorly documented. Moreover, each of these areas is a relatively minor component of more extensive regional karst landscapes. The eastern karst extends both southward beneath the glacial deposits of Illinois and also arcuately northeastward through the Garden Peninsula of Upper Michigan and the Huron Peninsula of Ontario to New York's Niagara Falls and beyond. The convex sweep of the Niagara Escarpment is the most obvious surface expression of this carbonate sequence, but there are more extensive, although generally unexposed carbonates farther west, including the subdued escarpments in the Sinnipee Group limestones. Additionally, Silurian carbonates and carbonate-cemented shales extend centripetally beneath the Great Lakes, with considerable regional hydrologic significance (Torrey 1976; Cherkauer and McKereghan 1991). Approximately 37,566 km^2 of carbonate rocks underlie Lake Michigan, the waters of which are chemically carbonate dominated (Mauel, pers. comm. 2016; Torrey 1976; USGS 2016).

Similarly, the southwestern karst is but a portion of the broader Upper Mississippi karst developed in Paleozoic carbonates and clastic rocks within the four-state area that extends into southeastern Minnesota, northwestern Illinois, and northeastern Iowa (Hedges and Alexander 1985; Delong 2005; Doctor 2020, Chap. 1 of this volume). Again, the regional hydrologic significance of this karst is far greater than its relatively restricted and understated surface expression might suggest (Day et al. 1989; Reeder and Day 1993; LePain et al. 2005).

Three Paleozoic carbonate rock sequences host Wisconsin's karst and their subsurface pattern describes a broad V-shape across the southern half of the state, with a subsurface area of some 37,000 km^2 (Fig. 6.1). The Upper Ordovician Sinnipee Group carbonates are the most extensive in the subsurface, with an area of nearly 14,000 km^2 beneath both the eastern and western halves of southern Wisconsin, but their surface exposure is essentially restricted to southwestern Wisconsin, south of the Wisconsin River and only 3500 km^2 is within 1.5 m (5 ft) of the surface.

Silurian and restricted Devonian carbonates underlie far eastern Wisconsin, with a subsurface area of 12,300 km^2, but only about 1200 km^2 of this is within 1.5 m (5 ft) of the surface, so their exposure is severely limited. By contrast, the Early Ordovician Prairie du Chien Group carbonates have a subsurface area of just over 11,000 km^2 beneath much of south-central and southwestern Wisconsin, but of this nearly 4000 km^2 is within 1.5 m (5 ft) of the surface, particularly in the western part of the state. Thus, the Silurian and Sinnipee Group carbonates have greater subsurface areas but relatively little surface expression, whereas the Prairie du Chien Group carbonates are rather less extensive in the subsurface but have relatively greater surface expression. For this reason, the caves and karst of the Prairie du Chien are proportionally better known than those of the Sinnipee Group and Silurian carbonates, although their significance is not necessarily greater. The karst and caves of west-central Wisconsin, in Trempealeau, Buffalo, Pepin, Pierce, St Croix, Polk and Dunn counties (Fig. 6.1) are very poorly documented, although it appears to be quite restricted and, like eastern Wisconsin, is largely obscured by Pleistocene glacial sediments.

Additionally, there are in Wisconsin numerous caves and other karstic features that have developed in noncarbonate rocks, particularly Paleozoic sandstones. In general, these features have attracted relatively little attention, but the rather sparse literature is reviewed here for completeness and to illustrate the potential for additional inquiry.

Historically, studies of Wisconsin's karst date well back into the nineteenth century (Brick 2020, Chap. 2 in this volume) with particular emphasis on the formerly economically important lead and zinc deposits in the Sinnipee Group carbonates southwest of the Wisconsin River (Murrish 1871; Strong 1877; Dockal 2020, Chap. 7 in this volume). Geomorphic studies began with attention to caves and sinkholes (Lange 1909; Bretz 1938) and were encouraged by Lawrence Martin's influential writings about Wisconsin's physical geography (Martin 1916, 1932, 1974). Hydrologic and geomorphic research has gathered momentum since the middle of the twentieth century and the Wisconsin Speleological Society (WSS) was particularly influential in compiling a series of county-wide cave inventories in the 1960s and 1970s (Frater 1966; Peterson 1968; Gietkowski 1972; Hennings et al. 1972; Cronon and Frater 1974). The history of the WSS is interesting in itself (Day and Soule 1988) and the caves are not without their challenges (Day and Kueny 1990). Appreciation of the broader karstland management implications has lagged, coming to the forefront only recently in the context of competing water supply interests,

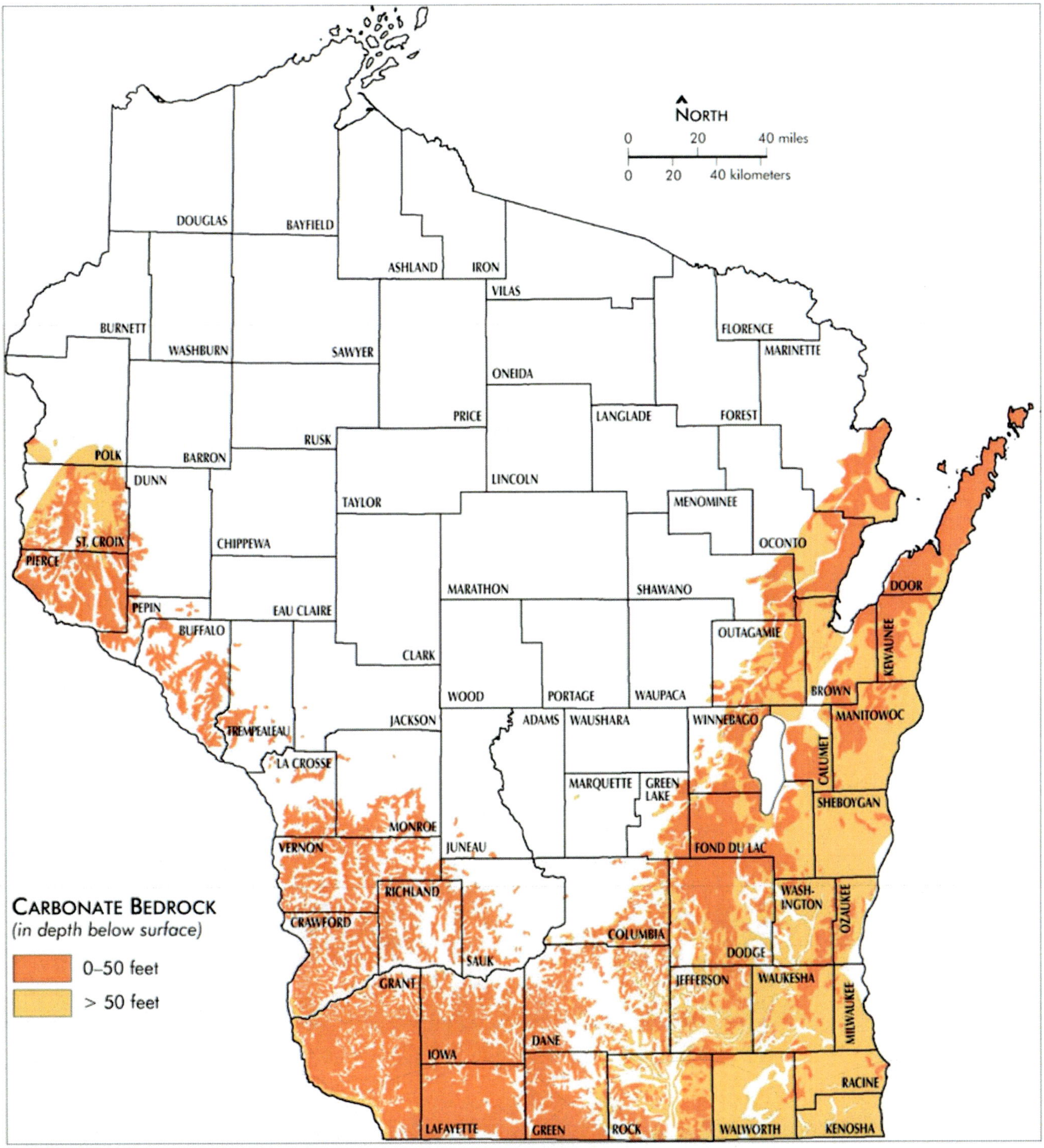

Fig. 6.1 Karst and shallow carbonate bedrock in Wisconsin (WGNHS 2009)

changing climate, and the potential environmental impacts of proposed shifts in agricultural and industrial land use (Juckem et al. 2008; Crawford Stewardship Project 2016). The Wisconsin Geological and Natural History Survey is now documenting karst features throughout the state via a voluntary reporting program using a standard Wisconsin Interagency Karst Feature Reporting Form (available at https://www.tfaforms.com/209523) (Pearson 2001).

6.2 Southeastern and South-Central Wisconsin—Glaciated and Buried Karst in Silurian and Ordovician Carbonates

In eastern Wisconsin, northward from the southern state line through the two tiers of counties immediately west of Lake Michigan and beyond the Door Peninsula, karst is well developed within the Silurian carbonates that are most prominent as the Niagara Escarpment (Fig. 6.2).

Within the state, surface expression of this karst decreases southward from the Door Peninsula, largely because of increasing depths of overlying Pleistocene glacial deposits. Carbonate exposure numbers and surface karst area decline steadily from Door County southward through Kewaunee,

Fig. 6.2 Niagara Dolomite escarpment, west side of Door County, east side of Green Bay. Photo by Michael J. Day

Manitowoc and Sheboygan counties, with near-complete glacial burial by up to 45 m of drift at the latitude of Milwaukee (Fig. 6.1). Devonian carbonates occur only in very small areas of Ozaukee and Milwaukee counties adjacent to Lake Michigan and for convenience they are grouped here with the Silurian formations.

West of the Silurian outcrop, older Ordovician carbonates, although more extensive in subsurface area, are almost entirely buried by Pleistocene glacial drift, with only localized surface expression near the crests of the Sinnipee Group and Prairie du Chien Group cuestas. As such, surface karst is of very limited extent and caves are few and largely undocumented. The overlying Maquoketa Shale, which is presumed to have been more extensive prior to Pleistocene glaciation, may have effectively isolated the underlying Ordovician dolostones from aggressive surface drainage or groundwater (Luczaj and Stieglitz 2008).

Pleistocene glaciation wrought the substantial transformation of the surficial eastern Wisconsin karst through glacial erosion and deposition. It was estimated previously that glacial erosion removed carbonate rock to a depth of some 30–60 m (Martin 1974) but this is now regarded as an overestimation (Colgan et al. 2002). As elsewhere, many preexisting cave systems survived glaciation, although many were filled by glacial sediments and there has been sufficient time for the development of postglacial caves in locations where meltwaters produced by deglaciation stimulated speleogenesis in carbonates not buried deeply beneath carbonate-rich glacial sediments (Ford and Williams 2007; Cooper and Mylroie 2015). The effects of glaciation on karst are numerous and varied, with the establishment of permafrost, ice advance, ice sheet hydrology, deglacial seismicity, and rebound and meltwater impacts all being significant. There are clear parallels between the Paleozoic limestones of eastern Wisconsin and similarly aged limestones in the northeastern USA, where glacial effects were comparably dramatic (Cooper and Mylroie 2015).

The Silurian carbonates are predominantly white to buff gray-colored dolostones, lithologically biomicrites, fossiliferous, medium to coarse grained, and with carbonate purities in the 85–95% range (Kluessendorf and Mikulic 1989; Stieglitz 1990). Bedding is mostly thin, although more variable and increasing in thickness upward and dip is generally southeastward at less than five degrees, with pronounced jointing along orientations at 25, 70, and 155° from north (Kluessendorf and Mikulic 1990; Rosen and Day 1990). Cyclic paleokarst formation is evident within the Silurian sequence (Kuessendorf and Mikulic 1990). Mechanically, the limestones are quite competent, with Schmidt Hammer hardness generally within the range between 35 and 45 (Day and Goudie 1977; Day 1980; Goudie 2006).

Fig. 6.3 Open joint, Door County. Photo by Michael J. Day

Scientific documentation of the surface karst landscape in eastern Wisconsin is essentially limited to that in Door County, where there are extensive open joints (Fig. 6.3), sinkholes, sinking streams, karren, caves, and springs (Johnson and Stieglitz 1990; Rosen and Day 1990).

Dry valleys are not a major component of this landscape, but there are prominent east-facing ledges and steps, many with small karst pavements (Fig. 6.4) which have been identified as components of a glaciokarstic landscape incorporating elements of both karst development and subsequent glacial modification (Rosen et al. 1987). In Door County and elsewhere where the drift is thin and patchy, there are restricted exposures of glaciokarst, including characteristic glacial benches with steps and risers (*schichttreppenkarst*) and dolostone pavements (Rosen and Day 1990; Johnson and Stieglitz 1990). Staircases and pavements are predominantly south- and east-facing, as predicted by the general model of northwest–southeast Pleistocene ice movement. Many of the smaller features are postglacial, although shallow features may have an

Fig. 6.4 Jointed glacial bench/pavement, Door County. Photo by Michael J. Day

important inherited component and the larger sinkholes and caves may antedate Wisconsinan glaciation (Rosen and Day 1990).

Many of the most obvious karst landforms on the Door Peninsula are on the western side, where the drift is thinnest (Rosen and Day 1990), the hydraulic gradient is steepest (cf. Cowell and Ford 1983), and joints are dilated as a result of unloading and/or ice wedging (Stieglitz et al. 1980). Additionally, there is some evidence of preferential karst formation within the Burnt Bluff Group, which outcrops at the higher elevations on the west side of the peninsula (Barden 1980a).

6.2.1 Sinkholes

Sinkholes, the diagnostic karst landforms (Ford and Williams 2007) are numerous in Door County, although most are small (Fig. 6.5). At Brussels Hill, 99% of enclosed depressions are structurally controlled, 58% occur at three-way joint intersections, and 71% are elongated along joints at 25, 70, and 155° (Rosen et al. 1987). Depression widths are generally in the range from 0.6 to 12.0 m, with depths between 0.15 and 3.0 m (Rosen and Day 1990; Rosen et al. 1987). Large scattered examples are often prominent in agricultural fields, where many have been infilled or otherwise modified. By contrast, in wooded areas, most depressions are grouped in high-density lattice networks that reflect the closely spaced joint sets. At Brussels Hill, densities are up to 8.7/100 m^2 and at Ledge View depressions occupy 95% of a 170 m^2 area (Rosen and Day 1990).

Dissolutionally enlarged joints are perhaps the most numerous karst features in Door County, where they represent both agricultural and engineering hazards and potential entryways to the groundwater system for contaminants such as agricultural chemicals (Sherrill 1978; Barden 1980a; Johnson and Stieglitz 1990) (Fig. 6.3). These enlarged joints, otherwise known as cutters and grikes, are aligned in accordance with locally dominant joint orientations. Most are probably postglacial, reflecting glacial unloading, isostatic rebound, and near-surface dissolution, but some larger examples may have preglacial origins analogous to those on limestone pavements elsewhere (Vincent 1995; Wilson et al. 2012). Enlarged joints are most numerous where surficial deposits are less than 0.6 m thick, ranging up to more than 10 m in length and 0.8 m in width (Rosen and Day 1990).

6.2.2 Caves

Some 25 caves of dissolution origin have been documented in Door County (Barden and Soule 1980; Hennings et al. 1972; Soule, pers. comm. 2017) and the Door Peninsula contains several significant caves, including Horseshoe Bay Cave, the longest known cave in eastern Wisconsin at 946 m, Paradise Pit (576 m) and Brussels Hill Pit Cave, which is one of the state's deepest and most paleontologically significant caves (Brozowski and Day 1994). There are also about 25 sizeable contemporary and relict littoral caves, for example, the actively forming caves at Cave Point and the relict Wellever Cave near Egg Harbor and Eagle Cave in Eagle Bluff in Peninsula State Park, plus many smaller examples (Soule, pers. comm. 2017).

Horseshoe Bay Cave (Fig. 6.6), also known as Tecumseh Cave (among other names), is the best-known cave in the Door Peninsula and Wisconsin's longest noncommercial cave at 946 m (3103 ft) (Soule 2014). Like many caves, the subject of extensive folklore, the cave was first documented in 1896 when a stream was noted draining from its (then obscured) entrance near the base of the Niagara Escarpment near the west coast settlement of Egg Harbor (Soule 2014). The cave has been recognized as Wisconsin's most

Fig. 6.5 Small Door County sinkholes and field joints. 22 June 2008 Google Earth photo

challenging because much of the passageway is low and water filled, particularly in the further reaches, although there is also considerable vadose walking passage and several large rooms, including the largest two cave rooms in the state, in the more accessible section. Historically, much of the drainage through the caves are thought to have derived from Plum Bottom, a large enclosed depression located above and beyond the cave which contains several plugged outlets and in which an ephemeral lake was present until the 1920s, since when it has remained dry (Soule 2014; Alexander et al. 2008). The cave has been explored and its known length extended considerably since the 1960s by WSS and other cavers. Located on privately owned land used for fruit growing and as a country club through the twentieth century, the cave entrance area was incorporated into the Frank E. Murphy County Park in 2010 and a management plan that involves opening up the most accessible section for visitors was developed and adopted in 2014 (Aleson et al. 2014; Soule 2014).

Also, in Door County near Sturgeon Bay, Paradise Pit Cave, with 576 m of passage mapped, is notable particularly because since its discovery by Gary Soule in 1968, it has been the focus of extensive excavation, including the digging of two new entrance pits by members of the WSS (Soule 1986). Like Horseshoe Bay Cave, Paradise Pit shows evidence of both phreatic and vadose development and includes 47 separate rooms (Soule 1986).

Dorchester Cave was discovered in 1972 during blasting for the construction of a basement beneath the Dorchester nursing home in Sturgeon Bay (Soule 2010). With over 100 m of mapped passage, the cave extends both north and south from the basement and appears to be unique in Door County, being the result of mechanical stresses, perhaps resulting from tectonic stress or glacial unloading, with obvious lateral and vertical displacement creating a markedly rectangular passage along a joint oriented at 340° (Soule 2010). Access via the nursing home is strictly controlled.

Brussels Hill Pit Cave is 28 m deep and aligned along a joint trending at 62° east of north. The cave contains a postglacial sediment sequence covered by later sediments and organic material of Holocene age, with the cave sediments being markedly different from surface materials (Brozowski and Day 1994). Lower coarse sands and gravel may date to early stages of glacial recession, with higher finer sands and silts reflecting subsequent decreasing transport potential with the increasing recession. The cave itself predates the sediment fill and the sinkhole entrance collapse probably occurred relatively recently (Brozowski and Day 1994).

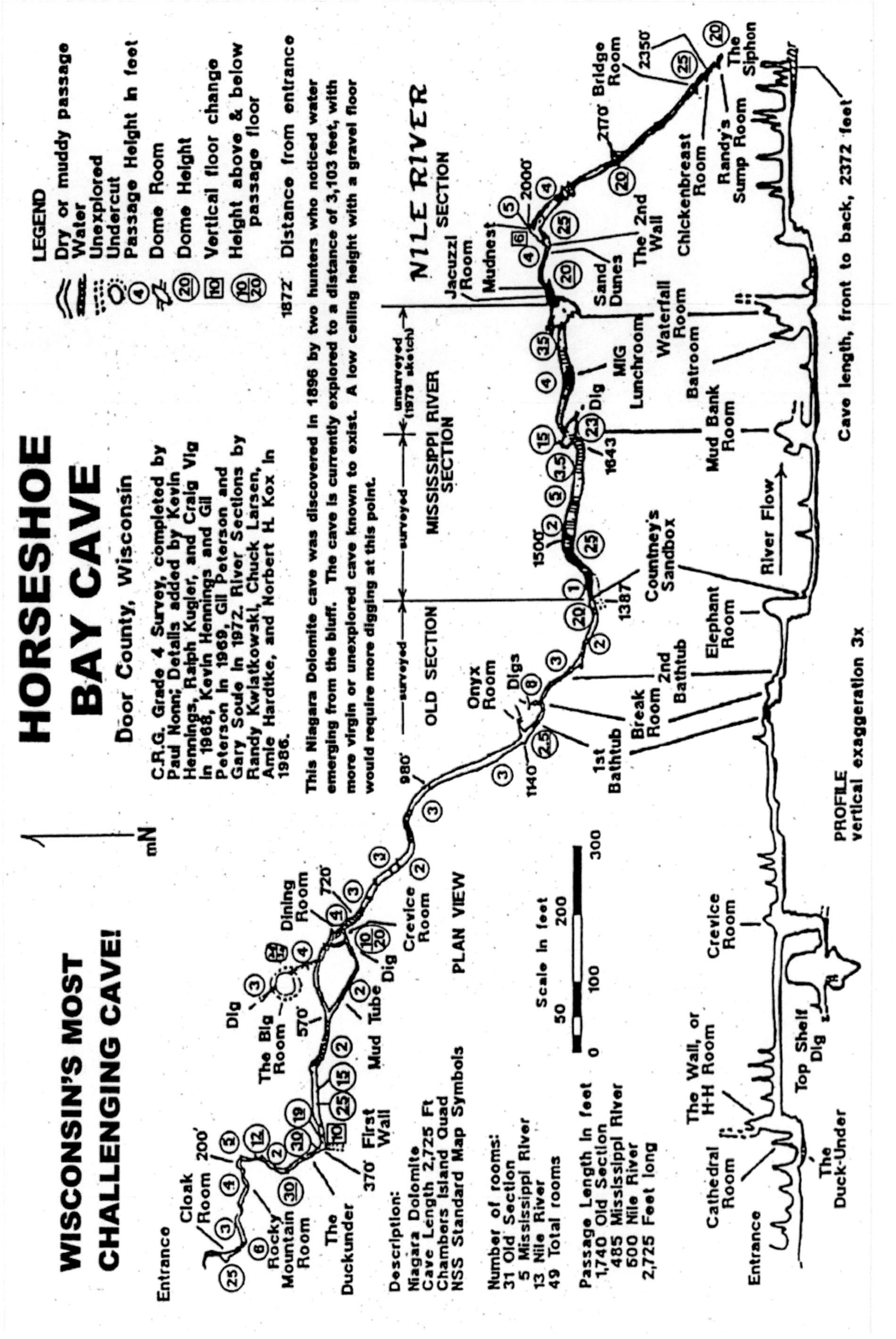

Fig. 6.6 Horseshoe Bay Cave. Map provided by Gary Soule

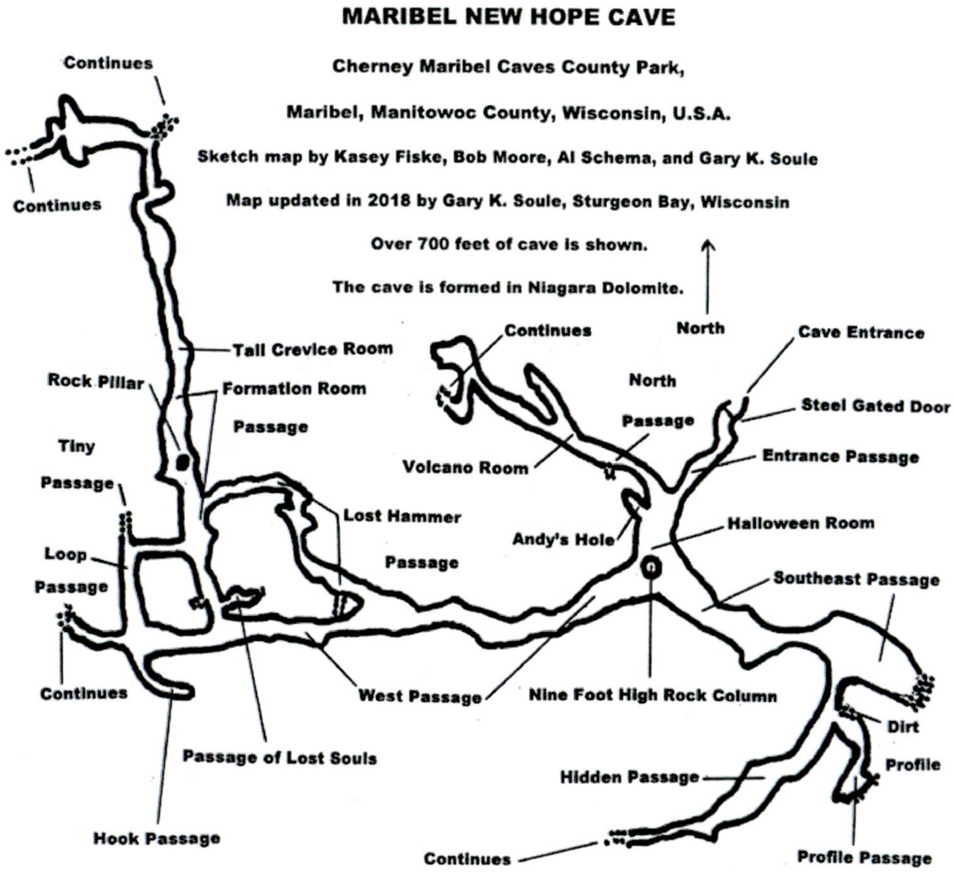

Fig. 6.7 Maribel New Hope Cave map. Map provided by Gary Soule

In Manitowoc County, the Cherney Maribel Caves, particularly Maribel New Hope Cave (Fig. 6.7), have been excavated extensively by members of the Wisconsin Speleological Society and are incorporated into the Cherney Maribel Caves County Park (CMCCP 2013). Three similarly excavated caves, Carolyn's Cavern, Montgomery Cave, and Mother's Cave, are located in Ledge View Nature Center, near Chilton in Calumet County.

Maribel New Hope Cave contains an impressive sequence of fluvial sediments with a wide array of sedimentary structures (Luczaj and Stieglitz 2008). Three main sedimentary packages have been identified, requiring that at least some of the sediment be 5,590–5,740 years old. What have been interpreted as water-escape structures and high-flow regime sedimentary structures are consistent with glacial activity in the region. The lower sediment unit displays an unusual orange iron oxide banding and is likely considerably older than the other two units, possibly even pre-Pleistocene (Luczaj and Stieglitz 2008).

There has been no suggestion to date that the eastern Wisconsin karst and caves have anything other than an epigenic origin, with dissolution essentially proceeding downward from the surface as a result of the gravitational movement of meteoric water. This conventional interpretation may be correct, but alternative hypotheses arising from hypogenic cave development theory (Palmer 1991; Klimchouk 2007, 2017; Klimchouk et al. 2017), in which dissolution is caused by ascending groundwater may merit future investigation, particularly in light of indications of hydrothermal diagenesis along with the Wisconsin Arch (Luczaj 2006) and the suggestion of similar possibilities elsewhere in the region (Pipes and Day 2006; Barr et al. 2008).

6.2.3 Hydrogeology

The karst hydrology of the Door Peninsula was poorly understood until late in the twentieth century, when concerns were raised about potential contamination of the shallow karst aquifer associated with increasing levels of construction and agricultural activity (Sherrill 1975, 1978; Wiersma et al. 1986). Subsequently, karstic hydrologic connections between sinkholes, crevices, springs, and groundwater have been confirmed (Johnson and Stieglitz 1990; Alexander et al. 2008), a regional Karst Task Force was established (Erb and

Stieglitz 2007) and appropriate land-use practices have been instituted (Erb et al. 2015). Similar studies have been conducted in adjacent counties, such as Brown and Calumet (Fermanich et al. 2006) and subsequent agricultural and domestic groundwater contamination has resulted in similar recognition and land-use regulation throughout northeastern Wisconsin. Appropriate land-use regulations are now in effect regionally, for example, in Brown County (2017), Manitowoc County (2017), and Kewaunee County, which is one of the state's leading dairying counties and is also underlain by shallow karst (WGNHS 2009; WisconsinWatch 2014).

The Niagaran aquifer to the north of Milwaukee, particularly in Door County, is extensively karstified and exhibits high levels of secondary permeability, with a spectrum of flow styles and velocities including rapid, discrete flow along joints, bedding planes, and other discontinuities. Muldoon et al. (2001) characterize the aquifer as being dual porosity, with the lower permeability matrix providing storage capacity but higher permeability fractures transmitting the majority of the water. Pre-Pleistocene caves and karst have been infilled by glacial deposits and are rarely exposed except along "the ledge"—the discontinuously exposed surface of the Niagara Escarpment.

The economy of Door County includes a significant tourism element, with the Niagara Escarpment itself playing an important role in locations such as Peninsula State Park (WDNR 2017a). Regionally, the Niagara Escarpment has been recognized as a significant ecological habitat and cultural icon (WDNR 2010; NERN 2017).

Surface karst south and west of Door County has been only poorly documented, but possibly warrants greater attention, for example, in the vicinity of the Maribel-Cherney caves (Manitowoc County), east of Lake Winnebago in High Cliff State Park (Calumet County) and in and around Ledge County Park (Dodge County) (NERN 2017). Open joints and small surface collapses are the most numerous karst features here, but springs are not uncommon and several natural bridges have been documented in the Silurian dolostones in Brown and Fond du Lac counties (Paull 1992). The natural bridge at Fonferek Glen, in Brown County, appears to be of fluvial origin, but those at Oakfield Ledges, in Fond du Lac County, may have originated as caves, subsequently modified by mechanical processes (Paull 1992). Further illustrating the potential both for karst landform development and for environmental and land-use problems, unauthorized dredging in 2017 of a navigable stream near Appleton in Outagamie County resulted in the excavation of a sinkhole that subsequently pirated surface drainage and potentially contaminated local groundwater supplies; remediation was estimated to cost $100,000 (Appleton Post-Crescent 2017). Elevated radon levels have also been detected in local housing (Hawk et al. 1993).

There is essentially no surface karst landscape in the Milwaukee area because Pleistocene glaciation has eroded the preexisting karst surface and deposited up to 45 m of carbonate-rich drift. Nonetheless, incipient epikarst occurs in the upper weathered zone of the dolostone and groundwater recharge is largely saturated with respect to carbonates (Rovey and Cherkauer 1994) and there remains the potential for karstification within the aquifer that predates Pleistocene attenuation.

6.2.4 The Silurian Aquifer

The hydrology of the Silurian aquifer in the context of regional water supplies has been recognized explicitly (Rovey and Cherkauer 1994; Muldoon et al. 2001) but its karstic nature has been and continues to be underappreciated, particularly south of Door County where there is little surface expression of the karst (Day 2004). This became manifest during construction of a deep tunnel stormwater management network in Milwaukee in the 1980s and 1990s, when boring encountered dissolution cavities and underground conduits, the drainage of water from which caused unexpected construction delays and necessitated extensive grouting (Day 2004). A particularly critical component of Milwaukee's $2.8 billion Water Pollution Abatement Program is a 31.2 km inline storage system comprising three 5–10 m diameter deep tunnel sections that were bored between 1984 and 1993 at a depth of 80–100 m within the Niagara dolostone.

Construction of these tunnels by the Milwaukee Metropolitan Sewerage District (MMSD) proved more difficult and expensive than estimated because of the karstic nature of the dolostone, particularly its hydrology, had not been fully appreciated. Rock collapse, subsidence, and groundwater intrusion necessitated remedial grouting and lining of about 45% of the tunnels, costing some $50 million above estimates and delaying completion by nine months. Tunneling encountered bedding planes, joints, and faults enlarged by dissolution (Figs. 6.8, 6.9), voids not apparently related to rock mass discontinuities and collapse structures (Burke 2002). Some voids were open and water-filled, but others were filled partially or completely with maroon or green siltstone or mudstone. The collapse structures included dolostone blocks 1.5–3 m in diameter and surrounded by a siltstone/mudstone matrix and larger collapse zones spanning up to 30 m (Burke 2002).

The main karstic problem encountered was groundwater intrusion via open fissure flow intercepted during boring. Inflows between 10 and 50 gpm were numerous and, during 1988, construction of the North Shore tunnel, which had previously experienced inflows averaging 51 l/s (800 gpm) was halted for 4 months because of flooding, necessitating a

Fig. 6.8 Open void in the wall of 5.2 m diameter MMSD tunnel. Photo HU35-7, © 2020 MMSD records

Fig. 6.9 Clean fractures in wall of 5.2 m diameter MMSD tunnel. Photo HU29-21, © 2020 MMSD records

$1.5 million grouting campaign. Despite grouting efforts, discrete groundwater intrusion in some sections reached 220 l/s (3500 gpm). In 1992, after grouting, the MMSD measured inflow into the North Shore tunnel at 4300 m^3/day (Cherkauer 1996). Water inflow also reached 190 l/s (3000 gpm) in the eastern section of the Crosstown Tunnel, with additional grouting taking 11 months; four companies working on the Crosstown Tunnel subsequently sued MMSD for nearly $12 million, charging that they had been misled by indications that groundwater flow into the tunnel would be considerably less. Similar water infiltration problems plagued the Kinnickinnic/Lake Michigan tunnel (Day 2004).

Tunnel performance since completion continues to be controversial. Among the concerns are those of groundwater flow into the deep tunnels and potential contamination of the regional aquifer by sewage leaking from them. Cherkauer (1996) showed that the North Shore tunnel induced recharge into the dolostone and overburden aquifer of at least 1600 m^3/day from Lake Michigan and 534 m^3/day from the Milwaukee River. Groundwater inflow into the Crosstown Tunnel, which is about 20% lined, is considerably greater than into the North Shore Interceptor, which is about 50% lined.

There is also concern about the potential for tunnel leakage from unlined sections and contamination of adjacent groundwater, with at least one apparent, although disputed case, in which a well, used since 1949 by the Red Star Yeast Company, some 152 m from the Crosstown Tunnel was contaminated by fecal coliform bacteria and abandoned in March 1999. MMSD has asserted that leakage, when the tunnels are full, is returned to them as they empty and a spokesman also asserted that cracks formed during tunnel flooding during storms are subsequently sealed by calcite deposition. Subsequent to tunnel construction, the anisotropy inherent to both the dolostones and the overlying glacial deposits have been recognized more explicitly (Carlson 2001).

6.3 Southwestern Wisconsin—Karst Development in a Cambrian–Ordovician Carbonate/Clastic Sequence

The karst of southwestern Wisconsin is better known than its eastern counterpart, in part because it is integral to the larger and more extensively studied karst of the Upper Mississippi Driftless Area (Hedges and Alexander 1985). In contrast to eastern Wisconsin, a portion of southwestern Wisconsin (and extreme northwestern Illinois) has never been glaciated (the area enclosed in the yellow dashed line in Fig. 1.13 and shown in Fig. 1.12 in Doctor 2020, Chap. 1 of this volume). The rest of the “classic” Driftless Area in Wisconsin, Minnesota, Iowa, and Illinois has been glaciated at least once, but those areas were not covered by the Wisconsinan ice sheets. The true Driftless Area of Wisconsin escaped the direct erosional and depositional ravages of Pleistocene continental glaciations and that karst area does not exhibit as much recent disruption. This is a matter of degree, however, rather than an absolute distinction. The conventional interpretation has been that the southwestern Wisconsin karst has evolved more-or-less continuously since the Paleozoic, in parallel with the broader regional fluvial landscape (Day et al. 1989; Baker et al. 1998) although this may not be entirely correct, underestimating hypogenic contributions, long-term environmental episodicity, and recent Pleistocene influences.

Land use in the southwestern Wisconsin karst is dominated areally by agriculture, forestry, and recreation (Fig. 6.10). Ridgetops, particularly the more extensive uplands south of the Wisconsin River, are a focus for regional dairying, including grazing and fodder crop cultivation, although there is considerable agricultural diversity, with sheep, goats, pigs, elk, bison, camelids, and horses locally important and a mix of hay, corn, soybeans, and other crops on arable land. There is an increasing emphasis on organic farming, cooperative, and farm-to-market business, with Organic Valley Inc., headquartered in LaFarge, Vernon County, and the largest US organic agricultural cooperative, serving as a significant regional catalyst. There is a strong Amish community too, for example, around Cashton in Monroe County. While the relationship between agriculture and the karst is generally harmonious, sinkhole formation and water supply problems may cause inconvenience and farm effluents, herbicides, and other materials may represent threats of groundwater contamination (Day et al. 1989; Reeder and Day 1993, 1994; Coxon 2011). Agricultural chloride and nitrate contamination of surface and groundwater involve distinct seasonality, with base levels in the summer, winter storage above or close to the ground surface, and contaminant release, often as distinct slugs, during spring snowmelt (Reeder and Day 1993, 1994).

The Paleozoic rocks in southwestern Wisconsin consist of approximately 400 m of mainly sandstones and dolostones, generally dipping at less than 10° to the south, into the Illinois basin and southwest, into the Forest City Basin (Barden 1980b). The major cavernous formations are the Ordovician carbonates: the dolomitic Oneota formation of the Prairie du Chien Group and the Platteville and Galena dolostones of the Sinnipee Group. The carbonates were folded and fractured during the Paleozoic and these initial structures have been accentuated by subsequent dissolution and slumping (Hedges and Alexander 1985). Karst formation may have commenced early in the Paleozoic (Hedges and Alexander 1985) and hydrothermal activity affected the Platteville Formation and Galena Dolomite around the same time (Heyl et al. 1970). The dolostones have undergone extensive and complex diagenesis (Smith and Simo 1997) and there is a considerable paleokarstic legacy (Mai and Dott 1985; Smith et al. 1993, 1996).

Fig. 6.10 Dry Valley, forest, and contour farming, Richland County. Photo by Michael J. Day

The carbonate–siliciclastic sediments of the Prairie du Chien Group were deposited under shallow marine conditions across the Wisconsin Arch and the adjacent Michigan Basin during the Ordovician (Smith et al. 1993). The Prairie du Chien Group has two major depositional components—the Oneota Dolomite and Shakopee Formation—both of which are bounded by type 1 sequence boundaries which across the Wisconsin Arch are associated with karst development and silicification of the underlying carbonates, indicating the development of an unconformity during prolonged subaerial exposure (Smith et al. 1993).

The Oneota Dolostone is the lowest formation in the Ordovician Prairie du Chien Group and attains a thickness of nearly 60 m. Barden (1980b, p 8) describes it as "…a thick or poorly bedded, medium crystalline, saccharoidal dolostone with minor amounts of chert and shale, cavernous zones of poorly preserved algal stromatolites and, in many areas, large secondary calcite crystals." Day (1979, 1984) describes the Prairie du Chien overall as a medium-textured impure sandy dolostone, with quartz contents in excess of 10% and clay averaging 2.1%. Mean insoluble residue is 12.17% by weight, although ranging locally from 1 to 26%; porosity is estimated at 10% and recrystallization is evident, often with silica-replacing carbonates (Day 1979, 1984). Compressive strength, as measured by Schmidt Hammer hardness, is highly variable but averages about 31 (Day 1984).

Dolostones of the Platteville Formation (Fig. 6.11) reach 25 m in thickness. The lowest member, the Pecatonica, is composed of "Fine to medium crystalline, buff to blue-gray, fossiliferous, thick-bedded dolostone with medium to coarse grains of well-rounded quartz and phosphatic nodules in the lower 1 m." (Barden 1980b, p 8). Overlying the Pecatonica is the McGregor Member, a fine-grained, light grey or buff, fossiliferous, nodular or wavy bedded limestone, and dolomitic limestone. The upper member, the Quimbys Mill, is a buff to blue–grey, fine-textured, fossiliferous, thick-bedded limestone, or dolostone. The Galena Dolomite—the upper unit of the Sinnipee group—includes three members, in ascending order, the Dunleith, Wise Lake, and Dubuque, which are thin to thick-bedded, buff, shaley, dolostones, locally with chert, and shale bands. Total thickness of the Galena is nearly 80 m (Barden 1980b).

Fig. 6.11 Incipient sinkhole in road cut in the Platteville Formation, Iowa County. Photo provided by Michael J. Day

Karst, perhaps best characterized as covered fluviokarst (Sweeting 1972), is a significant component of the upland landscape of southwestern Wisconsin's Driftless Area, with a wide array of dry valleys, sinkholes, caves, and springs (Day et al. 1989). Although dissolution may have been sluggish, the area was spared the apparent ravages of Pleistocene glaciation (Mickelson et al. 1982; Syverson and Colgan 2004), which has allowed the persistence of the spatially restricted, essentially relict karst dissected by fluvial valley incision. Pleistocene glacial influences were not, however, totally insignificant. Permafrost and periglacial processes are implicated in much of the contemporary surface geomorphology (Clayton et al. 2001; Iannicelli 2010), glacial loess overlies much of the karst landscape and infiltrates caves (Stiles and Stensvold 2008; Knox et al. 2011; Day 1988), glacial modification of drainage systems influenced karst hydrology and glacial meltwaters had both local and regional impacts on surficial karst landforms and speleogenesis (Hobbs 1999), although these were perhaps not as wholesale and traumatic as in glaciated eastern Wisconsin (Johnson and Stieglitz 1990) and in the northeastern USA (Cooper and Mylroie 2015). Although there remains uncertainty as to precisely why the Driftless Area remained ice-free during the Pleistocene, it has been suggested that the terrain itself was a major factor in that "The dissected limestones and sandstones of the Paleozoic Plateau prevented the spread of glacial ice onto the Driftless Area, because they allowed dewatering at the base of the ice sheet" (Hobbs 1999, p 101).

North of the Wisconsin River, the karst increases in area and thickness from east to west but is manifest largely as ridgetop Prairie du Chien Group (Fig. 6.12) carbonate caps above the older noncarbonates into which long-established fluvial valley systems are incised. Because of the limited upland catchment areas this karst is functionally constrained, although there is considerable surface karren development and ridge-end castellations and surface collapse dolines provide access to shallow cave systems of limited length. This ridgetop karst is normally dry, but valley systems become active following snowmelt or during particularly heavy rains, such as those of 1978 (Hughes et al. 1981), 2007, 2008 (Fitzpatrick et al. 2008) and 2016. The karst is dominantly forested, although some ridgetops remain cleared for agricultural use. Lower in the landscape channel flow is more perennial, particularly where it is maintained by discharge from springs at or near the base of the carbonates.

Fig. 6.12 Prairie du Chien quarry with incipient sinkhole, Vernon County. Photo by Michael J. Day

Although there are outcrops of the Prairie du Chien Group immediately south of the Wisconsin River in Grant and Iowa counties, most of the karst, there is formed in the Platteville–Galena carbonate formations of the Sinnipee Group. South of the Wisconsin River, the landscape is less deeply dissected and the karst is areally more extensive. Here dry valley systems are visually dominant, particularly where the landscape is in agricultural use, although there are also sinkholes, caves, and springs.

Fig. 6.13 Dry valley ephemeral stream bed, Crawford County. Photo by Michael J. Day

Fig. 6.14 Dry valley contour cropping, Richland County. Photo by Michael J. Day

Overall, dry valleys, ephemerally active streams, and spring-fed streams dominate the Driftless Area karst (Day et al. 1989) (Figs. 6.10, 6.13) but they have received little scientific attention and certainly merit additional study. One interesting question concerns the meteorological and hydrologic thresholds at which stream flow in the valleys becomes activated. Spring snowmelt appears to generate at least some flow in most dry valley segments, but otherwise it seems from preliminary observations that 24-hour rainfall totals of about 50–100 mm are sufficient to produce stream discharge, suggesting that these "dry valleys" may be active more regularly than generally recognized. These flood events have proven very disruptive, resulting in stormflow events, localized flooding and damage to fields, houses, bridges, culverts, roads, and other infrastructure. Even then, there may be considerable loss of discharge into channel beds; during an August 1982 storm an ephemeral stream tributary to Tainter Creek in Crawford County lost 20% of its 500 gpm discharge into the stream bed over a distance of only 200 m (Day et al. 1989) (Fig. 6.14).

There have been few process studies in the southwestern Wisconsin karst and the general consensus has been that the karst is relatively inactive, even essentially relict (Day 1986a, b, c). By contrast, a nearly 40-year investigation of weathering rates employing carbonate erosional weight loss tablets (Trudgill 1977; Gunn 2013) has demonstrated that the karst remains geomorphologically active (Day 1979, 1984, unpublished). Although there are limitations of weight-loss tablets, especially when using different carbonates, single sites and short time periods (Day et al. 1980; Gunn 2013), initial results suggested that dissolution of the Prairie du Chien dolostones is relatively sluggish but that rates at a given depth were consistent over sites of up to 10 m^2 and increased with increasing soil depth (Day 1979, 1984). Preliminary results (1977–1978) suggested that the Prairie du Chien tablets dissolved more slowly than "control" tablets of Jamaican and Yugoslavian limestones but that variability within the single lithology was minimal and that mechanical processes were also significant (Day 1979). Medium-term (1977–1984) results generally supported these initial conclusions and emphasized the significance of soil depth and mechanical processes (Day 1984). Results from the emplacement of 80 Prairie du Chien tablets from 1979 until 1984 showed mean weight losses at the surface and at soil/sediment depths of 5 cm, 15 cm, and 25 cm of 1.02, 2.21, 3.18, and 2.47 $\mu g/cm^2/day$ (Day 1984). By international standards, these rates were low and lower than those for the Jamaican and Yugoslavian "control" tablets by at least 50% (Day 1984).

Longer term unpublished data (1978–2017) using only Prairie du Chien dolostone tablets (n = 55) suggest that dissolution ratesover nearly 40 years are 15–20% higher than those measured previously, but with soil depth and site selection being more significant than previously acknowledged. For 1978–2017, mean weight losses at the surface and at comparable soil/sediment depths of 5 cm, 15 cm, and 25 cm were 1.18, 2.61, 3.82, and 2.90 $\mu g/cm^2/day$. These overall apparently increased weight loss rates should be interpreted with caution because relationships with soil depth are more varied than in the earlier studies, nearly 40% of tablets were broken (by freeze–thaw?) and because inter-site variability (four sites) is greater. Despite this, the longer term results may reflect any combination of several variables: increased emplacement time combined with reduced disturbance; changes in meteorological and hydrologic conditions; and/or increasing surface area to volume ratios as tablets are broken by freeze–thaw and other mechanical weathering processes (Day unpublished).

Slope processes within the karst have received insufficient attention and warrant greater scrutiny. Carbonate rockfalls pose occasional hazards to local roads, houses, and other structures but are rarely of great magnitude (LaCrosse

Tribune 2011). The landscape position of the carbonates results in slope materials experiencing considerable mixing, which complicates potential studies, but there is evidence of granular disintegration, rockfall, rock collapse, and occasional slab failure, with active and abandoned quarries providing suitable although disturbed sites for monitoring.

Narrow ridges and slopes are dominantly wooded, with a mix of hardwood tree species that have considerable economic value and represent important wildlife habitat, particularly for the burgeoning deer herd. Hunting apart, the karst provides a range of recreational opportunities, exemplified by state parks at Wyalusing (Grant County), Tower Hill (Iowa County), and Wildcat Mountain (Vernon County) (WDNR 2017b).

6.3.1 Sinkholes

Although dry valleys dominate the karst numerically and visually, sinkholes or dolines are diagnostic of karst terrain (Ford and Williams 2007) (Fig. 6.15). More than 250 sinkholes have been documented north of the Wisconsin River (Day and Reeder 1989) and it is estimated that there are in excess of 500 in southwestern Wisconsin (Day et al. 1989). The majority of these sinkholes are small: mean diameter is 6.8 m and mean depth is 1.1 m, with 78% of individuals less than 10 m in diameter and 51% less than 5 m wide (Day et al. 1989). The largest recorded sinkhole is the Reynolds sinkhole in Crawford County, which exceeds 50 m in length, 23 m in width, and 18 m in depth (Oh et al. 1993). Sinkhole densities are generally low, less than 0.5/km^2, but there are several clusters with >20/km^2 (Day et al. 1989). Few sinkholes are significantly elongated along structural lines and most are near-circular in plan (Day and Reeder 1989). Most sinkholes are on the tops and sides of interfluvial ridges, with suffosion and surface collapse both operative mechanisms and up to 10 sinkholes forming yearly (Day et al. 1989).

The sinkholes pose problems for human activities, mostly associated with agricultural inconvenience and small examples or new occurrences are often infilled (Day and Reeder 1989). Land-use strategies designed to minimize established sinkhole problems include avoidance, encouragement of tree growth, fencing, and infilling (Day and Reeder 1989). Some sinkholes provide natural entrances to caves, such as Star Valley Cave, Pops Cave, Haines Cave, and Boscobel Bear Cave. The entrance sinkhole to John Gray Cave (Fig. 6.16) was in the nineteenth century a visitor attraction in its own right (Day and Kueny 2000). It is possible that some enclosed depressions, particularly in the Sinnipee Group carbonates, are actually borrow pits resulting from early lead mining or maybe sinkholes that were excavated by miners, although the presence of spoil materials around the periphery is often indicative (Day et al. 1989).

Fig. 6.15 Sinkhole dumping, Crawford County. Photo by Michael J. Day

Fig. 6.16 Sinkhole entrance to John Gray Cave. Photo by Michael J. Day

6.3.2 Caves

Caves in southwestern Wisconsin have formed throughout the Ordovician carbonates, particularly in the dolostones, of the Prairie du Chien Group, the Platteville Formation, and Galena Dolomite, which are increasingly dolomitized across the Wisconsin Arch. The carbonates were folded and fractured during the Paleozoic and the initial structures have been accentuated by subsequent dissolution and slumping (Hedges and Alexander 1985). Karstification may have commenced in the Paleozoic (Hedges and Alexander 1985) and hydrothermal activity affected the Platteville–Galena units around the same time (Heyl et al. 1970). The dolostones have undergone extensive and complex diagenesis and there is a considerable paleokarst legacy (Smith and Simo 1997).

Many southwestern Wisconsin caves are the subject of extensive folklore (Cronon 1980) and several caves are operated commercially, including Crystal Cave (Pierce County), Kickapoo Caverns (Crawford County), and Eagle Cave (Richland County). Lost River/Pokerville Cave (Iowa County) was operated commercially, although sporadically, until 1980. The St John Mine (Fig. 6.17) (Snake Cave) (Grant County) was added to the National Register of Historic Places in 1979 but is currently closed. In 1988, Cave of the Mounds (Dane County), which is notable for its impressive speleothems, was designated a National Natural Landmark by the US Department of the Interior and the National Park Service (Soule 1993). These commercial operations are economically significant, at least locally. Full- and part-time employment at the commercial caves totals nearly 100 people and revenues exceed $500,000 annually (Day et al. 1989). Cave of the Mounds has an annual visitor total of over 80,000 (Day et al. 1989) and plays an important regional role in educating young people about caves (Cave of the Mounds 2017).

Fig. 6.17 Mined passages, St John Mine, Grant County. Photo by Michael J. Day

The longest cave (1200 m) in southwestern Wisconsin is Crystal Cave, near Spring Valley in Pierce County (King 1972) (Fig. 6.18), which is also an excellent example of the multilevel maze caves that occur in the Prairie du Chien carbonates. With a depth exceeding 20 m, the lower level of the cave is formed within the Oneota Formation dolostone and the overlying dolomitic New Richmond Sandstone, while the upper levels are developed within the Shakopee Formation. Located beyond the periphery of the Driftless Area, Crystal Cave has an extensive infill of glaciofluvial deposits which are atypical of southwestern Wisconsin caves. Glacial sediment in Crystal Cave has been dated by optically stimulated luminescence (OSL) to around 24,000 BP and is overlain by loess aged between 20,000 and 17,000 BP (Bellomo et al. 2011).

There are over 250 individual documented carbonate caves in southwestern Wisconsin but most are shallow and with one exception all are less than 1 km in length. Their dimensions are constrained by slow dissolution rates, thin bedding, and the dismemberment of networks and reduction in catchment areas produced by valley incision (Day et al. 1989). Passages tend to be tubular, with some vadose trenching and they contain extensive breakdown deposits, together with extensive red-brown silts and clays produced by weathering of the carbonates and infiltration of loess (Day 1988). Passages are mostly near-horizontal, sometimes multilevel, with occasional large rooms and vertical pits. Entrances are typically through sinkholes or in quarry faces. These caves represent ridgetop or hillside remnants of formerly more extensive systems that were dismembered by long-term valley incision, although probably accelerated by Pleistocene glacial meltwater. The caves are essentially abandoned phreatic tubes and mazes with little vadose entrenchment and most are at least partially blocked by breakdown slabs, with active breakdown remaining a hazard in several caves (Day 1986a). Most cave passages are within 10 m of the ground surface (Day 1986a, b, c; 2009).

Despite their obscurity, the caves and karst of southwestern Wisconsin are not without broader scientific interest. Significantly, karst valleys and caves are integrated into the regional drainage system and provide inputs supporting springs and perennial drainage within the underlying non-carbonates (Reeder and Day 1993). The sediments and mineralogy of southwestern Wisconsin caves clearly merit additional investigation, as they can provide evidence not only about cave development but also about the overlying karst landscape and about broader regional paleoenvironmental conditions. The mineralogy of southwestern Wisconsin caves has been the object of very limited prior study, although most reports suggest that the mineral suites are calcite/aragonite dominated (Smith and Simo 1997). Detailed mineralogic studies are necessary in order to produce significant evidence about cave development processes.

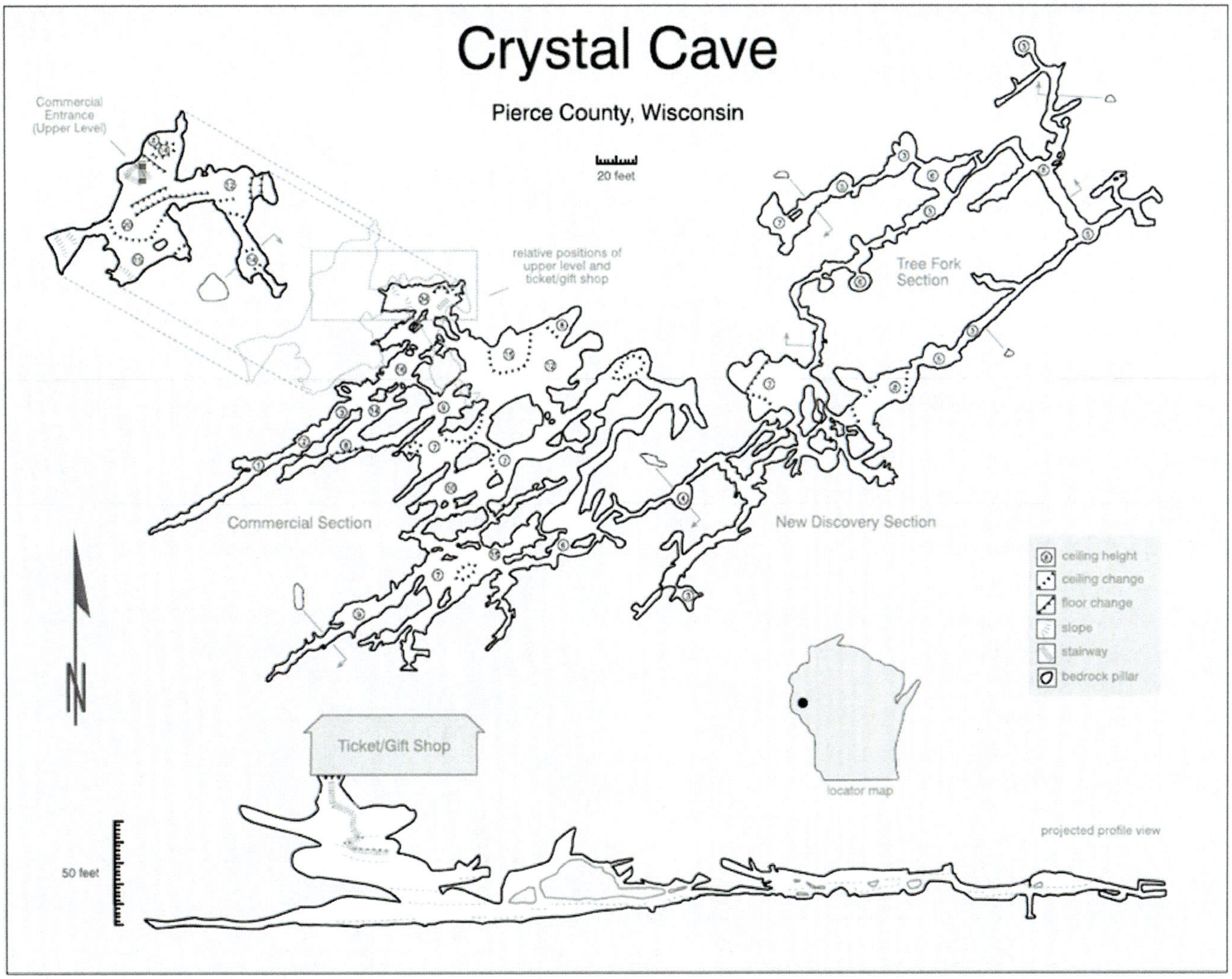

Fig. 6.18 Crystal Cave map by John Lovaas. Image provided by Crystal Cave

Sedimentological studies in the 1980s by Pete Bull and the author were never adequately documented because of medical and other exigencies and only a brief overview is available (Day 1988).Driftless Area caves contain a characteristic suite of red-brown silts and clays which are remarkably uniform both within and between individual caves, but sediments in the adjacent glaciated area are very different. The bulk of the Driftless Area cave sediment appears to be derived from windblown loess which has infiltrated the caves and which in many respects resembles surficial deposits, but there is no evidence of any materials of direct glacial origin. These studies require repetition and expansion, since similar regional studies have yielded valuable information (Milske et al. 1983; Bellomo et al. 2011). In particular, the evidence of loess-derived sediments in the caves should prove relevant in the context of ongoing debates about sediment sinks and budgets in the Driftless Area (e.g., Beach 1994; Hobbs 1999; Syverson and Colgan 2004). Brick (2013) reported nitrate-rich cave sediments from Lake Pepin dolomite crevices on both the Minnesota and Wisconsin sides of the river, which could have been used by French fur traders for the manufacture of saltpeter.

Cave and sinkhole sediments particularly provide evidence about former surface environments and karst development. Investigations of sinkhole sediments in the Reynolds and Seneca sinkholes in Crawford County indicate a range of inputs, some local, some exotic, but suggesting overall that the local lithological and landscape context is the most important factor influencing sediment accumulation (Oh and Day 1990, 1991; Oh et al. 1991, 1993). The sediment cores contain distinct layers, suggesting episodic inputs perhaps linked to Holocene climatic variability. Sediments adjacent to the Reynolds sinkhole floor have been dated to 6540 ± 90 yr BP, roughly coincident with the period of maximum Holocene aridity (Bartlein and Webb 1982; Bartlein et al. 1984). Silt components appear to be loess derived, although the relative contributions to immediate sinkhole accumulation by aeolian, fluvial, and other processes are uncertain. Potential contributions by gullying, slope wash, colluviation, and back-wasting are all possibilities and there is evidence of both permafrost conditions and freeze–thaw cycles (Oh et al. 1991, 1993). Sands in the sediments may be derived from the dolostone itself or from the overlying St. Peter Sandstone, with increasing sand contents toward the periphery of the sinkholes suggesting the latter (Oh and Day 1991). Clays are derived primarily from the carbonate bedrock, with occasional gravels perhaps derived from pre-Pleistocene gravels of the Windrow Formation (Clayton and Attig 1990). Sediments in sinkholes on ridgetops are generally thinner and less variable than those in hillside sinkholes (Oh et al. 1991) and it is suggested that at least some of the sinkholes adjacent to the interface of the carbonates and the overlying St. Peter Sandstone may be interstratal or subjacent (Oh et al. 1991).

Additionally, an array of calcite formations has developed in the caves and some of these may be amenable to uranium-series dating (Ford and Schwarcz 1981) or other dating techniques. In many of the noncommercial caves, however, many of the formations have been damaged by breakdown or vandalism and the remaining intact examples are small and accessible only with difficulty.

Other noteworthy caves in southwestern Wisconsin include Pops Cave (Richland County), with about 200 m of surveyed passage and John Gray Cave at 216 m. Pops is probably the most heavily visited noncommercial cave in the area and is characterized by its hilltop sinkhole entrance, extensive breakdown, bi-level phreatic passageway, rimstone dams, and consistent dripwater. Like Pops, John Gray is entered via a ridgetop sinkhole (Fig. 6.16) beneath which is an entrance room containing a talus pile and from which three small rounded passages radiate, leading to extensive "pancake" sections with breakdown. John Gray was one of the four caves studied by Lange (1909).

Bogus Bluff Cave (Richland County) is unusual in having three cliff-side entrances and extends primarily laterally to total 195 m in length. Much of the cave shows signs of excavation and the name is said to derive from a counterfeiting operation in the cave around 1900 (Peterson 1968).

Castle Rock Cave (Grant County) is perhaps southwestern Wisconsin's most confusing maze cave and is entered via a tight crawlway leading to a large room, beyond which are the maze sections. Total passage length is about 275 m.

Crawford County's Haines Cave (252 m), Boscobel Bear Cave (260 m) (Fig. 6.19), and Star Valley Cave are all entered through sinkholes. Haines consists of several pits with large rooms and passages at different levels and includes a maze and extensive breakdown. Boscobel Bear, studied by Lange (1909), is similar, with various levels, large rooms, and considerable breakdown. Star Valley (244 m) is perhaps best known for the vadose "keyhole" passage extending from the entrance but also contains some tight crawlways and a lower northern level which floods occasionally.

Fig. 6.19 Sinkhole entrance to Boscobel Bear Cave. Photo by Michael J. Day

Bear Creek Cave (Sauk County), discovered during blasting in 1954, is entered via a disused quarry and contains 290 m of passageway. Parts of the cave are well decorated and there are extensive "pancake" sections. Bear Creek cave has been an important bat hibernaculum, monitored for many years by the WDNR and biologists from UW-Madison.

Ages of the caves have yet to be established with any certainty, but it is entirely possible that their earliest development may have been initiated during the Paleozoic, immediately following lithification. It has certainly been suggested that they may be Mesozoic or earlier (Wopat 1974; Barden 1980a) but the Mesozoic argument rests on a dubious emphasis on paleoenvironmental stimuli, such as warm, wet climatic conditions (Day 1986a, b, c). Inception is generally considered to have occurred under shallow phreatic conditions prior to valley incision (Olmstead and Borman 1968; Milske et al. 1983). The following sequence proposed by Olmstead and Borman (1968) for Pop's Cave, in Richland County may represent a model of general utility. The cave originated through bi-level, joint-controlled phreatic solution contemporaneous with the commencement of valley incision. As valley downcutting progressed, a major phase of formation in the upper part of the phreatic zone occurred as increasingly aggressive water came through the cave. Further valley incision then stranded the bi-level caves in the vadose zone where breakdown dominated, through which the lower passage eventually intercepted the upper to produce the present configuration.

The relationship of the caves to regional drainage patterns at their time of formation is unknown, especially since the caves are fragmentary. Preincision they may have drained toward the proto-Mississippi River or its tributary the proto-Wisconsin, a hypothesis supported by a preliminary analysis of cave elevations (Patti Day, pers. comm.) and jointing influences may have been significant (Olmstead and Borman 1968; Barden 1980a). Structural trends notably fracture orientations and anticlinal and synclinal structures, control drainage patterns and ridge orientation (Heyl et al. 1959) and this also appears to apply to the caves since many passages follow N–S, NE–SW, or NW–SE alignments. Cave passage orientation data from 21 caves in Richland County were compared statistically with orientation data for adjacent valleys (Terlau and Day 1997). Rose diagrams suggested visually a close relationship between the orientation datasets and statistical testing showed a strong relationship in 80% of cases, suggesting that the orientation of caves and valleys may be influenced by a common set of bedrock fractures.

It has generally been assumed that the caves of southwestern Wisconsin developed primarily through "normal" epigenic dissolution of the carbonate bedrock by dilute carbonic acid solutions. Although this may well be the case, it is clear that hydrothermal processes have affected the caves south of the Wisconsin River, where lead and zinc ores were emplaced during the Paleozoic (Reeder and Day 1990; Dockal 2020, Chap. 7 in this volume) and it is also possible that similar hydrothermal modification may have occurred in the caves north of the Wisconsin River. Moreover, it is also possible that the caves were influenced, even possibly initiated, by hypogenic sulfuric acid dissolution processes, the broader significance of which has only recently been fully appreciated within the USA (Klimchouk et al. 2017) and/or by dissolution under saline–freshwater mixing conditions (Pipes and Day 2006). Extensive hydrothermal diagenesis has been indicated regionally along the Wisconsin Arch (Luczak 2006) and this undoubtedly influenced cave development.

Consideration of hydrothermal influences has previously been restricted to caves formed in the Platteville–Galena units in the lead and zinc district south of the Wisconsin River, but even here the assumption has usually been that the caves developed initially through "normal" processes and that hydrothermal ores were emplaced subsequently (Paull and Paull 1977; Day 1986a, b, c). There has been little previous investigation of hydrothermal speleogenesis possibilities north of the Wisconsin River where the Platteville–Galena is absent and the older carbonates crop out, although regional hydrothermal dolomitization is clearly indicated (Smith 1990; Smith and Simo 1997). Similarly, there has been no meaningful study of the similarities and/or differences in morphology, hydrology, or chemistry between the caves in the different formations.

What is certain is that tectonic conditions potentially influencing southwestern Wisconsin's carbonate rocks during the Paleozoic were such that widespread hydrothermal activity cannot be dismissed. Regional tectonism during the Paleozoic caused extensive fracturing (Holst 1982) and the Permian-aged lead and zinc ores in the Platteville–Galena appear to be controlled by a fold-related fracture pattern that provided access for mineral fluids (Paull and Paull 1977). The Prairie du Chien carbonates are older than the Platteville–Galena, so they were certainly in place at the same time as hydrothermal action introduced metal ores into the latter during the early Permian (Rowan et al. 1995). Similar copper deposits, together with minor occurrences of lead and zinc do occur within the Prairie du Chien carbonates (Paull and Paull 1977) and regional hydrothermal dolomitization is clearly indicated (Smith 1990; Smith and Simo 1997) but their relationship to cave development is currently unknown.

In terms of morphology, there is currently no definitive evidence to demonstrate that caves north and south of the Wisconsin River are fundamentally dissimilar. Several southwestern Wisconsin caves have large chambers, but these are not always associated with branching networks of ascending passages terminating in spherical pockets. Likewise, some southwestern Wisconsin caves have maze-like sections (e.g., Castle Rock Cave) but this is not consistent. According to Hedges and Alexander (1986: 45), "Many maze caves unrelated to modern topography also occur in the (Lower Ordovician) Oneota dolomite," which is also suggestive of other than "normal" development. Some southwestern Wisconsin caves do have natural entrances, although many have been discovered during quarrying.

As noted previously, the mineralogy of southwestern Wisconsin caves has been the object of very limited prior study, so there is no definitive evidence of hydrothermal mineral associations north of the Wisconsin River, although the Prairie du Chien carbonates do host copper, lead, and zinc deposits on the northern fringes of the lead–zinc district (Paull and Paull 1977). The abundance of breakdown in the caves has usually been ascribed to instability during drainage (Olmstead and Borman 1968) but it is possible that it is related to hydrothermal activity and it would be useful to compare the breakdown to the collapse breccias characteristic of hydrothermal caves elsewhere.

Cave air temperatures currently are about 10 °C but there is considerable fluctuation in temperature and humidity, particularly in shallow caves with multiple entrances (Mueller and Day 1997). Although much speleological literature stresses the temporal, particularly annual consistency of cave atmospheres, there is increasing evidence of seasonal and other variability (e.g., deFreitas and Littlejohn 1987) and this certainly applies to some of Wisconsin's caves. A year-long MS thesis study of atmospheric variability in several southwestern Wisconsin caves by Bob Mueller was unfortunately never written up, but the results of a 2-week study of John Gray Cave, Richland County in May–June 1990 were documented by Mueller and Day (1997). John Gray Cave is a typical shallow ridgetop cave with a single sinkhole entrance, an entrance lobby, and three radiating passages 49 m, 72 m, and 111 m in length. During the study period (May 26–June 8, 1990), daily measurements of temperature and relative humidity were made at the surface, in the lobby, and within the three passages. Mean external temperature was 19 °C, ranging from 10.0 to 23.9 °C and lobby temperatures were similar, ranging from 12.5 to 23.3 °C, with a mean of 18.6 °C. Temperatures in the three passages varied less but with ranges of 6.2–8.4 °C around station means between 14.1 and 16.4 °C. External relative humidity (RH) ranged from 32.0 to 90.0%, with a mean of 58.9%. Mean lobby RH was 52.2%, ranging from 41.0 to 69.2% and RH in the three passages ranged from 38.9 to 65.6%, with station means between 47.0 and 52.9% (Mueller and Day 1997). These readings, although covering only a single cave over a relatively brief period, indicate a dynamic cave atmosphere suggestive of considerable air exchange with the surface. Although unpublished, year-long results revealed similar fluctuations, particularly in Bogus Bluff Cave, which has three entrances.

White-nose syndrome, attributed to the fungus *Pseudogymnoascus destructans* (Gargas et al. 2009), was first diagnosed in bats in Grant County in 2014 and continues to afflict cave bat populations across Wisconsin, having now been identified in 14 counties and suspected in 10 others (WDNR 2016a,2017c). Wisconsin's bat population was estimated at 350,000 to 500,000 in 2015, but recent surveys reveal "catastrophic" population declines of between 30 and 100% at some hibernacula in the second and third years of infection (WDNR 2017c). Appropriate decontamination protocols were adopted statewide (WSS 2017) but were largely ineffective (White 2018; Cushman 2019; Echolocator 2020).

Several of the sandstone caves in southwestern Wisconsin have proven to be valuable **archaeological sites** yielding a variety of pre-European Native American artifacts. Much of Wisconsin's pre-European rock art is associated with sandstone caves and rockshelters in southwestern Wisconsin (Salzer 1987a, 1997; Birmingham and Green 1987; Stiles-Hanson 1987; Boszhardt 2003). Over 150 rock art caves have been recorded (Broihahn 2008). In particular, Arnold Cave contains an impressive array of recently documented pictographs (Steelman et al. 2001) and a famous petroglyph was discovered in the Gottschall Rock Shelter (Salzer 1987b). The Raddatz Rockshelter and the Durst Rockshelter are two other intensely excavated archaeological sites in the state (Wittry 1959a, b).

6.3.3 Springs

Springs are an integral component part of the karst geomorphology and hydrology (Day et al. 1989, 2004) (Fig. 6.20). Springs are also numerous among southwestern Wisconsin's karst landforms. A 1960s survey of a 900 square-mile (2330 km^2) study area south of the Wisconsin River documented 7210 springs, with a density of eight per square mile (3.1/km^2) with discharges between 0.25 and 30 gpm (0.02 to 1.9 l/sec) (De Geoffroy et al. 1970). 2,278 springs have been recorded in Grant County alone (Smith and Ball 1972) and surveys in the 1950s documented 586 springs in Richland County (WCD 1958) and 469 in Crawford County (McNurlin 1959). The carbonate aquifers are dominant of the diffuse flow type, which is typical of impure limestones and dolostones (White 1977). Recharge is primarily through infiltration and transmission velocities generally are low, mostly via pores, tight fractures, and small voids. In the saturated or phreatic zone, water is in dynamic storage in unconfined aquifers and slow gravity flow maintains moderate spring discharges. Portions of the aquifers appear to be of the free-flow type (White 1977) with higher velocity, turbulent flow in integrated conduits enlarged by dissolution (Reeder 1992; Reeder and Day 1993).

Fig. 6.20 Castle Rock Spring, Grant County. Photo by Michael J. Day

Paleoconduits, abandoned as the regional potentiometric surface was lowered, exist in several regional caves (Day 1986a, b, c).

In Richland and Crawford counties at least, the combination of regional geologic dip toward the southwest and joint orientation predominantly northeast–southwest is reflected both in the regional drainage pattern, with rivers tributary to the Wisconsin draining essentially northeast–southwest and in spring locations, which are dominantly at the base of slopes on the east and north sides of valleys (Day et al. 2004). Groundwater flow may reasonably be modeled initially as dominantly down-dip, i.e., toward the southwest, although there may be localized exceptions (Day et al. 2004).

With regional geologic dip is toward the southwest and joints oriented predominantly northeast–southwest, it might be expected that springs would be located primarily on valley sides with a southwestern aspect, at elevations declining southwestward and reflecting the elevation of the contact between the carbonates and underlying sandstones. The hypothesis was tested by field location, aspect, and elevation determination of 72 major springs in Richland and Crawford counties (Day et al. 2004) with aspect attributes generally supporting the hypothesis and 38 of the springs (52.8%) having aspects within the southwestern quadrant. Elevation data, however, revealed a trend contrary to that hypothesized, with spring elevations generally declining toward the northeast. These results may reflect the sampling design, measurement errors, complexities in the regional geologic structure, and/or incomplete understanding of the regional hydrology. Larger springs are located particularly on valley sides with a southwestern aspect, but elevations are less predictable, perhaps reflecting complexities in the regional geologic structure (Day et al. 2004). Historically, these springs have been important domestic and agricultural water sources, particularly as documented by historical documentation, although not substantiated by preliminary spatial analysis of maps dating from between 1842 and 1895 (Kemp and Day 1989).

Spring discharges are highly variable but 90% of those sampled in Iowa County between 1986 and 1988 were discharging less than 0.2 cfs (100 gpm) (Day et al. 1989), confirming previous estimates from Brynildson (1966). By contrast, also in Iowa County, Big Spring has a regular discharge of about 3.5 cfs (1600 gpm) and Arndt Spring discharges 2.7 cfs (1200 gpm). Similarly, about 83% of the 469 springs in Crawford County have discharges of less than 0.38 cfs (McNurlin 1959). The largest regional spring is probably Castle Rock Spring, in Grant County, which has a regular discharge of about 6.7 cfs (3000 gpm). Many higher elevation springs are intermittent or ephemeral (Reeder 1992; Reeder and Day 1993, 1994). Spring water temperatures measured between 1986 and 1988 ranged between 8.9 and 10.6 °C (48–51 °F), with a mean temperature of 9.7 °C (n = 185) (Day et al. 1989). Similarly, 12 springs sampled monthly in 1980–81 ranged between 9.1 and 10.3 °C (48.4–50 °F) (Heller, pers. comm. 1988). Springwater is typically calcium–magnesium–bicarbonate dominated and supersaturated with carbon dioxide. Heller (1988) measured total hardness (as $CaCO_3$) ranging from 327 to 373 mg/l and pH values from 7.2 to 8.0. Locally, springs are contaminated by agricultural chemicals (Reeder 1992; Reeder and Day 1993, 1994).

A survey of springs in Iowa County suggests that any loss of spring resources over the last 50 years has been minimal (Swanson et al. 2007), although preliminary investigations suggest that this may not always be the case north of the Wisconsin River (Day et al. 2004). Iowa County springs are associated with every major stratigraphic unit but are most numerous in association with the Sinnipee Group, near the upper contact of the St. Peter Sandstone, or near the upper contact of the Cambrian sandstones (Swanson et al. 2007). Spring waters discharging from different geologic units can be distinguished on the basis of major ion geochemistry and springs discharging from stratigraphically higher units with small recharge areas have more variable flow and are likely to be vulnerable to reduced discharge or desiccation as a result of pumping from ridgetop wells (Swanson et al. 2007).

Likewise, there has been only limited investigation of southwestern Wisconsin's karst hydrology, despite its significance to recreational trout streams (Graczyk 1993). Chemical analyses of spring waters were utilized extensively for identifying prospective zinc mining sites in areas south of the Wisconsin River in the 1960s (De Geoffroy et al. 1967, 1970) although it contributed little to an understanding of the karst itself. Chemical analysis of two tufa depositing springs in the Platteville Formation indicated bicarbonate levels in the 396–450 mg/l range, with Ca/Mg ratios between 1.06 and 1.54 (Heller 1988). Seasonal comparison revealed a summer decrease in pH, saturation with respect to calcite and dolomite and bicarbonate concentration, but a summer increase in temperature and pCO_2 as a result of carbonic-acid recharge during the growing season. As might be expected, carbon dioxide degassing during tufa formation caused an increase in carbonate saturation and pH, precipitation of calcite or aragonite and a subsequent decline in dissolved calcium, bicarbonate, total hardness, and conductivity. However, the greatest loss of calcium carbonate occurred in the winter and an increase in carbonate mineral saturation during the summer suggested that deposition was inhibited during that season, possibly because the magnesium ions retarded calcite precipitation at higher temperatures (Heller 1988).

County-wide hydrogeological studies would be appropriate and valuable contributions to understanding of the southwestern Wisconsin karst, with a study of Dane County providing an excellent model (Bradbury et al. 1999), but to date there have been only limited karst hydrogeological investigations, certainly in comparison to those in neighboring Minnesota (e.g., Green and Alexander 2015). The most pressing need is for data constraining the occurrence and function of karst in the Prairie du Chien Group (LePain et al. 2005). The balance of the flow regimes within the aquifer is currently "entirely a matter of speculation" (LePain et al. 2005, p 9), but it has great importance for managing the Prairie du Chien aquifer. Priorities include "characterizing the transience of springs and groundwater levels; identifying correlations between solution features, structural features and lithostratigraphy; characterizing the hydraulic relationships between the Prairie du Chien and adjacent units; and completing karst-appropriate studies of groundwater flow (e.g., using tracers to confirm point-to-point flow paths)" (LePain et al. 2005, p 9).

One difficulty hindering hydrologic studies is that the upland karst has very restricted surface catchment areas, so dye tracing from sinkholes and through caves to springs is possible only during unusually intense rainfall events, during snowmelt, or if water can be otherwise delivered. Even then, results have been inconclusive or ambiguous, as exemplified by an unpublished dye trace by Mike Neal and the author from Star Valley Cave, Crawford County in 1985. Following an instructional dye trace in Belize in 1984 (Day et al. 1987), leftover Rhodamine WT dye was introduced to spring snowmelt floodwater in the lower rear chamber of Star Valley Cave in April 1985 and the appropriate activated charcoal detectors (Smart and Brown 1973) were placed in several local springs, within a lateral distance of 5 km and up to 100 m below the cave. The dyed water took nearly 2 months to exit the cave through sediments in its floor and approximately monthly monitoring of the springs revealed no dye presence until 18 months later, when a very weak signal was detected at a spring 1.2 km away and approximately 100 m below the cave elevation. This investigation was never published because it was initially assumed that the detection was a false positive resulting from inadvertent contamination of the spring water by the researchers and because the geologic situation indicated that to reach the spring from the cave the water would have had to drain through the apparently nonkarstic Jordan Sandstone Formation. However, since subsequent studies that have demonstrated the hydrologic connectivity between the Oneota and the Jordan (Reeder and Day 1993, 1994), it is possible that this trace may have shown an actual connection, although this clearly needs confirmation using more sophisticated tracing methodology.

6.3.4 Mining

Lead ores were mined extensively in the Driftless Area of southwestern Wisconsin during the middle of the nineteenth century, when the Upper Mississippi Valley Lead District was one of the major lead-producing regions in the world (Reeder and Day 1990; Day and Reeder 2013). This topic is treated at greater length by Dockal (2020), Chap. 7 in this volume.

Quarrying and aggregate mining are significant historical and contemporary industries within the karst (Day and Bindl 2010–2011) (Fig. 6.21). The history of quarrying extends back to earliest nineteenth Century European settlement, when limestone and other rocks were quarried for local construction and there are estimated to be in excess of 1000 abandoned quarries in the region. Larger scale quarrying, primarily of limestone and dolostone for aggregate for road construction, dates to the early and mid-twentieth century and continues today, with about 200 quarries currently active on at least a temporary basis. Richland, Crawford, Vernon counties are in Wisconsin's Crushed Stone/Sand and Gravel District 5, whereas Sauk, Grant, Iowa, and Lafayette counties are in District 1 (USGS 2006). Statewide, these two districts rank consistently in the top three districts in crushed stone production, with outputs in 2005 of 7.26 million tones (8.00 million tons) and 6.97 million tonnes (7.68 million tons) respectively, valued at $46.0 million and $43.4 million, respectively (USGS 2005). Regionally, contemporary quarrying is dominated by the Kraemer Company, based in Sauk County. Aggregate production is currently about 14.5 million tonnes (16 million tons) annually. Quarrying currently employs only about 200 people directly, although hundreds more in distribution and thousands in product usage. It contributes nearly $90 million annually directly to the regional economy and considerably more indirectly.

Regional quarrying is not without controversy, but remediation and other efforts can significantly reduce environmental impact. Potential environmental impacts include partial or complete destruction of landforms, interference with groundwater flow and contamination of groundwater, dumping of overburden, and air pollution by dust (Ford and Williams 2007). Fortunately, avoidance of unnecessary impacts is usually possible without undue inconvenience, particularly if quarrying is restricted to small dispersed sites at distance from zones of groundwater recharge.

Quarrying may also cause or exacerbate subsidence and/or sinkhole collapse (Kastning 2003) but this problem has not been documented in southwestern Wisconsin. By

Fig. 6.21 Kraemer Company's Clockmaker aggregate Quarry near Westby in Vernon County. Google Earth photo

contrast, regional quarrying has destroyed several small cave systems without previously known entrances and this, while perhaps inevitable, is to be regretted. Contemporary operations are more sensitive to cave locations and the Kraemer Company strives to avoid cave destruction unless absolutely necessary, drawing caves disclosed by quarrying operations to the attention of the Wisconsin DNR (J. Kraemer, pers. comm. 2010). Although quarrying may destroy caves, it also exposes caves which were previously unknown, making them potentially accessible to scientific research and exploration. Among southwestern Wisconsin caves discovered through quarrying are Cave of the Mounds (1939), Bear Creek Cave (1954), and Eysenogel Hill Cave (1964) (Day 1986a, b, c).

Abandoned quarries also present challenges, particularly if they are water filled. Use of abandoned quarries for official and unofficial disposal of solid wastes in landfill sites poses even greater potential problems, including permanent groundwater contamination. Nearly 30% of the abandoned quarries located in Richland and Crawford counties in 1988 and 1989 were being used for agricultural, commercial, and industrial wastes, with variable amounts of refuse including paint, oil, batteries, vehicles, and household appliances (Day and Bindl 2010–2011). Natural processes may partially and slowly rehabilitate abandoned quarries through slope failures and vegetation growth and human remediation can replicate natural landforms and improve groundwater quality (Gunn et al. 1997).

Public understanding of the significance of karst in southwestern Wisconsin was extremely limited until recently, when proposed agricultural and industrial developments drew attention to the potential for significant land-use change and possible groundwater contamination. Land use in the karst of southwestern Wisconsin has been dominated by small-scale agriculture, which has adapted to the karst with minimal difficulty (Day and Reeder 1989). Two recent developments—the proposed construction in 2007 of a coal-fired power station fly ash disposal facility in Vernon County and the expansion since 2010 of confined animal feeding operations (CAFOs) in Crawford County—raise more critical issues about human interaction with the karst and highlight the potential for land-use conflicts and detrimental environmental impacts. Proposals for fly ash disposal in the karst have been discontinued following public opposition and scientific input, but several CAFOs are already in operation, although there is mounting public and professional opposition to proposed new facilities and to the existing state environmental assessment and permitting process. Even more recently, the rapid escalation of frack-sand mining in western Wisconsin has increased public awareness of the broader environmental challenges resulting from changing industrial priorities. This industry is focused currently on west-central Wisconsin (WDNR 2016b), in which context the relatively unknown karst of Trempealeau, Buffalo, and Dunn counties may have particular relevance.

In response to these and other developments, locally based environmental organizations such as the Crawford Stewardship Project have become increasingly prominent, enlisting professional assistance and organizing networks of volunteers to monitor variables such as air and water quality and disseminate information (CSP 2016). Regional karst science has yet to be incorporated adequately into planning mechanisms and there have been calls for a regional karst study similar to that carried out and applied to land-use planning in northeastern Wisconsin (Erb and Stieglitz 2007). Regardless of the outcomes of the current controversies, these issues have served to greatly increase public awareness of southwestern Wisconsin's karst, which is now a central theme for local non-government organizations (NGOs) and the regional press.

6.4 Caves and Karst in Sandstones and Other Lithologies

Over 77 or about 31% of the approximately 250 caves recorded and mapped in Wisconsin are developed in sandstones (Day 2001). Despite this, these sandstone caves have received scant attention, except from recreational cavers and little research into them has been conducted. To date, the most authoritative summary remains that by Cronon (1970) who provides a listing of the state's known sandstone caves, grouping them into two broad classes: collapse caves and erosional caves. The former number at least 10 and the latter at least 51 with an additional 16 unclassified. Of the erosional class caves, four are classified in a stream meander group and at least four in an "exterior erosion" group, which includes crevice caves formed by rock failure, but the remaining 43 or more are attributed to groundwater erosion, or speleogenesis.

Wisconsin's sandstone caves are formed within three geologic units. In the northern part of the state, a few caves are developed in Precambrian sandstones, but these are not considered in detail here since they are few, small, and produced primarily by littoral and other processes, rather than by dissolution. Littoral caves around the perimeters of the Bayfield Peninsula and the Apostle Islands have formed by lacustrine wave erosion of the Precambrian Devils Island Formation sandstones and are quite impressive, if under-studied (Thwaites 1912; Brick 2002). Notable littoral cave, arch, and stack locations include those near Squaw Bay on the mainland and on Sand Island and Devils Island. Both the caves and their winter ice formations are significant visitor attractions NPS (2016a, b).

Fig. 6.22 Small caves in the Tunnel City Formation, Sauk County. Photo by Michael J. Day

More significantly, caves in southwestern and central Wisconsin have developed in Paleozoic sandstones, particularly in the Cambrian-aged Jordan and Tunnel City Group sandstones (Fig. 6.22), which underlie the main carbonate cave host rock, the Early Ordovician dolostones of the Prairie du Chien Group and in the Middle Ordovician St. Peter Sandstone, which overlies the Prairie du Chien Group. Collapse caves are formed predominantly in the St. Peter Sandstone and dissolutional caves in the Jordan Formation.

Depositional patterns during the Cambrian reflect the influence of the Wisconsin Arch and adjacent basins, with five rhythmic transgressional sequences of sandstones, dolostones, and shales (Paull and Paull 1977). The basal Upper Cambrian formation is the 100–250 m thick shallow water Mount Simon Sandstone, which is overlain by the finer-grained impure sandstone of the Eau Claire Formation. Overlying this unconformably is the course, better-sorted sandstone of the Wonewoc Formation, which is up to 120 m thick and above this is the Tunnel City Sandstone, which is 30–60 m thick and lithologically similar to the Eau Claire. The dolostones of the St. Lawrence Formation cap this second transgressive cycle, which was followed by a period of erosion (Paull and Paull 1977).

The Jordan Sandstone is the youngest of the sequence of Cambrian sandstones and is a clean, well-sorted, white, medium-grained, high-energy sandstone about 6–46 m thick that was deposited during a third marine transgression onto the Wisconsin Arch (Paull and Paull 1977). As the transgression continued, increasing marine depths favored carbonate deposition and the Jordan graded into the overlying Oneota dolostone, which is the youngest of the Prairie du Chien Group. Regional uplift at the end of the Early Ordovician was followed by two further transgressions, during the second of which the St. Peter Sandstone was deposited. The St. Peter is typically a white, massively bedded, medium-grained, well-sorted quartz sandstone, 12–107 m thick, in places cross-bedded and in part of aeolian origin (Paull and Paull 1977).

The sandstones are integral components of the northernmost of the three westward-dipping cuestas that dominate Wisconsin's western uplands (Martin 1974). North of the Wisconsin River, the Jordan Sandstone typically forms 10 m high laterally extensive vertical cliffs beneath a Prairie du Chien dolostone caprock; further south and west the St. Peter outcrops above the Prairie du Chien. Further north and east the older Cambrian sandstones outcrop, with the Tunnel City Group forming particularly extensive valley-side cliffs.

In Sauk County, a small number of caves have developed in the Precambrian Baraboo Quartzite (Gietkowski 1972). Most of these are talus caves, enlarged joints, or cleavage fractures (Hendrix and Schaiowitz 1964). **Elephant** or **Devil's Cave**, at the north end of the east bluff at Devil's Lake State Park, is essentially a 10 m long enlarged joint cave but may have developed through the erosion of a conglomerate fill in either an open fracture or possibly as a littoral cave (Gietkowski 1972). Many smaller caves were formerly accessible in Cambrian Jordan sandstone at the Wisconsin Dells in Sauk County before being flooded by a power dam. These were formed by piping processes and include historically significant caves (Hedges and Alexander 1985; Brick 2007).

6.4.1 Caves in the St. Peter Sandstone

The caves in the St. Peter Sandstone represent subjacent karst development, since they have developed essentially by the collapse of the sandstones into cavities in the underlying Prairie du Chien dolostones. Although not numerous, these are regarded as some of the most interesting caves in the state (Cronon 1970). In neighboring Minnesota, many St. Peter caves are due to piping processes, forming pseudokarst (Brick 1997).

The progression of cavity migration from the dolostone into the overlying sandstone is outlined by Cronon (1970, p 85) and results in a variety of sinkhole and cave morphologies with variable carbonate–sandstone ratios. Most of these caves are single rooms entered through sinkhole bases and they are generally symmetrical, with circular plan profiles and ceilings arching upward toward the center. There is often a central pile of sand and sandstone rubble and the floors typically slope downward toward one of the edges. At least 10 of these collapse caves are cataloged by Cronon (1970) and several more are known.

Several of southwestern Wisconsin's best-known caves, including **Star Valley Cave** and **Viroqua City Cave**, have exposures of the St. Peter Sandstone in their ceilings and their sandy floors attest to gradual upward migration. In other cases, the upward migration has been such that the caves are now entirely within the overlying sandstone. Several pit caves occur in this category, including **E-Pit**, **Jones Cave**, **Cobb Cave,** and **Bridgeport Cave**, which contain the largest cave room in Wisconsin.

In addition to caves, the St. Peter Sandstone features a variety of striking residual fragile rock formations, such as Three Chimneys and Monument Rock, which may be in part dissolutional in origin. There are also localized boulder fields and trains, whose development has been inadequately studied and well developed clint and grike structures with obvious parallels to karstic features, particularly on ridges in Vernon and Crawford counties.

6.4.2 Caves in the Jordan and Other Cambrian Sandstones

Some of the better-known and more accessible caves within the Cambrian sandstones have developed through stream meandering, waterfall undercutting, or exterior erosion, including rock failure (Martin 1932; Cronon 1970) but at least 43 have formed through dissolution by groundwater and are thus of true speleogenic origin. These occur throughout the Cambrian sandstone sequence where the sandstones have higher carbonate contents, such as in the Lone Rock Formation of the Tunnel City Group and in the Jordan and St. Lawrence formations of the Trempealeau Group. They occur particularly within the upper Jordan Sandstone where groundwater flow is focused downward through the overlying fractured and karstified Oneota dolostone. The vertical continuity of the carbonate–clastic aquifer has been documented by the tracing of agricultural contaminant flushes to caves, springs, and wells within both lithologies (Reeder and Day 1993) and the Prairie du Chien-Jordan is widely regarded as a single distinct aquifer (Runkel et al. 2003; Tipping et al. 2006).

Some caves in the Cambrian sandstones are joint controlled, with tall narrow passages, while others are enlarged into gently sloping "pancake" passages and rooms along bedding planes. Overall, their morphology and orientation are similar to that of regional carbonate caves (Cronon 1970; Day et al. 1989; Terlau and Day 1997). In particular, they slope generally downward toward their entrances, indicating water egress (Cronon 1970). One notable difference, however, is that the sandstone caves contain very little of the silt–clay sediment infill, which characterizes the dolostone caves (Day 1988), presumably because the sediment has been retained within the latter rather than transported down into the underlying sandstones.

Notable caves in the Cambrian sandstones include **Anderson's**, **Grunt,** and **Hummel's Caves** in Richland County (Peterson 1968). Although the longest cave in the

Cambrian sandstones, **Autograph Cave**, in Juneau County, attains nearly 100 m in length, most are much smaller and a large number have not been recorded or mapped. For example, there are numerous small crevice caves in the Jordan Sandstone cliffs flanking the Kickapoo River Valley north of Viola, but only one, **Mount Nebo Cave**, is catalogued by Cronon (1970).

The development of these caves involves processes additional to dissolution, notably granular disintegration, the mechanical flaking of interior wall and ceiling surfaces, and the development of breakdown. Freeze–thaw may play an important role around entrances, where sand piles and vegetative debris accumulate and cavities may be initiated or expanded by tree root growth or animal burrowing. Mass wasting of slopes, for example, through rock toppling or rockfall may further disrupt entrances.

6.5 Natural Bridges and Other Fragile Rock Formations

The absence of Pleistocene glaciation has permitted the development and persistence within the sandstones and dolostones of the Driftless Area of numerous fragile rock formations, some of which have at least a partial speleogenic origin. Two natural bridges occur in the Tunnel City Group Sandstone, one at Pier Natural Bridge Park in Rockbridge, Richland County (Fig. 6.23), the other at Natural Bridge State Park in Sauk County. The former is essentially of fluvial origin, but the latter, known as the Leland Natural Bridge (Fig. 6.24) and the largest natural bridge in Wisconsin, may well have originated as a cave (Day 2008).

Fig. 6.23 Pier Natural Bridge, Rockbridge, Richland County. Photo by Michael J. Day

Located northwest of Sauk City, Leland Natural Bridge appears to be a remnant of a promontory or residual interfluvial ridge extending approximately northeast–southwest into the tributary valley of the North Branch of Honey Creek, a tributary of the Wisconsin River. The bridge has three distinct components: a basal plinth or foundation, two abutments or ribs, and a broadly rectilinear coping, or span (Fig. 6.24). The rock span is oriented northeast–southwest, with the opening of the bridge oriented WNW to ESE. The bridge has an internal span of 14.2 m (46.2 ft) at the base of the abutments and 8 m (26 ft) immediately beneath the coping. The internal height is 5.4 m (17.5 ft) and the coping is 4.8 m (15.6 ft) deep at its center with a mean upper width of 1.4 m (4.6 ft). Below and adjacent to the bridge is the Raddatz Rockshelter (Dott and Attig 2004), which is one of the most intensely excavated archaeological sites in the state

Fig. 6.24 Leland Natural Bridge, Natural Bridge State Park, Sauk County. Photo ©AdamMartin.SPACE

and which was occupied by Native Americans during the Archaic (9000 to 2500 BP) and later Woodland (2500 to 850 BP) periods, "as indicated by sequences of fire pits, artifacts and projectile points and bone debris" (Paull and Paull 1977, p 131), with earliest human occupation dating to perhaps 11,000 BP (WDNR 2008).

The bridge is formed in sandstones of the Mazomanie Formation of the Late Cambrian-aged Tunnel City Group, which in Sauk County is 30 m (97 ft) to 45 m (146 ft) thick and is a clayey or shaley bedded, partly calcareous, or dolomitic-cemented fine-grained quartizitic sandstone, with trace amounts of glauconite (Paull and Paull 1977; Clayton and Attig 1990). The sandstone is typically cross-bedded and, significantly, "somewhat calcareous, perhaps beginning to transition into the overlying dolomitic St. Lawrence Formation" (Dutch 2008, p 4).

Day (2001) suggested that many of southwestern Wisconsin's fragile rock formations may actually be cave remnants and regional speleogenesis may at least partially be hypogenic, involving upward dissolution by sulfuric acid solutions (Pipes and Day 2006), in which case cave formation may have also occurred within noncarbonate rocks underlying the regional limestones and dolostones, including Cambrian clastic rocks such as the Tunnel City Group. There is also independent evidence for laterally extensive, high permeability zones, and karst conduit flow in the Tunnel City Group, suggesting that stratigraphically controlled heterogeneities like contrasts in lithology or bedding-plane fractures may influence the flow of groundwater and promote speleogenesis (Swanson et al. 2006; Swanson 2007).

The carbonate contents of the bridge components, at about 40%, confirm the potential for karstic dissolution and sampled seepage water hardness at 78 ppm is suggestive of active dissolution within the rock formation. Interestingly, the mean hardness of the natural bridge waters (78 ppm) is very similar to that of the Wisconsin River at Muscoda (80 ppm) (Hem 1959), suggesting that carbonate dissolution may play a larger role in the regional landscape than is sometimes recognized.

The morphology of the Leland Natural Bridge itself is consistent with it being a cave roof remnant and the presence of both the adjacent rock shelter and a small arch within the coping show that cave-like features exist locally as well as elsewhere in the Tunnel City Formation. The orientation of the bridge broadly accords with the karst landform orientations elsewhere in southwestern Wisconsin (Terlau and Day 1997) and with regional groundwater flow models (Day et al. 2004). Also, the bridge is not an isolated landform element and it is possible that the adjacent features are unroofed remnants of cave passages or rooms, or the remnants of adjacent collapsed sinkholes.

Fig. 6.25 Five column rock (table rock) near Readstown in the Kickapoo Valley (Martin 1916, Pl 8)

Another particularly striking rock formation is Five-Column Rock (Fig. 6.25), also known as Table Rock, in Vernon County (Day and Kueny 1999). The rock is formed at the transition from the Jordan Sandstone to the overlying Oneota dolostone and has a basal sandstone plinth, a set of columns enclosing "windows" and a tabular dolostone summit, the entire structure being over 6 m high. The morphology of the feature, its stratigraphic context and its juxtaposition to extant cave passage all point to a speleogenic origin, which may have broader significance for the development of similar features throughout the region. In particular, the columns are developed within the transitional Sunset Point Member of the lower Prairie du Chien Group, which may represent a significant locus of speleogenesis adjacent to the sandstone–carbonate contact. Reno Cave, in neighboring Minnesota, appears to represent an earlier stage in the genesis of a columned rock, its columns still confluent with one another, forming the walls of a cave (Greg Brick, pers. comm. 2020).

Fragile rock formations in the St. Peter Sandstone include Elephant Trunk Rock (Fig. 6.26), Monument Rock, Maiden Rock, Steamboat Rock, Spook Rock, and the Three Chimneys, none of which has been the subject of detailed

Fig. 6.26 Elephant Trunk Rock, Jordan Formation, Richland County. Photo by Michael J. Day

geomorphological study. Rock castellations in the Jordan Sandstone are numerous, especially at the tapering extremities of the interfluvial ridges, but these too have not been studied in detail (Day 2001).

Acknowledgments I am grateful to Steve Mauel of the Wisconsin Geological and Natural History Survey particularly for providing data on the statewide areas of carbonate bedrock and carbonate groundwater extent. I also acknowledge the assistance provided over the years by Steve and other UW-Milwaukee graduate students, who drew my attention back to Wisconsin's karst when other, more alluring areas distracted me.

References

Aleson E, Redell J, Schuster W, Thomas G (2014) Management plan for Horseshoe Bay Cave. https://map.co.door.wi.us/parks/Horseshoe%20Bay%20Cave%20Management%20Plan%20(Approved%20June%2024,%202014).pdf

Alexander EC Jr, Green JA, Alexander SC (2008) Plum Bottom closed depression groundwater trace final report. https://www.researchgate.net/profile/Scott_Alexander2/publication/266528432_Plum_Bottom_Closed_Depression_Groundwater_Trace_Final_Report_prepared_by/links/55818a8c08ae6cf036c168e2.pdf. Accessed 22 Jan 2017

Appleton Post-Crescent (2017) Greenville karst obliterated, https://www.postcrescent.com/story/news/local/2017/04/07/greenville-pulls-nasty-obliterating-karst/99682980/. Accessed 25 Feb 2017

Baker RW, Knox JC, Lively RS, Olsen BM (1998) Evidence for early entrenchment of the upper Mississippi River valley. In: Patterson CJ, Wright HE (eds) Contributions to quaternary studies in Minnesota. Minnesota geological survey report of investigations 49, pp 113–120

Barden MJ (1980a) Geology of Door County. In: Alexander EC Jr (ed) An Introduction to Caves of Minnesota, Iowa and Wisconsin, NSS convention guidebook 21, pp 136–139

Barden MJ (1980b) Geology field trip guide. In: WSS Hodag hunt guidebook section 2

Barden MJ, Soule GK (1980) Caves of Door County, Wisconsin. In: Alexander EC Jr (ed) An Introduction to Caves of Minnesota, Iowa and Wisconsin, NSS convention guidebook 21, pp 136–141

Barr K, Klimchouck A, Alexander EC Jr (2008) Hypogenic karst and its implications for Minnesota hydrogeology. In: Yuhr LB, Alexander EC Jr, Beck BF (eds) Proceedings of the eleventh multidisciplinary conference on sinkholes and the engineering and environmental impacts of karst. American Society of Civil Engineers, Reston, pp 43–53

Bartlein PJ, Webb T (1982) Holocene climatic changes estimated from pollen data from the northern Midwest. In: Quaternary history of the driftless area. Wisconsin geological and natural history survey, vol 5, pp 67–82

Bartlein PJ, Webb TI, Fleri E (1984) Holocene climatic change in the northern Midwest: pollen-derived estimates. Quat Res 22(3):361–374

Beach T (1994) The fate of eroded soil: Sediment sinks and sediment budgets of agrarian landscapes in southern Minnesota. Ann Assoc Am Geogr 84(1):5–28

Bellomo L, Hanson PR, Jennings CE, Alexander EC Jr (2011) Optically stimulated luminescence (OSL) dating of glacial sediment in Crystal Cave, WI. Paper 182-2 presented at the 2011 GSA annual meeting, Minneapolis, 9–12 October 2011

Birmingham RA, Green W (eds) (1987) Wisconsin rock art. Wis Archeol 68(4): 273–477

Boszhardt RF (2003) Deep cave rock art in the upper Mississippi Valley. Prairie Smoke Press, St. Paul, MN

Bradbury KR, Swanson SK, Krohelski JT, Fritz AK (1999) Hydrogeology of Dane County, Wisconsin. USGS Open-File Report 4

Bretz JH (1938) Caves in the Galena formation. J Geol 46:828–841

Brick GA (1997) Classification and Morphology of St. Peter Sandstone Piping Caves in Minnesota. Journal of Cave and Karst Studies 59 (3): 168

Brick GA (2002) Joseph LeConte's account of the caves of the Apostle Islands in 1844. Minn Speleol Mon 34(8):89–91

Brick GA (2007) Recycled imagery: the Minnesota alter ego of a Wisconsin Cave. Hollow Earth News 14(7):4

Brick GA (2013) The nitrate deposits of rock crevices in the Upper Mississippi Valley. Minneapolis: University of Minnesota dissertation, 165p. University of Minnesota Digital Conservancy. https://hdl.handle.net/11299/157750

Brick GA (2020) Exploration and institutional history (Chap. 2, this volume)

Broihahn JH (2008) No crevice too narrow; no tunnel too deep. Wis Speleol 2008:3–6

Brown County WI (2017) Brown County ten year land and water management plan. https://www.co.brown.wi.us/i/f/land_conservation/draft%20LWP%20-%20finalpdf. Accessed 10 Jan 2017

Brozowski J, Day MJ (1994) Development of Brussels Hill pit cave, Door County, Wisconsin: evidence from flowstone and sediment. Trans Wis Acad Sci Arts Lett 82:1–16

Brynildson C (1966) Spring survey of Iowa County. Wisconsin Conservation Department Report
Burke J (2002) Deja-vu completion for Milwaukee Deep Sewer. World Tunn 15(10):471–474
Carlson DA (2001) Dependence of conductivities and anisotropies on geologic properties within the near-surface aquifier in Milwaukee, Wisconsin. Dissertation, University of Wisconsin-Milwaukee
Cave of the Mounds (2017). https://www.caveofthemounds.com/. Accessed 10 Jan 2017
Cherkauer DS (1996) Impact of tunnel dewatering on surface water bodies in Milwaukee County: determination of hydrogeologic controls and the efficiency of monitoring arrays. Groundwater Research Report WRC GRR 9603, UW System Groundwater Research Program, Madison
Cherkauer DS, McKereghan PF (1991) Ground-water discharge to lakes: focusing in embayments. Ground Water 29(1):72–80
Clayton L, Attig JW (1990) Geology of Sauk County, Wisconsin. Wisconsin Geological and Natural History Survey Information Circular 67
Clayton L, Attig JW, Mickelson DM (2001) Effects of late Pleistocene permafrost on the landscape of Wisconsin, USA. Boreas 30(3):173–188
CMCCP (2013) Cherney Maribel Caves County Park. https://www.maribelcaves.com/
Colgan PM, Bierman PR, Mickelson DM, Cafee M (2002) Variation in glacial erosion near the southern margin of the Laurentide Ice Sheet, south-central Wisconsin, USA: implications for cosmogenic dating of glacial terrains. Bull Geol Soc Am 114(12):1581–1591
Cooper MP, Mylroie JE (2015) Glaciation and speleogenesis: interpretations from the northeastern United States. Springer, New York
Cowell DW, Ford DC (1983) Karst hydrology of the Bruce Peninsula, Ontario. J Hydrol 61:163–168
Coxon C (2011) Agriculture and karst. In: Van Beynen P (ed) Karst management. Springer, New York, pp 103–138
Cronon W (1970) The sandstone caves of Wisconsin. Wis Speleol 9 (3):53–99
Cronon W (1980) An introduction to Wisconsin caves. In: Alexander EC Jr (ed) An Introduction to Caves of Minnesota, Iowa and Wisconsin, NSS convention guidebook 21, pp 105–108
Cronon W, Frater N (1974) The caves of Iowa County. Wis Speleol 13 (3):68–95
CSP (2016) Crawford Stewardship Project. https://crawfordstewardshipproject.org/. Accessed 1 Oct 2016
Cushman W (2019) After 5 Years, white-nose syndrome is devastating Wisconsin Bats, WisContext. https://www.wiscontext.org/after-5-years-white-nose-syndrome-devastating-wisconsin-bats
Day MJ (1979) Preliminary results of an investigation of current rates of carbonate erosion in the Wisconsin karst. Wis Speleol 16(2):11–23
Day MJ (1980) Rock hardness: field assessment and geomorphic importance. Prof Geogr 32(1):72–81
Day MJ (1984) Carbonate erosion rates in southwestern Wisconsin. Phys Geogr 5(2):142–149
Day MJ (1986a) Cave studies in southwestern Wisconsin: implications and importance. Wis Speleol 19(3):1–21
Day MJ (1986b) Caves in southwestern Wisconsin, USA. In: Proceedings of 9th international speleology congress, vol 1, pp 155–157
Day MJ (1986b) Caves in the driftless area of southwestern Wisconsin. Wis Geogr 2:42–51
Day MJ (1988) The origin of cave sediments in southwestern Wisconsin. Geo^2 15:8–9
Day MJ (2001) Sandstone caves in Wisconsin. In: Proceedings of 13th international speleology congress, vol 1, pp 89–92. Reprinted in Wis Speleol 25(1):18–22 (2002)
Day MJ (2004) Karstic problems in the construction of Milwaukee's deep tunnels. Environ Geol 45(6):859–863
Day MJ (2008) A speleogenic origin for the Leland Natural Bridge, Wisconsin? Wis Geogr 23:50–66
Day MJ (2009) Wisconsin. In: Palmer AN, Palmer MV (eds) Caves and karst of the USA. National Speleological Society, Huntsville, pp 143–145
Day MJ, Goudie AS (1977) Field assessment of rock hardness using the Schmidt Test Hammer. Br Geomorphol Res Group Tech Bull 18:9–29
Day MJ, Kueny JA (1990) Some Wisconsin caving hazards. Wis Speleol 21(1):3–19
Day MJ, Soule GK (1988) A geographical analysis of Wisconsin Speleological Society membership, 1960–1985. Wis Speleol 20 (3):43–50
Day MJ, Reeder PP (1989) Sinkholes and landuse in southwestern Wisconsin. In: Beck BF (ed) Engineering and environmental impacts of sinkholes and karst. AA Balkema, Rotterdam, pp 107–113
Day MJ, Kueny JA (1999) A speleogenic origin for five-column rock, Wisconsin? J Cave Karst Stud 61(3):141–144
Day MJ, Kueny JA (2000) The history of John Gray, Pleasant Ridge and John Gray Cave, Richland County, Wisconsin. Wis Speleol 24 (1):2–10
Day MJ, Bindl M (2010–2011) Quarrying in the Karst of southwestern Wisconsin: History, significance and prospects. Wis Geogr 24:71–86
Day MJ, Reeder PP (2013) Lead mine caves in southwestern Wisconsin, USA. In: DeWaele J, Forti P (eds) Mine Caves. Italian Institute of Speleology, Series 2, vol 28, pp 213–228
Day MJ, Leigh C, Young A (1980) Weathering of rock discs in temperate and tropical soils. Zeit Fur Geomorph NF Supp 35:11–15
Day MJ, Gaertner P, Neal MK (1987) A dye trace at the Blue Hole in Central Belize. Wis Speleol 20(2):1–12
Day MJ, Reeder PP, Oh JW (1989) Dolostone karst in southwestern Wisconsin. Wis Geogr 5:29–40
Day MJ, Kueny JA, Parrish A, Tenorio R (2004) Testing a preliminary model of spring location in the karst of southwestern Wisconsin. Wis Geogr 20:29–34
de Freitas CR, Littlejohn RN (1987) Cave climate: assessment of heat and moisture exchange. Int J Clim 7(6):553–569
De Geoffroy J (1969) Geochemical prospecting by spring sampling in the Southwest Wisconsin Zinc Mining Area. Wisconsin Geological and Natural History Survey Information Circular 10
De Geoffroy J, Wu SM, Heins RW (1967) Geochemical coverage by spring sampling method in the southwest Wisconsin zinc area. Econ Geol 62(5):679–697
De Geoffroy J, Wu SM, Heins RW (1970) Some observations on springs in southwest Wisconsin. Water Resour Res 6(4):1235–1238
Delong MD (2005) Upper Mississippi River Basin. In: Benke AC, Cushing CE (eds) Rivers of North America. Elsevier, New York, pp 327–374
Dockal JA (2020) Lead-Zinc crevices of Iowa, Illinois, and Wisconsin (Chap. 7, this volume)
Dott RH, Attig JW (2004) Roadside geology of Wisconsin. Mountain Press, Missoula
Dutch S (2008) Natural Bridge State Park. https://www.uwgb.edu/dutchs/geolwisc/geostops/NaturalBridge.htm. Accessed 30 Sept 2016
Echolocator (2020). https://wiatri.net/Inventory/Bats/News/2020Mar-Echolocator.pdf
Erb KR, Stieglitz RD (2007) Final Report of the Northeast Wisconsin karst Task Force, University of Wisconsin—Extension, Madison
Erb KR, Ronk E, Koundinya V, Luczak J (2015) Groundwater quality changes in a karst aquifer of northeastern Wisconsin, USA:

Reduction of brownwater incidence and bacterial contamination resulting from implementation of regional task force recommendations. Resources 4:655–672

Fermanich K, Zorn M, Stieglitz R, Waltman CS (2006) Mapping and characterization of springs in Brown and Calumet Counties. Report to Wis DNR, Bureau Drinking Water and Groundwater

Fitzpatrick FA, Peppler MC, Walker JF, Rose WJ, Waschbusch RJ, Kennedy JL (2008) Flood of June 2008 in Southern Wisconsin. USGS Scientific Investigations Report

Ford DC, Schwarcz HP (1981) Uranium series disequilibrium dating. In: Goudie A (ed) Geomorphological techniques. Allen and Unwin, London, pp 284–287

Ford DC, Williams PW (2007) Karst hydrogeology and geomorphology. Wiley, Chichester

Frater N (1966) The caves of Dane County. Wis Speleol 5(3):82–93

Gargas A, Trest MT, Christensen M, Volk TJ, Blehert DS (2009) Geomyces destructans sp. nov. associated with bat white-nose syndrome. Mycotaxon 108:147–154

Gietkowski R (1972) Caves of Sauk County. Wis Speleol 11(1):1–25

Goudie AS (2006) The Schmidt Hammer in geomorphological research. Prog Phys Geogr 30(6):703–718

Graczyk DJ (1993) Surface-water hydrology and quality and macroinvertebrate and smallmouth bass populations in four stream basins in southwestern Wisconsin, 1987–90. USGS Water Resources Investigation Representative 934024

Green JA, Alexander EC Jr (2015) Creation of a map of Paleozoic bedrock springsheds in southeast Minnesota. In: Doctor DH, Land L, Stephenson JB (eds) Proceedings of 14th multidisciplinary. conference on sinkholes and the engineering and environmental impacts of Karst. National Cave and Karst Research Institute, Carlsbad, NM, pp 211–222

Gunn J (2013) Denudation and erosion rates in karst. In: Frumkin A (ed) Treatise on geomorphology 6: karst geomorphology. Academic Press, San Diego, pp 72–81

Gunn J, Bailey D, Handley J (1997) The reclamation of limestone quarries using landform replication. Department of Environment, Transport and Regions, London

Hawk K, Stieglitz RD, Norman JC (1993) Radon concentrations in homes in an area of dolomite bedrock: Door County, Wisconsin. Environ Int 19(3):241–252

Hedges J, Alexander EC Jr (1985) Karst related features of the Upper Mississippi Valley region. Stud Speleol 6:41–49

Heller SA (1988) Seasonal geochemistry of two tufa-depositing springs in southwestern Wisconsin. Geosci Wis 12:77–83

Hem JD (1959) Study and interpretation of the chemical characteristics of natural water. USGS Water-Supply Paper 1473

Hendrix TE, Schaiowitz M (1964) Gravitational structures in the Baraboo Quartzite near Baraboo, Wisconsin. Geol Soc Am Bull 75 (10):1045–1050

Hennings F, Soule G, Peterson G (1972) Caves of Door County, Wisconsin. Wis Speleol 11:63–129

Heyl AV, Agnew AF, Lyons EF, Behr CH (1959) The geology of the upper Mississippi Valley zinc-lead district. USGS Prof Paper 309

Heyl AV, Broughton WA, West WS (1970) Guidebook to the Upper Mississippi Valley base metal district. Wisconsin Geological and Natural History Survey Information Circular 16

Hobbs H (1999) Origin of the driftless area by subglacial drainage – a new hypothesis. In: Mickelson DM, Attig JW (eds) Glacial processes past and present. Geological Society of America Special Paper 337, pp 93–102

Holst TB (1982) Regional jointing in the northern Michigan basin. Geology 10:273–277

Hughes PE, Hannuksela JS, Danchuk WA (1981) Flood of July 1–5, 1978 on the Kickapoo River, southwestern Wisconsin. USGS Hydrol Atlas 653

Iannicelli M (2010) Evolution of the driftless area and contiguous regions of Midwestern USA through Pleistocene periglacial processes. Open Geol J 4(1):35–54

Johnson SB, Stieglitz RD (1990) Karst features of a glaciated dolomite peninsula, Door County, Wisconsin. Geomorphology 4(1):37–54

Juckem PF, Hunt RJ, Anderson MP, Robertson DM (2008) Effects of climate and land management change on streamflow in the driftless area of Wisconsin. J Hydrol 355(1–4):123–130

Kammerer PA (1981) Ground-water-quality atlas of Wisconsin. US Geological Survey Information Circular 39

Kastning E (2003) Quarrying in Karst: geotechnical estimation of environmental risk. In: Beck BF (ed) Sinkholes and the environmental impact of Karst. American Association of Civil Engineers Geotechnical Special Publication, vol 122, pp 704–713

Kemp SL, Day MJ (1989) Karst springs as water sources in southwestern Wisconsin: an historical perspective. Wis Geogr 13–14:29–40

King T (1972) Crystal Cave, Pierce County, Wisconsin. Wis Speleol 11 (2):55–60

Klimchouk A (2007) Morphogenesis of hypogenic caves. Geomorphology 106(1–2):100–117

Klimchouk A (2017) Types and settings of hypogene karst. In: Klimchouk A, Palmer AN, De Waele J, Auler AS, Audra P (eds) Hypogene karst regions and caves of the world. Springer, New York, pp 1–39

Klimchouk A, Palmer AN, De Waele J, Auler AS, Audra P (eds) (2017) Hypogene karst regions and caves of the world. Springer, New York. https://doi.org/10.1007/978-3-319-53348-3

Kluessendorf J, Mikulic DG (1989) Bedrock geology of the Door Peninsula. In: Palmquist JC (ed) Wisconsin's Door Peninsula: a natural history. Perin Press, Appleton, pp 12–31

Kluessendorf J, Mikulic DG (1990) An early Silurian sequence boundary in Illinois and Wisconsin. In: Witzke BJ, Ludvigson GA, Day J (eds) Paleozoic sequence stratigraphy: views from the North American Craton. Boulder Geological Society of America Special Paper 306

Knox JC, Leigh DS, Jacobs PM, Mason JA, Attig JW (2011) Introduction: Part 2: quaternary loess lithostratigraphy in Wisconsin. In: Syverson KM, ClaytonL, Attig JW, Mickelson DM (eds) Lexicon of Pleistocene stratigraphic units in Wisconsin. Wis Geol Nat Hist Survey Technical Report 1:14–17

LaCrosse Tribune (2011) Boulder slams into bluffside home, 24 Feb 2011.

Lange EG (1909) Original work on the caves of the driftless area of southwestern Wisconsin. Thesis, University of Wisconsin

LePain DL, Bradbury KR, Cobb MK (2005) Hydrostratigraphy of West-Central Wisconsin: a new approach to groundwater management. University of Wisconsin, Water Resources Institute

Luczak JA (2006) Evidence against the Dorag (mixing-zone) model for dolomitization along the Wisconsin Arch—a case for hydrothermal diagenesis. Am Assoc Petrol Geol Bull 90:1719–1738

Luczak JA, Stieglitz RD (2008) Geologic history of New Hope Cave, Manitowoc County, Wisconsin. Wis Speleol 2008:7–17

Mai H, Dott RH (1985) A subsurface study of the St. Peter sandstone in southern and eastern Wisconsin. Wisconsin Geological and Natural History Survey Information Circular 47

Manitowoc County (2017) Manitowoc County karst Inventory. https://www.co.manitowoc.wi.us/departments/q-z/soil-and-water-conservation/groundwater-programming/karst-inventory/. Accessed 22 Jan 2017

Martin L (1916) The physical geography of Wisconsin. Wis Geol Nat Hist Surv Bull 36

Martin L (1932) The physical geography of Wisconsin, 2nd edn. University of Wisconsin Press, Madison

Martin L (1974) The physical geography of Wisconsin, 3rd edn. University of Wisconsin Press, Madison
McNurlin T (1959) Spring survey of Crawford County. Wisconsin Conservation Department Report
Mickelson DM, Knox JC, Clayton L (1982) Glaciation of the driftless area: an evaluation of the evidence. In: Quaternary history of the driftless area. Wisconsin Geological and Natural History Survey Field Trip Guide Book, vol 5, pp 155–169
Milske JA, Alexander EC Jr, Lively RS (1983) Clastic sediments in Mystery Cave, southeastern Minnesota. Nat Speleol Soc Bull 45:55–75
Mudrey MG, Brown BA, Greenberg JK (1982) Data accompanying M078: bedrock geology map of Wisconsin. Wisconsin Geological and Natural History Survey. https://wgnhs.uwex.edu/pubs/download_m078di/. Accessed 25 Feb 2017
Mueller RJ, Day MJ (1997) Daily atmospheric variation within caves in southwestern Wisconsin, U.S.A. In: Proceedings of 12th International Speleology Congress, vol 1, pp 215–217
Muldoon MA, Simo JA, Bradbury KR (2001) Correlation of high-permeability zones with stratigraphy in a fractured-dolomite aquifer, Door County, Wisconsin. Hydrogeol J 9(6):570–583
Murrish J (1871) Report to the State of Wisconsin on the Geological Survey of the Lead Regions, pp 21–22
NPS (2016a) National Park Service, Apostle Islands Mainland Sea Caves. https://www.nps.gov/apis/learn/nature/caves.htm. Accessed 11 Apr 2016
NPS (2016b) National Park Service, Apostle Islands Mainland Sea Caves—Ice Caves. https://www.nps.gov/apis/mainland-caves-winter.htm. Accessed 11 Apr 2016
NERN (2017) Niagara Escarpment Resource Network. https://www.escarpmentnetwork.org/
Nicholson SW, Dicken CL, Foose MP, Mueller JAL (2007) Preliminary integrated geologic map databases for the United States: Minnesota, Wisconsin, Michigan, Illinois and Indiana. USGS Open-File Report 2004-1355. https://pubs.orgs.gov/of/2004/1355/. Accessed 30 Sept 2016
Oh J, Day MJ (1990) Loess-derived karst soils and sediments in the southwestern Wisconsin Driftless Area. J Reg Dev (Korea) 15:29–43
Oh J, Day MJ (1991) Sediments of the Seneca sinkhole in the southwestern Wisconsin karst. Wis Geogr 7:25–39
Oh J, Day MJ, Gladfelter B, Huppert G, Fredlund G, Kolb M (1991) Potential sources of the sinkhole sediments in the Wisconsin driftless area. Kyung Hee Geogr Rev 19:31–58
Oh J, Day MJ, Gladfelter B (1993) Geomorphic environmental reconstruction of the Holocene sinkhole sediments in the Wisconsin driftless area. In: Proceedings of Korean Science Engineering International Conference (Geoscience Section), pp 390–397
Olmstead R, Borman R (1968) The origin and development of Pop's Cave. Wis Speleol 7:57–66
Palmer AN (1991) Origin and morphology of limestone caves. Bull Geol Soc Am 103(1):1–21
Paull RA (1992) First report of natural bridges in eastern Wisconsin. Trans Wis Acad Sci Arts Lett 80:139–148
Paull RK, Paull RA (1977) Geology of Wisconsin and Upper Michigan. Kendall/Hunt, Dubuque
Pearson R (2001) Wisconsin interagency karst feature reporting form. In: Beck BF and Herring JG (eds) Geotechnical and environmental applications of karst geology and hydrology. CRC Press, pp 95–100
Peterson GN (1968) Caves of Richland County. Wis Speleol 7(3):78–108
Pipes T, Day MJ (2006) Some alternative hypotheses of cave formation in southwestern Wisconsin. Wis Geogr 21:82–94
Reeder PP (1992) Groundwater contaminant pathways in a fractured dolostone-clastic aquifer: Richland County, Wisconsin. Dissertation, University of Wisconsin-Milwaukee
Reeder PP, Day MJ (1990) Lead mining from the caves in the upper Mississippi Valley: an historical perspective. J Spelean Hist 24 (1):14–22
Reeder PP, Day MJ (1993) Seasonality of chloride and nitrate contamination in the southwestern Wisconsin karst. In: Beck BF (ed) Applied karst geology. AA Balkema, Rotterdam, pp 53–61
Reeder PP, Day MJ (1994). Groundwater contamination pathways in the karst terrain of Southwestern Wisconsin. Water Resources Center, University of Wisconsin Madison, Technical Report WIS WRC, pp. 94–101
Rosen CJ, Day MJ (1990) Glaciated karst terrain in the Door Peninsula of Wisconsin. Trans Wis Acad Sci Arts Lett 78:39–44
Rosen CJ, Day MJ, Piepenburg K (1987) Glaciokarst depressions in the Door Peninsula, Wisconsin. Phys Geogr 8(2):160–168
Rovey CW, Cherkauer DS (1994) Relation between hydraulic conductivity and texture in a carbonate aquifer: observations. Ground Water 32(1):53–62
Rowan ELK, Goldhaber MB, Hatch JR (1995) Duration of mineralization in the upper Mississippi Valley lead-zinc district: implications for the thermal-hydrologic history of the Illinois basin. Geol Soc Am Abs Prog 27:A-328
Runkel AC, Tipping RG, Alexander EC Jr, Green JA, Mossler JH, Alexander SC (2003) Hydrogeology of the Paleozoic bedrock in southeastern Minnesota. Minnesota Geological Survey, Report of Investigations 61
Salzer RJ (1987a) Introduction to Wisconsin rock art. Wis Archeol 68 (4):277–286
Salzer RJ (1987b) Preliminary report on the Gottschall site. Wis Archeol 68(4):419–472
Salzer RJ (1997) Wisconsin rock art. Wis Archeol 78(1/2):48–76
Sherrill MG (1975) Ground-water contamination in the Silurian dolomite of Door County, Wisconsin. Ground Water 13(2):209–213
Sherrill MG (1978) Geology and ground water in Door County, Wisconsin, with emphasis on contamination potential in the silurian dolomite. USGS Water-Supply Paper 2047. US Government Printing Office, Washington DC
Smart PL, Brown MC (1973) The use of activated charcoal for the detection of the tracer dye Rhodamine WT. In: Proceedings of 6th International Speleology Congress, vol 4, pp 285–292
Smith GL (1990) Dolomitization of the Lower Ordovician Prairie du Chien Group in southern Wisconsin and southeastern Minnesota—a case for confined and unconfined aquifer systems. Am Assoc Petrol Geol Bull 74:766–775
Smith TD, Ball JR (1972) Surface Water Resources of Grant County. Wisconsin Conservation Department Report
Smith GL, Byers CW, Dott RH (1993) Sequence stratigraphy of the Lower Ordovician Prairie du Chien Group on the Wisconsin Arch and in the Michigan Basin. Am Assoc Petrol Geol Bull 77(1):49–67
Smith GL, Byers CW, Dott RH (1996) Sequence stratigraphy of the Prairie du Chien Group, Lower Ordovician, Midcontinent, USA. In: Witzke BJ, Ludvigson GA, Day MJ (eds), Paleozoic sequence stratigraphy: views from the North American Craton. Geological Society of America Special Paper, vol 306, pp 23–33
Smith GL, Simo JA (1997) Carbonate diagenesis and dolomitization of the Lower Ordovician Prairie du Chien Group. Geosci Wis 16:1–16
Soule GK (1986) Beneath the Door Peninsula: the story of Paradise Pit Cave. NSS News 1986:240–246
Soule GK (1993) A hidden treasure—cave of the mounds. NSS News 1993:178–182
Soule GK (2010) Dorchester cave—an unusual urban discovery. NSS News 2010:18–24

Soule GK (2014) Door County's Legendary Horseshoe Bay Cave (Tecumseh Cave) Egg Harbor, WI. NSS Spelean History Series 22. https://www.wisconsincaves.org/gallery/door-county-horseshoe-bay-cave

Steelman KL, Rowe MW, Boszhardt RF, Southon JR (2001) Radiocarbon age determination of a rock painting at Arnold/Tainter Cave, Wisconsin. Midcont J Archaeol 26(1):121–131

Stieglitz RD (1990) The geologic foundations of Wisconsin's Door Peninsula. In: Hersbell KB (ed) Door County and the Niagara Escarpment: foundations for the future. Wisconsin Academy of Sciences, Arts, and Letters, Madison

Stieglitz RD, Moran JM, Harris JD (1980) A relict geomorphological feature adjacent to the Silurian escarpment in northeastern Wisconsin. Trans Wis Acad Sci Arts Lett 68:202–207

Stiles CA, Stensvold KA (2008) Loess contribution to soils forming on dolostone in the driftless area of Wisconsin. Soil Sci Soc Am J 72 (3):650–659

Stiles-Hanson C (1987) Petroglyphs and pictographs of the Coulee Region. Wis Archeol 68(4):287–340

Strong M (1877) Geol Wis 2:661–662

Swanson SK (2007) Lithostratigraphic controls on bedding-plane fractures and the potential for discrete groundwater flow through a siliclastic sandstone aquifer, southern Wisconsin. Sed Geol 197:65–78

Swanson SK, Bahr JM, Bradbury KR, Anderson KM (2006) Evidence for preferential flow through sandstone aquifers in Southern Wisconsin. Sed Geol 184(3–4):331–342

Swanson SK, Bradbury KR, Hart DJ (2007) Assessing the ecological status and vulnerability of springs in Wisconsin. University of Wisconsin, Water Resources Institute

Sweeting MM (1972) Karst landforms. Macmillan, London

Syverson KM, Colgan PM (2004) The Quaternary of Wisconsin—a review of stratigraphy and glaciation history. In: Ehlers J, Gibbard PL (eds) Quaternary glaciations—extent and chronology: part II: North America. Elsevier, Amsterdam, pp 295–311

Terlau CA, Day MJ (1997) A comparison of the orientation of cave passages and surface tributary valleys in the karst of southwestern Wisconsin, U.S.A. In: Proceedings of 12th International Speleology Congress, vol 1, pp 133–136

Thwaites FT (1912) Sandstones of the Wisconsin coast of Lake Superior. State of Wisconsin, Madison

Tipping RG, Runkel AC, Alexander EC Jr, Alexander SC, Green JA (2006) Evidence for hydraulic heterogeneity and anisotropy in the mostly carbonate Prairie du Chien Group, southeastern Minnesota, USA. Sed Geol 184(3–4):305–330

Torrey MLS (1976) Chemistry of Lake Michigan. Argonne National Laboratory, Springfield, VA

Trudgill ST (1977) Problems in the estimation of short-term variations in limestone erosion processes. Earth Surf Proc 2(2–3):251–256

USGS (2005) The Mineral Industry of Wisconsin. In: USGS Minerals Yearbook 2005. https://minerals.usgs.gov/minerals/pubs/state/2005/myb2-2005-wi.pdf. Accessed 25 Feb 2016

USGS (2006) The Mineral Industry of Wisconsin. In: USGS Minerals Yearbook 2006. https://minerals.usgs.gov/minerals/pubs/state/2006/myb2-2006-wi.pdf. Accessed 25 Feb 2016

USGS (2016) Geological map of North America resources. https://ngmdb.usgs.gov/gmna/. Accessed 10 Jan 2016

Vincent PJ (1995) Limestone pavements in the British Isles: a review. Geog J 161(3):265–274

White JP (2018) White-nose syndrome continues to ravage cave bat populations. https://dnr.wi.gov/news/Weekly/article/?id=4254

White WB (1977) Conceptual models for carbonate aquifers: revisited. In: Dillamarter RR, Csallany CS (eds) Hydrologic problems in karst Regions, Western Kentucky University, Bowling Green, pp 176–187

Wiersma JH, Stiegltiz RD, Cecil DL, Metzler GM (1986) Characterization of the shallow groundwater system in an area with thin soils and sinkholes. Environ Geol Water Res 8(1):99–104

Wilson P, Barrows TT, Lord TC, Vincent PJ (2012) Surface lowering of limestone pavement as determined by cosmogenic (^{36}Cl) analysis. Earth Surf Proc Landf 37(14):1518–1526

WCD (1958) Spring survey of Richland County. Wisconsin Conservation Department Report

WDNR (2008) Natural Bridge and Rockshelter. Wisconsin Department of Natural Resources. https://dnr.wi.gov/topic/parks/name/naturalbridge/history.html

WDNR (2010) A look at the Ledge. Wisconsin Department of Natural Resources. https://dnr.wi.gov/wnrmag/2010/10/ledge.htm. Accessed 11 Apr 2017

WDNR (2016a) Saving Wisconsin's bats. Wisconsin Department of Natural Resources. https://dnr.wi.gov/topic/Wildlifehabitat/bats.html. Accessed 11 Apr 2017

WDNR (2016b) Industrial sand mining in Wisconsin: strategic analysis for public review. Wisconsin Department of Natural Resources, Madison

WDNR (2017a) Peninsula State Park. Wisconsin Department of Natural Resources. https://dnr.wi.gov/topic/parks/name/peninsula/nature.html. Accessed 25 Feb 2017

WDNR (2017b) Wisconsin State Parks. Wisconsin Department of Natural Resources. https://dnr.wi.gov/topic/parks/. Accessed 25 Feb 2017

WDNR (2017c) Bat disease takes its toll. Wisconsin Department of Natural Resources. https://dnr.wi.gov/news/Weekly/?id=587#art1. Accessed 11 May 2017

WGNHS (2009) Karst and shallow carbonate bedrock in Wisconsin. Wisconsin Geological and Natural History Survey. https://wgnhs.uwex.edu/pubs/000905/. Accessed 1 May 2017

WisconsinWatch (2014). https://wisconsinwatch.org/2014/08/in-states-karst-area-even-good-farming-may-pollute-groundwater/. Accessed 1 May 2017

Wittry WL (1959a) The Raddatz Rockshelter, Wisconsin. Wis Archeol 40:33–90

Wittry WL (1959b) Archeological studies of four Wisconsin rockshelters. Wis Archeol 40:137–267

Wopat MA (1974) The karst of southeastern Minnesota. Wis Speleol 13:1–47

WSS (2017) Wisconsin WNS Caving Provisions. https://docs.google.com/a/wisconsincaves.org/viewer?a=v&pid=sites&srcid=d2lzY29uc2luY2F2ZXMub3JnfHdzc3xneDo2NTlkMWI0OGZjM2M1OTY2

Lead-Zinc Crevices of Iowa, Illinois, and Wisconsin

7

James A. Dockal

Abstract

Lead-zinc mineralized crevice caves developed within the Galena Group (Ordovician) of the Upper Mississippi Valley lead-zinc district (UMVLZD) result from both hypogene and epigene processes. Far-field tectonic movements occurring in the late Paleozoic resulted in extensive fracturing and consequent hydrothermal circulation in the late Paleozoic between the Proterozoic basement and the carbonate strata of the Galena Group. This caused significant coeval solution of the limestone, dolomitization, and sulfide mineral precipitation, especially where fractures intersected strata of a specific lithofacies characterized by the trace fossil *Thalassinoides*. At these intersections pod-shaped masses of sulfide minerals formed. During the Cenozoic, erosion of strata overlying the Galena Group resulted in the oxidation of the pod-shaped sulfide mineral deposits with resultant production of sulfuric acid and consequent solution of surrounding dolostone forming in some cases substantial caves and cave systems. Historically these caves were a major source of the mineral galena (PbS) with a utilization time line reaching back to Archaic (3500–1000 BCE).

7.1 Introduction

The Upper Mississippi Valley lead-zinc district (UMVLZD) is a Mississippi Valley Type (MVT) stratabound, sulfide mineralization. The district covers approximately 39,000 km^2 in northeastern Iowa, northwestern Illinois, and southwestern Wisconsin (Fig. 7.1). The western boundary of the district is defined by a rapid transition from limestone in the west to dolostone to the east. The northern and eastern boundaries in Wisconsin and Illinois represent the present erosional limit of the Galena Group. The southern boundary in Illinois represents a weakly constrained limestone to dolostone transition. The area where the caves occur covers approximately 14,000 km^2 and consists of that portion of the UMVLZD where the Silurian strata have been removed by erosion. The district and the caves have been a significant source of the mineral galena from pre-European contact times almost up to the present. There is an extensive body of literature on the UMVLZD; however, quality descriptive information can be found in: (Daniels 1854; Percival 1855, 1856; Whitney 1858, 1862, 1866; Strong 1877; Leonard 1896; Calvin and Bain 1900; Bain 1905, 1906; Grant 1906; Cox 1914; Willman et al. 1946; Bradbury 1959; Heyl et al. 1959, 1970; Panno et al. 2015, 2017). In addition to these, there are 13 USGS reports covering 19 7.5-min geologic quadrangles: (Brown and Whitlow 1960; Carlson 1961; Agnew 1963; Whitlow and Brown 1963; Allingham 1963; Taylor 1964; Klemic and West 1964; Mullens 1964; Whitlow and West 1966a, 1966b, 1966c; West and Blacet 1971; West et al. 1971; West and Heyl 1985).

Caves developed within the Galena Group (Ordovician) carbonate strata of the UMVLZD are unusual in that they resulted from a two-step process of formation. The first or hypogene step involved two episodes of limestone dissolution, dolomitization of limestone, and precipitation of the sulfide minerals galena, marcasite (FeS_2), pyrite (FeS_2), and sphalerite (ZnS) all in association with hydrothermal fluids moving through coeval fractures. This step occurred at the end of the Paleozoic. The second or epigene step involved the oxidation of the sulfide minerals, either by inorganic processes or bacterial modulated processes which results in additional carbonate rock dissolution and consequent formation of a network of interconnected solution-expanded fractures. This step was initiated probably in the early Cenozoic and continues to the present.

J. A. Dockal (✉)
Department of Earth and Ocean Science, University of North Carolina, Wilmington, NC 28403, USA
e-mail: jimdockal@gmail.com

G. A. Brick and E. C. Alexander Jr. (eds.), *Caves and Karst of the Upper Midwest, USA*,
Cave and Karst Systems of the World, https://doi.org/10.1007/978-3-030-54633-5_7

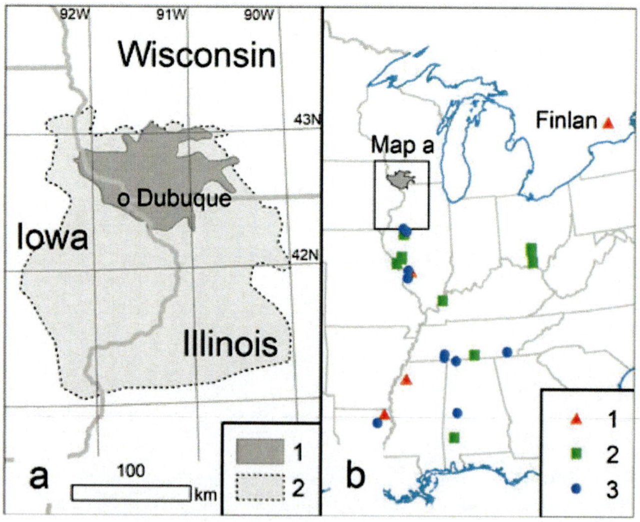

Fig. 7.1 Location Maps. **a** Upper Mississippi Valley lead-zinc district area. (1) Area referred to as the productive area of the Upper Mississippi Valley District (Heyl et al. 1959). (2) Mineralized region as delineated by extensive dolomitization in the Galena Group as indicated from well logs. Considered here to represent the true extent of the UMVLZD. **b** Mississippi Valley area of the USA. (1) Archaic to Early Woodland Native American sites where galena was found that was derived from the UMVLZD area. (2) Middle Woodland and (3) Mississippian sites were galena was found that was derived from the UMVLZD. Rectangular box denotes area of Map a

The remainder of this chapter is divided into two parts. The first is a discussion of human utilization of the caves that includes a synopsis of the physical character of the caves as revealed at Dubuque, Iowa. The second presents an interpretation of the geologic development of the caves.

7.2 History of Human Utilization and Physical Character of the Caves

Human utilization of the caves extends back several thousand years. The primary interest from ancient through historic time was in mining the mineral wealth associated with the caves. Recent use is chiefly recreation as at present there are no operating lead or zinc mines.

7.2.1 Pre-contact Native American

The caves and associated lead and zinc mineral deposits have played a prominent role in the human history of North America. Earliest indication of human interest in the caves comes from the discovery of galena at Finlan, an Archaic (3500–1000 BCE) site near Lake Erie, Ontario (Fig. 7.1b). That galena has a lead isotope composition that matches that of the southwestern portion of the UMVLZD (Farquhar and Fletcher 1980, 1984). Galena, identified as likely coming from the UMVLZD, has also been found in widely scattered burial and occupation sites belonging to the Woodland and Mississippian cultures (Ghosh 2008; Walthall et al. 1980; Walthall 1981). Native Americans were probably using the galena for face paint, ceremonial powder, or glaze on ceramics (Walthall et al. 1980; Dockal 2011).

If pre-contact Native Americans were making use of galena and trading it widely across North America, then one might speculate on whether they were making use of surface finds or going underground in caves to mine such as they did for muscovite in North Carolina (Kerr 1875; Holmes 1919; Margolin 2000) or copper in Michigan (Baldwin 1871; Martin 1995; Whitney 1858). Kerr (1875) noted that the pre-contact mica miners in North Carolina made "hundreds of old pits and connecting tunnels," these were constructed in a "very systematic, skillful way," and that "some have been followed for fifty [15 m] and a hundred feet [30 m] and upwards." Whitney (1858, p. 424) noted that the pre-contact copper miners of the Lake Superior region worked to depths of 15 m and in rock that was much more difficult to deal with than that of the UMVLZD.

7.2.2 French Interest in the Caves

Recorded French advances into the Upper Mississippi Valley region begin with the arrival of Jean Nicolet in the autumn of 1634 (Thwaites 1895). At that time, the Illiniwek (Illini) were in possession of the territory. Between 1658 and 1660 Pierre-Esprit Radisson and Médard Chouart des Groseilliers made a reconnaissance into the area and made the first known record by a European of galena occurrence there. Circa 1680 a member of the Illiniwek people brought a large piece of galena to a trading post located near the future site of Peoria, Illinois. Nicolas Perrot, who may have been at that trading post, apparently saw or heard of this find and quickly headed to the lead region and was engaged in lead mining there by 1682 and opened a trading post on the bank of the Mississippi River at the future site of East Dubuque, Illinois (Keyes 1912). The earliest written record that provides a geographically verifiable location of Perrot's mines, that of Pierre-Charles LeSueur who on August 25, 1700 recorded:

> River of the Mine. It comes from the north at its mouth, then it turns off toward the northeast. Seven leagues [about 27 km] farther upstream, to the right, on this river, there is a lead mine in a meadow a good three miles [4.8 km] inland. But the river, with the exception of the first 9 miles [14.5 km], is navigable only at the time of high water, that is from spring until the month of June. Made 5½ leagues/1 ½ leagues. Lead mine on the slope of a bluff [mountain] about half mile [0.8 km] from the river, toward the west. We made some lead there. (Translated here from the French by Wilkie (1987, Fig. 1.10, p. 20).)

The first location is at present day Galena, Illinois and the second is most likely the bluff to the west of downtown Dubuque, Iowa and known as Kelley's Bluff, the site of the mine of Tom Kelley circa 1830. In another passage LeSueur notes "We found on the right and on the left some lead mines, which today are still called the Mines of Nicolas Perrot, which is the name of the one who found them" (Wilkie 1987, p. 21).

Many of the eighteenth-century maps of the region, such as that of Guillaume Delisle 1702 or Jacques-Nicolas Bellin circa 1755 (Wood 2001 Plates 4D, 7, and 9) note at various locations *mine de plumb* which could be taken literally to mean a lead mine, or it could denote a location where lead (galena) could be found and not implying actual mining operations. The Illiniwek prior to circa 1710 and then later, sometime prior to 1733, the Meshkwahkihaki (Meskawki or Fox) having gained knowledge of firearms, were extracting and smelting galena to use as ammunition and in trade with the French, English, Spanish, and eventually Americans. The *mine de plumb* on the maps being the places from which they occasionally obtained the galena.

During the eighteenth century the primary use of lead, west of the Appalachians, would have been for ammunition: musket balls and bird shot. At that time in North America there were only two areas capable of producing significant amounts of lead, southeast Missouri and the UMVLZD. Galena was discovered in southeastern Missouri circa 1720 but the mining of such did not become a regular business until 1798 (Whitney 1858).

Mining on a regular, continuing, and systematic basis began in the UMVLZD in 1788 with the arrival of Julien Dubuque. It has been passed down that the wife of Peosta, a Meshkwahkihaki warrior, in 1780 discovered a galena deposit at the future site of Dubuque (Schoolcraft 1821). At about that time Julien Dubuque arrived at the confluence of the Wisconsin and Mississippi rivers, present day Prairie du Chien, Wisconsin, and was engaged in the fur trade industry. On September 22, 1788 Julien Dubuque held a council with representatives of the Meshkwahkihaki which resulted in a written instrument in French, signed by Dubuque, representatives of the Meshkwahkihaki, and some of Dubuque's associates. This document allows Dubuque (Shiras 1902, p. 323):

> to work at the mine as long as he shall please, and to withdraw from it, without specifying and term to him; moreover, that they sell and abandon to him all the coast and the contents of the mine discovered by the wife of Peosta, so that no white man or Indian shall make any pretension to it without the consent of Mr. Julien Dubuque; and in case he shall find nothing within, he shall be free to search wherever he may think proper to do so, and to work peaceably without anyone hurting him, or doing him any prejudice in his labors.

This document gives a clue to the nature of the site found by Peosta, mainly the phrase "he shall find nothing within" implies a cave and not surface diggings. The actual location of that cave is unknown; however, there is a map supposedly drawn in 1804 by Dubuque that shows the sites where he found or mined galena (Wilkie 1987, p. 76, and Dockal 2011). The "Diagram of the Lead-bearing Crevices Near Dubuque, Iowa" of Whitney (1858) indicates Dubuque Cave as being in the south half of Section 14 which does roughly correspond to a location on the 1804 map. Local lore points to a cave, also known as Haugen Cave, located in that area, as being that of Dubuque (Figs. 7.2 and 7.3). This cave is developed along an EW striking fracture. There is a short side passage at the east end which is developed on a NNW striking fracture. The cave is developed on two levels where the uppermost is at the contact between the Dubuque and Stewartville formations and the second is 5 m lower, about the middle of the Stewartville. The overall layout of this site is consistent with the type of mines operated in the area between 1830 and 1860. However, there are two portions of this site that Dubuque could have worked. The westernmost shaft and adit is a cleft in the hillside. That area and that immediately east could have been worked without the use of lighting, blasting, or ventilation. The shafts east of there are dug and blasted and hence probably a later mining effort. The area on the surface near the eastern end may also have been a site worked by Dubuque. When I examined the site in 1966 there was a shallow trench-like depression there. Elevation of the surface there corresponds to the elevation at the west end and the soil there may have contained float galena (pieces of galena mixed in the soil). Henry Schoolcraft visited the future site of Dubuque on August 7, 1820 and examined the mines then being worked there by the Meshkwahkihak (Schoolcraft 1821, 1855). Given the fact that he only had a few hours to do this in, the site that he visited was most likely either this site or another close by. Schoolcraft notes that: "no shafts are sunk, not even the simplest kind" and "They run drifts into the hills so far as then conveniently go, without the use of gun-powder." Schoolcraft (1821, p. 346) noted that the Meshkwahkihak were also exploiting three other sites in addition to the site he had visited: The *Mine au Fevre* which was situated at the present site of Galena, Illinois; the *Mine of Maquanquitons* which was situated near present day Durango, Iowa; and the *Sissinaway Mines* which he noted were at the confluence of the Mississippi and Sinsinawa rivers. Meeker (1872 p. 282), who arrived at Mine au Fevre in November 1822 and continued in the mining business there and in Wisconsin for many years noted that:

> The Indian women proved themselves to be the best as well as the shrewdest miners. While Col. Johnson's men were sinking their holes or shafts, in some instances the squaws would drift under them and take out all the mineral or ore they could find. When the men got down into the drift made by the women, the latter would have a hearty laugh at the white man's expense.

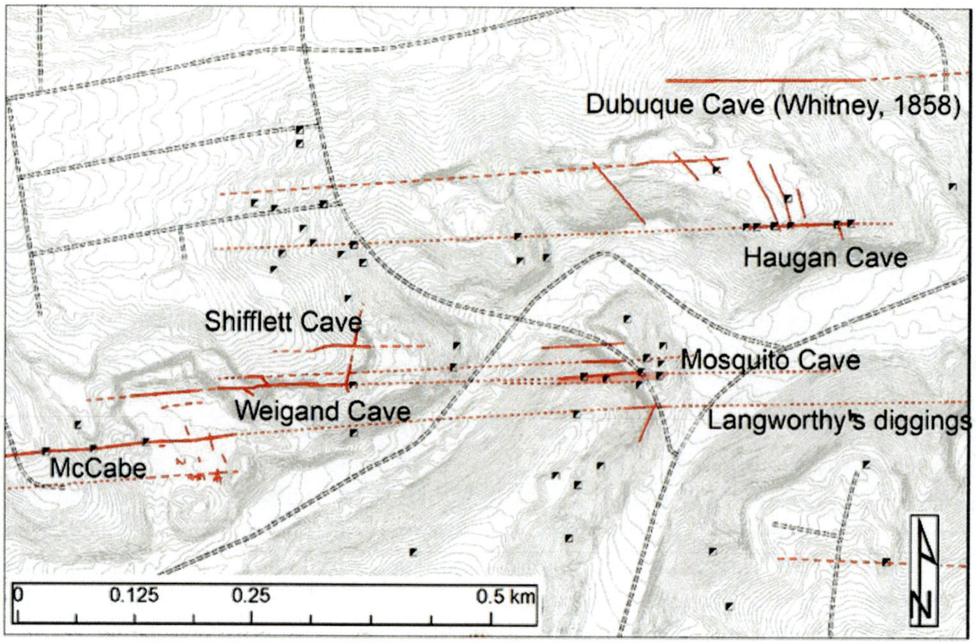

Fig. 7.2 Map of the crevices and caves of Langworthy Hollow (Kaufman Avenue area of Dubuque). Solid reds lines denote known locations of caves and crevices; dashed lines denote inferred extensions of known crevices into areas with little or no information. The southern grouping of crevices, including those of Weigand Cave, is known as Langworthy Range. Background is a 2-foot (0.6 m) contour topographic map derived from Lidar data. The map covers most of the southern 1/3 of Section 14 , T 89 N, R 42 W

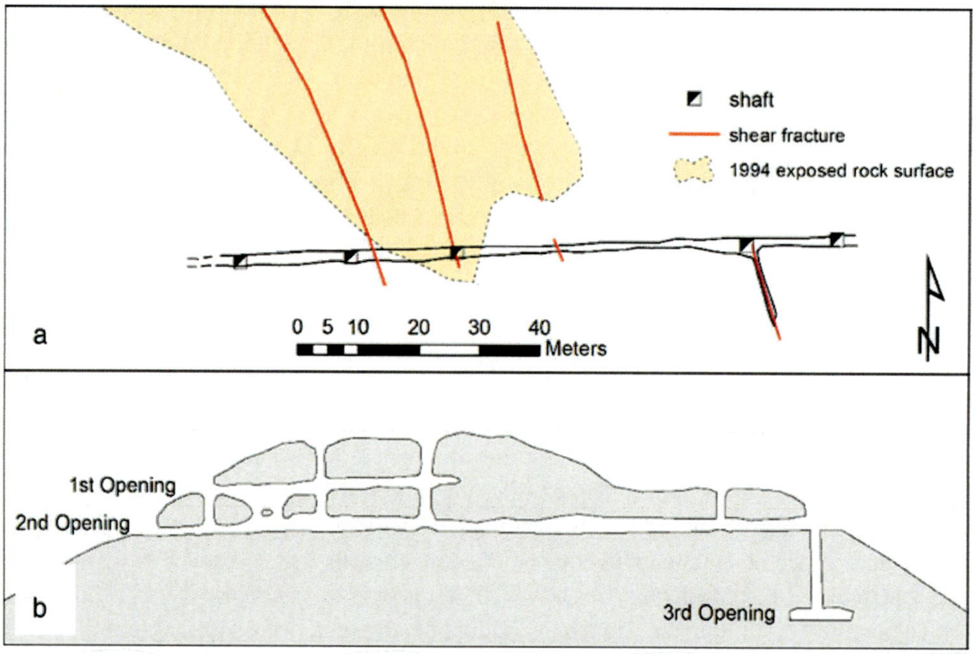

Fig. 7.3 Map of Haugen Cave also known as Dubuque's Cave. **a** Map view shows the extent of the passages or drifts on the 2nd Opening. **b** Cross section view, conforms to the map view in position and scale. The presence of the 3rd Opening is inferred from a near-by closed adit and apparent drift located in the valley floor

These statements imply that the Meshkwahkihak were not only working a surface deposit in 1820 at Dubuque but they were more than able to work underground.

7.2.3 Early Exploitation by American Miners

After the death of Dubuque in 1810 the Meshkwahkihak continued to mine for lead, controlled the region both east and west of the Mississippi River and did not allow any intruders with the exception that they allowed traders to operate on the islands of the Mississippi where they exchanged galena for trade goods. This situation changed in 1819 with the establishment of a mining camp by Jesse W. Shulls at the future site of Shullsburg, Wisconsin (Bain 1905) and the arrival in 1822 of a mining venture lead by Col. James Johnson at the future site of Galena, Illinois (Meeker 1872). In the fall of 1824, John Bonner in searching old Indian diggings discovered galena at the present day Hazel Green, Wisconsin. In the first day of his endeavor he extracted 3,175 kg of galena. The same summer a significant deposit was opened at the present site of New Diggings, Wisconsin. By 1829 American miners had discovered most of the major mining areas of the UMVLZD in northwestern

Illinois and southwestern Wisconsin (Heyl et al. 1959, Plate 9). The puzzling question here is: how did the American miners discover deposits of galena so quickly over such a broad area in so short of time? Meeker (1872, p. 290) provides a clue to this noting "The Indians generally were very fond of whisky" and to obtain it they would "promise to show, for a given number of bottles of whisky, where mineral could be found." What the American miners got for their whisky were sites like Haugen Cave described above, trench mines in the Mines of Spain Recreation Area south of Dubuque (Dockal 2011), or the Indian Walk-in Mine on the south side of Dubuque (Fig. 7.4).

There are number of terms used by American miners and early geologic reports that are unique to the area or are applied in a manner that differs from common modern usage.

A *crevice* refers to any vertical fracture that is to some extent expanded beyond being a simple crack through the action of mineralizing solutions or weathering. Most of the galena mined in the nineteenth century occurred within a crevice. The big discoveries occurred in crevices that were oriented roughly east to west; these they called east-west crevices or *east-wests*. Crevices that trended roughly north to south were *north-souths*. Crevices that trended roughly 30–60 degrees from the trend of the local east-wests were called *quarterings* or *swithers*. If several parallel east-west crevices were closely spaced, a few meters of one another, this was called a *range*. Whitney (1862, p. 335) noted that Polkinghorn range near Potosi, Wisconsin is: "a long series of crevices in one range; there are three parallel continuous ones, from eight [2.4 m] to ten feet [3 m] apart." Fractures inclined 45–80 degrees were all called *pitches* and fractures that were horizontal or parallel to bedding were known as *flats*. The galena usually occurred as the thin seam within the crevice, pitch, or flat; this the miners referred to as *sheet ore*. Whitney (1858) recognized that these sheet ore deposits were not like veins described elsewhere in the world as they had limited depth, only occurring within the Galena Group strata and not passing deeper, hence he coined term *gash vein*. Sheet ore could fill the entire width of a crevice or it could be surrounded on either side by *ochre* where ocher referred to a reddish colored material that consisted of goethite (FeO(OH)) sand-like dolomite crystals, and pieces of dolostone. The galena in some crevices was scattered throughout this sediment. These loose pieces of galena were called *cog ore*. Cog ore consisted of euhedral to subhedral galena crystals and crystal clusters that ranged in size from less than a centimeter to several meters in diameter. The 3,175 kg of galena noted above as dug by John Bonner in a single day came from one of these larger pieces of cog ore. Similar pieces of galena found in the soil above bedrock were called *float ore*.

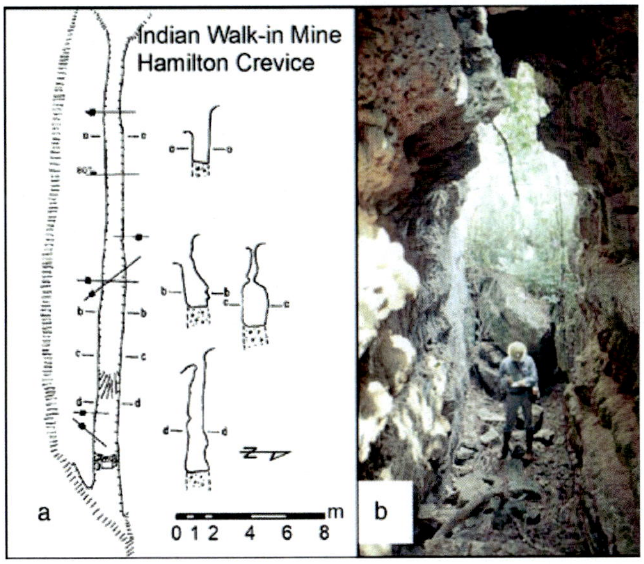

Fig. 7.4 Indian Walk-in Mine, probable Native American mine site, 42.476°N 90.682°W. **a** Sketch map by Dockal, 1973. **b** Photograph looking east, camera position about at the position of section c-c

Some of the crevices at Dubuque extended laterally for several kilometers (Calvin and Bain 1900). The width of a crevice varied considerably both vertically and laterally. For most of the vertical extent of a crevice, its width would be less than a centimeter, however, where the crevice cut certain favorable stratigraphic beds it could be expanded to several meters in width. These favorable stratigraphic beds were usually between one and three meters thick and frequently traceable laterally for several kilometers. The body fossil *Receptaculities oweni* was usually found in association with these favorable beds, consequently, it was sometimes referred to as the *lead fossil* (Leonard 1896, p. 23).

At the various mining camps in the UMVLZD, these favorable beds occurred at differing stratigraphic levels within the Galena Group, furthermore in any area there could be more that one of these sets of favorable beds (Fig. 7.3). Each set of favorable beds was called an *opening* and was given a semi-formal name, the use of which was limited to the area of the particular mining camp.

For most of the area within the City of Dubuque one of these occurs just below the contact between the Dubuque and Stewartville: this was the *1st Opening* at Dubuque. A second set of favorable beds occurs near the middle of the Stewartville: the *2nd Opening* of the Dubuque miners. The *3rd Opening* occurs just above the Stewartville-Prosser contact and the *Upper Flint Opening* occurs about 10 m below that contact. The miners at Beetown, Wisconsin recognized two openings, the *12 Foot* and the *65 Foot*. The 12 Foot, which was the uppermost and occurred within the upper portion of the Prosser, was 12 feet (3.7 m) thick, hence the name, while the 65 Foot occurred "65 feet" (19.8 m) below the 12 Foot.

The presence of an opening only provided a potential for a crevice to be expanded. If a crevice was traced laterally at the stratigraphic position of say the 1st Opening at Dubuque, at some points it would be expanded and at others it was not. Generally, in three dimensions the area of expansion had a pod-like shape which had a height equivalent to or less than the thickness of the favorable beds; a width greater than a few centimeters, typically one to two meters, but could exceed 10 m; and a length generally less than 200 m but could go on for ten times that. A single pod-shaped area of expansion was unfortunately also called an *opening* but used in an informal sense.

Stratigraphically equivalent areas along the crevice where it was not expanded were sometimes referred to as *bars* or barren ground. The area between the walls of an expanded portion of a crevice prior to mining would consist of any combination of: rock that was degraded by mineralizing processes; primary minerals deposited by the mineralizing processes especially galena, sphalerite and marcasite; sediment formed in situ from the weathering of the preceding; sediment that was transported into the feature from outside the crevice; calcite in the form of large scalenohedrons or as flowstone, stalactites, and stalagmites; and physically open space.

Miners would explore along the trend of a crevice by first finding an indication of the crevice then finding the elevation of, say, the 2nd Opening. If the proper elevation was at or near the land surface, they would dig pits along the trend of the crevice searching for float ore. These pits were called variously *badger holes*, *sucker holes*, *mineral holes*, or *prospects*. If those pits uncovered a significantly expanded portion of the crevice, then they struck an opening. If the opening trended into a hill side, they would run a *drift* into the hillside following the crevice. Running a drift basically involved removing the sediment from the opening, i.e., the sediment between the walls of the crevice leaving a tunnel as wide as the crevice and as tall as the opening.

If the width of the opening was too narrow and not enough mineral present to make further progress worthwhile they would move some distance further along the crevice following its compass trend and sink a shaft down to the level of the stratigraphic opening that they were pursuing and, if there was an opening there, they would run drifts from the bottom of the shaft along the crevice. Some shafts were sunk deep enough to explore more than one stratigraphic opening (Fig. 7.3). If a shaft or a drift encountered a naturally occurring physical open space that was big enough to allow a boy to enter and work, this, they called a *cave*. Mostly caves were small, less than a half meter tall, meter wide, but could run laterally along the crevice for tens or hundreds of meters. Flatbelly Cave, an unmined opening encountered in a quarry near Murphy Park in Dubuque, has a ceiling height 30–50 cm throughout most of its extent and a width of one to one and a half meters. Some caves were quite large; Stewart's cave in Dubuque was 24 m wide and 61 m long, the opening was 12–15 m high but nearly filled to the roof with "fallen rock" (Calvin and Bain 1900, pp. 538–539).

Later in the twentieth century, the term cave as used in the region of UMVLZD took on a more expanded meaning to include any underground open space big enough for a person to enter regardless of origin. Fizzle Cave west of Dubuque consisted of three drifts, two head west from the valley floor and one driven north-south to connect them. All three drifts were hand drilled and blasted in hard rock following crevices, but at the wrong elevation of the locally productive opening. Kemling Cave (Brown and Whitlow 1969, Fig. 2, p. 11 and Leonard 1896, Fig. 2, p. 30) is a series of drifts, some portions of which were considered a cave in the nineteenth-century sense. Snake Cave at Potosi, Wisconsin, a commercial cave open to the public, is an abandoned lead mine. Whitney (1862) reported that:

> The Cave range makes a large opening beneath the caprock; and, on its east end, terminates in a cave, known as the "Snake Cave," from which the diggings here were originally called the "Snake diggings When this cave was first discovered, it was found to contain an incredible number of rattlesnakes-many thousands, it is said."

The term *diggings* as Whitney uses here refers to surface workings, badger holes, along one or more crevices.

In summary there are three terms: crevice, opening, and cave that are critical to understanding the historical literature of the caves of the UMVLZD. A crevice refers to any solution-expanded vertical through-going fracture regardless of the origin or orientation of that fracture. All vertical through-going fractures in the UMVLZD display some degree of mineralization and/or solution expansion. The term opening has two related usages: (a) opening can refer to a specific stratigraphic horizon where a crevice could be expected to show significant mineralization and/or solution expansion or (b) an opening can refer to a specific pod-shaped expansion of a crevice. The term cave also has two related usages: (a) a cave may refer to any open space that existed by natural processes prior to any mining activity or (b) a cave may refer to any open space regardless of the processes of formation of the open space, but generally caused or facilitated by human activity. Caves of the first usage are referred to in an informal sense; i.e., Stewart's cave, while those of the second usage is referred to in a formal sense; e.g., Crystal Lake Cave. The term *crevice cave* will embody both usages in this paper.

7.2.4 Nineteenth-Century Lead Mining of the Crevice Caves at Dubuque, Iowa

American miners first crossed over the Mississippi River and attempted to work sites in the Dubuque area in 1830. Among this first group of miners was Lucius H. Langworthy who mined an area known as Langworthy diggings, located about 300 m southeast of the area of Haugen's Cave (Fig. 7.2). Langworthy diggings are crossed by several closely spaced east-west crevices which were probably all mineral rich. This grouping of crevices was known as Langworthy range (Fig. 7.2). The range was worked by several different individuals; just west of Langworthy diggings were those of Mathias Ham and a Mr. Sims. These workings consisted of several drifts and shafts along east-west fractures. A road cut on North Grandview Avenue, just east of its intersection with Kaufman Avenue, exposed a portion of these workings (Fig. 7.2). This site was investigated by Dockal (1973) to develop a geochemical exploration method for UMVLZD type crevice deposits. Five east-west crevices are exposed in the cut; the northern most is expanded to a width of 5 cm with the gap filled with reddish colored mixture of individual weathered dolomite euhedra and goethite.

The next crevice south has a small opening developed at the top of the 3rd Opening in which M. Ham ran a short, 30 m long, less than a meter wide, and a meter-high drift known today as Mosquito Cave (Fig. 7.5). The walls of the drift are very rough, consisting of a very porous, friable sandstone-like, dolostone that contains a considerable amount of goethite and traces of smithsonite ($ZnCO_3$). When examined in July 2008 there was a white powdery substance on the walls which was tentatively identified as goslarite ($ZnSO_4 \cdot 7H_2O$). The gash vein is evident in the ceiling of the drift where it contains goethite, small amount of galena, traces of cerussite ($PbCO_3$), and smithsonite. At the far end of the drift the gash vein forms a sheet of galena, but that sheet is less than 0.5 cm wide; which, at that time, would have been considered non-economic to work further (Whitney 1858).

Fig. 7.5 View looking west into the Mosquito Cave. This is what most of the passages or drifts in the Dubuque area look like after miners removed the sediment. At this point it is roughly three quarters meter wide and one meter tall. The white material is tentatively identified as goslarite ($ZnSO_4 \cdot 7H_2O$) Scale bar divided into 0.1 m units

Twelve meters south of that crevice lies an area of broken rock that apparently was once a large cave which suffered a ceiling collapse. The opening there joins two parallel east-west crevices, 15 m apart, and appears to have developed on the entire thickness of the 3rd Opening and possibly extending down to the Upper Flint Opening. The fifth east-west crevice lies 26 m south of the south side of the collapsed cave. The road cut at that point exposes nearly the entirety of the 3rd Opening yet there is only minor solution expansion of the crevice. According to Whitney's 1858 map this is the North Langworthy crevice. The five crevices that are exposed in the road cut are traceable today to the west. The northern-most crevice is exposed in a small quarry where an opening at the top of the Stewartville, the 1st Opening, was mined out and now called Jim Shifflett's Cave (Fig. 7.2). This east-west crevice is intersected by several NNW striking fractures and a north striking fracture which the miners explored for short distances via less than meter-high drifts.

The crevice on which Mosquito Cave lies correlates across Kaufman Avenue to a crevice that forms a portion of Weigand Cave (Fig. 7.2). Weigand Cave apparently was discovered by miners sinking a shaft on the trend of the crevices of the collapsed cave exposed in the road cut. That shaft missed the target crevice by 6 m but luckily encountered a mineralized north-south fracture, the same fracture as exposed in Jim Shifflett's Cave. The miners followed it northward by drifting in the opening striking the target east-west crevice where they encountered an opening at the level of the 1st Opening which they worked eastward 8 m until the opening pinched to barren rock and a trace of a gash vein. That same opening was worked 140 m westward where again the miners encountered barren rock. At the western end two winzes were sunk to the 2nd Opening were the miners encountered an opening which they developed for a short distance.

West of Weigand Cave the same crevice was encountered in the Becker Quarry where it was still barren of galena and consisting of a less than meter wide opening filled with a reddish, unstratified, sediment, or ocher. At Becker Quarry a second well developed east-west crevice was exposed 45 m south of the crevice that Weigand Cave is developed upon. When first encountered by quarrying operations there was a drift 1 m wide by 1.5 meters high. Spelunkers in the 1960s named this Becker Quarry Cave. This crevice is the same as the southernmost crevice exposed in the road cut, the North

Langworthy crevice, however, from Becker Quarry on west it is known as Levens crevice.

Possibly the greatest discovery in the area was Levens's cave located 820 m west of Becker Quarry on Levens crevice. It was discovered by Tom Levens in 1850 and by 1852 had produced 1,600,000 kg of galena. Upon discovery, it consisted of an open room 40 m long, 6-10 m wide, with a domed or arched ceiling with a maximum height of 6 m and containing many large blocks of galena, one of which had an estimated mass of 12,000 kg. (Whitney 1858). Once the galena and associated sediments were excavated from this opening it formed a room 150 m long, 8–12 m wide, and with a ceiling 10–12 m high (Calvin and Bain 1900). The opening is 27 m below the surface and developed on the 1st Opening. Later development work resulted in discovery of another sulfide-bearing opening 12 m below the cave (Calvin and Bain 1900). At the west end of the Levens's cave, the opening pinches but then after about 800 m opens into another "great cave" (Calvin and Bain 1900, p 541). North Langworthy crevice (Levens crevice) is traceable for 4.8 km, the longest traceable crevice in the region. During the period 1830 to 1865 an estimated 8,800,000 kg of galena was extracted from the various openings on this crevice (Calvin and Bain 1900).

Approximately 200 m south of the Langworthy range is a pair of parallel east-west crevices with a substantial opening that was mined by Booth and Carter circa 1840 and produced an estimated 1,600,000 kg of galena (Whitney 1858) (Fig. 7.6). The main room in the mine is developed at the 1st Opening, it is 6–10 m wide, near the western end it had a ceiling height of almost 3 m., and the accessible portion of the opening was 190 m in length.

On the south side of the main room near its western end there is a drift that follows a crevice southward (Fig. 7.6). This drift leads into a maze of crevices and associated openings, some of which have drifts dug on them. These drifts are tight, some less than a half meter square. There is a single route through the maze that eventually intersects another major east-west crevice and associated opening. That crevice is the Level crevice and the now mined-out opening along it is called Level Crevice Cave. The initial mining was done by McBride circa 1840–1850. Level Crevice Cave extends westward from the maze entry at least for 1000 m making it the longest crevice cave that was accessible in recent times. For most of the cave's length it is from one to three meters wide. However, where the cave is crossed ENE striking fractures it is usually much wider. At those points rich pockets of galena were encountered (Calvin and Bain 1900). A short distance east of the maze entry point, the crevice enters barren ground or a bar and drifting stopped.

This barren ground extends for 170 m and then opens into Glom Cave, a name applied in the 1960s to a circa 1860s mine. Glom Cave is 180 m long and near the east end the Rose Shaft extends from the surface to the level of the 3rd Opening. Glom Cave is on the 1st Opening. There was no opening developed at the 2nd Opening. At the 3rd Opening there is a drift which to the west is blocked by a concrete dam, iron pipe, and valve. That drift continues westward along the Level crevice for 524 m (according to annotations added in 1927 to an unpublished 1899 map by Cole of the developments on the crevice). To the east the drift runs 400 m to an exit at the valley floor.

The entire drift, both east and west of Rose Shaft, is a drainage gallery, but known by the miners as a *level* and its function was to dewater the lower sulfide-bearing horizons of Level crevice and the adjacent area. Later the City of Dubuque used its outflow as the source of potable water for the city. The estimated flow in 1965 through the pipe and valve when wide open was 7,500 m^3/day (Morehouse 1968). East of the valley, the Level crevice disappears in the *patch diggings*, an area populated with closely spaced NNE striking shear fractures that bore thin veins of galena.

The workings on Langworthy or Levens crevice, in the Booth and Carter mine and along Level crevice are largely typical of the mine workings within the UMVLZD for the nineteenth century. They are also typical of the character of the crevice caves of the UMVLZD. Lead mining in the UMVLZD peaked in the late 1840s at which time the district was contributing 85–90% of the lead consumed in the United States (Agnew et al. 1953). Thereafter it declined due to a combination of low price, the working out of the easy-to-reach ore, and the exodus of the miners to the newly discovered gold fields in California.

7.2.5 Zinc Mining at Dubuque

Many of the shallower lead mines also encountered smithsonite known by the miners as *dry bone* while the deeper mines encountered sphalerite known as *black jack*. Both were tossed on the spoil heaps as waste material of no value. This changed in 1860 with the opening of a zinc smelter at La Salle, Illinois. However, the first zinc ore from Dubuque was shipped in the fall of 1880 to a plant in Benton, Wisconsin (Leonard 1896). This started a zinc mining boom at Dubuque. Spoil piles were worked over for the smithsonite and sphalerite, old mines reopened, old shafts deepened, new deep shafts sunk, and large water pumps installed.

Smithsonite was reported to occur at Ewing's diggings on Timber range near Durango, Iowa as early as 1839 but was discarded (Owen 1840). Ewing's diggings lie at the east end of Timber range which was probably the Mine of Maquanquitons of Schoolcraft (1821). Owen (1840) described the site as "the crevice or cave from which the ore was taken is, in places, thirty feet [9.1 m] wide." Whitney (1858) visited the

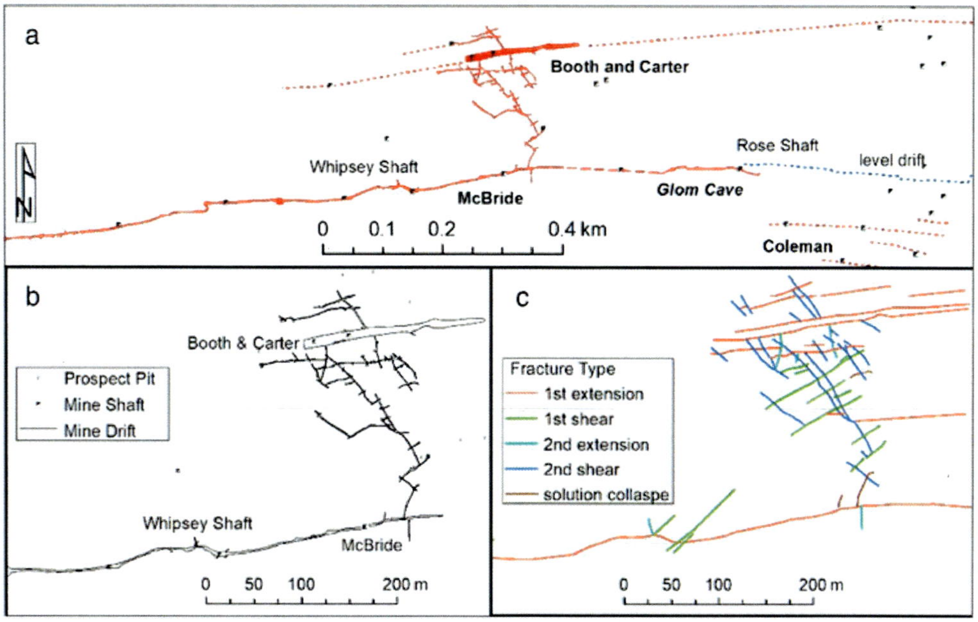

Fig. 7.6 Maps of the Booth and Carter Level Crevice area. **a** Map of the extent of the workings on the two crevices; (solid red lines) developed drifts on the 1st Opening; (red dashed line) inferred extensions of drifts or approximate positions of workings; (blue dashed line) drift on the 3rd Opening initially used to drain the lower areas of Level Crevice and later as a water supply for the city. **b** Detailed map of the enterable (circa 1973) drifts. **c** Map of the fracture types that control the openings. 1st and 2nd refer to the first and second episode of fracturing (Sect. 7.3.3)

site in 1856 when it was in disuse and noted the workings were inaccessible. In 1894 the site had become a very productive zinc mine known as the Durango Drybone Mine complete with an open cut, "caverns of immense size." The mine had a special rig for concentrating the ore (Leonard 1896).

There are three aspects to the site: a very large cave, the open cut, and many open shafts. The cave lies at the eastern end of the site. It trends almost due west and is 108 m from drip line to drip line. The eastern portal is a gaping hole 13 m wide and about 8 m high. This rapidly narrows to 4 m for the next 15 m but then opens into an elliptically shaped room 18 m wide and 36 m long with a ceiling 3 to 5 m high. West of there the roof is unstable and partially collapsed and then the cave becomes a drift 3 m wide all the way to the western portal. This was probably a cave prior to mining but was extensively worked out.

The original cave probably extended further east, but the roof rock was mined for its smithsonite content leaving an open cut and substantial mine dump. The main open cut lies 150 m west of the west portal of the cave and is well described by Leonard (1896) and Heyl et al. (1959). The cut is 75 m long, 18 m at its widest point, and 10 m high. The north wall is an EW striking pitch that dips 45° north, the south wall is a similar striking pitch that dips south 40–60°; these are cross-cut by a series NNE striking shear fractures some of which show a slight vertical offset (Heyl et al. 1959, Plate 21). The open cut was a cave prior to mining, portions of which remain. The open cut and the cave are developed in the lower portion of the Stewartville. There are a series of shallow shafts just beyond the south wall of the open cut that lead to workings below the cut.

The eastern-most shaft leads to Spiral Cave which was named for the confusing nature of the low narrow drifts which seem to pass downward in a spiral fashion penetrating well into the Prosser, possibly 20 to 30 m below the floor of the open cut. The rock is rotten, consisting of buff-colored dolostone and nodules of chert surrounding pockets of reddish brown very friable dolostone or dolomite sand. The nineteenth-century miners in the UMVLZD called such rock *calico* rock. Galena occurs as scattered cerussite encrusted euhedra throughout. The next shaft to the west leads to Maze Cave, named for its maze-like character. The rock is identical to that found in Spiral Cave. A shaft at the west end of the cut leads to Skull Cave which is also like Spiral Cave. It appears that the entire open cut is underlain by *calico* rock.

The Dubuque Lead Mining Company deepened Sloan Shaft on Level crevice hoping to discover sphalerite ore. What they encountered was a mineralized area of galena and marcasite one to two meters wide along a fracture developed at the Upper Flint Opening (Leonard 1896; Bain 1906). Leonard (1896) described the galena as "scattered through the rock, sometimes in particles of considerable size." Bain (1906) refers to the ore as a mixed galena-pyrite *honeycomb* ore. It is also referred to as a breccia for its outward appearance but on closer inspection it is not a breccia. The local mining term for this and similar deposits is *brangle* (Wheeler 1908). The major zinc mines of the Dubuque area all produced a mixed sphalerite-marcasite brangle ore, including the Dubuque and Superior Mining Company and their mines west of Dubuque, the Avenue Top Mining Co. and their mines within the city in an area long worked for galena (Heyl et al. 1959) and the Goosehorn Mining Co. which operated the Goosehorn Mine, the last formal zinc mine in Dubuque, which shut down in 1912. Unfortunately, these mines are not accessible and there are no descriptions of the workings.

However, in 1973 there remained a sizable pile of brangle ore next to Beadle shaft which was one of the mines operated by The Dubuque and Superior Mining Co. A 7-kg random grab sample of the ore consists of a breccia appearing gray cherty dolostone that has large vugular pores (1.5–5.5 cm, average 2.8 cm, in maximum width) with no regular shape or distribution but whose volume was variously filled with pyrite, sphalerite, galena, calcite, smithsonite, goethite, and gypsum ($CaSO_4{\cdot}2H_2O$). The rock composition by volume: 60% dolostone and chert; 40% vug fillings consisting of sphalerite (15% total sample volume), pyrite (5%), galena (<1%), calcite (5%), and empty pore space (15%). The smithsonite was a thin gray coating covering portions of some of the samples; the goethite and gypsum were very minor superficial coatings probably the result of 60 years of exposure on the surface. The brangle represents the un-oxidized sulfide mineralized rock of the opening.

7.3 Geologic Formation of the Crevice Caves

The crevice caves formed as the result of oxidation of sulfide minerals deposited within fractures and within dolomitized strata immediately adjacent to the fractures. The geometry of the caves is a direct result of the spatial geometry of the sulfides, the timing of sulfide precipitation, and the intimated association of sulfide minerals with coeval calcite dissolution and dolomitization. Bethke (1986) postulated that the source of Pb and Zn for the ores was the Illinois Basin where topographically driven basinal fluids transported the metals to sites of mineralization in the UMVLZD. This transport of ore precursor elements through deep basins to basin margins where they result in formation of sulfide deposits is a common theme in the discussion of MVT deposits.

Such a transport model fails to account for the volumetrically significant coeval calcite dissolution and dolomitization of calcite associated with the UMVLZD ores. Fluid traveling hundreds of kilometers through a basin like the Illinois Basin to the basin margin will have encountered enough calcite and dolomite along the fluid pathway to have reached chemical saturation with respect to both calcite and dolomite regardless of the initial origin of that fluid. Spirakis (1995) and Spirakis and Heyl (1996) challenged the basin fluid flow model for the sulfide mineralization and offered a model based upon thermal convection between basement rocks and overlying strata, where thermal convection brought Pb and Zn, sourced in Precambrian rock that underlie the UMVLZD, to mineralization sites in the Galena Group. The problem with that model is that it relies upon much of the UMVLZD being underlain by rock that is unusually rich in K, Th, and U. Here I propose a variant upon that model which also accommodates calcite dissolution and dolomitization and does not rely upon thermal convection.

7.3.1 Stratigraphy

Sedimentary strata of the UMVLZD are underlain by 1.7–1.4 Ga Precambrian Proterozoic rocks that are nowhere exposed within or near the UMVLZD. Those basement rocks and are known only through drillers logs and a few petrographic studies of cuttings and core from a small number of wells (Fig. 7.7). Though the rocks are much older than the formation of the crevice caves, they had a significant impact on the formation of these mineralized crevice caves. Significant features of the Precambrian basement rocks relative to this study include:

1. The rock is felsic in composition consisting of quartz, K-feldspar, Na-plagioclase (oligoclase), and biotite (Anderson 1950; Eckstein et al. 1983; Lidiak and

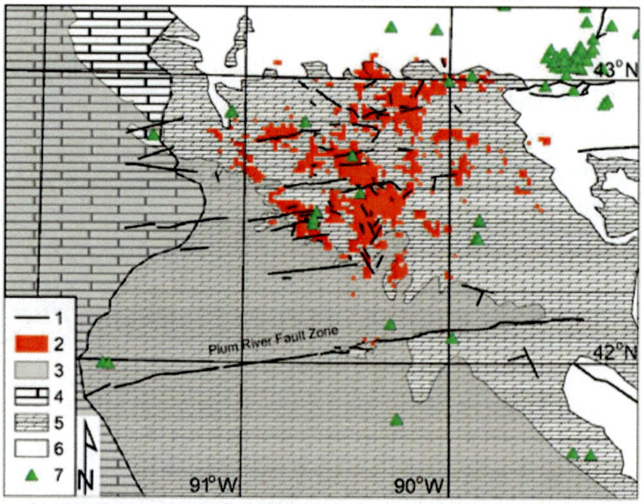

Fig. 7.7 Geologic map. (1) Probable shear zone. (2) Areas of documented mining activity. (3) Area where the Galena Group lies under significant cover of younger Paleozoic strata. (4) Galena Group strata predominately limestone. (5) Galena Group strata predominately dolostone. (6) Area where the Galena Group has been removed by erosion. (7) Well that reached the Precambrian basement

Denison 1983; Vitaliano et al. 1986). To the west of the UMVLZD the Precambrian rock is mafic, tracolite or olivine gabbro, and contains the more calcium rich plagioclase, labradorite (Yagapur 1979; Jiras et al. 2006). The boundary between the felsic and mafic rock, as delineated by areal gravity and magnetic surveys, corresponds to the western boundary of the dolomitized area (Fig. 7.7).

2. The rock is fractured, the fractures strike EW, NS, NNW, and NE (Haimson and Doe 1983) which is like the orientation of the dominant sets of fractures in the Paleozoic strata. Fractures are generally open and coated with chlorite (Daniels et al. 1983).
3. Doe et al. (1983) describe a Pb depletion event in the Proterozoic rock occurring at 260 ± 35 Ma and suggested that a large amount of Pb was lost and that it was highly radiogenic like the Pb found in the ore deposits of UMVLZD. Interestingly Brannon et al. (1992) obtained a 87Rb-86Sr age of 269 ± 4 Ma on sphalerite from a mine within the Galena Group within the UMVLZD and paleomagnetic dating by Pannalal et al. (2004) found regional dolomitization in the UMVLZD during the early Permian, 282 ± 10 Ma.

Approximately 600 m of Cambrian and Ordovician sandstone and dolostone separate the Precambrian basement from the strata of the Galena Group that are directly involved in the formation of the crevice caves. The nature of the Galena Group within the UMVLZD is well documented (Agnew et al. 1956; Willman and Kolata 1978; Ostrom 1987).

The crevices caves are developed within the Prosser and Stewartville Formations. The caves only form within the rock of a specific lithofacies that is characterized by the body fossil *Receptaculites oweni* (Hall) (Fig. 7.8a) and the trace fossil *Thalassinoides* (Fig. 7.8 b–d) which forms a three-dimensional network of tubular sediment filled burrows (Kendall 1977; Morrow 1978; Palmer 1978). The burrows give the rock a distinctive mottled texture. The material between the burrows prior to dolomitization was fossiliferous fine-grained limestone (wackestone to packstone sensu Dunham 1962) (Fig. 7.9a). The infillings of the burrows, when evident, is a coarser grained limestone (grainstone) consisting of fragments of bioclasts; but mostly this sediment was dolomitized very early, possibly penecontemporaneously with deposition (Morrow 1978). This dolomite is very distinct in character consisting of

Fig. 7.8 Galena Group, *Thalassinoides* lithofacies lithology. **a** *Receptaculites oweni* (Hall) preserved here as a steinkern in dolostone. **b** Mottled texture of limestone, *Thalassinoides* lithofacies, Galena Group (lower portion of Prosser Fm.) exposed in fresh outcrop along the Northwest Arterial, Dubuque, IA. (42.539°N, 90.695°W). The darker areas are the dolomite filled burrows of the trace fossil *Thalassinoides*. The lighter areas are limestone (fossiliferous wackestone and packstone). Red/white ranging bar divided into 0.1 m segments. **c** Slab of rock from same locality stained with alizarin red-s to differentiate the limestone portion (red or dark colored) from the dolomitized *Thalassinoides* burrows (tan or light colored); scale divided into centimeters. **d** Mottled texture of dolostone, Thalassinoides lithofacies, upper portion of the Stewartville Fm. from Dubuque's Monument, Mines of Spain Recreation Area, Dubuque, Iowa. Dolomitized burrows, lighter areas, dolomitized limestone (wackestone), darker areas

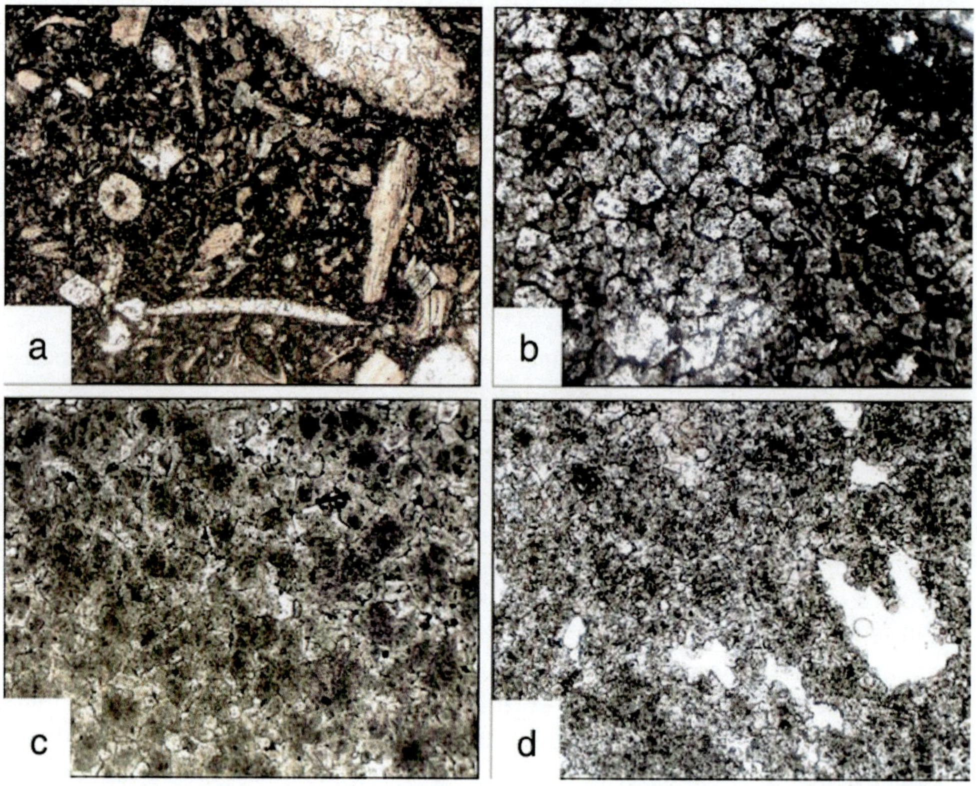

Fig. 7.9 Photomicrographs, plane polarized light, from the *Thalassinoides* lithofacies, width of photomicrographs = 3.0 mm. **a** Unaltered limestone adjacent to a *Thalassinoides* burrow, Prosser Fm. Guttenberg, IA, stained with alizarin red-s. **b** Dolomite replacing the fill of a *Thalassinoides* burrow, Prosser Fm., Dubuque, IA **c** Typical dolomite adjacent to burrows, Stewartville Fm., Dubuque, IA. **d** Dolomitized rock with moldic pore after an echinoderm, Stewartville Fm., Dubuque, IA

poorly formed dolomite euhedra that contain abundant semi-opaque brownish inclusions (Fig. 7.9b). The burrows possess a porosity and permeability that is greater than that of the surrounding limestone as can be observed by wetting polished slabs of the rock taken from outside the dolomitized area of the UMVLZD and observing how water is rapidly absorbed into the burrow but remains standing upon the non-burrow portion of the rock. Individual beds of this lithofacies in the study area range from a decimeter to a little over a meter in thickness with the boundary sometimes marked with what appears to be a sub-marine corrosion surface upon which *Receptaculites* occurs in growth position. A bed may grade laterally through a progressive loss of *Thalassinoides* into a laminated, weakly sorted packstone or into a totally unsorted and thoroughly bioturbated grainstone. These beds can be stacked one upon another to give the lithofacies a maximum thickness several meters. The lateral extent of the lithofacies ranges from a few tens of meters to possibly as much as a kilometer. The Thalassinoides lithofacies can occur at any stratigraphic position within the Prosser and Stewartville formations; however, in the Dubuque area it is particularly prevalent at the top of the Stewartville, near the middle of the Stewartville, at the top of the Prosser, and near the base of the Prosser. These stratigraphic positions are equivalent to the 1st Opening, 2nd Opening, etc., of the miners. The significance of this lithofacies is that it has a porosity and horizontal permeability that is much greater than all the other strata between the top of the Maquoketa and the top of the St. Peter Sandstone.

By the end of the Paleozoic, the UMVLZD had approximately 2.5 km of strata deposited upon the Galena Group (Pannalal et al. 2004; Spirakis and Heyl 1996; Sangster et al. 1994). Most of that sediment would have been Upper Carboniferous sandstone and shale which was later eroded in the Cenozoic leaving at present few scattered outliers in the southern portion of the UMVLZD.

7.3.2 Structural Geology

Strata within the region dip southwest 2.6 m/km. The only structure of note in the UMVLZD is the Plum River Fault Zone (Kolata and Buschbach 1976; Ludvigson et al. 1978; Bunker et al. 1985) (Fig. 7.7). Historical mapping revealed few other faults and most of those are of limited extent and offset. Folding is limited to minor, gentle undulations where structural closure is usually less than 25 m. The significant structural feature is fractures. The region is cross-cut by an array of four types of fractures:

1. Vertical extension fractures which are through-going in that each cut across all stratigraphic layers, are vertical in their gross aspect but show moderate variation in dip both laterally and vertically, when not significantly

Fig. 7.10 Outcrop exposures of the various fracture types found in the UMVLZD. **a** EW striking vertical extension fracture in the upper Prosser Fm. (42.472°N, 90.650°W). **b** ENE striking vertical shear fracture in the Dubuque Fm. (42.514°N, 90.851°W). **c** Bedding parallel tension fracture in the Stewartville Fm. (42.526°N, 90.823°W). **d** Inclined tension fractures in the lower Stewartville Fm. (42.474°N, 90.685°W)

widened by solution display walls that are mirror images of one another, and their strike within an area is uniform but with some deviation or wanderings (Figs. 7.10a, 7.11a, b).

2. Vertical shear fractures which are also through-going, their dip is without much variation both vertically and laterally, and the walls are smooth except where modified by solution (Figs. 7.10b, 7.11a–c). The gap between the walls is usually filled with gouge or secondary mineral deposits. Infrequently the space between the walls is occupied by lithologies different from that of the wall rock, especially sandstone that is identical to that of the St. Peter Sandstone but occurring stratigraphically above the position of the St. Peter.
3. Bedding parallel tension fractures or the flats of the miners occur in all stratigraphic units of the Galena Group but are most commonly observed in the Decorah (Fig. 7.10c).
4. Non-through-going vertical to inclined tension fractures which at the outcrop scale display a uniform strike and dip but in map view at slightly larger scales are arcuate (Figs. 7.10d, 7.11b).

7.3.3 Hypogene Sulfide Mineralization and Dolomitization

At some point, after deposition of the Silurian strata, possibly late Paleozoic, the region was subjected to far-field stress associated with distant tectonic events. This resulted in the strata of the UMVLZD being cut by two sets of vertical extension fractures and four sets of vertical shear fractures. Fracturing occurred in two episodes.

7.3.3.1 The First Fracturing Episode

The first fracturing episode produced a pervasive array of approximately EW striking vertical extension fractures. These occur throughout the UMVLZD and to the west of the UMVLZD. They range in length from one hundred meters to several kilometers and are typically spaced 50–100 m apart. Occurring with these are west-northwest and ENE striking vertical shear fractures and similarly trending zones of shearing (Fig. 7.7). The Plum River Fault Zone is most obvious of the latter mainly because it has a significant vertical offset. A similar fault zone, but lacking significant vertical offset, passes through southern Wisconsin (Fig. 7.11a).

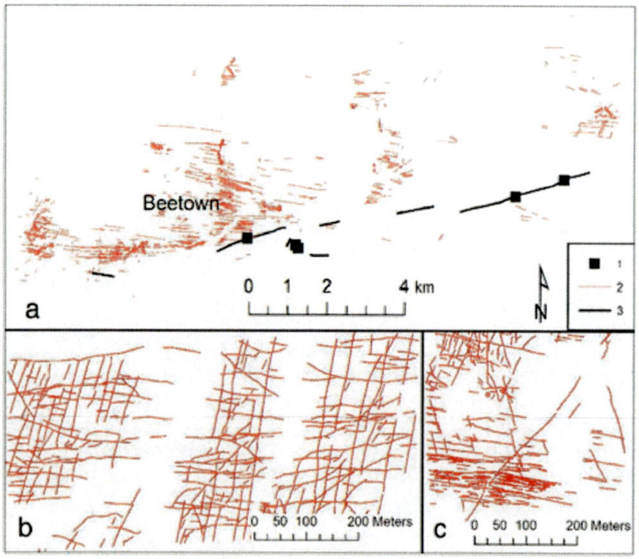

Fig. 7.11 Fracture patterns form various localities in the UMVLZD. **a** Fracture pattern at Beetown, Wisconsin revealed by abandoned mine workings along crevices. Pattern compiled from hillshade effect Lidar DEM from 2010, observation from 1940 air photography, Heyl et al. (1959, Plate 7) **b** Fractures revealed by croplines in hay fields observed on 2012 Google Earth photography (Panno et al. 2015, 2017) Located near Shullsburg, Wisconsin (42.583°N, 90.177°W). Visible are EW extension fractures, NNE shear fractures, and arcuate subsidence fractures. **c** Croplines located near Meekers Grove, Wisconsin (42.641° N, 90.288°W). Visible include WNW shear fractures, NNW shear fractures, and ENE shear fractures

Another is apparently associated with the Meekers Grove Anticline (Fig. 7.11c). The pattern of these fractures implies a principal stress is oriented east to west, intermediate stress oriented vertically, and minor stress oriented north-south. Fracturing would have been an ongoing process for the duration of the episode which could have lasted for thousands of years and resulted in thousands of minor earthquakes.

These fractures, once formed, collectively gave the sub-Maquoketa strata of the region a significant vertical permeability and east-west oriented horizontal permeability. Several processes occurred coevally with fracturing. Saline connate water trapped within the sub-Maquoketa sediment column circulated and intermixed with water entrapped within the Precambrian felsic basement through the formation of differential pressures associated with movement on the shears and opening of the extension fractures (Oliver et al. 2006 for a discussion of fluid flow between the sedimentary strata and basement rock). Prior to fracturing, water within the Paleozoic strata would have been at chemical equilibrium with the minerals contained within those strata, principally calcite, dolomite, and quartz. Likewise, within the basement felsic rock the water there would have been in chemical equilibrium with quartz, biotite, plagioclase, and K-feldspar. During fracturing cool water from the sediment column was brought into contact with the minerals of the basement rock resulting in the alteration of biotite to produce chlorite, sericite, vermiculite, or other clay minerals while adding F^-, Fe^{2+}, Mg^{2+}, K^+, and Zn^{2+} to circulating solutions. Similar alteration of plagioclase produced epidote, sericite, or clay minerals including smectite while releasing Pb^{2+} to solution. Ca^{2+} was taken up from the solution in the formation of epidote, fluorite, and as interlayer cations in the smectite. The resultant solutions, which would been at the temperature of the surrounding basement rock, then circulated back to the sub-Maquoketa strata, in particular the Galena Group. That returning water was depleted in Ca^{2+} and enriched in Mg^{2+} relative to water that was initially within the Galena Group. These fluid interactions resulted in the dissolution of calcite and the dolomitization of calcite (Dockal 1988). This entire flow system was effectively a closed chemical system confined by the overlying shale beds of the Maquoketa and by the limits of the depth of fracturing in the basement rock. Taking the depth of recent earthquake hypocenters in the Midwest USA between latitudes 38° and 45° and longitudes 84° and 94° as a guide, that depth could have been as great as 18 km.

The magnesium enriched, calcium depleted water moved vertically and laterally through the fracture network and moved laterally through the strata between fractures. Strata with low relative permeability such as micritic limestone beds would have experienced calcite dissolution at the face of fractures that bears chemically aggressive water, water that was undersaturated with respect to calcite.

Dolomitization of the calcite also would have occurred within the rock adjacent to fractures since that water also would been supersaturated with respect to dolomite. The dolomite thus formed was stable, there is no evidence that these solutions at any point were undersaturated with respect to dolomite. Further dissolution of the calcite adjacent to the newly formed dolomite created some porosity around the dolomite and brought the solution closer to saturation with respect to calcite. Eventually the solution reached saturation with respect to calcite but if it was still supersaturated with respect to dolomite additional dolomitization of calcite would have occurred. The result, starting at the fracture face and moving laterally away from that face and into the rock was: (1) solution expansion of the fracture; (2) a zone of rather porous dolostone (Fig. 7.12b); (3) a zone of dolostone with progressively fewer pores (Fig. 7.9d); (4) dolostone without pores (Fig. 7.9c); (5) limestone with only partial dolomitization (Fig. 7.12a); and (6) eventually unaltered limestone. The width of the altered zone depended upon the permeability of the rock, the duration of the flow event, and the chemical aggressiveness of the solution. The lateral extent of the various alteration zones noted above is dependent upon two competing but linked processes, the dissolution of calcite and the dolomitization of calcite. Once the rock is completely dolomitized no further porosity development is possible except that formed by fracturing.

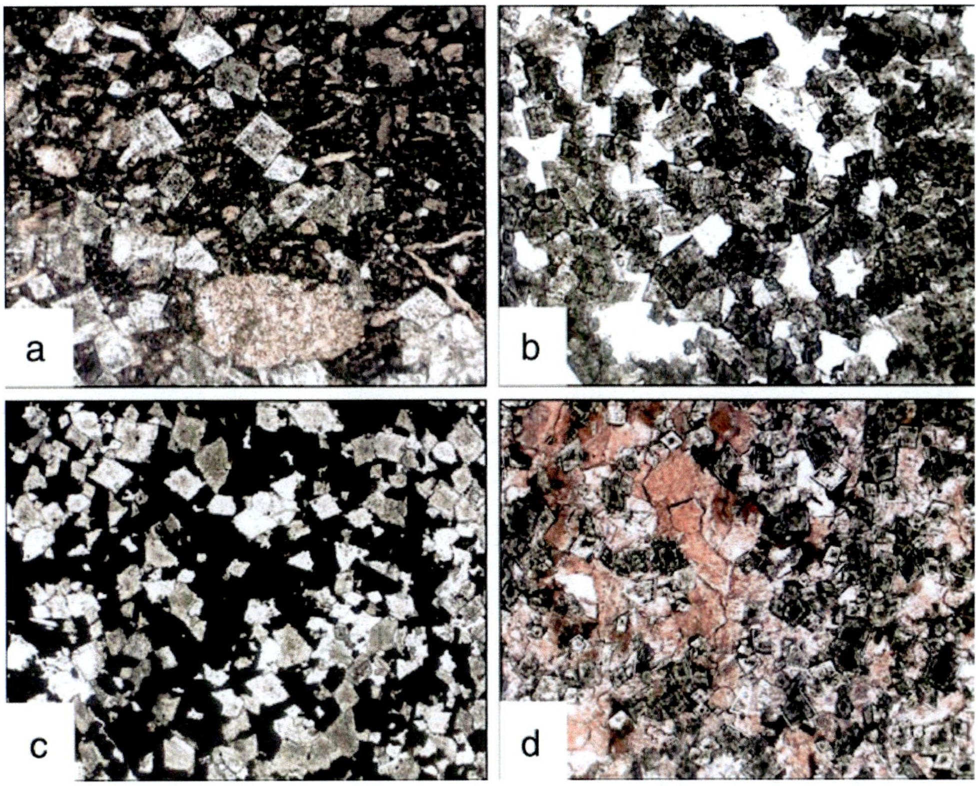

Fig. 7.12 Photomicrographs of *Thalassinoides* lithofacies strata, width of photomicrographs = 3.0 mm. **a** Partially dolomitized limestone. **b** Dolostone with abundant intercrystalline porosity. **c** Dolostone with intercrystalline porosity completely reduced by marcasite. **d** Dolostone with intercrystalline porosity completely reduced by blocky calcite

The *Thalassinoides* lithofacies presents a special case in that the dolomitized burrows are very permeable relative to the encasing limestone. As in the case micritic limestone, at the fracture face the limestone portion dissolved widening the fracture but the dolomite of the burrows was unchanged. The aggressive solution also penetrated the burrow resulting in dissolution of calcite adjacent to the burrow and dolomitization including syntaxial overgrowths on the dolomite crystals at the edge of a burrow. Microscopically the resulting altered rock is like that described above; but macroscopically it looks different. Starting at the fracture face and moving into the rock: (1) there is a zone characterized by the burrows surrounded by open pore space (Fig. 7.13); (2) this progresses to where the pores are smaller, the pore space is less, and the dolomitized burrows are enchased in dolomitized limestone; (3) a rock that is a dolostone consisting of minor porosity, dolomitized limestone, and the dolomitic *Thalassinoides* burrows; (4) to eventual unaltered *Thalassinoides* lithofacies limestone. Additionally, strata overlying or underlying the *Thalassinoides* lithofacies are altered by solution that first passed through the *Thalassinoides* lithofacies and then spread upwards and downwards. There is a good accessible example of this situation in a road cut along the Northwest Arterial in Dubuque (42.539°N, 90.695°W). There a fracture cuts the Prosser where it has not been entirely dolomitized; 3 meters from the fracture the rock is limestone (*Thalassinoides* lithofacies) with minor disseminated dolomite rhombohedra.

The important aspect here is the aggressiveness of the circulating water to the dissolution of calcite relative to the speed which can cause the dolomitization of the calcite. Conceivably the water could dissolve the calcite with little if any dolomitization—which would create a void space with just the dolomitized burrows present. But structurally they are not strong enough to be free standing, so they would collapse forming a rubble at the bottom of the void. Even

Fig. 7.13 Weathered east-west vertical extension fracture face illustrating the typical character of the *Thalassinoides* lithofacies. The raised areas are the dolomite burrow fills. The cavities are where calcite was dissolved during mineralization. The entire rock is dolostone. The exposure is of the upper portion of Prosser Fm. located at the mouth of Catfish Creek immediately below the Julien Dubuque Monument (42.468°N, 90.645°W). Rock hammer is 0.8 m long

with some dolomitization occurring around the burrows there could still be some settling forming a void above a breccia like deposit. Some of these voids are large enough to be considered caves.

The same aggressive water entering the St Peter Sandstone dissolved or dolomitized any calcite within further enhancing the porosity and permeability. Water within the St. Peter also interacted with the limestone beds immediately above the St Peter causing their dissolution and/or dolomitization. The action was like the action of the solution upon a fracture face except here it was acting upon a bedding surface. If there was enough volume loss on the bedding surface, the local stress field of the overlying strata would change such that the vertical oriented stress became the weaker or minor stress. That provided the impetus for bedding plane parallel tension fractures (the flats of the miners) to form. Once formed these new fractures became fluid pathways resulting in additional calcite dissolution and dolomitization around additional bedding surfaces. If dissolution in this situation was significant enough, then settling of the overlying strata may have occurred with the additional development of vertical to inclined tension fractures, the pitches, and thus created even more new pathways for fluid movement.

The circulating water, fresh from the basement, would have been hot; fluid inclusion studies indicate 80–220 °C (McLimans 1977; McLimans et al. 1980; Rowan and Goldhaber 1996). Furthermore, water circulating from depths of 5 to 18 km would have been heated to temperatures of 80 to 250 °C, respectively; using a thermal gradient of:

$$\text{Temp}(^\circ\text{C}) = 0.0138\text{X} + 26.2 \quad (7.1)$$

where X is the depth in meters below the surface. This equation is based upon an analysis of bottom hole temperatures from 600 wells drilled throughout Illinois. Heat from that water is transferred to the sedimentary strata, in particular, it heats the organic rich beds of the Decorah, Dubuque, and the lowermost beds of the Maquoketa resulting in the thermal alteration of organic material, specifically type IIs kerogen, and the release of sulfur to the circulating water. If sufficient sulfur and metallic ions were available then the pore spaces or fractures would have been filled at the same time by the open space precipitation of galena, marcasite, pyrite, and sphalerite (Fig. 7.12c); if not then they remained as open or void spaces (Fig. 7.12b) which could later be filled by minerals. Where the rock adjacent to a fracture was of low permeability the resultant sulfide body would have been a thin vertical vein, the sheet ore of the miners. Where the rock adjacent to the fractures was of the *Thalassinoides* lithofacies and there was a significant amount of secondary solution porosity developed and those pores mostly filled with sulfides, this would have been the *brangle ore* of the miners (Wheeler 1908) or a *honeycomb run* (Bain 1906, p. 61) or *solution breccia* (Heyl et al. 1959, p. 133); the terms are synonymous.

7.3.3.2 The Second Fracturing Episode

A second, later episode of deformation resulted in the formation of north-south vertical extension fractures, NNE and NNW vertical shear fractures. These orientations imply that during this second episode the principal stress was oriented north-south, the intermediate was oriented vertically, and the minor oriented east-west. The north-south extension fractures are generally less than 200 m in length and frequently terminating against a fracture from the first episode. They are not evenly distributed throughout the region and do not have the even spacing that the east-west fractures of the first episode. The shear fractures occur throughout the region, but their frequency also varies where in some areas only north-northeast fractures are present (Fig. 7.11b), other areas just the north-northwest, while also both can occur together (Fig. 7.11c). The shear fractures clearly cross-cut fractures of the first episode but they rarely show any significant displacement. Some of the inclined tension fractures from the earlier episode may have been reactivated by this second episode to form reverse faults, especially those which had an EW strike.

With formation of these fractures, additional fluid circulation occurred between the basement Precambrian felsic rock and the Paleozoic strata, producing further diagenesis of the strata and sulfide mineral precipitation. If the rock had been dolomitized during the prior episode of fracturing, then the only new open space formed would be that of the fractures otherwise everything else would have been the same as in the first episode. Sulfide precipitation, as earlier, occurred in any open space, including precipitation upon pre-existing sulfide deposits of the earlier episode and precipitation between the fracture planes of all fractures. This second episode of fracturing may have been more significant in terms of sulfide mineralization. In Level Crevice Cave noted earlier, the major galena deposits occurred where second episode shear fractures crossed first episode east-west expansion fracture.

At the culmination of the second episode of deformation, most of the fractures may have had sulfide mineral deposited to some extent along their lengths. These deposits range from thin sheets of mainly marcasite or pyrite on fracture faces to meters wide zones of extensive secondary porosity completely filled by combinations of galena, marcasite, pyrite, and sphalerite.

Furthermore, all the Dubuque, Stewartville, and most of the Prosser within an extensive region was completely dolomitized (Figs. 7.1a, 7.7). At the western boundary of that region the transition from dolostone to limestone occurs over a short distance (Leonard 1905). In comparing the dolomitized region to the limestone region to the west:

(1) there is no significant change in the character of the Galena Group except for the dolomitization; (2) there is no significant change in the basal beds of the Maquoketa except the carbonate beds therein display a dolostone to limestone transition like that of the Galena Group; and (3) the same fracture sets are present and cut across the transitions area into the limestone area. The only significant difference between the limestone area and the dolostone area is the apparent lithology of the underlying basement rock as indicated by a very small number of wells and analysis aeromagnetic and gravity surveys (Anderson 2006; Sims et al. 2008), felsic under the dolostone area and mafic under the limestone area. The mafic rock contained the more calcic plagioclase; apparently when altered by circulating fluid the alteration of the plagioclase provided sufficient calcium for the other alteration products, hence the Precambrian basement there did not provide a sink for the calcium liberated by dissolution and dolomitization of the Paleozoic strata and the circulation fluid remained at equilibrium with respect to calcite.

7.3.3.3 Calcite (Iceland Spar) Precipitation

Calcite was deposited as Iceland spar in some of the remaining pore spaces well after the second episode of fracturing. In a few places, calcite was the only mineral deposited in the pores and such that it preserved for observation the character of brangle (Figs. 7.12d, 7.14). Calcite brangle is well exposed in a road cut along U.S. 151 on the south side of Dubuque. (42.484°N, 90.663°W). At that locality there is a 4 m wide brangle zone developed in the upper strata of the Prosser. At the top of the brangle there is a small cave lined with 2–5 cm long scalenohedrons of calcite. The cave results from the apparent settling of the brangle due to minor solution collapse prior to calcite precipitation.

There is a similar locality in the Mines of Spain Recreation Area (42.463°N, 90.636°W) where quarry operations exposed a brangle zone develop along an east-west expansion fracture near the top of the Prosser; now exposed on the quarry floor. During an attempt to expand the quarry a small, 9 m long, 1–2 m wide, 1 m tall, calcite lined cave was encountered; Mar Jo Quarry Cave. The calcite was Iceland spar, scalenohedrons of calcite covered the entire interior of the cave. At the same site, within the quarry there is a flat, a calcite mineralized bedding plane tension fracture with small 1–2 m wide solution pockets scattered along its exposed length. North of Dubuque near the town of Sageville, Iowa, a calcite brangle is exposed in a road cut along Sherrill Road (42.569°N, 90.723°W). This is the best locality to view a calcite brangle. It is developed in the upper Prosser. Interestingly, it is cut by several fractures that postdate the calcite.

Fig. 7.14 Calcite brangle exposed in a road cut on Sherrill Road, northwest of Sageville, Iowa (42.569°N, 90.723°W)

7.3.4 Epigene Processes

Sometime after calcite precipitation, probably during the Cenozoic, the UMVLZD was uplifted and the subsequent erosion of the strata overlying the Galena Group eventually resulted in the erosion of the Maquoketa Fm. over a large portion of the UMVLZD (Fig. 7.7). Once that barrier is removed the Galena Group was no longer in a chemically closed system. Oxygenated groundwater could enter the system and react with the sulfide minerals, especially marcasite and pyrite, and cause the dissolution of dolomite and calcite. Morehouse (1968) investigated this process in Level Crevice Cave where he noted the presence of sulfide oxidizing bacteria. These were later identified by Peck (1986) as *Gallionella ferrunginea* and *Leptothrix* sp. Morehouse (1968) remarked that sulfide weathering can proceed with or without the aid of bacteria with the outcome being the same in terms of resultant chemical species.

General marcasite or pyrite oxidation reaction:

$$2FeS_2 + 2H_2O + 7O_2 \rightarrow 2Fe^{2+} + 4SO_4^{2-} + 4H^+ \quad (7.2)$$

The Fe^{2+} can be further oxidized and precipitated as goethite:

$$4Fe^{2+} + O_2 + 4H^+ + 10H_2O \rightarrow 4Fe^{3+} + 12H_2O \quad (7.3)$$

$$4Fe^{3+} + 12H_2O \rightarrow 4Fe(OH)_3 + 12H^+ \text{[goethite precipitation]} \quad (7.4)$$

The overall outcome is that for every mole of pyrite or marcasite oxidized two moles of SO_4^{2-}, one mole of goethite, and four moles of H^+ are produced. Sphalerite can be oxidized either by dissolved oxygen or Fe^{3+} (Heidel et al. 2011 and citations therein).

$$ZnS + 2O_2 \rightarrow Zn^{2+} + SO_4^{2-} \quad (7.5)$$

$$ZnS + 8Fe^{3+} + 4H_2O \rightarrow Zn^{2+} + 8Fe^{2+} + SO_4^{2-} + 8H^+ \quad (7.6)$$

The overall outcome is that for every mole of sphalerite oxidized one mole each of Zn^{2+} and $SO^{2-}{}_4$ are produced.

There is no net change in H^+ regardless of the oxidant since every mole of Fe^{3+} required, one mole of H^+ was consumed in the oxidation of Fe^{2+} (Eq. 3). Galena can be oxidized in a similar fashion, but the outcome will be anglesite ($PbSO_4$) which will coat the surface of the galena and impede further oxidation. The H^+ produced by marcasite or pyrite oxidation may react with calcite or dolomite and cause their dissolution:

$$CaCO_3 + H^+ \rightarrow Ca^{2+} + HCO_3^- \text{[calcite dissolution]} \quad (7.7)$$

$$CaMg(CO_3)_2 + 2H^+ \rightarrow 2HCO_3^- + Ca^{2+} + Mg^{+2} \text{ [dolomite dissolution]} \quad (7.8)$$

If all the H^+ produced by pyrite or marcasite oxidation reacted with calcite, then one mole of pyrite or marcasite could cause the dissolution of 4 mol of calcite or similarly 2 mol of dolomite. Given the following molar volumes: marcasite (24.58 cm^3/mole), pyrite (23.94 cm^3/mole), sphalerite (23.83 cm^3/mole), calcite (36.934 cm^3/mole), dolomite (64.340 cm^3/mole), goethite (20.7 to 29.2 cm^3/mole) (Robie et al. 1979). One cubic meter of marcasite, if totally oxidized, will result in the formation of 0.8 to 1.2 m^3 goethite and can cause the dissolution of up to 6.0 m^3 of calcite or 5.2 m^3 of dolomite. One cubic meter of brangle consisting of 50% dolomite, 25% marcasite, and 25% sphalerite, when totally oxidized, will result in dissolution of 1.3 m^3 of dolomite and the formation of about 0.2 m^3 of goethite, or combined would result in a cave with an open space volume of 1.6 m^3 and a total volume (open space + sediment) of 1.8 m^3. The sediment in a typical crevice cave is primarily a very porous mixture of dolomite fragments and goethite where the dolomite results from incomplete dissolution of portions of the brangle and surrounding wall rock. The result is that the open space is much less than the ideal where the sediment was only goethite. Furthermore, some of the Zn^{2+} can combine with bicarbonate to form smithsonite further reducing open space volume. Certain other ephemeral secondary minerals may form including: epsomite ($MgSO_4{\cdot}7H_2O$), goslarite ($ZnSO_4{\cdot}7H_2O$), gypsum ($CaSO_4{\cdot}2H_2O$), and melanterite ($FeSO_4{\cdot}7H_2O$).

If an opening was initially filled with a marcasite-sphalerite brangle then after complete oxidation of the sulfides the outcome would be a pod-shaped cave floored by an unstratified sediment consisting of goethite, smithsonite, and loose dolomite crystals and dolostone fragments. Fractures that are without brangle deposits but containing a gash vein will have their sulfides oxidized with the result that the fracture is solution widened through the loss of sulfides and the dissolution of the wall rock. The H^+ charged solution will also penetrate the wall rock at stratigraphic levels where there is enough porosity and permeability, such as the *Thalassinoides* lithofacies, and attack the dolomite there. The collective result is the formation of a pod-shaped mass of goethite and dolomite fragments surrounded by a greatly weakened wall rock with substantial intercrystalline pore space at the intersection of the gash vein and the *Thalassinoides* lithofacies. If the gash vein was thick enough a natural cave could have formed. Cave systems like that of Level-Booth and Carter (Fig. 7.6) or Crystal Lake Cave (Fig. 7.15) were initially east-west oriented pod-shaped bodies of brangle that were cross-cut by numerous gash veins that formed along the shear fractures. Oxidation of the sulfides produced open spaces where there was brangle or sufficient width to the gash vein. These open spaces were floored by a sediment consisting of goethite, galena, smithsonite, and weathered dolostone debris. These were connected by the weakened and goethite impregnated dolomite along the gash veins.

Once the sulfide minerals marcasite, pyrite, and sphalerite are oxidized within the crevice cave and immediately surrounding rock the chemistry of the system shifts to a more typical phreatic/vadose situation chemically characterized by the action of carbonic acid charged meteoric water which would cause additional solution of dolomite. Many of the natural cave spaces prior to mining held carbonate speleothem deposits that were no different than one would expect in a typical cave formed in limestone or dolostone. The miners considered speleothems of no real or aesthetic

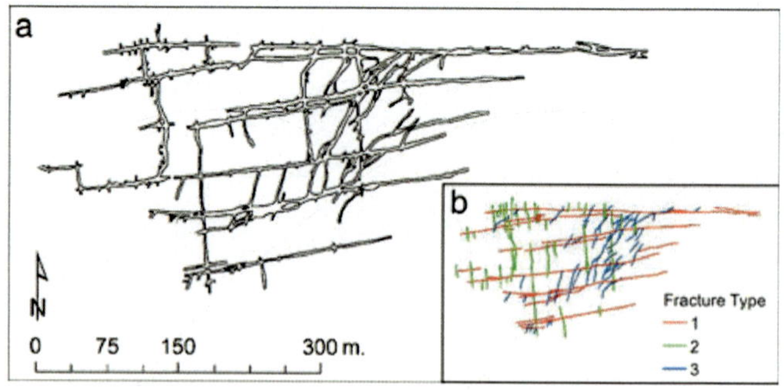

Fig. 7.15 Crystal Lake Cave or Rice's cave (42.434°N, 90.620°W). **a** Map of surveyed drifts redrawn from Iowa Grotto, Iowa Cave Book 1958 which was based upon the 1936 survey of Luckhardt, Bertholf, Jones, and Monk and the 1956 survey of Vevitt, Lillhuse, and Knockel. **b** Map of the interpreted fracture type controlling the various drifts and passages. (1) East-West vertical extension fracture (2) North-South vertical extension fracture (3) vertical to inclined subsidence tension fracture

value and removed them to the spoils piles. Rice's cave, now known as Crystal Lake Cave, is one of the few former mines where speleothems remain.

7.4 Summary and Conclusions

There are many contributing factors to the formation of the crevice caves. (1) The basement rock that underlies the region where they occur is felsic, not mafic. This rock supplied the Fe, Pb, and Zn for the sulfide mineralization and provided the source of the Mg and sink for the Ca of the dolomitization. Mafic rock, though it potentially could provide Fe, Pb, and Zn for the sulfide minerals and Mg for dolomitization, provided a poor sink for Ca due to the presence of the more calcic variety of plagioclase. (2) Far-field tectonic stress resulted in the formation of an extensive fracture system that provided pathways for fluid circulation between the basement rock and overlying Paleozoic strata and provided much of the energy to drive that circulation. The two episodes of deformation, each with very different stress orientation, resulted in an interconnected grid of fractures. (3) The dolomitized *Thalassinoides* burrows that occur in various layers of the Prosser and Stewartville formations provided a pathway for the circulating fluid to penetrate the strata thus exposing portions of the strata further from a fracture face to chemical interaction with the fluid. Strata without the *Thalassinoides* burrows or strata with burrows that were not dolomitized were only exposed to chemical interaction at the fracture faces. (4) The timing and relative rates of the subsequent chemical reactions, the dolomitization of calcite and the dissolution of calcite, dictated the volume and character of the pore spaces formed for the subsequent but coeval deposition sites for the sulfide minerals. (5) The thermal energy transferred from the basement rock to the type IIs kerogen found in the Decorah, Dubuque, and lower portion of the Maquoketa formations provided the sulfur for the sulfide minerals. Variation in the quantity of sulfide minerals present across the UMVLZD probably reflects variation in the degree of thermal heating of the kerogen and/or variation in the quantity of type IIs kerogen present. (6) The physical formation of a crevice cave relies upon the oxidation of marcasite and pyrite which occurs only when the strata overlying the Galena Groups are removed by erosion. The oxidation of those sulfide minerals produces the H^+ ions which dissolve the dolostone. The volume of dolostone dissolved depends upon the available quantity of marcasite and pyrite. The location of the dolostone dissolution depends upon the ability of the solutions evolving from the oxidation reactions to contact the dolomite; proximity and permeability being the key factors. The product of these factors is a cave consisting of tubular shaped passages that formed at a specific stratigraphic horizon that follows grid defined by the various local sets of fractures.

Acknowledgements This paper is dedicated to David F. Morehouse (1943–2014) who introduced me to the geology of the UMVLZD. The structural analysis aspect of this study relied heavily upon the online resources of the Illinois State Geological Survey, Iowa Geological Survey, and the Wisconsin Geological & Natural History Survey, especially the availability of the various well logs, historical aero photography, and Lidar data. The clarity and content of the paper was greatly improved through the comments of Calvin Alexander, Greg Brick, and James E. Liles. I am indebted to reference librarian Katherine Burright for her efforts in locating obscure publications.

References

Agnew AF (1963) Geology of the Platteville Quadrangle, Wisconsin. USGS Bulletin 1123-E, pp 245- 277

Agnew AF, Flint AE, Allingham JW (1953) Exploratory drilling program of the USGS for evidences of zinc-lead mineralization in Iowa and Wisconsin, 1950–51. USGS Circular 231, 37 p

Agnew AF, Heyl AV, Behre CH, Lyons EJ (1956) Stratigraphy of Middle Ordovician rocks in the Zinc-Lead District of Wisconsin, Illinois, and Iowa. USGS Professional Paper 274-K, p 251–312

Allingham JW (1963) Geology of the Dodgeville and Mineral Point quadrangles, Wisconsin. USGS Bulletin 1123-D, p 169–244

Anderson KE (1950) Basement complex biotite granite at Dubuque, Iowa. Proceedings of the Iowa Academy of Sciences 57:241–244

Anderson RR (2006) Geology of the Precambrian surface of Iowa and surrounding area. Iowa Geological Survey Open File Map, OFM 06–7

Bain HF (1905) Zinc and lead deposits of northwestern Illinois. U.S Geological Bulletin 246, 56 p

Bain HF (1906) Zinc and lead deposits of the Upper Mississippi Valley. USGS Bulletin 294, 155 p

Baldwin JD (1871) Ancient America, Notes on American Archaeology, Harper & Brothers, New York, 299 p

Bradbury JC (1959) Crevice lead-zinc deposits of northwestern Illinois. Illinois State Geological Survey Report of Investigations 210, 49 p

Brannon JC, Podosek FA, McLimans RK (1992) Alleghenian age for the Upper Mississippi Valley zinc-lead deposit determined by Rb-Sr dating of sphalerite. Nature 356:509–511

Brown CE, Whitlow JW (1960) Geology of the Dubuque South Quadrangle Iowa-Illinois: USGS Bulletin 1123-A, p 1–93

Bunker BJ, Ludvigson GA, Witzke BJ (1985) The Plum River Fault Zone and the structural and stratigraphic framework of eastern Iowa. Iowa Geological Survey Technical Information Series No. 13, 125 p

Calvin S, Bain HF (1900) Geology of Dubuque County. Iowa Geological Survey Annual Report 10:379–622

Carlson JE (1961) Geology of the Montfort and Linden quadrangles, Wisconsin. USGS Bulletin 1123-B, p 95–138

Cox GH (1914) Lead and zinc deposits of northwestern Illinois. Illinois State Geological Survey Bulletin No. 21, 120 p

Daniels E (1854) First Annual Report of the Geologic Survey of the State of Wisconsin. 84 p

Daniels JJ, Olhoeft GR, Scott JH (1983) Interpretation of core and well log physical property data from drill hole UPH-3, Stephenson County, Illinois. J Geophys Res 88:7346–7354

Dockal JA (1973) Bedrock geochemical sampling near lead-zinc mineralizations, Dubuque, Iowa. Master's thesis, Iowa State University, Ames, Iowa, 178 p

Dockal JA (1988) Thermodynamic and kinetic description of dolomitization of calcite and calcitization of dolomite (dedolomitization). Carbonates Evaporites 3:125–141

Dockal JA (2011) History of Lead (Galena) mining at Dubuque, Iowa, with reference to the mining features in the Mines of Spain Recreation Area. in: Dubuque: Living with the Rocks and the River, Association of American State Geologists 2011 Annual Meeting-Dubuque, Iowa, Tuesday, June 14, Field Trip, p 8–32

Doe BR, Stuckless JS, Delevaux MH (1983) The possible bearing of the granite of the UPH deep drill holes, Northern Illinois, on the origin of Mississippi Valley ore deposits. J Geophys Res 88:7335–7345

Dunham RJ (1962) Classification of carbonate rocks according to depositional texture. in: Ham WE (ed), Classification of Carbonate Rocks. American Association of Petroleum Geologists Memoir 1, p 108–121

Eckstein Y, Dahl PS, Vitaliano CJ (1983) Petrographic and Physical Factors Controlling Thermal Conductivity of Granitic Rocks in Illinois Deep Holes UPH-1, -2, and -3. J Geophys Res 88:7381–7385

Farquhar RM, Fletcher IR (1980) Lead Isotope Identification of Sources of Galena from Some Prehistoric Indian Sites in Ontario. Canada. Science 207(4431):640–643

Farquhar RM, Fletcher IR (1984) The Provenance of Galena from Archaic/Woodland Sites in Northeastern North America: Lead Isotope Evidence. Am Antiq 49(4):774–785

Ghosh S (2008) Heavy Stable Isotope Investigations in Environmental Science and Archeology: Doctoral dissertation. Florida State University, Tallahassee, p 132

Grant US (1906) Report on the lead and zinc deposits of Wisconsin, with an atlas of detailed maps. Wisconsin Geological Survey Bulletin 14, 100 p

Haimson BC, Doe TW (1983) State of stress, permeability, and fractures in the Precambrian granite of northern Illinois. J Geophys Res 88:7355–7371

Heidel C, Tichomirowa M, Breitkopf C (2011) Sphalerite oxidation pathways detected by oxygen and sulfur isotope studies. Appl Geochem 26:2247–2259

Heyl AV, Agnew AF, Lyons EJ, Behre CH (1959) The geology of the Upper Mississippi Valley zinc-lead district. U. S. Geological Survey Professional Paper 309, 310 p

Heyl AV, Broughton WA, West WS (1970) Geology of the Upper Mississippi Valley Base-metal District. University of Wisconsin-Extension Geological and Natural History Survey Information Circular 16, 45 p

Holmes WH (1919) Mica Mines. Handbook of Aboriginal American Antiquities, Part I, Bureau of American Ethnology Bulletin 60:242–252

Jirsa MA, Miller JD, Severson MJ (2006) Geochemical data from Phase 2 of the study geology, geochemistry, and PGE potential of mafic-ultramafic intrusions in Minnesota excluding the Duluth Complex. Minnesota Geological Survey Open- File Report OF-06–1

Kendall AC (1977) Origin of dolomite mottling in Ordovician Limestones from Saskatchewan and Manitoba. Bull Can Pet Geol 25:480–504

Kerr WC (1875) Report of the Geological Survey of North Carolina 1. Josiah Nichols, Printer, Raleigh, North Carolina, p 325

Keyes C (1912) Spanish Mines: an episode in primitive American lead-mining. Annals of Iowa 10:539–546

Klemic H, West WS (1964) Geology of the Belmont and Calamine Quadrangles, Wisconsin. USGS Bulletin 1123-G, p 361–435

Kolata DR, Buschbach, TC (1976) Plum River Fault Zone of Northwestern Illinois. Illinois State Geological Survey Circular 491, 20 p

Leonard AG (1894) Satin Spar from Dubuque. Iowa Academy of Science Proceedings 1(4), 52–55 Leonard AG (1896) Lead and zinc deposits of Iowa. Iowa Geological Survey, Annual Report 6:11–66 Leonard AG (1905) Geology of Clayton County. Iowa Geological Survey Annual Report 16:217–309

Lidiak EG, Denison RE (1983) Petrology and physical properties of granites from the Illinois deep hole in Stephenson County. J Geophys Res 88:7287–7299

Ludvigson GA, Bunker BJ, Witzke BJ, Bounk MJ (1978) A field guide to the Plum River Fault Zone in east-central Iowa. 42nd Annual Tri-state Geological Field Conference Guidebook, Trip 1

Margolin PR (2000) The Sink Hole at Bandana: an historic Blue Ridge mica mine reveals its past. North Carolina Archaeology 49:43–58

Martin SR (1995) The State of Our Knowledge About Ancient Copper Mining in Michigan. Michigan Archaeologist 41:119–138

McLimans RK (1977) Geological, fluid inclusion and stable isotope studies of the Upper Mississippi Valley zinc-lead district, southwest Wisconsin. Doctoral dissertation, Pennsylvania State University, College Station, Pennsylvania, 175 p

McLimans RK, Barnes HL, Ohmoto H (1980) Sphalerite stratigraphy of the Upper Mississippi Valley Zinc-Lead District. Southwest Wisconsin, Econ Geol 74(3):351–361

Meeker M (1872) Early history of the Lead Region of Wisconsin. Wisconsin Historical Collections VI (Madison, 1872): 271–296. Online facsimile at: http://www.wisconsinhistory.org/turningpoints/search.asp?id=895; Visited on: 10/12/2017

Morehouse DF (1968) Cave development via the sulfuric acid reaction. National Speleological Society Bulletin 30(1):1–10

Morrow DW (1978) Dolomitization of lower Paleozoic burrow-fillings. J Sediment Petrol 48:295–306

Mullens TE (1964) Geology of the Cuba City, New Diggings, and Shullsburg quadrangles, Wisconsin and Illinois. USGS Bulletin 1123-H, p 437–531

Oliver NHS, McLellan JG, Hobbs BE, Cleverley JS, Ord A, Feltrin L (2006) Numerical models of extensional deformation, heat transfer, and fluid flow across basement-cover interfaces during basin-related mineralization. Econ Geol 101:1–31

Ostrom ME (1987) Middle Ordovician rocks at Potosi Hill, Wisconsin. Geological Society of America Centennial Field Guide-North-Central Section, p 201–204

Owen DD (1840) Report of a geological exploration on part of Iowa, Wisconsin, and Illinois, 1839. Congressional Documents, 26th Congress, 1st Session, H.R. Executive Document No. 239

Palmer TJ (1978) Burrows at certain omission surfaces in the Middle Ordovician of the Upper Mississippi Valley. J Paleontol 52:109–117

Pannalal SJ, Symons DTA, Sangster DF (2004) Paleomagnetic dating of Upper Mississippi Valley zinc-lead mineralization, WI, USA. J Appl Geophys 56:135–153

Panno SV, Luman DE, Kolata DR (2015) Characterization of karst terrain and regional tectonics using remotely sensed data in Jo Daviess County, Illinois. Illinois State Geological Survey Circular 598, 29 p

Panno SV, Luman DE, Kelly WR, Larson TH, Taylor SJ (2017) Karst of the Driftless Area of Jo Daviess County, Illinois. Illinois State Geological Survey Circular 586, 39 p

Peck SR (1986) Bacterial deposition of iron and manganese oxides in North American caves. Bull Natl Speleol Soc 48:26–30

Percival JG (1855) Annual Report of the Geological Survey of the State of Wisconsin for 1854. 101 p Percival JG (1856) Annual Report of the Geological Survey of the State of Wisconsin for 1855. 111 p

Robie RA, Hemingway BS, Fisher JR (1979) Thermodynamic Properties of Minerals and Related Substances at 298.15 K and 1 Bar (105 Pascals) Pressure and at Higher Temperatures. USGS Bulletin 1452, 456 p

Rowan EL, Goldhaber MB (1996) Fluid inclusions and biomarkers in the Upper Mississippi Valley zinc-lead District- Implications for the fluid-flow and thermal history of the Illinois Basin. USGS Bulletin 2094-F, 34 p

Sangster DF, Nowlan GS, McCracken AD (1994) Thermal comparisons of Mississippi Valley-type lead-zinc deposits and their host rocks using fluid inclusions and conodont color alteration index data. Econ Geol 89:493–514

Schoolcraft HR (1821) Narrative journal of travels through the northwestern regions of the United States, Albany, 419 p

Schoolcraft HR (1855) Summary narrative of an exploratory expedition to the sources of the Mississippi River in 1820 resumed and completed by the discovery of its origin in Itasca Lake, in 1832. Grambo, and Co., Philadelphia, Lippincott, p 596

Shiras OP (1902) The Mines of Spain. Annals of Iowa 5(5) 3rd Series, p 321–334

Sims PK, Saltus RW, Anderson ED (2008) Precambrian basement structure map of the continental United States-An interpretation of geologic and aeromagnetic data. USGS Scientific Investigations Map 3012

Spirakis CS (1995) Problems with applying topographically driven flow to genesis of the Upper Mississippi Valley zinc- lead district and to fluid flow in the Illinois Basin. USGS Bulletin 2094-C, 12 p

Spirakis CS, Heyl AV (1996) Fracturing radioactive basement rocks as the energy source to drive mineralization in the Upper Mississippi Valley zinc-lead district. Society of Economic Geologists Special Publication No. 4:390–399

Strong M (1877) Geology and topography of the lead region. Geology of Wisconsin 2:643–752

Taylor AR (1964) Geology of the Rewey and Mifflin quadrangles, Wisconsin. USGS Bulletin 1123-F, p 279–360

Thwaites RG (1895) Notes on early lead mining in the Fever (or Galena) River Region. Collections of the state historical society of Wisconsin 13 (Madison, 1895), p 271–292. Online facsimile at: http://www.wisconsinhistory.org/turningpoints/search.asp?id=896. Visited on: 10/22/2017

Vitaliano CJ, McCormick GR, Dahl PS, Eckstein Y (1986) Petrochemistry of the Granitic Rocks from Illinois Deep Drill Hole UPH-3. J Geol 94:594–599

Walthall JA (1981) Galena and Aboriginal Trade in Eastern North America. Illinois State Museum Scientific Papers 17, 66 p

Walthall JA, Stow SH, Karson MJ (1980) Copena Galena: Source Identification and Analysis. Am Antiq 45:21–42

West WS, Blacet PM (1971) Geologic Map of the Lancaster Quadrangle, Grant County, Wisconsin. USGS Map GQ-949

West WS, Heyl AV (1985) Geologic Map of the Hurricane Quadrangle Grant County, Wisconsin. USGS Map GQ-1582

West WS, Whitlow JW, Brown CE, Heyl AV (1971) Geologic Map of the Ellenboro Quadrangle Grant County, Wisconsin. USGS Map GQ-959

Wheeler WH (1908) Brangle ore. Engineering and Mining Journal, January 27, p 1303–1304

Whitlow JW, Brown CE (1963) Geology of the Dubuque North Quadrangle, Iowa-Wisconsin-Illinois. USGS Bulletin 1123-C, p 139–168

Whitlow JW, West WS (1966a) Geology of the Potosi Quadrangle, Grant County Wisconsin, and Dubuque County, Iowa. USGD Bulletin 1123-I, p 533–571

Whitlow JW, West WS (1966b) Geologic Map of the Dickeyville Quadrangle Grant County, Wisconsin. USGS Map GQ-488

Whitlow JW, West WS (1966c) Geologic Map of the Kieler Quadrangle Grant County, Wisconsin and Jo Daviess County, Illinois. USGS Map GQ-487

Whitney JD (1858) Chemistry and economical geology (of Iowa). in Hall J, Whitney JD, Report on the Geological Survey of the State of Iowa, Iowa Geological Survey 1. Pt. 1:324–472

Whitney JD (1862) Report of the lead region of Wisconsin. Geological Survey of Wisconsin 1, p 73-420 Whitney JD (1866) Geology of the lead region. Geological Survey of Illinois 1:153–207

Wilkie WE (1987) Dubuque on the Mississippi. Loras College Press, 505 p

Willman HB, Kolata DR (1978) The Platteville and Galena Groups in northern Illinois. Illinois State Geological Survey Circular 502, 75 p

Willman HB, Reynolds RR, Herbert P (1946) Geological aspects of prospecting and areas for prospecting in the zinc-lead district of northwestern Illinois. Illinois Geological Survey Report of Investigations 116, 48 p

Wood WF (2001) An Atlas of Early Maps of the American Midwest: Part II. Illinois State Museum Scientific Papers, v. XXIX, Springfield Illinois, 26 p

Yaghubpur A (1979) Preliminary geologic appraisal and economic aspects of the Precambrian basement of Iowa, Doctoral dissertation, Iowa University, Iowa City

Paradigmatic Studies in Midwestern Cave Science

8

Greg A. Brick

Abstract

Ten fields of cave science in the Midwestern USA are examined from the standpoint of the dominant paradigms in those fields: Karst, geology, hydrogeology, sedimentology, meteorology, biology, spring ecology, paleontology, archeology, and rock art. The paradigm chosen for examination in each field generally has roots in the nineteenth century, so as to give it some historical depth, but not so remote as to have little relevancy to ongoing, contemporary research. It was found that the dominant paradigms arose within the Midwest itself in some cases, diffusing to the outside, while in others it was imported from surrounding areas.

8.1 Introduction

When writing cave science, one may take the strictly chronological approach, as, for example, Trevor Shaw's *History of Cave Science to 1900*, covering the history of each category of cave feature (Shaw 1992) or one may employ a more historiographic approach, as will be done here.

This account of Midwestern cave science is by no means all inclusive. Indeed, literature reviews, which are often mistaken for historical introductions, seek to be exhaustive, whereas historical narratives seek to exclude much material so as to be able to tell a story. For literature reviews on each topic, consult the relevant chapters in this book.

The following brief essays address the top 10 paradigmatic aspects of cave science in so far as they manifest themselves in the Midwest. And while the notion of a paradigm in history of science studies is connected with Kuhn's *The Structure of Scientific Revolutions*, the paradigms discussed here are informal, not the "hard" paradigms of Kuhn, with their mutual incommensurability (Kuhn 1962).

As there are often several paradigms in each field, the one chosen for examination generally has roots in the nineteenth century, so as to give it some historical depth, but not so far back as to have little relevancy to ongoing, contemporary research.

8.2 The Karst Paradigm

The word *Karst* is of ancient derivation, referring to a stony place, associated with the classic Karst of Slovenia (Gams 1993). At least as early as Hacquet (1778), the solutional aspects of the formation of karst landforms were suggested. Although Lyell (1839) described how carbon dioxide solutions in water created caves in the Chalk of England, he did not use the word karst. The more generalized application of the word to signify a solutional landscape with underground drainage, characterized by caves and springs, grew with Cvijic's *Das Karstphanomen* (1893). On the other hand, his contemporary, the great French caver Edouard Martel, avoided the term karst in his writings, according to Hromnik (2001).

The classic Karst area of Europe is a complicated kind of karst, especially because of the superimposed tectonics, while the American Midwest, with its flat-lying, comparatively undisturbed limestones, might have afforded a better general model for some aspects of karst science, as emphasized by Ford (2006). But although caves and springs were known in America from the earliest times, there does not seem to have been a home-grown karst concept. So when did the karst concept come to be applied by name in this part of the world?

While Lapham (1867), Murrish (1871), and Strong (1877, 1882) give preliminary accounts of karst features in

G. A. Brick (✉)
Minnesota Department of Natural Resources, 500 Lafayette Road, St. Paul, Minnesota, 55155, USA
e-mail: drgregbrick@gmail.com

G. A. Brick and E. C. Alexander Jr. (eds.), *Caves and Karst of the Upper Midwest, USA*, Cave and Karst Systems of the World, https://doi.org/10.1007/978-3-030-54633-5_8

Wisconsin, Lange (1909), whose bachelor's thesis gives the first systematic account of the caves of the Driftless Area, uses the word Karst, but only as a European place name. He explains cave formation by carbonate dissolution along joints. Lawrence Martin (1932), in his classic *The Physical Geography of Wisconsin,* cites Lange (1909) as his cave authority, but although he describes the peneplains of Wisconsin in some detail, he does not integrate them into a theory of cave formation as Davis (1930) did before him. He attributes the paucity of Wisconsin caves outside the Driftless Area "to sculpture by ice during the Glacial Period" (Martin 1932).

Jillson (1924), in his article on "American Karst Lands," writes that "one of the most interesting physiographic regions in our country is that which has, within recent years, come to be designated as the 'Karst' region of Kentucky." From there the term spread out to the periphery. The term "karst topography" first appears in the Upper Mississippi Valley (UMV) region descriptive of the Silurian Niagaran dolomite outcrops in the vicinity of Kankakee, IL (Ekblaw 1925). Ekblaw noted three episodes of karstification, two in the Paleozoic and one in the Pleistocene. Ekblaw derives his terminology from an English version of Cvijic (Sanders 1921), commenting that "Many karst areas occur in the United States and other countries," showing that by this time the word had achieved the full generality we are familiar with today.

The pioneering Minnesota geologist Newton H. Winchell (1839–1914), who wrote so prolifically on that state, rarely mentioned caves, and did not even use that word in his report on cave-rich Fillmore County (Winchell 1884). However, he did recognize the existence of subterranean drainage (Brick 2018). It was not until Thiel (1944) that "a karst type of topography" is described for Minnesota. The term karst did not seem to stick however, not being used by Hogberg and Bayer (1967), until the 1970s came along. Herak and Stringfield (1972) included the Upper Midwest as one of the "important karst regions of the Northern Hemisphere" and Wopat (1974) applied the term "fluviokarst," derived from Sweeting (1973), to southeastern Minnesota. From then on, the word karst is used consistently in Minnesota.

In Iowa, Prior (1976) also spoke of "karst topography" but substituted the term "Paleozoic Plateau" for the Iowa portion of the Driftless Area. Hedges (1972) explained the formation of expanded joints on the Niagara Escarpment by means of dolomite blocks sliding on the underlying shale layer, aided by ice wedging. Hansel (1976) used this to make a distinction between the "mechanical karst" of the Niagara Escarpment, where expanded joints allow lines of sinkholes to form along ridges collecting limited surface drainage, from solutional karst proper.

The entire subject of karst in the UMV has been succinctly summarized by Hedges and Alexander (1985).

8.3 Geological Paradigms: Speleogenesis

Shaw (1992) has described the early history of theories of cave formation. Strangely, caves seemed to demand some explanation as a secondary landscape feature, how they got there, while other common landscape features were taken as given by the casual observer. By the time of UMV region settlement, cave formation had largely settled into a debate between physical and chemical theories.

Speleologically speaking, the UMV region has been peripheral to Mammoth Cave (Kentucky) on the east and to Carlsbad Caverns (New Mexico) on the west, both of from which concepts of cave science have diffused toward it. But in one regard, the natural sandstone caves of Minnesota were in the vanguard, and that is cave formation in pseudokarst (non-solutional forms that mimic true karst), even though the term pseudokarst was not used in print till 1906 (Halliday 2007).

Two natural sandstone caves, Carver's and Fountain, in what is now St. Paul, Minnesota, have served as landmarks for European explorers along the Mississippi River during the past quarter millennium and will elucidate this paradigm. Despite the ongoing confusion between these two caves among such explorers, it's fairly easy to tell them apart in the historical literature even when they have been misnamed: Carver's Cave is the short cave downstream from St. Paul containing a lake, whereas Fountain Cave is the long cave above town containing a stream (Woolworth et al. 1987). Because of this intertwined history, they share many of the same literature references and are discussed together here. The best reference for these caves is Brick (1995, 2004b, 2006, 2009b, 2014d). To prevent overwhelming of the text with heavy-duty scholarly apparatus, please note that the citations from the primary sources in this section are contained in Brick (2014a, b). See Figs. 2.1 and 2.2 in Chap. 2 of this volume for historic images of the caves.

While both caves formed in the Ordovician-age St. Peter Sandstone in postglacial times by the same mechanical process of "piping," involving the washing away of sand grains by groundwater, there were very different historical perceptions of what was going on, geologically speaking, at each cave. Why was one cave overloaded with speleogenetic interpretation, even by casual visitors, while the other was not?

Native Americans refer to **Carver's Cave**, in St. Paul, Minnesota, as *Wakan Tipi*, the Dwelling of the Great Spirit. Captain Jonathan Carver visited what he called the "Great Cave" in 1766 and again in 1767, and it became the earliest Minnesota cave in the published literature when the first edition of Carver's best-selling *Travels through the Interior Parts of North America* (1778).

While it can be shown from the comparison of cave surveys that Carver's Cave has changed little over long stretches of time, and thus approaches stasis, the overall impression you get from some historical accounts is exactly the opposite. For example, when Major Stephen H. Long, U. S. Corps of Topographical Engineers, who had ascended the Mississippi River in a "six-oared skiff" to find a location for what would become Fort Snelling, visited Carver's Cave on July 16, 1817, he had this to say:

> "Two miles [3.2 km] above the village [Kaposia] on the same side of the [Mississippi] river is Carvers Cave, at which we stopped to breakfast. However interesting it may have been, it does not possess that character in a very high degree at present. We descended it with lighted candles to its lower extremity. The entrance is very low and about 8 feet [2.4 m] broad, so that a man in order to enter it must be completely prostrate. The angle of descent within the cave is about 25°.... In shape it resembles a Bakers oven. The cavern was once probably much more extensive. My interpreter informed me that since his remembrance the entrance was not less than 10 feet [3 m] high, & its length far greater than at present."

Another account, by ethnologist Henry R. Schoolcraft, who would discover the "true head" of the Mississippi River in 1832, dates from 1820, not long after Long's first visit. We read in his *Narrative Journal of Travels* (1821) for August 2, 1820, that:

> "The cave itself, appears to have undergone a considerable alteration since [Carver's] period.... As the rock is of a very friable nature, and easily acted upon by running water, it is probable that the lake has been discharged, thus enlarging the boundaries of the cave."

These early perceptions of change at Carver's Cave were illusory, however, because both travelers had almost certainly confused the cave with other, nearby caves. In Long's case, it has been alleged that he confused Carver's with Dayton's Cave—a much smaller cave—while it's known with more certainty that Schoolcraft confused Carver's with Fountain Cave, a much larger cave. Either way, the cave appeared to the bewildered traveler to have undergone a veritable metamorphosis in the half century since Carver's original visit (Brick 2009b).

Even more bizarre, however, was the account of the eccentric Italian traveler Giacomo C. Beltrami. He confounded Carver's and Fountain caves in his 1828 memoirs, *A Pilgrimage in America*, creating the hybrid "Cave of Trophonius," referring to a famous Greek cave that contained an underground river where an oracle was consulted. The physical description of Beltrami's cave undeniably belongs to Fountain Cave, yet he attributed to its Native American ceremonies and the "hieroglyphics" associated with Carver's Cave, even using the name "*Whakoon-Thiiby*" (*Wakan Tipi*) which pertains to the latter cave (Brick 2014c).

But Carver's Cave was certainly undergoing some change, regardless of Long's and Schoolcraft's misperceptions. The French scientist Joseph N. Nicollet, after a distinguished mathematical career in France, immigrated to the United States in 1832, devoting the rest of his life to mapping the Upper Mississippi, determining for the first time accurate altitudes, latitudes, and longitudes. His great cartographic work is considered the "mother map" of Minnesota. In his *Report Intended to Illustrate a Map of the Hydrographical Basin of the Upper Mississippi River* (1843), he describes digging Carver's Cave open in 1837. Nicollet wrote:

> Its entrance has been, for more than thirty years, closed by the disintegrated debris of the limestone capping the sandstone in which it is located.... I saw enough to satisfy myself of the accuracy of Carver's description.

Here also we see a change of emphasis from internal changes at Carver's Cave to that of its entrance. Edward D. Neill, founder of Macalester College in St. Paul, reported in 1851 that "The cave has since then been materially altered by the tools of time, frost, air, and water. Many years ago, the roof of the cave fell in, thus exposing to the light the side walls." In the following year, Daniel S. Curtiss, in his *Western Portraiture and Emigrants' Guide* (1852), was more emphatic:

> "Among the most singular or attractive curiosities in Minnesota, beside the great Falls, are the caves, or subterranean lakes and creeks. Carver's Cave is one of some note; but it can rarely ever be explored, as the entrance to it is constantly changing and being obstructed by sliding rocks and earth, which frequently fill up the orifice, so that there is no access for several days, till the little stream issuing from it bursts out again, leaving a passage, sometimes, through which a man can enter and explore, though it is a hazardous experiment, not often attempted; yet, within the cave there is a beautiful crystal lake, with shining rock walls and inclosures."

By 1867, the centenary of Carver's purported treaty with Native Americans at the "Great Cave" (as Carver himself called it) there's another change of emphasis. The secretary of the Minnesota Historical Society, J. Fletcher Williams, who reported the festivities in 1872, said that "The cave remained unchanged in appearance for over a century." He goes on to say: "Within the past two years, however, sad changes have taken place. The St. Paul and Chicago Railroad, having condemned for their use the strip of land along the river bank, including the 'bluff,' or cliff in which is the cave, have dug it down and nearly destroyed it." At this point in history, the minimal geological change seems overwhelmed by human alterations.

Amid the perceptions, what's the reality? The best summary is given by the antiquarian Theodore H. Lewis in the *Macalester Monthly* in 1898, of which I quote only the length statistics: "About the year 1857 Dr. Edward D. Neill had a survey made of the cave...total length of the cave, 117 feet [35.7 m]…. My own measurements, made October 27, 1878, are as follows…total length inside, 113 feet [34.4 m]." This throws doubt on the claim that about 22 feet (6.7 m) of the cave entrance was removed by the railroads in 1869; unfortunately, what was carved away held most of the petroglyphs. "According to Dr. Neill," Lewis continued, "a portion of the roof had fallen in many years previous to the date of his survey, and the cavity thus formed was called the 'dome'." This dome, no longer in existence, can easily be seen near the entrance in Illingworth's 1870 photograph taken inside the cave. So the cave has been shortened by at least that much since 1870.

St. Paul druggist Robert O. Sweeny drafted a map of Carver's Cave, presumably about the time of this centenary (which he attended) and the passage configuration is similar to when the cave was resurveyed in 1981, except that the modern cave is 26 m in length. Overall, the amount of change has been more modest than sometimes claimed.

During a quarter of a millennium, then, piping was not directly observed (or at least not reported) at Carver's Cave, even though that's the process to which the cave is commonly attributed. This contrasts with the next cave to be described.

Unlike other UMV region caves before and since, almost every visitor to **Fountain Cave** had something to say about its genesis. While some discussed the cave's origin, per se, others addressed the cave's ongoing development. No other cave even comes close in the sheer volume of references—a selection of which is included below.

Major Long explored and named Fountain Cave later on the same day that he visited Carver's Cave, reporting that "instead of a stagnant pool and only one accessible room of a very different form, this cavern [Fountain Cave] has a brook running thro' it and at least 4 rooms in succession one after the other. Carver's Cave is fast filling up with sand so that no water is now to be found in it, whereas this from the very nature of the place must be enlarging, as the fountain will carry along with its impact all the sand that falls into it from the roofs & sides of the Cavern." Schoolcraft wrote that "the rock is of a very friable nature, and easily acted upon by running water…thus enlarging the boundaries of the cave." George W. Featherstonhaugh, the first person to hold the title "United States Geologist," in his *Canoe Voyage up the Minnay Sotor* (1847) wrote for September 12, 1835, "Like many other caves, this appears to have a reservoir of water in it arising from springs, that in long periods of time have effected the excavations in the rock, which is so soft and incoherent as to be easily cut by a knife." A Canadian visitor, Peter Garrioch, in his diary for November 16, 1837, described "narrow passages formed by the force of the stream of water running through the cave and washing away the sand from between the contiguous and more consolidated rocks. The apartments diminish in size, however, as they approach the head or termination of the cavern. The water running through the cave, and which doubtless has brought it to its present form, is a beautiful, crystal stream, and as pleasant to the taste as any water I ever tasted."

Nicollet gave the most elaborate account of the genesis of Fountain Cave—the stream piracy theory. Writing about Fountain Cave, he stated that, "It owes its formation to the dislocation and decomposition of the upland [Platteville] limestone, which have left sloughy places; the waters of which have penetrated into the [St. Peter] sandstone, wearing it away, and giving origin to the streamlet [Fountain Creek] that issues from it. The location of this cave is on my map designated as the *new cave*." Two subsequent accounts are quite similar to Nicollet's, perhaps because of borrowing. E. S. Seymour, in his *Sketches of Minnesota, The New England of the West* (1850) wrote, "This cave is probably produced by the action of this stream of water, which has broken through the strata of superincumbent limestone, and worn a passage through the sandstone. The latter is constantly crumbling off, and is carried away by the current." The artist Henry Lewis, in his *Valley of the Mississippi Illustrated* (originally published in German in 1854) wrote that "This cave is, doubtless, of comparatively recent formation and owes its origin to the stream of water breaking through the fissures of the plate of limestone which forms the roof, disintegrating and washing out the stratum of soft sandstone beneath."

A letter to the *Congregationalist* of Boston, for September 19, 1856, signed "H," describes "A small stream, a rill of water, has worn itself a channel through this bed of sand-stone, and thus formed the cave." Robert Watt, a Danish visitor, wrote in 1871 (as translated by Jacob Hodnefield in 1929) that "A stream from within apparently has hollowed it out, and some maintain that a person could penetrate a couple of miles [3.2 km] beneath the surface either by canoe or by picking his way along the narrow white sand edges of the stream." James Davenport, in his *Minnesota Tourist's and Traveller's Guide* (1872) wrote, "Here a stream of water, which empties into the [Mississippi] river a short distance below, has hollowed out a large cave over one hundred feet [30.5 m] in length, while a narrow passage extends still further into the bowels of the earth, and is said to have been explored for a quarter of a mile [0.4 km] by some adventurous persons some years since."

Fountain Cave largely disappears from the literature after 1880, when it became a convenient sewer for the overlying Omaha Railroad shops. But in 1932, St. Paul landscape architect George L. Nason completed our present understanding of the cave when he described how the ravine at the cave's entrance—"the beautiful little valley," as he called it—was "formed by the caving in of the roof at various times."

In 1960, the original, natural entrance to Fountain Cave, the one used by so many famous explorers, was sealed by the highway department. The construction of Shepard Road along the river bluffs, begun years earlier, was intended to create "a fast route downtown" from the Minneapolis-St. Paul Airport. The cave ravine lay directly in the path of the new highway, whose engineers were looking for a good place to dump "surplus excavated material."

Anticipating the grading crews, Mayor Joseph E. Dillon of St. Paul, and the city engineer, the eponymous George M. Shepard, went searching for the cave on June 16, 1959. "Historic Saloon Eludes Officials" was the *Pioneer Press* headline the next day. "Dillon and Shepard could find no sign of that cave," it was reported. "Shepard said that quite likely falling rock, sand and debris have hidden the mouth."

The reason was not far to seek. Back in 1878, the pioneer Minnesota geologist Newton H. Winchell in his *Geology of Ramsey County* (Winchell 1878) had reported that "The water that issues at Fountain Cave, St. Paul, is that of a creek which disappears in the ground about half a mile [0.8 km] distant" and this is shown clearly on the oldest and only complete map of Fountain Cave known to exist, dating to 1880. Judging from this map, Fountain Cave is the longest natural sandstone cave in Minnesota, about 1,100 feet (335 m). Fountain Creek is shown flowing on the surface, disappearing into a sinkhole near the former Omaha shops, and then flowing through the cave and out to the Mississippi. In 1923, a railroad spur servicing the former Ford Motor Company plant in Highland Park was laid over this sinkhole so that the "upper entrance" to the cave, if ever humanly enterable, was now sealed. Other early maps indicated that the ultimate source of the surface stream itself was the old Fort Road wetland to the west, long since paved over. The cave's water supply was thus cut off and cliff debris began to accumulate at the cave entrance—debris that ordinarily would have been flushed away by the cave stream. Which explains why Dillon and Shepard "could find no sign of that cave" in 1959.

These early ideas about the genesis of Fountain Cave involved "piping," the washing away of sand grains by groundwater. Cause and effect were juxtaposed at Fountain Cave as nowhere else. The sight of the flowing stream in contact with the loose sandstone made the conclusion obvious. It was not so obvious at the other major sandstone cave in the vicinity—Carver's Cave—which contained a stagnant pool, rather than a stream. But no one used the term "piping" back then; the word was borrowed from engineers by soil scientists and geologists who applied it to caves. Hogberg and Bayer's *Guide to the Caves of Minnesota* (1967) gave currency to the term piping locally. The word "pipe" itself, as used in geology, was of venerable antiquity, having been used, for example, by Lyell (1839) among others.

Indeed, since some of the caves described in the present volume are artificial, I would like to clarify the meaning of the term "natural cave." Among the general public, I have found that unlined, artificial caves dug into bedrock will often be called "natural caves" because the avowedly natural rock surface is plainly visible. Whereas to most geologists, and here in this book, natural refers to the space itself, not the walls.

In fact, a new category of cave has been proposed based on its prevalence or prominence in the St. Peter Sandstone in the Twin Cities, namely, the anthropogenic cave, which is neither natural nor artificial, but an unintentional washout of the easily excavated sandstone, owing to nearby man-made tunnels (Brick 2000, 2009a, 2017c, p 7).

When the topic of cave genesis was next taken up seriously in the UMV region, in the early twentieth century, it was with regard to the chemical solution of limestone caves. The Harvard geomorphologist William M. Davis (1850–1934) published his classic paper, "Origin of Limestone Caverns," in the *Bulletin of the Geological Society of America* (Davis 1930) which dealt heavily with Mammoth Cave, and within a dozen years four other important papers on speleogenesis appeared. While Davis only briefly touched on UMV region caves, which were peripheral to Mammoth Cave, one of Davis's followers, the University of Chicago geologist J Harlen Bretz (1882–1981) published his own paper, "Caves in the Galena Formation," in the *Journal of Geology* in 1938, dealing with the genesis of Niagara Cave in Minnesota, Crystal Lake Cave in Iowa, and Smith's Cave in Illinois (Bretz 1938). Bretz and Harris (1961) extended the Davisian interpretation of speleogenesis to the remainder of Illinois. Among speleologists, Bretz seems to have been the chief figure in centrifugal diffusion of cave science to the Midwestern periphery.

Hedges and Darland (1963) extended the Davisian-Bretzian paradigm to Iowa caves, relating them to peneplains. Hedges (1967) used the work of Bogli (1964) on mixing corrosion to explain features of Worden's Cave in Iowa. Bounk (1983) described the frequently observed association between the presence of caves and the *Cyclocrinites* beds of the Hopkinton dolomite, in Iowa. He concluded that most of the caves of northeastern Iowa formed since the last glaciation, as they are related to the current topographic-hydraulic gradient. Bounk and Bettis (1984) developed this explanation further. Even after process-based geomorphology gained favor and led to a de-emphasis on peneplains, there are traces of Bretzian influences today, as, for example, the author's own *Iowa Underground: A Guide to the State's Subterranean Treasures*, where it provided a convenient vehicle for popular exposition (Brick 2004a).

Another concept that had roots in the UMV region was the sulfuric acid reaction for cave development, at Level Crevice Cave in Iowa, where pyrite grains were suspected

(Morehouse 1968). According to Egemeier (1987), hydrogen sulfide speleogenesis explained the origin of Carlsbad Caverns in New Mexico, a model that was subsequently applied to nearby Lechuguilla Cave. But in these latter cases the development was hypogenic, with hydrogen sulfide rising from oilfield brines and oxidizing above the water table to form sulfuric acid, rather than from oxidation of pyrite in rock layers above the cave (Jagnow et al. 2000).

Mystery Cave was explained as a subterranean meander cutoff that utilized a pre-existing maze of rectilinear rock joints (Milske et al. 1983). The origin of the maze itself has latterly been attributed to hypogenic recharge (Barr et al. 2008) and the model extended to sandstone caves in the Twin Cities (Barr and Alexander 2009).

8.4 Hydrogeological Paradigms: Groundwater Concepts

While the UMV region was on the periphery of other great cave systems and often a late adopter of concepts from elsewhere, in other fields it was ahead of the rest of the country. Certainly that was the case for some aspects of groundwater geology.

Thomas Chrowder Chamberlin (1843–1928) taught at three distinguished Midwestern institutions (Beloit College, University of Wisconsin, University of Chicago) and did important work in several fields, promoting the concept of multiple glaciation and proposing the planetesimal theory of Earth formation. In groundwater, his U.S. Geological Survey monograph of 1885 included cross-sections of artesian systems in the comparatively flat conditions of the UMV region (Chamberlin 1885).

The importance of Chamberlin's groundwater research to hydrogeology generally, rooted in the UMV region, is attested by numerous accolades over the years. Tolman (1937, p 360) wrote that "The classic paper of Chamberlin is the first comprehensive and detailed geological description of artesian flow in stratiform aquifers with gentle dip.… His cross sections have been reproduced in every treatise on ground water." According to Bredehoeft et al. (1982), it was "the first classic paper on regional flow." "The study of artesian conditions was so important that Chamberlin's 1885 report is generally recognized as the beginning of hydrogeology in the United States," according to Back and Herman (1997). "This was the first hydrogeologic report published by the USGS…. Chamberlin recognized that water occurred in both fractured and porous media" (Fetter 2004).

The basic visual vocabulary of groundwater movement was also developed in the UMV region (Brick 2013a). Meinzer (1934) wrote that "The simple concept of the water table has developed rather tardily, although a good contour map of the water table was published by Gustave Dumont in 1856" in France. The U.S. Geological Survey eventually took up the innovation: Gilbert's (1897) Pueblo, Colorado, Folio showed by contours the depth to water (rather than elevations), whereas King (1899) added flow arrows to his elevation-based contours in the UMV region, on the shores of Lake Mendota on the University of Wisconsin campus in Madison, which, used together with cross-sections, is a crucial innovation.

8.5 Sedimentology: The Case of Grain-Size Analysis

It is not merely caves but the deposits in them that are of interest. Chemical (dripstone) and clastic deposits allow scientists to answer many questions about how the cave formed, its age, and the reconstruction of past surface climates—the subject matter of paleoclimatology. Herman et al. (2012) offer a comprehensive recent review of clastic sedimentology in fluviokarst aquifers.

Early cave sediment studies in the United States used generalized descriptions of sediments, such as clay or dripstone, in so much as it was relevant to the Davisian paradigm. Following Davis (1930), observations of cave sediments were often tied to proving the two-cycle theory of cave formation: phreatic residual clays, derived from insoluble limestone residues, were deposited while the cave was forming under the peneplain, followed by vadose dripstone deposits after the cave had drained. Bretz (1942) added a third, intermediate epoch of clay filling, but the clays in this epoch derived from surface soils through dolines and fissures. Bretz and Harris (1961) applied this distinction to cave genesis in Illinois. Hedges (1963) performed a reconnaissance of cave sediments in the Maquoketa River Valley of Iowa. His study of the stratigraphy of five caves revealed, in the simplest example, a pattern of unctuous reddish residual clay which he attributed to phreatic conditions, overlain by a layer of breakdown which he attributed to draining of the cave, and an uppermost layer of subaerial speleothems.

But Davis and Bretz spoke of clays generally; no granulometric analyses were performed. The classic sedimentological techniques in geology, employing grain-size analysis and so forth, were developed by J. A. Udden and C. K. Wentworth in the Midwest between 1890 and 1920 (Krumbein 1932; Law 1980). The "Chicago Group" (Law 1980) added further refinements to the classic methods, with Friedman (1961) advancing inferences about depositional environment from sediment characteristics. Davies and Chao (1959) applied these techniques at Mammoth Cave, Kentucky, but in the UMV this came to fruition in Milske et al.'s (1983) study of Mystery Cave sediments, which made full use of grain-size methods, reporting a preglacial basal silt overlain by stream gravels of glacial origin, capped by flowstone (which is datable). It was indeed the first study to relate clastics to uranium/thorium dating. Brozowski and Day (1994) applied this approach in Wisconsin: the sediments of Brussels Hill Pit Cave, a 28 m deep pit, the deepest in the state, were used to reconstruct cave development. Several articles by the Korean soil scientist Jong-woo Oh investigated the karst sediments of southwestern Wisconsin. Oh et al. (1993) applied statistical inference and mineral provenance to the sediments at Reynolds Sinkhole, the largest known sinkhole in southwestern Wisconsin. Such studies show that cave sediments in the Driftless Area are often silt-size and derived from wind-blown loess, differing from cave sediments outside this area (Day 1988). Hobbs (1994) continued this work in Minnesota.

Other characteristics of cave sediments have been examined. Christiansen et al. (1961), studying the springtails of Hunter's Cave, Iowa, considered the impact of sedimentological characteristics on microarthropod ecology. Josephs (2007) applied rigorous micromorphological (thin section) methods to the study of cave and rock shelter sediments in Iowa. At Bogus Cave, he demonstrated an uppermost, Holocene culture-bearing layer in which modern artifacts were thoroughly mixed with prehistoric ones. Below that was a layer of stones interpreted as cryoclastic (freeze-thaw) breakdown from the late Pleistocene, a pattern he had observed in other nearby caves (Josephs 2002) which correlated with accelerated mass wasting throughout the Midwest during late Pleistocene times.

Bradbury (1959) published cross-sections of the Herman Smith Mine in Illinois, showing detailed stratigraphy. At Fogelpole Cave, redeposited glacial sediment was dated with carbon 14 and it was inferred that the cave began forming near the end of the Illinois glacial episode (Panno et al. 2001). Isotopic analysis of carbon in cave sediments allowed Panno et al. (2004) to infer the presence of C3 or C4 plants growing on the land surface above, hence whether it was woodland or prairie, respectively.

Sediments at cave entrances are more varied (Ford 1976). Nitrate deposits in Midwestern Caves were of military interest as late as World War I. Brick (2012, 2013b, c) examined more than 100 caves and rock shelters in all four UMV states, finding cave sediments enriched in nitrate, up to 3.5% by weight, in the entrance zones of two-thirds of them. This tapered to zero in the deep cave environment. Stable isotopes and other evidence point to an organic origin for this nitrate.

Layered clastic sediments, especially in the finer fractions, may contain magnetic minerals that recorded the orientation of the Earth's magnetic field during deposition (Herman et al. 2012). Schmidt (1982) conducted the first large-scale study using magnetostratigraphy of sediments in Mammoth Cave, Kentucky, finding that the upper levels of that cave dated back about 2 million years based on the magnetic polarities encountered. Webb et al. (2010) applied this technique to laminated sediments in the Garden of the Gods passage in Mystery Cave, Minnesota, but the sediments postdated the latest magnetic reversal, 780 ka BP. No UMV caves that have been studied to date record this reversal.

In some cases, the interpretation of cave deposits has even called for the revision of commonly received recent human history (Brick 2007a).

8.6 Cave Meteorology and the Cave of Paradox

Much of the subject matter of cave meteorology deals with air movement, temperature, and humidity (Wigley and Brown 1976). The distinction between cave meteorology and cave climatology is less valuable underground than it is on the surface because the cave environment is more nearly constant (Wigley and Brown 1976, p 330). Nonetheless, some rudimentary understanding of cave microclimates has been important in the cultural use of caves, as for beer lagering, cheese ripening, storage, and so forth, as seen so well in the St. Peter Sandstone caves of Minnesota (Brick 2009a).

Air movement in caves was noticed as early as the cave of Aeolus in the Classical world. Athanasius Kircher (1602–1680) described air circulating in the caves of the Classical Karst in 1678 (Cigna 2017). But it was the Austrian scientist Hermann Bock (1882–1968) who in 1913 first applied a mathematical treatment to air circulation in Alpine caves with various entrance configurations (Cigna 2017).

The most important air movement in multiple-entrance caves is the chimney effect, whereby air flows upward in winter and downward in summer (Wigley and Brown 1976, p 330). This basic notion was behind an important new paradigm in cave meteorology which originated in the UMV. Alois Kovarik (1880–1965) was a native of Iowa who became a Yale physicist. During his early, Iowa years, in the late 1890s, he investigated the Decorah Ice Cave, the largest glacière in North America east of the Black Hills, a topic thoroughly researched by Hedges and Knudson (1975). It was dubbed the "Cave of Paradox" in the "Ripley's Believe It or Not" newspaper column in 1932 because it contained ice during the summer but not in winter. Kovarik established that the convective air flow chilled the walls of the cave in winter, but that ice formation could only begin in spring when the ground above became unfrozen, providing a source of water. This explanation for certain kinds of ice caves became widely adopted elsewhere.

There are two forms of cave breathing, short-period resonance and long-period barometric breathing (Wigley and Brown 1976, p 336). Iowa again was the field site for some innovative work on this topic with the research of Warren Lewis at Coldwater Cave. Early studies by Faust (1947) dealt with Breathing Cave in Virginia, with its short-period breathing attributed to resonance. Lewis (1981, 1991) found that Coldwater Cave breathed through its artificial access shaft, but he also detected two hitherto unrecognized air patterns, for which he could not find an internal explanation. He concluded that they were gravity acoustic waves, some of which originated in the ionosphere while the others were generated by the jet stream.

Kambesis et al. (2013) investigated temperature fluctuations at Coldwater Cave, which is worth quoting at length: "This eight-year study demonstrated that both resurgences, and in-cave sites proximal to surface recharge points displayed significant variation in water and air temperature hourly, daily and seasonally as well as during storm events. In-cave sites that were located farthest from surface recharge showed very little fluctuation in water and air temperature and corresponded to the mean annual temperature of the area. These results offer important implications in terms of the study of aquifer vulnerability to surface contaminants, cave ecosystems, speleothem development, and thermodynamic controls on subterranean karst processes."

Moreover, "Temperature variations are also caused by recharge events. The period between March and October have the most precipitation and this is reflected in the spiky nature of the cave stream temperature graphs during that time span. Even the North Snake Passage, which shows the least effect from surface influences displays a little noise in the March through October time period. Though the period between November and March did not receive storm event precipitation, the cave temperature graphs showed fluctuations during this time period because of the diurnal freeze-thaw affect. There are significant temperature variations between different parts of the cave system. The diurnal and seasonal variations in surface temperature, are reflected in the cave stream temperature graphs for passages that are located less than 500 m from surface stream sink points." In neighboring Wisconsin, Mueller and Day (1997) likewise established that cave temperatures are much more variable than had been hitherto appreciated.

Beyond these conventional methods, De Freitas et al. (1982) used inert tracer gases to study air movements in caves, a method deployed in Minnesota by Lively and Krafthefer (1993) to gain insight into the radon distribution patterns of Mystery Cave.

8.7 Biological Paradigms: The Case of Adaptation

By an oddity, few other groups of organisms have been so closely described by their adaptation to a particular environment, than that of cave fauna. We don't usually speak of aquaphiles or montanophiles. Yet we speak of troglophiles and so forth. How did this come about?

The Danish zoologist J.C. Schiodte (1815-1884) recognized shade animals, twilight animals, cave animals, and stalactite animals in 1849, distinctions that were observed by Charles Darwin and late-nineteenth-century American biospeleologists such as Packard and Putnam (1872). The Schiner (1854) and Racovitzka (1907) ecological classification, dividing cavernicoles into trogloxenes, troglophiles, and troglobites, was early recognized as a possible evolutionary sequence. The chief paradigm of biospeleology had been established and was widely used (e.g., Vandel 1965).

But pioneers before this did not use an ecological classification. In 1845, the German physician Theodor Tellkampf (1812–1883), for example, studying the fauna of Mammoth Cave, Kentucky, described the blind fishes according to their degree of adaptation to caves, as bearing eyes, rudiments of eyes, or being eyeless (Brick 2017b).

The ambitious *Animalium Cavernarum Catalogus* (1934–1938) was an attempt to list cave species worldwide. But the first comprehensive troglobitic checklist of the United States in the twentieth century was by Nicholas (1960). While pioneering ecologist Stephen A. Forbes (1844–1930), the founder of the Illinois Natural History Survey, had published on blind cave and spring fishes (Forbes 1881, 1882), Leslie Hubricht (1908–2005) was the first to devote himself to the troglobites of the UMV region, as at Morrison's Cave in Illinois (Mackin and Hubricht 1940, Hubricht 1943). Peck and Lewis (1978) examined the biogeography of cavernicoles in Illinois, showing that the Pennyroyal fauna characteristic of the Mammoth Cave region reaches the southern tip of Illinois, while the remainder of the state is characterized by the depauperate drift fauna, except for the northwestern corner. It was Peck and Christiansen (1990) who first applied the ecological classification to the UMV region's Driftless Area in a robust way. They identified several endemic troglobitic species which had evolved there. They distinguished troglobites (terrestrial) from phreatobites (aquatic).

8.8 Spring Ecology Paradigms

How does the ecology of springs get done? In the UMV region, we see at least two ecological traditions at play: the sphere of discharge ecology (Bornhauser 1913) involving habitats and species lists, and the influence of Odum (1957) where energetics and trophic levels are the focus of discussion.

Geologists usually classify springs by their point of discharge. As an example, many UMV springs are "contact springs," i.e., occurring at the geologic contacts between rock layers. Whereas the sphere of discharge terminology, widely used by ecologists, originally began with the three terms *limnocrene, helocrene*, and *rheocrene*. First proposed by Bornhauser (1913) for the springs of Basel, Switzerland, this Basel Nomenclature was revived more recently by Hynes (1970) and is currently in use by various entities conducting spring inventories, such as the National Park Service, and many consulting firms in the Midwest.

Springer et al. (2008, 2009) expanded the categories to twelve. An example of a confusing situation is their category of "cave spring." The common acceptation of this term in the Minnesota Spring Inventory (Brick 2017d) and among geologists generally is where a stream exits a spring cave. According to sphere of discharge terminology, however, a cave spring would be called a "gushette." On the other hand, a cave spring in sphere of discharge terminology means a spring resurging into a cave chamber; and so by definition, the groundwater has not reached the Earth's surface, as most definitions of springs require. *Hypocrene* is another category of "spring" without direct surface expression, important in deserts, but outside most spring definitions.

Howard T. Odum (1924–2002) conducted the first large-scale ecological study in the United States at Silver Springs in Florida (Odum 1957). The Odum method was applied to Cone Spring in Iowa by Tilly (1968). However, the ecosystems of these springs have a significant input of direct solar energy in their energy budget, whereas the ecosystem of cave springs was touched on by Brick (2009b) at Carver's Cave, with its subterranean lake, where the direct solar inputs are negligible, but the populations of *Gammarus pseudolimnaeus,* pigmented isopods, white flatworms, and various gastropods, seem to be supported by the leaf litter detritus that blows into the entrance. Carver's Cave also serves as an overwintering refuge for frogs (*Lithobates clamitans*) and mosquitoes. Beavers have been observed inside the cave amassing stick caches (Brick 2007b). An unverified report exists of blind white crayfish (Anonymous 1913).

There are few other Odum-style studies of springs in the UMV. Muck and Newman (1992) examined Minnesota springs but from the narrow aspect of using Spearman rank correlations to account for the distribution of two amphipod species. Webb et al. (1995, 1996) examined the biodiversity of Illinois springs but did not investigate community structure. Swanson (2013) applied the sphere of discharge terminology to Wisconsin springs.

Glazier and Gooch (1987) described the sharp dichotomy between insect- and crustacean-dominated springs in Pennsylvania. Perennial springs that have higher alkalinities tend to be characterized by amphipods, whereas springs with less dependable flow and lower alkalinities have insects which can more readily disperse to them. However, few springs in the UMV states have alkalinities sufficiently low, although a search was made by this writer among the oligotrophic springs of northeastern Minnesota.

8.9 Cave Paleontology: Buckland's Dilemma

Taphonomy is the study of how bones got where they did and this is an important theme in vertebrate cave paleontology. Early on, the English geologist William Buckland (1784–1856) struggled to account for how various bones got into the bone caves he examined in Yorkshire, UK. Some bones were washed in, some were dragged in by scavengers, some were animals who lived in the cave, and some were trapped in pitfalls (Buckland 1822). We are still grappling with some of these same scenarios two centuries later.

The earliest report of fossils in a UMV region cave was in 1804 during the Lewis and Clark expedition up the Missouri River, when fish fossils were noted from a cave near what is now Council Bluffs, Iowa. But the speculation is that the fossils were part of the wall rock and not a separate, more recent cave deposit (Brick 2010). A Mississippian-age fissure filling at Delta, Iowa, contains the remains of the earliest known North American tetrapods, including labyrinthodonts and lungfishes (Lundelius 2006).

Historically, some of the best early documentation for multiple glaciation in North America came from the UMV. Evidence included the discovery of interstadial forest beds, like those at Two Creeks, Wisconsin. Despite this, the UMV has a limited and late record of cave vertebrate fossils and sites relative to surrounding regions. The UMV has no good Irvingtonian-age cave sites, for example, such as the abundant faunal remains found at the Port Kennedy Cave of Pennsylvania, which records a warmer setting, with tapirs, from the early Pleistocene (Daeschler et al. 1993). Perhaps the very fact of multiple glaciations has played a role, obliterating possible early sites. If that's the case, we would expect to find the best sites in the Driftless Area.

In Iowa County, Wisconsin, at the eastern edge of the Driftless Area, Moscow Fissure has a Last Glacial Maximum (LGM) small-mammal assemblage, including tundra rodents, dated to 17 ka BP, while the nearby Lost River Sink, 4 ka BP, has a fauna sympatric with that of today (West and Dallman 1980). Brussels Hill Pit Cave, in Door County, Wisconsin, contains early Holocene mammalian remains (Brozowski and Day 1994). Palmer (1974) reported an extinct peccary in association with Archaic artifacts at Castle Rock Cave, Grant County, Wisconsin. In Jackson County, Iowa, Duhme Cave is a natural trap cave with micromammal faunules from the late Pleistocene and late Holocene (Jans-Langel and Semken 2003). Josephs (2005) has described caribou (*Rangifer tarandus*) bones from Bogus Cave, Jones County, dating to 20 ka BP. As an interesting historical comparison, the bones of this species were so plentiful in European caves as to lead French paleontologist Edouard Lartet (1801–1871) to propose the Reindeer Age in an 1858 memoir (Boule and Vallois 1957, p 19).

Holocene assemblages are well represented in some UMV caves. In Monroe County, Illinois, Parmalee (1967) excavated the Meyer Cave fissure, finding a quarter of a million vertebrate remains in this pitfall trap. All of the remains were of extant species (except for the passenger pigeon) from the Holocene, although several species were extralimital taxa, beyond their present ranges. But no one limestone fissure in the UMV records a full glacial cycle of change, from Pleistocene through Holocene.

In Fillmore County, Minnesota, Widga et al. (2012) reported scimitar cat (*Homotherium serum*) and elk-moose (*Cervalces* sp.) bones from Tyson's Spring Cave, dated to 26.9 ka BP, a mere 60 km from the LGM ice front. So, was the scimitar cat using the cave for a den? Or were the bones washed in? Chris Widga (pers. comm., 2016) has indicated the case is still unclear at present. Frasz (1983) reports on Couch Cave, Monroe County, Illinois, containing Babar's Bone Room, with its mastodon bones. As with the foregoing case, it's unclear to this day how they got there (Widga, pers. comm., 2016). Buckland's dilemma remains as relevant as ever.

8.10 Archeology's First Paradigm

Finally, we arrive at the topic of that postglacial newcomer, *Homo sapiens*, as represented by archeological remains. Passing over the prevalence of the Mound Builder mythology in nineteenth-century American archeology, which also has ties to caves (Brick 2004c, 2009c), we segue to subsequent research.

One thread in the history of archeology deals with the divisions of archeological time. Christian Jurgensen Thomsen's (1788–1863) three-age system in Denmark in 1819 was initially used for classifying museum artifacts and subsequently applied in the field by JJA Worsaae (1821–1885) (Willey and Sabloff 1974, p 13). This was the celebrated division into Stone, Bronze, and Iron ages, which Rodden (1981) described as "archaeology's first paradigm." The English antiquarian Sir John Lubbock (1834–1913) divided the Stone Age into Paleolithic and Neolithic in his 1865 book *Pre-Historic Times* (Lubbock 1865, p 2). Gabriel de Mortillet (1821–1898) in France subdivided the Paleolithic into four epochs, Acheulian, Mousterian, Solutrean, and Magdalenian, named for type sites. He cited the Scandinavian three-age system as the inspiration for his own chronology, which he had also developed for museum collections (Chazan 1995).

In North America, Native Americans were living in the Stone Age at Contact and during the period following Lubbock's work, an attempt was made to define an "American Paleolithic" (Wilmsen 1965). The European subdivisions of the Paleolithic did not apply, however, so a different way of dividing up American archeological time was sought. In Europe, Lartet (as noted above) proposed a paleontological chronology in 1858, with his four ages of the Great Cave Bear, Elephant and Rhinoceros, Reindeer, and Aurochs, respectively (Boule and Vallois 1957, p 19). Wood et al. (1941) proposed the North American Land Mammal Ages (NALMAs) for the Tertiary, a scheme which Barnosky et al. (2014) extended into the Holocene, "to highlight some major human alterations to the Earth system that preceded industrialized Anthropocene times," but none of these played the robust role that Lartet's scheme had done.

During the first half of the twentieth century a relative chronology prevailed. The American archeologist Nels C. Nelson (1875–1964), observing the layer-by-layer excavation of Spanish caves in 1913, employed stratigraphic excavation at Mammoth Cave in Kentucky in 1917 (Browman and Givens 1996). But intriguingly, this method had precursors in the UMV. An early stratigraphy was logged at Samuel's Cave in Wisconsin 1879 by Dr. John Rice, who excavated five feet (1.5 m) before encountering the water table while "making a cross section." Four layers of alternating ashes and white sand, with included artifacts, allowed archeologists to assign the cave to the Oneota culture (Boszhardt 2003, p 85–86).

Will C. McKern (1892–1988) was a curator at the Milwaukee Public Museum and devised his famous Midwest Taxonomic System as a museum classification (McKern 1939, Davison 1989), somewhat as Thomsen had done in Europe more than a century earlier. The prehistoric Native American cultures of the Midwest were classified with a five level system (focus, aspect, phase, pattern, and base) and either Woodland or Mississippian if they could not be tied to existing peoples. McKern referred to published research before this, like those of Newton H. Winchell, as "pre-classification reports." While strictly applicable to the Midwest, it grew to cover eastern North America (Stoltman et al. 1978). Expanding the system, Roberts (1940) used the term Paleo-Indian to refer to the oldest finds. Ford and Willey (1941) referred to the lengthy interval between these two as Archaic, using the term in a different context than was in vogue among Meso-American archeologists.

Here are some examples of the McKern system as applied to Minnesota sites. The archeologist Albert Jenks (1869–1953), who established the Department of Anthropology at the University of Minnesota, regarded southeastern Minnesota almost as an unglaciated region somewhat analogous to the Dordogne region of southern France (Wilford 1937). As for excavated Minnesota caves, La Moille Cave (Brick 2002), a sandstone rock shelter, contained a 15 feet (4.6 m) thickness of sediments that bridged the Archaic to Woodland transition (Lukens 1964). Lloyd Wilford (1894-1982) became the most prolific rock shelter excavator from the 1930s to the 1950s, especially at the Tudahl Rock Shelter near the town of Peterson, where extensive Oneota pottery was found (Wilford 1937). The entrance alcove of Petrified Indian Cave, overlooking the South Branch of the Root River, near Mystery Cave, was excavated by Oothoudt (1980), revealing Woodland potsherds.

Early efforts at complementing the taxonomic method with a temporal method involved dendrochronology, or tree-ring dating, carried out in southern Illinois by the University of Chicago (Guthe 1952). But it was the advent of radiocarbon dating—itself developed in the Midwest, by Willard F. Libby (1908–1980) at the University of Chicago, in the late 1940s—which transformed the cultures of the McKern system into "periods" (Taylor 2000). James Griffin (1905–1997) was the first to integrate radiocarbon dates in a landmark American archeology textbook, sometimes dubbed the "Green Bible," published by the University of Chicago (Griffin 1952). Carbon-14 was the radioactive "yeast" that leavened "the flat view of the past" (Stoltman et al. 1978) into deep time. The advent of accelerator mass spectrometry (AMS) in the 1970s allowed radiocarbon dating on even "milligram amounts" of sample (Taylor 2000). In summary, a "three-age system" of a different sort is widely accepted in North America today, assigning artifacts to Paleo-Indian, Archaic, and Woodland periods, with a fourth, Mississippian/Oneota period in some regions (Theler and Boszhardt 2003).

Three Wisconsin sites are often discussed together. The Raddatz Rock Shelter in Sauk County was excavated to a depth of 9.5 feet (2.9 m) by Wittry (1959) and although the deposits fell wholly within the Archaic period, the site is notable as an example where a geologist, Robert F. Black, investigated the stratigraphy, suggesting correlations of the rock shelter deposits with glacial events (Black 1959, Bettis et al. 1985). Wittry (1959) reported that the nearby Durst Rock Shelter, in Sauk County, had six zones ranging from the Middle Archaic up through the Late Woodland periods. Storck's study of Mayland Cave in Iowa County concluded that it served "as a base camp for family groups in the winter and as a temporary hunting station for small parties of male hunters during the remainder of the year" (Storck 1972, p 373). These people were members of Late Woodland Effigy Mound culture (Theler and Boszhardt 2000). Indeed, all three of these sites "appear to have been intermittently occupied campsites intended primarily for deer hunting" (Storck 1972, p 383).

The amateur archeologist Paul Sagers (1909–1982) excavated numerous rock shelters in eastern Iowa in the 1920s and 1930s, enough to fill a special museum of the artifacts (Brick 2004a, p 59). Charles R. Keyes (1871–1951) and his colleague Ellison Orr (1857–1951) investigated rock shelters in the 1930s and 1940s which contained Woodland period artifact assemblages (Schermer et al. 1996). In the 1970s and 1980s, excavation of Rock Run Shelter and Hadfields Cave employed modern excavation methods including precise stratigraphic controls and microartifact recovery (Schermer et al. 1996).

Excavations at the Modoc Rock Shelter in Illinois beginning in the 1950s have revealed the lengthiest cultural history in the UMV region, with 9.3 m of radiocarbon-dated layers comprising the early, middle, to late Archaic period in the Midwest, from 9000 to 4000 years BP (Ahler 1993). A possible Paleo-Indian cave site with bones of the extinct giant beaver (*Castoroides ohioensis*), discovered in St. Paul, Minnesota, in 1938, turned out to be spurious, except for the beaver fossil itself, which was radiocarbon dated to 10.3 ka BP in 1967 (Brick 2017a, p 20).

8.11 Rock Art Paradigms

As early as 1673, the French explorers Marquette and Joliet described pictographs of "painted monsters" on the bluffs at what is now Alton, Illinois, as they descended the Mississippi River in canoes. This Piasa monster was a hybrid creature, but often called a bird (Thwaites 1954, p 356–357). Early explorers in the UMV region, upon encountering rock art, usually described it within the hieroglyphics paradigm as late as the 1830s (e.g., Featherstonhaugh 1836, p 135) implying that each symbol was to be "deciphered" just as Egyptian hieroglyphics had been. Indeed, some asserted that the Mound Builders were in fact descended from the Egyptians, making the comparison even more apt!

The first recorded use of the more purely descriptive word "pictograph" was by the ethnologist Henry Rowe Schoolcraft (1793–1864) in 1851 (*Oxford English Dictionary*, 2nd ed., s.v. "pictograph"). He employed the term with reference to the Chippewa Indians of the Midwest. But by the last quarter of the nineteenth century most investigators were describing two distinct forms of rock art, namely, petroglyphs (carvings or sculptures on rock) and pictographs proper (paintings). Once it was concluded that rock art was not an alphabet or writing system, rock art studies were relegated to obscurity for many years (Finney 2006). In our own times, Rajnovich (1989) could assert that the Canadian Shield pictographs she studied "are not primarily art, rather they are picture writing."

In the UMV region, early studies focused on recording rock art along the Mississippi River in all four states. Mostly these have been carved in soft rock types such as the Jordan and St. Peter sandstones. The work of the Hill and Lewis Northwestern Archeological Survey (1881–1895) was important in documenting these (Finney 2006). Alfred Hill (1833–1895) provided financial backing, while Theodore Lewis (1856–1930) conducted fieldwork, most importantly the latter's 1888 rock art ramble, hiking along the Mississippi River from Carver's Cave in St. Paul to Allamakee County in northeastern Iowa, sketching the drawings from the river bluffs and seven caves. Newton H. Winchell's book *Aborigines of Minnesota* devoted a chapter to the topic, with reproductions of the Lewis petroglyphs (Winchell 1911, p 560–568).

In Minnesota, Native American petroglyphs had been reported from Carver's Cave as early as 1766. Nearby, Dayton's Bluff Cave was like a smaller replica of Carver's Cave but its walls were even more densely covered with petroglyphs. In southeastern Minnesota, petroglyphs have been documented from La Moille Cave, Mazeppa Cave, and Reno Cave, among others (Brick 2017a). Most recently, Big Room Cave was examined by modern methods for the documentation of rock art (Sprengelmeyer 2006).

Dudzik (1995) gives a helpful shortlist of the possible "site functions" of rock art. One problematic example in the UMV will be given. In recent years, Carver's Cave has been reinterpreted as a sort of prehistoric planetarium, with the petroglyphs representing the Dakota constellations. After all, there was a natural dome-shaped ceiling in the front room of the cave before that part was destroyed by railroad construction. Perhaps it represented the dome of the heavens? In this scenario, the rattlesnake petroglyphs in the ceiling are seen as bands of stars. While an attractive notion, comparison with known Native American planetarium caves elsewhere does not support this interpretation. In the known examples, stars are represented as small crosses inside the outline of the petroglyph (Patterson 1992, p 157, 191), whereas the Carver's Cave snakes are empty outlines or contain the snake's chevron pattern. No stars are depicted among the petroglyphs at Carver's Cave (Brick 2017a, p 34).

Garrick Mallery's *Pictographs of the North American Indians* (1886) supplies a more plausible explanation that can be applied to many (but not all) petroglyphs seen at UMV caves, namely, that they are clan or personal totems, whereby Native Americans registered their visits to the caves. Mallery documented how the Dakota Indians themselves did this at the famous Pipestone Quarry in western Minnesota, into historical times (Mallery 1886, p 23).

Southwestern Wisconsin is well endowed with petroglyph caves, such as Samuel's Cave, near La Crosse, with its shaman pictograph (Brown 1926). But it has only been recently that rock art has been discovered in the dark zone of UMV region caves: in 1998, Tainter Cave, and in 1999, Larsen Cave, both in Crawford County, were found to possess extensive rock art. The black pictographs were charcoal paintings while the red paintings used hematite from the sandstone itself. Innovative techniques, such as the use of infrared and ultraviolet light to examine drawings, and AMS radiocarbon dating of rock art pigments were used to assign one of the Tainter drawings to AD 690 (Boszhardt 2003, p 73). One of the rock art panels in Tainter Cave seems to reflect a Native American cosmology of sorts, a natural bedding plane in the sandstone walls being employed by the artist to separate earth and sky symbols (Boszhardt 2003, p 72).

Ritzenthaler (1950) proposed a tentative cultural classification of Wisconsin petroglyphs based on associated artifacts found at various sites. Salzer (1997) described the motifs, themes, compositions, and contexts of Wisconsin rock art, deriving a "provisional generalized chronology." At Gottschall Rock Shelter in Iowa County, whose 5.5 meters of stratified deposits span from the late Archaic to late Oneota periods, Salzer (2005) interpreted a rock art panel as reflecting the so-called Red Horn legend recorded by ethnographers among contemporary Ho-Chunk Indians, thus making a connection with living peoples.

Acknowledgements The author would like to thank Marie Dvorzak, C. K. Leith Library of Geology and Geophysics, University of Wisconsin-Madison; Jean Prior, Iowa Geological Survey; Aldemaro Romero, Dean and Professor, Weissman School of Arts and Sciences, Baruch College, City University of New York; Constance Arzigian and Kathy Stevenson, Mississippi Valley Archeology Center, University of Wisconsin-La Crosse; and Chris Widga, Associate Curator of Geology, Illinois State Museum. And certainly the faculty and fellow students in his many graduate seminars in the History of Science program at the University of Minnesota, where he earned a minor in the subject.

References

Ahler SR (1993) Stratigraphy and radiocarbon chronology of Modoc Rock Shelter, Illinois. American Antiquity 58(3):462–489

Anonymous (1913) Blind crayfish in Carver's Cave. St. Paul Pioneer Press, November 13

Back W, Herman JS (1997) American hydrogeology at the millennium: An annotated chronology of 100 most influential papers. Hydrogeol J 5(4):37–50

Badino G (2010) Underground Meteorology—"What's the Weather Underground?" Acta Carsologica 39(3):427–448

Barnosky AD, Holmes M, Kirchholtes R, Lindsey E, Maguire KC, Poust AW, Stegner MA, Sunseri J, Swartz B, Swift J, Villavicencio NA, Wogan GOU (2014) Prelude to the Anthropocene: Two new North American Land Mammal Ages (NALMAs). The Anthropocene Review 1(3):225–242

Barr K, Alexander EC Jr (2009) Examples of Hypogenic Karst Collapse Structures Twin Cities Metropolitan Area, Minnesota. In: Stafford KW, Land L, Veni G (eds) Advances in Hypogene Karst Studies: NCKRI Symposium 1. National Cave and Karst Research Institute, Carlsbad, NM, p 51–59
Barr K, Klimchouk A, Alexander EC Jr (2008) Hypogenic Karst and its Implications for Minnesota Hydrogeology. In: Yuhr LB, Alexander EC Jr, Beck BF (eds) Sinkholes and the Engineering and Environmental Impacts of Karst, Proceedings of the 11th Multidisciplinary Conference. ASCE/GI Geotechnical Special Publication No. 183, Amer. Soc. Civil Eng., Reston, VA, p 43–53
Bettis EA III, Benn DW, O'Brien MJ (1985) The Interaction of Archeology and the Earth Sciences in the American Midwest. American Archeology 5(2):120–126
Black RF (1959) Geology of Raddatz Rockshelter, Sk5, Wisconsin. Wisconsin Archeologist 40(2):69–82
Bögli A (1964) Mischungskorrosion—Ein beitrag zum verkarstungsproblem. Erdkunde 2:83–92
Bornhauser K (1913) Die Tierwelt der Quellen in der Umgebung Basels. Internationalen Revue der gesamten Hydrobiologie und Hydrographie, Biologische Supplement 5:1–90
Boszhardt RF (2003) Deep Cave Rock Art in the Upper Mississippi Valley. Prairie Smoke Press, St. Paul, MN, 94 p
Boule M, Vallois HV (1957) Fossil Men. Dryden Press, New York, 535 p
Bounk MJ (1983) Some factors influencing phreatic cave development in the Silurian strata of Iowa. Proceedings of the Iowa Academy of Sciences 90(1):19–25
Bounk MJ, Bettis EA III (1984) Karst development in northeastern Iowa. Proceedings of the Iowa Academy of Sciences 91(1):12–15
Bradbury JC (1959) Crevice Lead-Zinc Deposits of Northwestern Illinois. Report of Investigations 210. Illinois State Geological Survey, Urbana, IL, 49 p
Bredehoeft JD, Back W, Hanshaw BB (1982) Regional ground-water flow concepts in the United States: Historical perspective. Geol Soc Am Spec Pap 189:297–316
Bretz JH (1938) Caves in the Galena Formation. J Geol 46(6):828–841
Bretz JH (1942) Vadose and phreatic features of limestone caverns. J Geol 50(6):675–811
Bretz JH, Harris SE Jr (1961) Caves of Illinois. Report of Investigations 215. Illinois State Geological Survey, Urbana, IL, 87 p
Brick GA (1995) What Happened to Fountain Cave—The Real Birthplace of St. Paul? Ramsey County History 29(4):4–15
Brick GA (2000) Anthropogenic Caves in St. Peter Sandstone at Minneapolis, Minnesota. Journal of Cave and Karst Studies 62 (3):194
Brick GA (2002) La Moille Cave Exists!! Minnesota Speleology Monthly 34(1):1, 4
Brick GA (2004a) Iowa Underground: A Guide to the State's Subterranean Treasures. Trails Books, Black Earth, WI, 223 p
Brick GA (2004b) A Piping Voice: Theories of Cave Genesis in Minnesota Prior to 1880. Journal of Spelean History 38(1):22–23
Brick GA (2004c) Cave Hoaxes and Nineteenth Century Archaeological Theory. Journal of Cave and Karst Studies 66(3):116
Brick GA (2006) Historical perceptions of the rate of geological change at Carver's Cave. Minnesota Speleology Monthly 38(11&12):110–112
Brick GA (2007a) A new interpretation of Minneapolis history based on cave deposits. Minnesota Speleology Monthly 39(5):52–53
Brick GA (2007b) Beavering into Cave History. Minnesota Speleology Monthly 39(9):108
Brick GA (2009a) Subterranean Twin Cities. University of Minnesota Press, Minneapolis, 226 p
Brick GA (2009b) History and Geology at Carver's Cave. Ramsey County History 44(1):16–20
Brick GA (2009c) The Myth of the American Cave Man. Proceedings of the 15th International Congress of Speleology, Kerrville, TX 3, p 1955–1956
Brick GA (2010) Lewis and Clark's reference to a cavern on the Missouri River, 1804. Journal of Spelean History 44(2):8–9
Brick GA (2012) Le Sueur's Saltpeter Caves at Lake Pepin, Minnesota, and Wilderness Gunpowder Manufacture. Minnesota Archaeologist 71:7–20
Brick GA (2013a) "Water Flowing Uphill is One of the Commonest Facts of Nature": The Emergence of a Visual Language for Groundwater Geology. Ground Water 51(6):968–972
Brick GA (2013b) The nitrate deposits of rock crevices in the Upper Mississippi Valley. Ph.D. thesis, University of Minnesota, Minneapolis, 165 p. University of Minnesota Digital Conservancy, http://hdl.handle.net/11299/157750
Brick GA (2013c) French saltpeter caves in Minnesota, 1700. NSS News 71(11):22–23
Brick GA (2014a) Wakan Tipi Cave (Carver's Cave). Spelean History Series No. 26. Speece Productions, Altoona, PA, 24 p
Brick GA (2014b) Fountain Cave. Spelean History Series No. 27. Speece Productions, Altoona, PA, 24 p
Brick GA (2014c) Beltrami's Trophonian Cave of 1823. Journal of Spelean History 48(1):8–10
Brick GA (2014d) Carver's Cave, Fountain Cave, and Sandstone Speleogenesis. NSS News 72(9):8–10
Brick GA (2017a) Minnesota Caves: History and Lore. Arcadia/The History Press, Mount Pleasant, SC, 144 p
Brick GA (2017b) The prehistory of cave fauna paradigms. Proceedings of the 17th International Congress of Speleology, Sydney, Australia, I: 414
Brick GA (2017c) A third kind of cave in the world: Anthropogenic. Proceedings of the 17th International Congress of Speleology, Sydney, Australia, II: 248
Brick GA (2017d) Minnesota Spring Inventory Guidance Document. Minnesota Department of Natural Resources, Division of Ecological and Water Resources, County Geologic Atlas Project, St. Paul, 32 p
Brick GA (2018) "Well-known Subterranean Passages": The Prehistory of Mystery Cave. Minnesota. Journal of Spelean History 52(1):4–5
Browman DL, Givens DR (1996) Stratigraphic excavation: The first "New Archaeology". Am Anthropol 98(1):80–95
Brown CE (1926) Indian caves in Wisconsin. Wisconsin Archeologist 5(1):7–29
Brozowski J, Day MJ (1994) Development of Brussels Hill Pit Cave, Door County, Wisconsin: Evidence from flowstone and sediment. Transactions of the Wisconsin Academy of Sciences, Arts, and Letters 82:1–16
Buckland W (1822) Account of an assemblage of fossil teeth and bones of elephant, rhinoceros, hippopotamus, bear, tiger, and hyaena, and sixteen other animals; discovered in a cave at Kirkdale, Yorkshire, in the year 1821: with a comparative view of five similar caverns in various parts of England, and others on the Continent. Philos Trans R Soc Lond 112:171–236
Chamberlin TC (1885) The Requisite and Qualifying Conditions of Artesian Wells. Fifth Annual Report, US Geological Survey, Washington, DC, p 125–173
Chazan M (1995) Conceptions of time and the development of Paleolithic chronology. Am Anthropol 97(3):457–467
Christiansen K, Wilson M, Tecklin J (1961) Preliminary study of the microarthropod ecology of Hunter's Cave. NSS Bulletin 23:63–70
Cigna AA (2017) Hermann Bock, a forgotten precursor of cave meteorology. In: Moore K, White S (eds), Proceedings of the 17th International Congress of Speleology, Sydney, Australia, I:423–425
Cvijic J (1893) Das Karstphanomen. Versuch einer morphologischen Monographie. Geograph. Abhandl. 5(3):218–329

Daeschler E, Spamer EE, Parris DC (1993) Review and new data on the Port Kennedy local fauna and flora (Late Irvingtonian), Valley Forge National Historical Park, Montgomery County, Pennsylvania. Mosasaur 5:23–41
Davies WE, Chao ECT (1959) Report on sediments in Mammoth Cave, Kentucky, prepared for US National Park Service by US Geological Survey, 117 p
Davis WM (1930) Origin of limestone caverns. Bull Geol Soc Am 41:475–628
Davison M (1989) Will C. McKern. Lore 39(2):19–21
Day MJ (1988) The origin of cave sediments in southwestern Wisconsin. Geo2, Section of Cave Geology & Geography, NSS, 15 (1–3):8–9
De Freitas CR, Littlbjohn RN, Clarkson TS, Kristament IS (1982) Cave climate: Assessment of airflow and ventilation. J Climatol 2(4):383–397
Dudzik MJ (1995) Visions in stone: The rock art of Minnesota. Minnesota Archaeologist 51–54:99–108
Egemeier SJ (1987) A theory for the origin of Carlsbad Caverns. NSS Bulletin 49(4):73–76
Ekblaw GE (1925) Paleozoic Karst Topography. Transactions of the Illinois State Academy of Science 17:208–212
Faust B (1947) An unusual phenomenon. NSS Bulletin 9:52
Featherstonhaugh GW (1836) Report of a Geological Reconnaissance made in 1835, from the Seat of Government, by the way of Green Bay and the Wisconsin Territory, to the Coteau de Prairie, an elevated ridge dividing the Missouri from the St. Peter's River. Senate Document 333. Gales and Seaton, Washington, DC, 168 p
Fetter CW Jr (2004) Hydrogeology: A short history, Part 2. Ground Water 42(6):949–953
Finney FA (2006) The archaeological legacy of T. H. Lewis. Wisconsin Archeologist 87(1/2):1–48
Forbes SA (1881) A rare fish in Illinois. Am Nat 15(3):232–233
Forbes SA (1882) The blind cavefishes and their allies. Am Nat 16 (1):1–5
Ford DC (2006) Karst geomorphology, caves and cave deposits: A review of North American contributions during the past half century. GSA Special Paper 404
Ford JA, Willey GR (1941) An interpretation of the prehistory of the eastern United States. Am Anthropol 43(3):325–363
Ford TD (1976) The Geology of Caves. In: Ford TD, Cullingford CHD (eds) The Science of Speleology. Academic Press, London, p 11–60
Frasz G (1983) Couch Cave, Monroe County, Illinois. Windy City Speleonews 23:50–53
Friedman GM (1961) Distinction between dune, beach, and river sands from their textural characteristics. J Sediment Petrol 31(4):514–529
Gams I (1993) Origin of the term "karst", and the transformation of the Classical Karst (kras). Environ Geol 21:110–114
Gilbert GK (1897) Pueblo Folio, Colorado. Geologic Atlas of the United States, US Geological Survey, Washington, DC
Glazier DS, Gooch JL (1987) Macroinvertebrate assemblages in Pennsylvania (USA) springs. Hydrobiologia 150(1):33–43
Griffin JB (1952) Radiocarbon dates for the Eastern United States. In: Griffin JB (ed) Archeology of Eastern United States. University of Chicago Press, p 365–370
Guthe CE (1952) Twenty-five years of archeology in the Eastern United States. In: Griffin JB (ed) Archeology of Eastern United States. University of Chicago Press, p 1–12
Hacquet B (1778) Oryctographia Carniolica. Gottlob und Breitkopf, Leipzig, 186 p
Halliday WR (2007) Pseudokarst in the 21st Century. Journal of Cave and Karst Studies 69(1):103–113
Hansel A (1976) Mechanically Induced Karst Development along the Silurian Escarpment in Northeastern Iowa. M.S. thesis, University of Northern Iowa, Cedar Falls, 102 p
Hedges JA (1963) A reconnaissance of cave sediments in the Maquoketa River valley, Iowa. Iowa Cave Book 7:D1–D16
Hedges JA (1967) Mixture-solution hollows in Worden's Cave. Iowa. Caves and Karst 9(5):41–44
Hedges JA (1972) Expanded joints and other periglacial phenomena along the Niagara Escarpment. Biul Peryglac 21:87–126
Hedges JA, Alexander EC Jr (1985) Karst-related features of the Upper Mississippi Valley Region. Stud Speleol 6:41–49
Hedges JA, Darland GW (1963) The Scotch Grove strath in Maquoketa River valley, Iowa. Proceedings of the Iowa Academy of Sciences 70:295–306
Hedges JA, Knudson G (1975) The ice cave at Decorah, Iowa. Annals of Iowa 43(2):113–131
Herak M, Stringfield VT (1972) Karst: Important Karst Regions of the Northern Hemisphere. Elsevier, Amsterdam, 551 p
Herman EK, Toran L, White WB (2012) Clastic sediment transport and storage in fluviokarst aquifers: An essential component of karst hydrogeology. Carbonates Evaporites 27:211–241
Hobbs HC (1994) Karst is a repository for old sediments. North-Central Section, GSA Abstracts with Programs 26:21
Hogberg RK, Bayer TN (1967) Guide to the Caves of Minnesota. Educational Series 4. Minnesota Geological Survey, Minneapolis, 61 p
Hromnik CA (2001) Karst, Kras or Karasattu: Whence the name? Cave and Karst Science 28(2):79–88
Hubricht LH (1943) Studies on the Nearctic freshwater Amphipoda III. Notes on the freshwater Amphipoda of Eastern United States with descriptions of ten new species. Am Midl Nat 29:683–712
Hynes HBN (1970) *The Ecology of Running Waters*. University of Toronto Press, 555 p
Jagnow DH, Hill CA, Davis DG, DuChene HR, Cunningham KI, Northup DE, Queen JM (2000) History of sulfuric acid theory of speleogenesis in the Guadalupe Mountains, New Mexico. Journal of Cave and Karst Studies 62(2):54–59
Jans-Langel CM, Semken HA Jr (2003) Paleoecological interpretation of late Holocene and late Pleistocene micromammal faunules from Duhme Cave, eastern Iowa. In: Schubert BW, Mead JI, Graham RW (eds) Ice Age Cave Faunas of North America. Denver Museum of Nature & Science, p 119–148
Jillson WR (1924) American Karst Lands. Pan Am Geol 42:37–45
Josephs RL (2002) A late Pleistocene ceiling collapse in Bogus Cave, Jones County, Iowa: A potential relationship to coeval accelerated mass wasting events across the central Midwest. Journal of Cave and Karst Studies 64(3):175–179
Josephs RL (2005) A 20,000 CALYBP caribou mandible from a cave site in east-central Iowa. Current Research in the Pleistocene 22:73–75
Josephs RL (2007) Micromorphology of a late Woodland cave site, east-central Iowa. Midcontinental Journal of Archaeology 32(1):71–87
Kambesis PN, Lovaas J, Lace MJ (2013) High resolution temperature sampling of cave climate variation as a function of allogenic recharge, Coldwater Cave, Iowa, USA. Proceedings of the 16th International Congress of Speleology, Cave Climate and Paleoclimate Record, p 413–416
King FH (1899) Principles and Conditions of the Movements of Ground Water. Nineteenth Annual Report, Part 2, US Geological Survey, Washington, DC, p 59–294
Krumbein WC (1932) A history of the principles and methods of mechanical analysis. J Sediment Petrol 2(2):89–124
Kuhn TS (1962) The Structure of Scientific Revolutions. University of Chicago Press, 172 p
Lange EG (1909) Caves of the Driftless Area of Southwestern Wisconsin. University of Wisconsin thesis, Madison, 81 p
Lapham IA (1867) A geological, topographical, and subterranean map of the Brigham Lead Mines near the Blue Mounds. Manuscript map at the Wisconsin Historical Society, Madison, Wisconsin

Law J (1980) Fragmentation and investment in sedimentology. Soc Stud Sci 10(1):1–22
Lewis WC (1981) The "breathing" of Coldwater Cave. Proceedings of the 8th International Congress of Speleology, Bowling Green, KY, p 89–90
Lewis WC (1991) Atmospheric pressure changes and cave airflow: A review. NSS Bulletin 53(1):1–12
Lively RS, Krafthefer BC (1993) Radon Concentrations, Activities of Radon Decay Products, Meteorological Conditions and Ventilation in Mystery Cave, Final Technical Report, LCMR Mystery Cave Resources Evaluation, 135 p, https://www.leg.state.mn.us/docs/2015/other/155232/finaltechnicalrepot.pdf
Lubbock J (1865) Pre-Historic Times, as Illustrated by Ancient Remains, and the Manners and Customs of Modern Savages. Williams and Norgate, London, 512 p
Lukens PW Jr (1964) The mammals and molluscs from the La Moille Rock Shelter. Minnesota Archaeologist 26(2):50–54
Lundelius EL Jr (2006) Cave site contributions to vertebrate history. Alcheringa: An Australasian Journal of Palaeontology 30(S1):195–210
Lyell C (1839) On the tubular cavities filled with gravel and sand called "sand-pipes," in the chalk near Norwich. London and Edinburgh Philosophical Magazine and Journal of Science 15(96):257–266
Mackin JG, Hubricht L (1940) Descriptions of seven new species of Caecidotea (Isopoda, Asellidae) from central United States. Transactions of the American Microscopical Society 59(3):383–397
Mallery G (1886) On the Pictographs of the North American Indians. Fourth Annual Report of the Bureau of Ethnology. Government Printing Office, Washington, DC, p 12–264
Martin L (1932) The Physical Geography of Wisconsin, 2nd edn. University of Wisconsin Press, Madison, 608 p
McKern WC (1939) The Midwestern Taxonomic Method as an aid to archaeological culture study. Am Antiq 4(4):301–313
Meinzer OE (1934) The history and development of ground-water hydrology. J Wash Acad Sci 24(1):6–32
Milske JA, Alexander EC Jr, Lively RS (1983) Clastic sediments in Mystery Cave, southeastern Minnesota. NSS Bulletin 45:55–75
Morehouse DF (1968) Cave development via the sulfuric acid reaction. NSS Bulletin 30(1):1–10
Muck JA, Newman RM (1992) The distribution of amphipods in southeastern Minnesota and their relation to water quality and land use. Journal of the Iowa Academy of Sciences 99(1):34–39
Mueller RJ, Day MJ (1997) Daily atmospheric variation within caves in southwestern Wisconsin, USA. Proceedings of the 12th International Congress of Speleology, La Chaux-de-Fonds, Switzerland, Symposium 7: Physical Speleology, Vol 1, p 215–217
Murrish J (1871) Report on the Geological Survey of the Lead Regions. Wisconsin Geological, Mineralogical, and Agricultural Survey, Madison
Nicholas G (1960) Checklist of troglobitic organisms of the United States. Am Midl Nat 64(1):123–160
Odum HT (1957) Trophic structure and productivity of Silver Springs. Florida. Ecological Monographs 27(1):55–112
Oh J, Day MJ, Gladfelter B (1993) Geomorphic environmental reconstruction of the Holocene sinkhole sediments in the Wisconsin Driftless Area. Proceedings, Korean Scientists and Engineers International Conference, p 390–397
Oothoudt JW (1980) Petrified Indian Cave: The Legend and the Archaeology. In: Alexander EC Jr (ed), An Introduction to Caves of Minnesota, Iowa, and Wisconsin, p 11–29
Packard AS Jr, Putnam FW (1872) The Mammoth Cave and its Inhabitants. Naturalists Agency, Salem, MA, 72 p
Palmer HA (1974) Implications of an extinct peccary–early Archaic Artifact association from a Wisconsin Cave. Wisconsin Archeologist 55:218–230
Panno SV, Curry BB, Wang H, Hackley KC, Liu C-L (2001) Implications for climate change and speleogenesis in southern Illinois based on the age of redeposited glacial sediment in Fogelpole Cave. GSA Abstracts with Programs 33(6):24
Panno SV, Curry BB, Wang H, Hackley KC, Liu CL, Lundstrom C, Zhou J (2004) Climate change in southern Illinois, USA, based on the age and δ 13 C of organic matter in cave sediments. Quatern Res 61(3):301–313
Parmalee PW (1967) A recent cave bone deposit in southwestern Illinois. NSS Bulletin 29(4):119–147
Patterson A (1992) A Field Guide to Rock Art Symbols of the Greater Southwest. Johnson Books, Boulder, CO, 256 p
Peck SB, Christiansen K (1990) Evolution and zoogeography of the invertebrate cave faunas of the Driftless Area of the Upper Mississippi Valley of Iowa, Minnesota, Wisconsin, and Illinois, USA. Can J Zool 68:73–88
Peck SB, Lewis JJ (1978) Zoogeography and evolution of the subterranean invertebrate faunas of Illinois and southeastern Missouri. NSS Bulletin 40(2):39–63
Prior J (1976) A Regional Guide to Iowa Landforms. Iowa Geological Survey Educational Series 3. Iowa City, 71 p
Racovitzka EG (1907) Essai sur les problemes biospeologiques. Arch Zool Exp Gen, 4e Serie, 6:371–488
Rajnovich G (1989) Visions in the quest for medicine: An interpretation of the Indian pictographs of the Canadian Shield. Midcontinental Journal of Archaeology 14(2):179–225
Ritzenthaler RE (1950) Wisconsin petroglyphs and pictographs. Wisconsin Archeologist 31(4):83–128
Roberts FHH (1940) Developments in the problem of the North American Paleo-Indian. Essays in Historical Anthropology in North America, Smithsonian Miscellaneous Collections 100:51–116
Rodden J (1981) The development of the Three Age System: Archaeology's first paradigm. In: Daniel G (ed) Towards a History of Archaeology. Thames and Hudson, London, p 51–68
Salzer RJ (1997) Wisconsin rock art. Wisconsin Archeologist 78 (2):48–76
Salzer RJ (2005) The Gottschall Site: 3,500 years of ideological continuity and change. Ontario Archaeology 79(80):109–114
Sanders EM (1921) The cycle of erosion in a karst region (after Cvijic). Geographical Review 11(4):593–604
Schermer SJ, Anderson M, Lillie RM (1996) Field Trip Guide to Iowa Rock Shelters. 54th Annual Plains Anthropological Conference, Iowa City, IA
Schiner JR (1854) Fauna der Adelsberger, Lueger und Magdalener-Grotte. In: Schimdl A (ed) Die Grotten und Hoehlen von Adelsberg, Lueg, Planina und Laas. Braunmueller, Wien, p 231–272
Schmidt VA (1982) Magnetostratigraphy of sediments in Mammoth Cave, Kentucky. Science 217(4562):827–829
Shaw TR (1992) History of Cave Science: The Exploration and Study of Limestone Caves, to 1900, 2nd edn. Sydney Speleological Society, New South Wales, Australia, 338 p
Sprengelmeyer R (2006) The rock art of Big Room Cave. Minnesota Archaeologist 65:175–194
Springer AE, Stevens LE, Anderson DE, Parnell RA, Kreamer DK, Levin LA, Flora SP (2008) A comprehensive springs classification system: Integrating geomorphic, hydrogeochemical, and ecological criteria. In: Stevens LE, Meretsky VJ (eds) Aridland Springs in North America: Ecology and Conservation. University of Arizona Press, Tucson, p 49–75
Springer AE, Stevens LE (2009) Spheres of discharge of springs. Hydrogeol J 17:83–93
Stoltman JB, Brose DS, Brown IW, Dunnell RC, Klejn LS, Meacham W, Morse DF, Odell GH, Rivera MA, Starna WA (1978) Temporal models in Prehistory: An example from eastern North America. Curr Anthropol 19(4):703–746

Storck PL (1972) The Archaeology of Mayland Cave. PhD thesis, University of Wisconsin, 704 p
Strong M (1877) Geology of Wisconsin. Commissioners of Public Printing, Madison, WI 2:661–662
Strong M (1882) Geology of Wisconsin. Commissioners of Public Printing, Madison, WI 4(75–76):96–98
Swanson SK (2013) Wisconsin's spring resources: An overview. Geoscience Wisconsin 21(1):1–13
Sweeting MM (1973) Karst Landforms. Columbia University Press, New York, 362 p
Taylor RE (2000) The contribution of radiocarbon dating to New World archaeology. Radiocarbon 42(1):1–21
Theler JL, Boszhardt RF (2000) The end of the Effigy Mound Culture: The Late Woodland to Oneota transition in southwestern Wisconsin. Midcontinental Journal of Archaeology 25(2):289–312
Theler JL, Boszhardt RF (2003) Twelve Millennia: Archaeology of the Upper Mississippi River Valley. University of Iowa Press, Iowa City, 254 p
Thiel GA (1944) The Geology and Underground Waters of Southern Minnesota. Minnesota Geological Survey Bulletin 31. University of Minnesota Press, Minneapolis, 506 p
Thwaites RG (1954) The Jesuit Relations and Allied Documents. Vanguard Press, New York
Tilly LJ (1968) The structure and dynamics of Cone Spring. Ecological Monographs 38(2):169–197
Tolman CF (1937) Ground Water. McGraw-Hill Book Company Inc, New York, 593 p
Vandel A (1965) Biospeleology: The Biology of Cavernicolous Animals. Pergamon Press, Oxford, 550 p
Webb DW, Wetzel MJ, Reed PC, Phillippe LR, Harris MA (1995) Aquatic biodiversity in Illinois springs. J Kansas Entomol Soc 68 (2):93–107
Webb DW, Wetzel MJ, Reed PC, Phillippe LR, Young TC (1996) Karst springs in the sinkhole plain of Illinois: Their community diversity and hydrogeology. Technical Report 1996 (11) Illinois Natural History Survey Center for Biodiversity, 122 p
Webb S, Feinberg JM, Wang X, Calvin EC Jr (2010) Age constraints for laminated cave sediments using detrital remanent magnetization: Mystery Cave. Minnesota. AGUFM 2010: GP13A-0767
West RM, Dallman JE (1980) Late Pleistocene and Holocene vertebrate fossil record of Wisconsin. Geoscience Wisconsin 4:25–45
Widga C, Fulton TL, Martin LD, Shapiro B (2012) Homotherium serum and Cervalces from the Great Lakes Region, USA: Geochronology, morphology and ancient DNA. Boreas 41 (4):546–556
Wigley TML, Brown MC (1976) The Physics of Caves. In: Ford TD, Cullingford CHD (eds) The Science of Speleology. Academic Press, London, p 329–358
Wilford LA (1937) Minnesota Archaeology: With Special Reference to the Mound Area. Doctoral thesis, Harvard University, 319 p
Willey GR, Sabloff JA (1974) A History of American Archaeology. WH Freeman and Company, San Francisco, CA, 252 p
Wilmsen EN (1965) An outline of early man studies in the United States. Am Antiq 31(2):172–192
Winchell NH (1878) The geology of Ramsey County. The Geological and Natural History Survey of Minnesota, Sixth Annual Report, p 66–92
Winchell NH (1884) The geology of Fillmore County. The Geological and Natural History Survey of Minnesota, Final Report 1:268–324
Winchell NH (1911) Aborigines of Minnesota. Minnesota Historical Society, St. Paul, 761 p
Wittry WL (1959) Archaeological Studies of Four Wisconsin Rockshelters. Wisconsin Archeologist 40(4):143–263
Wood HEI, Chaney RW, Clark J (1941) Nomenclature and correlation of the North American continental Tertiary. GSA Bulletin 52:1–48
Woolworth AR, Alexander EC Jr, Hedges JA (1987) Fountain Cave: Birthplace of a Metropolis. Unpublished typescript
Wopat MA (1974) Part I, The Karst of Southeastern Minnesota. M.S. Thesis, University of Wisconsin-Madison, 68 p

Cave Faunas of the Upper Mississippi Valley Region

9

Stewart B. Peck, Julian J. Lewis, and John O. Whitaker

Abstract

Over 300 non-accidental animal species are recorded from subterranean (cave and groundwater) habitats in the Upper Mississippi River Valley (UMV) region. Most of these are trogloxenes or troglophiles and these are not restricted to nor morphologically specialized for subterranean habitats. Conspicuous examples are bats, fish, moths, camel crickets, and many kinds of flies. A rich fauna of some 56 species are seemingly obligatory inhabitants of subterranean habitats. Twenty-two species of these are terrestrial, termed troglobites, and 34 species are aquatic, called stygobites. The troglobites include pseudoscorpions, mites, spiders, millipeds, collembolans, bristletails, and a beetle. These generally have smaller distributional ranges, but some do occur in states outside of the UMV region. The stygobites include flatworms, snails, amphipods, and isopods. These generally have larger distributional ranges and most occur in southwestern Illinois, which was near to, but not covered by Pleistocene glacial ice. A subset of stygobites are known only from groundwaters in non-cavernous glaciated regions. Among the vertebrates, ten species of trogloxenic cave-inhabiting bats are known from the UMV. Other vertebrates include the Spring cavefish (*Forbesichthys agassizii*), three species of plethodontid salamanders, one frog, and a bird that nests in cave entrances. Included among the mammals, other than bats, are the White-footed mouse (*Peromyscus leucopus*), Eastern woodrat *Neotoma floridana*, and Raccoon (*Procyon lotor*).

9.1 Introduction

Karst regions and caves offer special, often rigorous, environments that support an often typical set of plants and animals. Caves themselves are characterized by the absence of light, a relatively uniform temperature, nearly saturated humidity, and scarcity of food. In the Upper Mississippi Valley (hereafter UMV) region cave temperatures approximate the average annual temperature of 8–11 °C (46–52 °F) in the north and 14 °C (54 °F) in southern Illinois.

Many different kinds of animals use caves as either temporary or permanent habitats. An ecological classification, based on the level of adaptation (if any) to life in caves, uses the terms listed in Table 9.1. Assignment of any species to one of these categories is a hypothesis, based on a summary of information about the species morphology, natural history, and distribution. Likewise, the assignment of a species name to an animal is a hypothesis which can be in error, or refined through time. Species names themselves are hypotheses of genetic similarity as actually or potentially interbreeding populations. Recent overviews of North American invertebrate cave faunas are Peck (1988, 1998) for Canada and Peck (1998) and Culver et al. (2000) for the United States.

The caves and karsts of the UMV region are broadly developed in Precambrian and Paleozoic dolomite and limestone bedrocks, and locally in Mesozoic gypsum in western Iowa (see earlier chapters). There is no indication that the animals are influenced by the kind and age of the rocks containing the air or water-filled subterranean spaces (simply termed "caves" from here on). It is the environmental conditions of the caves that are important: the

S. B. Peck (✉)
Biology Department, Carleton University, Ottawa, ON, Canada
e-mail: StewartPeck@cunet.carleton.ca

Entomology, Canadian Museum of Nature, 993 Normandy Crescent, Ottawa, ON K2C 0L3, Canada

J. J. Lewis
Biological Consulting LLC, 17903 State Road 60, Borden, IN 47106, USA

J. O. Whitaker
ISU Center for Bat Research Outreach and Conservation, Indiana State University, 200 North 7th Street, Terre Haute, IN 47809, USA

G. A. Brick and E. C. Alexander Jr. (eds.), *Caves and Karst of the Upper Midwest, USA*,
Cave and Karst Systems of the World, https://doi.org/10.1007/978-3-030-54633-5_9

Table 9.1 Terms and abbreviations for groupings of cave animals

Classification	Abbreviation	Definition
Troglobite	TB	Terrestrial, morphologically adapted and restricted to caves, must feed and reproduce in the cave environment
Troglophile	TP	Terrestrial, ± morphologically adapted to caves, not restricted to caves but can feed and reproduce in the cave environment
Trogloxene	TX	Terrestrial, not usually morphologically adapted to caves, usually leaving the cave to either feed or reproduce
Stygobite	SB	Aquatic, morphologically adapted and restricted to caves, must feed and reproduce in the cave environment; also including often smaller-sized interstitial species sometimes called phreatobites
Stygophile	SP	Aquatic, ± morphologically adapted to caves, not restricted to caves but can feed and reproduce in the cave environment
Stygoxene	SX	Aquatic, not usually morphologically adapted to caves, usually leaving the cave to either feed or reproduce
Parasite	PS	Occurs obligatorily on or in another organism
Accidental	AC	Falling or washing into caves, with no demonstrable regular affiliation with cave habitats

darkness, temperature, humidity, and low food availability. What does seem to be of some importance to the composition of the fauna of a cave is if it is located in lands previously glaciated. This is reflected in the fact that unglaciated temperate cave regions contain more obligatory subterranean species, showing a longer history of specialized life in cave habitats. This may be arguable because the aquatic crustaceans *Bactrurus mucronatus* and *Caecidotea lesliei* in the glaciated Midwest are eyeless and completely unpigmented and are identical in their degree of troglomorphic adaptations (e.g., characters associated with permanent cave life such as reduced or absent eyes, loss of pigment, elongated appendages, etc.) to their relatives in the unglaciated region.

The area discussed in this chapter is divided into four sub-regions for purposes of comparison (see the maps in earlier chapters). The first sub-region is the more northerly area, comprised of the adjacent parts of the states of Iowa, Minnesota, Wisconsin, and northwestern Illinois, with deeply carved river valleys and conspicuous limestone bluffs, called the "Driftless Area." All the 24,000 square miles (62,160 km^2) of the area was once thought to be unglaciated and free of glacial till deposits (drift). This area was in fact covered by earlier (pre-Illinoian) glaciations, but has remained unglaciated for the past some 790,000 years (Iannicelli 2010) (see maps in earlier chapters).

The second sub-region is in southwestern Illinois, and is comprised of a northern Lincoln Hills area and the more important southern area of the Salem Plateau section of the Ozarks. Most of this sub-region was very close to or perhaps partly overridden by the margin of the ice of continental Illinoian glaciation (about 130,000–190,000 yr BP) in the Pleistocene (Flint et al. 1959).

The third sub-region of caves occurs in southern Illinois in the Shawnee Hills section of the Interior Low Plateaus, and the caves are a few tens of km south of the limits of Illinoian glaciation (Flint et al. 1959). The east-west trending Cache Valley, forming the northern boundary of the Shawnee Hills, is interpreted as the former main channel of an earlier Ohio River (Alexander and Prior 1968) so there was at that time less of a river barrier between the Shawnee Hills and the Interior Low Plateaus of adjacent Kentucky.

The fourth sub-region is comprised of the flat till (drift) plains, the remnants of the Pleistocene glaciations. By area the till plains comprise the majority of the UMV region. Although soluble limestone or dolomite is present in many areas, here it is buried at varied depths under the glacial till and any caves present are inaccessible and likely sediment-filled as well as below the water table. These till fields present habitat for very small inhabitants of saturated soil interstices, but are seemingly unavailable for larger-bodied cave animals.

There have been few detailed studies of environmental conditions and cave faunas in the UMV region. Hubricht (1943) was the first to include specimens from the region's caves in his report on amphipods. Tilly (1968) presents a detailed study of the energetics and trophic dynamics of Cone Spring, a cold karst spring at Conesville, Louisa County, Iowa. *Gammarus pseudolimnaeus* Bousfield amphipods were responsible for almost one half of the primary consumer energy flow, but no obligate subterranean species were reported. Christiansen et al. (1961) conducted an ecological study of the invertebrates of Hunter's Cave, Iowa. Koch and Case (1974) gave details of a faunal survey of Coldwater Cave, Iowa. Peck and Lewis (1978) inventoried the invertebrate fauna of Illinois and adjacent Missouri, and reported 215 non-accidental species. Peck and Christiansen (1990) listed the invertebrate cave fauna of the Driftless Area, with 94 non-accidental species. Montz (1993) studied the aquatic invertebrates of Mystery Cave, Minnesota. Webb et al. (1993) summarized the faunas of 98 caves and other subterranean sites in 13 counties in Illinois, with 5810 invertebrate specimens and 215 genus level identifications in all animal groups, but much of the material was seemingly not identified. Webb et al. (1995) reiterated the spring fauna collection records of Peck and Lewis (1978) and added selected water chemistry data. Lewis et al. (2003) inventoried 71 sites in Monroe and St. Clair counties in southwestern Illinois for subterranean invertebrates, found 24 species to be obligate cave species and evaluated these from a conservation perspective. Webb et al. (1996, 1998b) listed the fauna of many Illinois springs and there were some records of subterranean species. All in all, the cave faunas of Illinois are the best studied and those of Minnesota and Wisconsin the least studied.

9.2 Summary of the Subterranean Faunas of the UMV Region

Comparisons show the cave faunas of the glaciated UMV region to be generally similar to those inventoried in more easterly Ontario, Canada (Peck 1988). Excluding accidentals, a large number of the species live in subterranean habitats as trogloxenes and troglophiles. The major differences in the unglaciated areas of southern Illinois are the larger numbers of troglobites and stygobites, similar in some ways to that described for the unglaciated karst of south-central Indiana (Lewis 2012). The cave fauna of southern Illinois is generally less diverse than that of adjacent southern Indiana, a fact attributed to the greater distance from the presumed Appalachian source area of the fauna. Thus, a diverse fauna of groups like the cave beetles of the genus *Pseudanophthalmus* or the cave milliped *Pseudotremia* is present in Indiana, but only single species of each genus are present in southern Illinois. Some of the species are still not formally named and described.

9.3 Invertebrates

The following is a summary mostly derived from Peck and Lewis (1978), Peck and Christiansen (1990), and Lewis et al. (2003). Over 215 invertebrate species have been recorded from caves in southwestern and southern Illinois and adjacent Missouri, and another nearly 100 species from caves in the Driftless Area of the UMV region. The following discussion briefly introduces mostly major and common groups of trogloxenes-stygoxenes and troglophile-stygophiles and notes some of their group features. The troglobites and stygobites are summarized in Table 9.2.

Platyhelminthes—Flatworms. The stygobite *Sphalloplana hubrichti* Hyman and the stygophile *Phagocata gracilis* (Haldeman) are reported from Illinois cave streams. An unidentified white flatworm has been reported from Carver's Cave, Ramsey County, Minnesota (Brick 2009a).

Annelida—Annelid worms. Oligochaeta. Earthworms (Clitellata) may occur in cave soils, especially in or near entrances with organic debris. Thirteen species are reported, but only *Allolobophora trapezoides* (Duges), *Allolobophora tuberculata* Eisen, and *Dendrobaena rubida* (Savigny) are known from more than one cave. As in surface habitats, the native earthworm fauna is being replaced by non-native species, largely introduced from Europe. Aquatic oligochaete annelids have been little studied, but could be expected in most cave streams. Wetzel and Taylor (2001) report 16 species (13 named to species) from cave streams in southwest Illinois and adjacent Missouri. Leeches (Hirudinea) occasionally occur in cave streams, where they may feed on crustaceans and aquatic insects.

Mollusca—Gastropoda—snails. Among the aquatic snails are physids variously identified as *Physa halei* Lea or *Physella* in caves in southwestern Illinois. These snails frequently have translucent shells that suggest troglomorphy, but their ecological status remains questionable.

The spring snails of the family Hydrobiidae are represented by *Fontigens antroecetes* (Hubricht), *F. aldrichi* (Call and Beecher), and *F. nickliniana* (Lea) (Hershler et al. 1990). The rarest of these is known only from the type-locality, Stemler Cave, St. Clair County, Illinois, and a few caves in adjacent Missouri. *F. aldrichi* is known from springs and caves, primarily in the Missouri Ozarks, but two localities are known from Fulton and LaSalle counties, Illinois. *F. nickliniana* is widespread in the eastern US, with many localities around the Great Lakes, including northern Illinois and Wisconsin. This species occurs in a variety of habitats, but in the UMV region it has a penchant for inhabiting emergent groundwater sites, such as at seeps, springs, and fens.

Land snails may occur in the twilight zone of cave entrances. *Discus macclintocki* Baker of Bixby Ice Cave, Clayton County, Iowa is in discontinuous cold habitats elsewhere (algific talus slopes, see chapter on Illinois) in the Driftless Area and the Appalachians. Lewis et al. (2003) reported terrestrial snails from Stemler Cave that were collected from sediment below a former sinkhole with input to the cave. An interesting insight was given by searching other caves and surface habitats, but this has failed to produce other records of some of these species, suggesting that they may represent the snail fauna of the late Pleistocene.

Table 9.2 A summary listing of species currently considered to be stygobites in aquatic (A) and troglobites in terrestrial (B) subterranean (cave and groundwater) environments in the UMV region. "?" indicates uncertainty of the status as a troglobite or stygobite. More details can be found in Peck and Lewis (1978), Peck and Christiansen (1990), Lewis et al. (2003), and Soto-Adames and Taylor (2013)

Higher taxa, genus and species name	UMV region State, and County with species records; and extra regional records
A. Aquatic subterranean habitats: stygobitic animals found only in cave streams, drip-pools, and groundwaters (these last usually sampled by wells and at seeps and drain tile outlets and which are sometimes also called phreatobites)	
Platyhelminthes, Turbellaria, Alloeocoelia	
Unnamed alloeocoel	WI: Richland
Turbellaria, Tricladida	
Kenkia cf glandulosa Hyman	IA: Jackson
Sphalloplana hubrichti (Hyman)	IL: Jackson, Monroe, St Clair, Union; also MO
Mollusca, Gastropoda, Pulmonata	
?*Physa halei* Lea	IL: Monroe (possibly in *Physella*; status of species taxonomy and ecological classification confused)
Gastropoda, Ctenobranchiata	
Fontigens antroecetes (Hubricht)	IL: St Clair; also MO
Arthropoda	
Crustacea, Cyclopoida	
Diacyclops yeatmani Reid	IL: Vermillion; also IN and other states
(= *C. clandestinus* Yeatman)	
Crustacea, Isopoda, Asellidae	
Caecidotea beattyi Lewis and Bowman	IL: southern and central Illinois into Indiana
Caecidotea bicrenata whitei Lewis and Bowman	IL: Union, Johnson; KY
Caecidotea kendeighi (Steves and Seidenberg)	IL: Champaign; many counties in IL, plus IA, IN, OH
Caedicotea packardi Mackin and Hubricht	IL: Adams, Monroe, Pike, St Clair; also MO
Caecidotea stygia Packard	IL: Hardin
Caecidotea spatulata Mackin and Hubricht	IL: St Clair; also MO
Crustacea, Amphipoda, Crangonyctidae	
Bactrurus brachycaudatus Hubricht and Mackin	IL: Calhoun, Green, Jackson, Johnson, Monroe, St Clair, Pike, Union; also MO
Bactrurus mucronatus (Forbes)	IL: Champaign, Fulton, Henderson, Knox, LaSalle, McDonough, McLean, Peoria, Saline, Vermillion, Warren; also IA, IN, OH, MI
Crangonyx packardi Smith	IL: Hardin, Johnson, Pike, Saline, Union
Stygobromus iowae Hubricht	IA: Fayette, Winneshiek; IL: Jo Daviess
Stygobromus lucifugus (Hay)	IL: Knox
Stygobromus putealis (Holmes)	WI: Dodge, Fond du Lac, Green
Stygobromus subtilis (Hubricht)	IL: Adams, Monroe, Jackson, Jersey, Union; also MO
Stygobromus sp 1	IL: Hardin
Stygobromus sp 2	IL: Pope
Gammaridae	
Gammarus acherondytes Hubricht and Mackin	IL: Monroe, St Clair

(continued)

Table 9.2 (continued)

Higher taxa, genus and species name	UMV region State, and County with species records; and extra regional records
B. Terrestrial subterranean habitats; troglobitic animals found only in sub-aerial (air-filled) subterranean habitats	
Arthropoda	
Arachnida, Pseudoscorpionida	
Apochthonius undescribed sp	IL: Hardin
Mundochthonius cavernicolous Muchmore	IL: Monroe; Not MO, contra Gardner (1986)
Arachnida, Aranea, Linyphiidae	
Phanetta subterranea (Emerton)	IL: Carroll, Hardin, Jackson, Johnson, Monroe, Saline, St Claire, Union; also widespread in over 12 eastern states; only in caves
Porrhomma cavernicolum (Keyserling)	IL: Jo Daviess, Monroe; IA: Jackson, also sporadic but widespread eastern US
Arachnida, Acarina, Rhagidiidae	
?*Robustocheles occulta* Zacharda and Pugsley	IA: Winneshiek
Myriapoda, Diplopoda, Nearctodesmidae	
Ergodesmus remingtoni (Hoffman)	IL: Adams, Hardin, Jersey, Monroe, Pike
Macrosternodesmidae (formerly Trichopolydesmidae)	
Chaetaspis (= *Antriadesmus*) sp.	IL: Monroe
Nemasomatidae	
Zosteractis interminata Loomis	IL: Pike
Insecta, Collembola, Arrhopalitidae	
Pygmarrhopalites ater (Christiansen and Bellinger)	IL: Monroe; also IN: Harrison, Washington
Pygmarrhopalites bimus (Christiansen)	IA: Delaware, Dubuque; also IN: Lawrence
Pygmarrhopalites carolynae (Christiansen and Bellinger)	IL: Monroe; also IN
Pygmarrhopalites dubius (Christiansen)	IA: Dubuque, Jackson; MN: Fillmore; also MI: Ingham; AR: Marion
Pygmarrhopalites fransjanssens Soto-Adames and Taylor	IL: St Clair
Pygmarrhopalites hirtus (Christiansen)	IA: Clayton, Jackson; IL: Monroe, Pike, Union; WI: Pierce, Richland, Sauk; also CA, KY, OH, OR, VA, PQ
?*Pygmarrhopalites incantator* Soto-Adames and Taylor	IL: St Clair
Pygmarrhopalites lewisi (Christiansen and Bellinger)	IL: Monroe; also IN
Pygmarrhopalites madonnensis (Zeppelini and Christiansen)	IL: Monroe
Pygmarrhopalites salemensis Soto-Adames	IL: Monroe, St Claire
Pygmarrhopalites sapo (Zeppelini and Christiansen)	IL: Monroe
Entomobryidae	
Pseudosinella sp 1 *argentea* complex	IL: Hardin, Monroe; also MO (Gardner 1986)
Pseudosinella sp 2	IL: Johnson
Hypogastruridae	
Ceratophysella cf. lucifuga (Packard)	IL: Monroe; IN: Crawford, Harrison

(continued)

Table 9.2 (continued)

Higher taxa, genus and species name	UMV region State, and County with species records; and extra regional records
Oncopoduridae	
Oncopodura iowae Christiansen	IA: Dubuque, Jackson; IL: Monroe, St Claire; Also MO, SD
Onychiuridae	
Onychiurus gelus Christiansen and Bellinger	IA: Winneshiek; WI: Richland
Onychiurus obesus Mills	IA: Clayton, Delaware, Dubuque, Jackson; WI: Pierce, Sauk; MN: Fillmore
Onychiurus pipistrellae Soto-Adames and Taylor	IL: Monroe, St Claire
Tullbergia hades Christiansen and Bellinger	IA: Jackson; MI: Wabasha
Tomoceridae	
Tomocerus (*Lethemurus*) *missus* Mills	IL: Jersey, Monroe, St Claire; also AL, CO, IN, KY, MO, TN, VA
Insecta, Diplura, Campodeidae	
?*Eumesocampa* sp	IL: Monroe, Pike, St Claire, Union; also MO
?*Haplocampa* sp	IL: Monroe, St Claire; also MO
?*Metriocampa* (*Tricampa*) sp	IL: Adams, Hardin, Saline
Insecta, Coleoptera, Carabidae	
Pseudanophthalmus illinoisensis Barr and Peck	IL: Hardin
Insecta, Diptera, Sphaeroceridae	
Spelobia tenebrarum (Aldrich)	IL: Calhoun, Hardin, Jackson, Monroe, Pope, Saline, St Clair, Union; also AL, AR, GA, IN, KY, MO, NY, PA, TN, WV
C. Species previously but incorrectly reported as troglobites or stygobites in the UMV region	
Arthropoda, Crustacea	
Isopoda, Asellidae	
Caecidotea tridentata Hungerford	IL: LaSalle (erroneously reported from Illinois, endemic to non-cave groundwater in Kansas)
Amphipoda, Crangonyctidae	
Crangonyx forbesi Hubricht and Mackin	IL: Calhoun, Gallatin, Hardin, Jackson, Jersey, Johnson, Monroe, Pike Randolph, St Clair; (this species was never considered as a stygobite; mostly occurs in springs and surface streams)
Arachnida, Aranea, Linyphiidae	
Bathyphantes weyeri (Emerton)	WI: Richland; also widespread in PA, VA, AR, KY, WV (known only from caves in the southern part of its range, possibly to be found in surface habitats)
"Myriapoda"	
Diplopoda, Trichopetalidae	
Scoterpes sp:	IL: Monroe; erroneously previously reported
Cleidogonidae	
Pseudotremia salisae Lewis	IL: Hardin (originally assumed to be a troglobite but this widespread species was recorded from a surface locality in Ohio; only in caves in Indiana and Illinois)
Insecta, Collembola, Onychiuridae	
Onychiurus reluctus Christiansen	IA: Jackson, Fayette counties; originally considered a troglobite but later declared a troglophile
Coleoptera, Leiodidae	
Ptomapahagus nicholasi Barr	IL: Monroe; described in error from mislabeled Kentucky specimens

9.3.1 Arthropoda—Arthropods

9.3.1.1 Crustacea

Cyclopoida—copepods. Lewis et al. (2003) and Lewis and Reid (2007) reported 11 species in Illinois cave waters, but most are little more than transient visitors. In contrast to the cave fauna in which no stygobites are known to occur in Illinois, *Diacyclops yeatmani* Reid (formerly called *Cyclops clandestinus* Yeatman) is an obligate groundwater inhabitant in the saturated interstices of soil and has been recorded from an agricultural field drain outlet in Vermillion County, Illinois. This species has also been found deep in stream gravel in Indiana and pools in caves in Tennessee (Reid 2004).

Isopoda—isopods. Seven aquatic isopod species of the genus *Caecidotea* are obligate inhabitants of subterranean waters in the region (Lewis and Bowman 1981; Lewis 1982) (Fig. 9.1). Of these, *Caecidotea stygia* Packard and *C. bicrenata whitei* Lewis and Bowman occur in caves of the Shawnee Hills while *C. packardi* Mackin and Hubricht occur in the Illinois Ozarks along the Mississippi River. None of these species is restricted to Illinois in distribution.

Four of the *Caecidotea* reported by Lewis and Bowman (1981) are obligatory groundwater inhabitants of saturated soil interstices, but rarely if ever occur in caves. All are depigmented with eyes vestigial or absent—in appearance these isopods look like their cavernicolous relatives. The majority of collections of these species are usually made from drain tiles placed to enhance drainage of farm fields. By far the most widespread of the saturated soil inhabiting species is *C. kendeighi* (Steeves and Seidenberg) that occurs from eastern Iowa across much of Illinois and Indiana to western Ohio (Lewis 2015). *Caecidotea beattyi* Lewis and Bowman was first found at Dixon Springs, Pope County, Illinois, but was subsequently found from eastern Missouri across Illinois to western Indiana. The rarest of this group of species is *C. lesliei* Lewis and Bowman known from a single drain tile outlet in McDonough County, Illinois. *Caecidotea spatulata* (Mackin and Hubricht) departs from the habitat norm since it has been reported from temporary pools, swales, and springs. With vestigial eyes and pigmentation it conforms in morphological appearance to the four species in this assemblage, but Mackin and Hubricht (1940) reported that *C. spatulata* had the ability to aestivate when their temporary pools dried.

Fig. 9.1 *Caecidotea kendeighi*, a stygobitic isopod that occurs in the UMV till plain in shallow groundwater habitats and occasionally occurs in caves, which may be secondary habitats.

Caecidotea brevicauda (Forbes) is generally an inhabitant of springs across its range, but in the caves of Monroe and St. Clair counties, Illinois, slightly troglomorphic populations occur and it generally greatly outnumbers the stygobite *C. packardi* Mackin and Hubricht with which it co-occurs. An unusual, slightly troglomorphic population of *Caecidotea intermedia* occurs in Niagara Cave, southern Minnesota. The occurrence of a somewhat troglomorphic population of this widespread epigean species is similar to the troglomorphic population closely resembling *C. forbesi* in a cave in the Lake Erie Islands, Ohio, that Lewis (2015) judged to be sufficiently isolated to merit description as *C. insula*. Unlike *C. insula,* found only on South Bass Island, the isopod population in Niagara Cave has no demonstrable barrier to dispersal that would suggest genetic isolation and speciation.

Four genera of terrestrial isopods (pill bugs or sow bugs) are known from Illinois caves, and these are primarily non-native species accidentally introduced from Europe.

Amphipoda—amphipods. The aquatic amphipod fauna is a rich one, with over four genera and 15 species, of which many are stygobites (see Table 9.2). The Illinois cave amphipod *Gammarus acherondytes* Hubricht and Mackin is of special conservation interest and is a federally listed endangered species (Webb et al. 1998a). Sampling conducted by Lewis et al. (2003) doubled the number of caves from which this unique amphipod is known. Panno et al. (2006) reported organic pollution with concomitant low dissolved oxygen levels in caves where the species occurs. Its metabolic rate was measured by Wilhelm et al. (2006). Lewis and Lewis (2014) have monitored the populations of the Illinois cave amphipod since its listing as an endangered species and reported that the only population in St. Clair County, in Stemler Cave, appears to have been extirpated by septic system pollution. Likewise, other populations in Fogelpole Cave and Illinois Caverns (Fig. 9.2) have been significantly reduced by nutrient enrichment from septic systems.

Besides *Gammarus acherondytes*, two other species of *Gammarus* are common in caves in western Illinois: *G. troglophilus* (Hubricht and Mackin) and *G. minus* Say. As

Fig. 9.2 Cavers conducting a transect for stream fauna through Illinois Caverns, Monroe County, Illinois.

the name implies, *G. troglophilus* is a common stygophile and tends to become the dominant amphipod species in cave stream communities where nutrient enrichment occurs. *G. minus* is a spring inhabitant that also occurs in cave entrances. *Gammarus pseudolimnaeus* Bousfield is common in Iowa caves (Peck and Christiansen 1990), inhabits the lake inside Carver's Cave (Brick 2000), is found in springs throughout southeastern Minnesota (Muck and Newman 1992) and is widely distributed in east central USA.

Amphipods of the genus *Crangonyx* were discussed in detail in a revision of the group by Zhang and Holsinger (2003). *Crangonyx packardi* Smith is a widespread stygobite occurring in caves, springs, and seeps from Indiana through southern Illinois, Missouri to western Kansas. In Illinois it has been recorded from caves and springs in Hardin, Johnson, Saline, Union, and Pike counties. *Crangonyx forbesi* Hubricht and Mackin is a stygophile with an even wider geographic range than that of *C. packardi*. In Illinois *C. forbesi* is recorded from eight counties, although the majority of records are from caves and springs in Monroe County. *C. anomalus* Hubricht, a species that mostly occurs in springs in Kentucky, Ohio, and Indiana, has been recorded in one spring on the western edge of its range, in Pope County, Illinois.

The large, widespread subterranean amphipod genus *Stygobromus* (Fig. 9.3) is represented in Illinois by *S. subtilis* (Hubricht) in caves and seeps in western Illinois, and two undescribed species in Dixon Springs, Pope County and a seep inside a sandstone cave in Hardin County. All of these species are rare. Similar to the situation found with the *Stygobromus* species in Illinois, two other species occur in the UMV region: *S. putealis* (Holmes) and *S. iowae* Hubricht. *S. putealis* is reported only from wells in Dodge, Fond du Lac, and Green Lake counties, Wisconsin (Jass 1994). Almost nothing is known of this species and only one population was known to be extant in 1994. Holsinger (1972) reported that *S. iowae* was known from two caves and a spring in northeastern Iowa and a mine in Jo Daviess County, Illinois. This species has been found recently in a cave in Minnesota (Warren Netherton, pers. comm., 2016).

Bactrurus brachycaudus Hubricht and Mackin is the largest subterranean amphipod in North America. This species is known primarily from caves, as well as seeps and springs in western Illinois and Missouri, and a few drain tile outlets in the west-central Illinois till plains (Koenemann and Holsinger 2001). *Bactrurus mucronatus* (Forbes) occurs primarily in saturated soil interstitial habitats of the glacial plains from eastern Iowa east to Ohio and as far north as Michigan.

Decapoda. Unlike adjacent states of Indiana and Missouri, where stygobitic crayfish occur, the occurrence of crayfish (*Cambarus* spp.) is relatively unusual in Illinois caves and none are stygobites. An unconfirmed and most unlikely newspaper report of "blind crayfish" exists for Carver's Cave, Ramsey County, Minnesota (Anonymous 1913). Webb et al. (1993) list Illinois cave crayfish and many associated ecto-commensal and epi-parasitic branchiobdellid worms, harpacticoid copepods and entero-cytherid ostracods.

Fig. 9.3 *Bactrurus mucronatus*, a stygobitic amphipod that occurs in the UMV till plain in shallow groundwater habitats and occasionally occurs in caves, which may be secondary habitats.

9.3.1.2 Arachnida

Pseudoscorpionida. The rare troglobitic pseudoscorpion *Mundochthonius cavernicolous* Muchmore is known from Saltpeter and Fogelpole caves, Monroe County, Illinois (Lewis 2003). An *Apochthonius* endemic to Brown's Hole, Hardin County, Illinois remains undescribed. Other pseudoscorpions (*Chthonius* spp. s. latu) are occasionally found around cave entrances.

Acarina. Mites are common, but poorly known. Many species have been found in many caves, probably mostly as accidentals (see details in Peck and Christiansen 1990; Peck and Lewis 1978). Some probably maintain populations in caves as troglophiles. The rhagidiid mite *Robustocheles occulta* Zacharda and Pugsley was reported from a cave in Iowa (Zacharda and Puglsey 1988) and could be considered a troglobite.

Opiliones. The common "daddy long-legs" (*Leiobunum* spp.) of forests use cave ceilings near entrances as daytime retreats. They may be found in clusters of dozens of individuals. Contrary to its name, the short-legged harvestman *Sabicon cavicolens* (Packard) occurs mostly in sinkhole floors and around cave entrances and it was reported from caves in Johnson and Monroe counties, Illinois (Lewis 2003).

Aranea. A diverse spider fauna exists in the caves of the UMV region. All are predators with a diverse array of webs used to capture prey. The most commonly encountered species is the large web-spinning *Meta ovalis* (Gertsch) (Fig. 9.4). North American populations were formerly called *Meta menardi* (Latreille). The webs and egg sacs of *Meta* are frequently obvious and abundant not far from entrances. In contrast, the troglobitic sheet-web spiders *Phanetta subterranea* (Emerton) and *Porrhomma cavernicola* (Keyserling) are tiny (2 mm) and cryptic, placing their webs and egg sacs on the undersides of rocks on cave floors. Both species occur in caves across the eastern US, with *Phanetta* occurring in the majority of caves within its broad range. Sheet-web spiders of the subfamily Erigoninae are mostly found at latitudes of cooler climate, but are frequently found in caves and sinkhole floors that provide similar cool habitats in areas of warmer climate (e.g., species of *Eperigone* and *Bathyphantes*). The dictynid meshweaver spiders *Cicurina* have troglobitic species in more southerly latitudes (e.g., Texas), but in the Midwest several species are trogloxenes or troglophiles. The hahniid spider *Calymmaria persica* (Heiss and Draney) (formerly *C. cavicola* Banks) is another occasional threshold trogloxene.

Fig. 9.4 A female spider *Meta ovalis* with egg sac. These most frequently occur near cave entrances where there are more flies as food items.

9.3.1.3 "Myriapoda"

Chilopoda. Centipedes are found in leaf litter in cave entrances in the region, but none appear to have any particular association with caves. The house centipede *Scutigera coleoptrata* (Linnaeus) was reported in abundance from artificial voids in sandstone around the foundations of buildings in the downtown business district of St. Paul, Minnesota (Brick 2009b, p 183).

Diplopoda. A diverse array of millipeds occurs in UMV region caves, with cavernicolous species representing four orders and numerous families. The first of these orders is Polydesmida, represented by the common forest millipeds of the genus *Pseudopolydesmus* that are occasional threshhold trogloxenes (Peck and Lewis 1978). *Ergodesmus remingtoni* (Hoffman) is a troglobite endemic to Illinois where it has been reported mostly from caves in the counties along the Mississippi River, and disjunctly from Cave Spring Cave, Hardin County (Lewis et al. 2003). This species is zoogeographically unique as the only representative of the Family Nearctodesmidae occurring outside of the Pacific Northwest (Shelley 1994). Peck and Lewis (1978) reported *Antriadesmus*, now placed in the genus *Chaetaspis* (Lewis 2002), from Pautler Cave, Monroe County, Illinois. Lewis et al. (2003) reported a collection of this milliped, from Pautler as well as two other caves in the same cave system, but a male is necessary for identification of the species and all specimens collected to date have been juveniles and females.

The second group of millipeds, Order Chordeumatida, contains numerous troglobites and troglophiles in eastern North America and is represented in the UMV region by several interesting species in the family Cleidogonidae. The large Appalachian genus *Pseudotremia* (Fig. 9.5) is known in the UMV region from Cave Spring Cave, Hardin County, Illinois. *Pseudotremia salisae* Lewis was described by Lewis (2000) from a cave in southern Indiana and was subsequently identified by Shear et al. (2007) as the species in Cave Spring Cave. Another millipede in the same family, *Cleidogona unita* Causey, is a trogloxene reported by Peck and Lewis (1978) from southern Illinois caves. This species is also known from epigean sites in southern Illinois, the Mammoth Cave area of Kentucky (Shear 1972), and unpublished sites in sinkhole floors in southern Indiana (Lewis in progress 2016).

In the Family Conotylidae, *Austrotyla specus* (Loomis) is a rather widespread troglophile occurring from southern Illinois and eastern Missouri through eastern Iowa to southern Wisconsin and Minnesota. Like other conotylids, *A. specus* has troglomorphic populations in the southern part of this range where it is found almost exclusively in caves. In the northern part of the range it has epigean populations where the millipeds are pigmented (Shear 1971). *Acheminides* (formerly *Conotyla*) *pectinatus* (Causey) is another conotylid endemic to the Driftless Area where it has been

Fig. 9.5 A troglobitic *Pseudotremia* sp. milliped feeding on decaying organic matter. © J. Lewis 2019. All Rights Reserved

found only in caves and mines (Peck and Lewis 1978). Shear (1971) cited the reduced ocelli and relatively large size as subterranean adaptations suggesting that the species may be troglobitic. *Tingupa pallida* Loomis (Family Tingupidae) is a troglobite that occurs in southern Illinois and southeastern Missouri (Shear and Shelley 2007). Among the trichopetalids, Peck and Lewis (1978) noted that Shear (1972) had suggested the presence of a milliped of the genus *Scoterpes* in Illinois localities, but these were unspecified and no populations have been discovered (Lewis et al. 2003) suggesting the record is an error.

The Order Julida is represented by *Zosteractis interminata* Loomis, a troglobite reported from Lost Creek Cave, Pike County, Illinois and other caves in adjacent Missouri (Peck and Lewis 1978). The final group, the Order Spirostreptida, contains no troglobites in the UMV region, but the widespread troglophile *Cambala minor* (Bollman) is reported from Illinois caves (Peck and Lewis 1978) where it is on the western edge of its range (Hoffman 1999). This species is frequently found in raccoon latrines in caves where it can be quite numerous.

9.3.1.4 Insecta

Aquatic insects of many orders (frequently the juvenile aquatic stages of dragonflies, mayflies, stoneflies, true flies, and water beetles) may be present in cave streams. These are usually washed into the caves and may possibly mature in caves, but do not form reproducing populations. Regardless, these organisms appear to be mostly important as a food source for other members of the ecosystem.

Collembola. Springtails are common and diverse, with over 50 taxa reported. The most in-depth report focused on cavernicolous collembolans was that of Soto-Adames and Taylor (2013), who reported 49 species in a survey of caves in southwestern Illinois. Although not focused solely on springtails, Peck and Lewis (1978) reported 39 taxa in the same general region.

Some troglophiles like *Pogonognathellus* (formerly *Tomocerus*) *flavescens* Tullberg, *P. bidentatus* Folsom, *Folsomia candida* Willem, and *Pygmarrhopalites pygmaeus* (Wankel) are widespread with ranges that include much of the UMV region.

Hundreds of individuals of *Folsomia candida* Willem were found on several occasions in heavily diesel contaminated observation wells (2.5 m deep) at the public works garage of Waterville, Le Sueur County, Minnesota, as identified by K.A. Christiansen in 2001 and reported by Brick (2006).

The status of many species, for example, the entomobryid *Pseudosinella argentea* Folsom, is uncertain. It is difficult to ascertain whether collembolans like *P. argentea* are one or a complex of closely related species. This makes ecological classification more or less impossible since *P. argentea* s. latu is reported from many caves, as well as a grave in the District of Colombia and a surface collection in Connecticut. There may be a complex of closely related troglobitic species inhabiting caves and surface populations of a troglophilic relative in more northerly latitudes, or alternately the taxon might be one widespread troglomorphic troglophile. The taxonomy of these microarthropods is difficult to formalize due to the phenotypic plasticity displayed by many taxa currently considered to be species.

The ecological classification of springtails is further complicated by the demonstrated presence of some species like *Sinella cavernarum* (Packard) and the aforementioned *Pseudosinella argentea* in edaphic (soil) habitats, as demonstrated by their collection from a grave along with dozens of caves over a wide area of the eastern US. Despite this, many species are troglomorphic and known only from caves and are likely true troglobites. Examples include the Iowa cave springtail *Oncopodura iowae* Mills or *Lethemurus missus* (Mills).

Zeppelini and Christiansen (2003) report numerous collections of the genus *Pygmarrhopalites* (formerly *Arrhopalites*, see Zeppelini et al. 2009) from caves in Illinois (19), Iowa (12), and other states. Some of the species of springtails of this speciose genus have been reported from relatively widespread sites, like *P. lewisi* (Christiansen and Bellinger) and *P. ater* (Christiansen and Bellinger) in Illinois and Indiana, or *P. carolynae* (Christiansen and Bellinger) and *P. hirtus* (Christiansen) with even wider ranges. These species occur along the southern margin of the Pleistocene Illinoian glacier and may represent widespread cave invasion of an epigean species that has not yet resulted in diverse speciation.

Diplura—Bristletails. Since all species in this order of soil-dwelling insects are colorless and without eyes, their degree of association with caves is difficult to assess. In caves of the Shawnee Hills in southern Illiniois diplurans are very rare, with only one undescribed species of *Metriocampa* known from Brown's Hole, Hardin County and Equality Cave, Saline County. This species is most closely related to a non-cavernicolous species from Colorado (Peck and Lewis 1978). In contrast, in the Ozark region of southwestern Illinois undescribed species of *Eumesocampa* (Fig. 9.6) and *Haplocampa* were reported from numerous caves and sometimes come to cheese bait by the dozens (Lewis et al. 2003). Two widespread endogean species of diplurans are known from two caves in the Driftless Area of Iowa, *Metriocampa vandykei* Silvestri from Dancehall Cave, Jackson County and *Campodea (C.) fragilis* Meinert from Horsethief Cave, Linn County.

Orthoptera. Cave (or camel) crickets, represented by 5 species of *Ceuthophilus*, are common as trogloxenes near cave entrances (Fig. 9.7). They leave the caves at night to feed as omnivores. Their droppings (guano) are an important food for other scavenging insects under their roosts on cave ceilings.

Coleoptera. About 50 species of beetles have been identified in caves of the UMV region, but many are probably accidentals. Around 20 species are regular troglophiles, of which some are the common staphylinid rove beetles *Quedius erythrogaster* Mannerheim, *Q. fulgidus* (Fabricius), and *Q. spelaeus* Horn (Peck and Thayer 2003). Small aleocharine rove beetles, e.g., *Aleochara lucifuga* (Casey), are well represented (Klimaszewski and Peck 1986). Notable is the presence of the leiodid *Ptomaphagus cavernicola* Schwarz in Hunter's Cave, Iowa. It is also found in 10 caves in southwestern Illinois and more commonly in caves in the Ozarks (in over 50 Missouri caves, Gardner 1986) and especially Texas (Peck 1982a). Otherwise there is one large Florida cave population and it is also known from two forest collections (Peck 1982b). The supposed troglobite *Ptomaphagus nicholasi* Barr, from Fogelpole Cave, Illinois, was found to be a synonym of *P. hirtus* based on mislabeled specimens actually collected in Kentucky (Peck 1984). The only troglobitic beetle in the region is the carabid *Pseudanophthalmus illinoisensis* Barr and Peck from Cave Spring Cave, Hardin County, Illinois. This is a large genus with over 200 species (many still unnamed) from Indiana and Kentucky to Virginia to Alabama with all but one

Fig. 9.6 A *Eumesocampa* sp. dipluran, a bristletail, on a rock coated with camel cricket dung.

Fig. 9.7 *Ceuthophilus gracilipes* and other species of camel crickets mostly occur in roosts on cave ceilings.

species in West Virginia limited to caves. The absence of *Pseudanophthalmus* in other parts of Illinois is somewhat of a mystery, considering the recent discovery of species in Missouri that remain undescribed.

Lepidoptera. Moths are mostly represented by the medium-sized reddish Herald moth *Scoliopteryx libatrix* (Linnaeus) (Noctuiidae), which uses caves as overwintering sites.

Diptera. Many species of flies use caves as daytime refuges, such as crane flies (Tipulidae), winter crane flies (Trichoceridae) (Fig. 9.8), and fungus gnats (Mycetophilidae). Fungus gnats are abundant in the dystrophic guanophile ecosystem of Schieks Cave, under downtown Minneapolis, Minnesota, flooded with raw sewage from leaking pipes (Brick 2009c). Inseminated female mosquitos (Culicidae, *Anopheles* and *Culex*) overwinter in caves. Small

Fig. 9.8 Trichocerid winter crane flies are one of the many kinds of two-winged flies which use caves as seasonal shelters.

Megaselia cavernicola Brues (Phoridae) and *Spelobia* species (Sphaeroceridae) are common scavenging troglophiles. *Spelobia tenebrarum* (Aldritch) is a troglobite known from hundreds of populations in at least 10 states, including Illinois, and occurs in most caves within its range (Marshall and Peck 1984). Larger reddish or yellowish *Aecothea specus* (Aldrich), *Amoebaleria defessa* (Osten Sacken), *A. sackeni* Garrett, and *Heleomyza brachyptera* (Loew) (all Heleomyzidae) overwinter in caves and occur there in the summer as well. Most heleomyzid and many other flies, as well as moths, are found on cave walls and ceilings and are part of what is called the "parietal" (wall and ceiling) fauna.

9.4 Vertebrates

9.4.1 Chiroptera—Bats

Bats are the most frequent vertebrates in caves, and are normally present during their winter hibernation. However, some individuals or male colonies (sometimes including non-breeding females) may be present in summer. In addition, Gray bats are true cave bats, meaning that they live in caves both summer and winter, although they use different caves during the two seasons. Mating of bats is usually in fall although some individuals may mate in winter or spring. Bats have delayed fertilization which means that the sperm stay in the reproductive tract and do not fertilize the eggs until spring. Mating very often occurs in swarms at cave entrances (Whitaker and Hamilton 1998).

Cave bats are facing a severe survival problem because of extensive mortality as a result of a fungal pathogen (*Pseudogymnoascus destructans*) which causes white nose syndrome (WNS). This syndrome leads to disruption in waking patterns during hibernation, starvation, and death. White nose syndrome was first found in Howe Caverns in New York State in 2006. It then spread rapidly to most of the northeastern states, southeastern Canada, and now even to Washington State. As of 2018, WNS had been verified in 33 states and seven Canadian provinces. Millions of bats of the genera *Myotis, Perimyotis*, and some *Eptesicus* have been killed by WNS. It is unclear whether Big brown bats have some immunity to WNS, or if their roosting behavior is such that they are less likely to be exposed. There is some sign that perhaps some of the bats in the northeast are developing some immunity to WNS and populations are starting a slow recovery.

Fig. 9.9 *Corynorhinus rafinesquii* (Lesson), Rafinesque's Big-eared bat.

Information on the ten species of bats living in caves in the UMV is given below in alphabetical order.

Corynorhinus rafinesquii (Lesson), Rafinesque's Big-eared bat (Fig. 9.9).

This is a forest species occurring especially in bottomland hardwoods and swamps. It usually roosts in small numbers (up to 70 or so) in summer in uninhabited dilapidated buildings and other structures. They are often in lighted areas. Colonies have occasionally occurred in caves, hollow trees, and under loose bark. Colonies are usually near water, and the same roosts are occupied yearly. These (and most other bats with big ears) feed heavily on moths. They hibernate in caves, old mines, and cisterns, often in the twilight zone. Summer and winter roosts are often close together. Rafinesque's bat occurs only in far southern Illinois, of the states considered here.

Eptesicus fuscus (Palisot du Beauvois), Big brown bat.

The Big brown bat is most often found in houses, followed by the Little brown bat. Big browns form maternity colonies mostly in buildings and other structures, but occasionally in trees. Big brown bats usually move from their maternity colonies to other buildings where they hibernate. Summer and winter roosts are often fairly close together. Maternity colonies number up to some 500 bats in clusters, although the colonies usually are much smaller. Big brown bats hibernating in buildings are fewer in number and spread out over the building, usually in the attic or in a steeple of a church. Also, many big brown bats hibernate in caves and mines, but are in lower numbers in winter to the south in Illinois. Whitaker and James B. Cope have seen large clusters hibernating in mines in Wisconsin. Big brown bats often fly outside in winter, but do not feed, at least in the north. Both maternity and hibernating colonies occur year after year in the same location. Big brown bats feed heavily on beetles and true bugs. This species is very common in all four UMV region states.

Lasionycteris noctivagans (LeConte), Silver-haired bat.

Silver-haired bats are highly migratory, with most moving south for the winter. Females form small maternity colonies, often in hollow limbs. Silver-haired bats often hibernate in caves, mines, or buildings. They feed on a variety of insects, but especially on moths, flies, beetles, and caddisflies. In southern Illinois, they have been found in abandoned silica mines (Layne 1958; Pearson 1962) but none were found in these silica mines by Whitaker (1975). This species occurs in winter in caves of Wisconsin, Illinois, and Minnesota.

Myotis austroriparius (Rhoads), Southeastern bat (state endangered in Illinois).

Southeastern bats are known only from the southern two tiers of counties plus Saline County in Illinois among the states covered by this work. In summer, maternity colonies are found in caves, hollow trees, and buildings. A few adult males may be found in maternity colonies, but most males in summer are in caves, mines, buildings, culverts, and under bridges. Southeastern bats travel short distances and hibernate in winter in small clusters in caves or mines. Unlike most *Myotis* which have one young per year, this species has two. It feeds on caddisflies, flies, moths, and some true bugs.

Myotis grisescens (Howell), Gray bat (federally endangered and state endangered in Illinois).

The Gray bat is a true cave bat since it spends both summer and winter in caves, but relatively few caves provide suitable conditions for this species. Gray bats migrate back and forth between summer and winter caves, sometimes up to several hundred kilometers. In summer, female gray bats live in large, warm, and humid limestone caves where they form very large clusters. Summer colonies are usually close to lakes or rivers. Male gray bats form separate bachelor colonies in summer, and non-breeding females may be found in these colonies. Gray bats sometimes roost in mines, storm sewers, or even buildings. The gray bat occurs in summer in Price and Monroe counties along the Mississippi River in southwestern Illinois, and in the southern two tiers of counties. In winter it occurs in LaSalle County in the north central part of the state, in Price, Adams, and Jackson counties along the Mississippi River in the southwestern part of the state and in Hardin County in the far southeastern part of the state. Gray bats feed heavily on aquatic insects such as mayflies and caddisflies, but also on beetles, moths, and stoneflies.

Myotis leibii (Audubon and Bachman), Eastern small-footed bat.

This is a small bat about which very little is known because it is rare and often difficult to find. It is found only in far southern Illinois in the states considered here. In summer it often occupies caves, but may occur under rocks or in openings under bridges. In winter it hibernates in caves, and is often the last species to enter hibernacula. It is often found hibernating under stones on the floor of the cave. Also, it often hibernates near the exit-entrance which is remarkable for such a small bat. It often feeds on Diptera, Hemiptera, Homoptera, and Coleoptera, as well as on other insects. Two Eastern small-footed bats were found in the late 1980s or early 1990s in Illinois, and two more were found under rocks in Johnson County. A small maternity colony was discovered in Pope County in the Shawnee National Forest.

Myotis lucifugus (Leconte), Little brown bat.

Most Little brown bats spend the summer in houses, barns, and other human structures where the females form large maternity colonies. They occasionally form maternity colonies in hollow trees, probably their original habitat. Some males spend the summer in the same buildings, but are usually lower in the buildings rather than in the attic or at the level of the maternity colonies. Most males occur in nearby buildings at that time. Both sexes migrate to caves and mines where they spend the winter in hibernation. Major foods of this species are Diptera, Lepidoptera, and Coleoptera. This species occurs in all four states considered here.

Myotis septentrionalis (Trouessart), Northern long-eared bat (federally endangered).

This is primarily a forest bat. In summer, maternity colonies are mostly in crevices, cavities, or under loose bark of live or dead trees. Colonies are usually small, seldom over 50 individuals. Roosts in summer may occasionally occur in buildings or under shutters, and under bridges. In fall and early winter these bats swarm at cave and mine entrances. Mating may take place there, and also to a lesser degree in spring. Most bats hibernating in caves are often obvious, but northern bats are usually in cracks and crevices or in other places where they are difficult to see, thus counts in caves of this species are unreliable. These bats feed on a variety of foods: Diptera, Coleoptera, and Lepidoptera and they often glean food items from foliage, tree trunks, and even the ground. They consume many spiders by gleaning. They occur in all four states considered here.

Myotis sodalis Miller and Allen, Indiana bat (federally endangered) (Fig. 9.10).

This species forms maternity colonies under loose bark of trees, often along the edges of woods. Females and males hibernate in winter in caves often in very large groups (several thousands). About 40,000 hibernated in Magazine Cave in Illinois prior to WNS. Males may spend the summer in bachelor colonies in the same caves in which they hibernated. Indiana bats mate during fall swarming, and one pup is born in spring after the bats return to the summer roost. Foods of this species are primarily Diptera, Lepidoptera, and Coleoptera. The Indiana bat occurs in Iowa (rarely) Wisconsin (rarely), and Illinois.

Fig. 9.10 *Myotis sodalis,* the Indiana bat, forms large colonies in a few caves in the UMV region. © S. Peck 2019. All rights Reserved

Fig. 9.11 *Perimyotis subflavus* (Cuvier), the common Eastern pipistrelle. © J. Lewis 2019. All rights Reserved

Perimyotis subflavus (Cuvier), Eastern pipistrelle (Fig. 9.11).

Many workers are currently using the common name "Tricolored bat" for *Perimyotis subflavus*, referring to the three bands of color of the fur. However, Whitaker et al. (2011) recommended that Eastern pipistrelle be retained as the "official" common name for this species as there is no compelling reason to change a long-used common name just because of a change in the scientific name, and also because there is already a tricolored bat (*Glyphonycteris sylvestris*) which occurs from Peru and Brazil north to parts of Mexico and also in Trinidad (Wilson and Reeder 2005). Eastern pipistrelles sometimes form maternity colonies in buildings, but most often a maternity colony is spread in several clusters of dead leaves, up to 3–4 bats per cluster. They may form a pre-maternity colony in a building (or perhaps a tree) prior to forming the maternity colony. The pre-maternity colony may give the bats protection before the leaves come out. This species feeds heavily on small Diptera, Lepidoptera, and Homoptera. Chute's Cave under Minneapolis, Minnesota, contains a large hibernaculum of pipistrelles (Brick 2009b, p 144). The Eastern pipistrelle occurs in all of the states considered here.

State distributions. The bats found in caves and mines in the four states of the UMV region are indicated below by state.

Illinois. There are about 400 caves in Illinois, mostly in the limestone region along the Mississippi River around St. Louis and southwards. Bats of Illinois include the same six species that occur in Wisconsin plus four additional species (Hofmann 2010; Feldhamer et al. 2015). They are the southeastern bat (*Myotis austroriparius*, the gray bat (federally endangered), *Myotis grisescens* (federally endangered), the eastern small-footed bat (*Myotis leibii*), and Rafinesque's big-eared bat (*Corynorhinus rafinesquii*).

Table 9.3 Data from Webb et al. (1993) on mammals in Illinois caves by county, cave, and species

Alexander –Silica Mine #42 *Myotis sodalis*
Carroll –Babe's Cave *Perimyotis subflavus* –Raccoon Den Cave *Procyon lotor*
Hardin –Crystal Cave *Perimyotis subflavus* –Mine #69 *Myotis austroriparius, M. sodalis* –Rich'Cave *Myotis lucifugus, Eptesicus fuscus, Perimyotis subflavus*
Hardin –Crystal Cave *Perimyotis subflavus* –Mine #69 *Myotis austroriparius, M. sodalis* –Rich'Cave *Myotis lucifugus, Eptesicus fuscus, Perimyotis subflavus*
Jackson –Ava Cave *Perimyotis subflavus* –Toothless Cave (or Bat Cave) *Myotis sodalis, M. grisescens, Eptesicus fuscus*
Joe Daviess –Jean's Cave *Peromyscus maniculatus* –Kevern's Cavern *Procyon lotor, Myotis lucifugus* –Kopper's Crevice *Myotis lucifugus, Perimyotis subflavus* –Tree Root Pit *Procyon lotor*
Johnson –Mason Cave #2 *Perimyotis subflavus* –Teal's Cave *Perimyotis subflavus*
Monroe –Fogelpole Cave (Lemonade Cave) *Myotis sodalis* –Fults-Saltpeter Cave *Perimyotis subflavus* –Illinois Caverns (and by several other names) *Perimyotis subflavus*
Running Spring Cave *Perimyotis subflavus*
Shivery Slither Cave *Perimyotis subflavus*
Pike –Cedar Cave *Procyon lotor* –Cloven Hoof Cave *Procyon lotor, Perimyotis subflavus* –Lost Creek Cave (Pearl Cave) *Perimyotis subflavus* –Slick Crawl Cave *Myotis septentrionalis, Perimyotis subflavus*
Pope –Big Grand Pierre Creek Cave *Perimyotis subflavus* –Brasher Cave *Myotis austroriparius, M. sodalis* –Lackey Cave *Castor canadensis* –Simmons Creek Cave #2 *Myotis sodalis, M. septentrionalis, Perimyotis subflavus* –Tube Cave *Perimyotis subflavus*
Randolph County –Indian Cave *Perimyotis subflavus*
Union –Guthrie Cave (Ephgrave Cave) *Eptesicus fuscus, Perimyotis subflavus*.

Table 9.4 Mammals in Iowa caves (courtesy of Joe Dixon) by county, cave, and species

Allamakee –Fungi Cave *Myotis lucifugus*
Cedar –Ax Head Cave *Procyon lotor*
Clayton –Black Cat Cave *Myotis lucifugus* –Toms Old Mine *Myotis lucufugus*
Delaware –Hunt's Cave *Myotis lucifugus*
Des Moines –Starr's Cave *Eptesicus fuscus, Myotis lucifugus, Peromyscus sp.*
Dubuque –Kemling Cave *Myotis lucifugus, Perimyotis subflavus*
Fayette –Sowards Annex *Eptesicus fuscus, Myotis lucifugus, Perimyotis subflavus* –Sowards Cave *Eptesicus fuscus, Perimyotis subflavus*
Floyd –Algal Mounds Cave *Procyon lotor* –Appleknocker's Cave *Procyon lotor* –Floyd Cave *Myotis lucifugus* –Jesse James Cave *Perimyotis subflavus*
Hardin –Wildcat Cave *Myotis lucifugus*
Jackson –Angry Chipmunk Cave *Tamias striatus* –Back Room Cave *Procyon lotor* –Black Cat Cave *Procyon lotor* –Blue Owl Cave *Perimyotis subflavus* –Bluff Mill Cave *Procyon lotor* –Bonehead Cave *Procyon lotor* –Brown Bat Cave *Eptesicus fuscus* –Burdock Cave *Procyon lotor* –Bushwack Cave *Procyon lotor* –Buzzard Nest Cave *Procyon lotor* –Cabin Cave *Procyon lotor* –Calamity Cave *Procyon lotor* –Castle Rock Cave *Procyon lotor* –Cedar Crevice Cave *Procyon lotor* –Century Cave *Procyon lotor* –Chelsea's Cave *Perimyotis subflavus* –Dancehall Cave *Eptesicus fuscus* –Decent Cave *Perimyotis subflavus* –Maquoketa Bend Cave *Myotis lucifugus* –Portal Cave *Eptesicus fuscus, Myotis lucifugus* –Scallop Cave *Perimyotis subflavus* –Shinbone Cave *Perimyotis subflavus* –Thumb Screwed Cave *Perimyotis subflavus* –Twin Arch Cave *Myotis lucifugus* –Werden's Cave *Eptesicus fuscus, Myotis lucifugus, Perimyotis subflavus* –Wide Mouth Cave *Perimyotis subflavus*
Jones –Abandoned Cave *Procyon lotor* –Alien Mustard Cave *Procyon lotor* –Angus Creek Cave *Procyon lotor* –Arena Cave *Procyon lotor* –Back Door Cave *Procyon lotor* –Bellowing Bovine Cave *Procyon lotor* –Bev's Cave *Bos taurus* (cows enter cave entrance for protection from weather) –Brainless Cave *Procyon lotor*

(continued)

Table 9.4 (continued)

–Brother's and Sister's Cave *Myotis lucifugus* –Indian Bluff Cave *Myotis lucifugus, M. septentrionalis, Perimyotis subflavus* –Lower Hitchhiker Cave *Myotis lucifugus* –Orchid Cave, *Perimyotis subflavus* –Sacrificial Alter Cave *Perimyotis subflavus* –Two Nest Cave *Perimyotis subflavus* –Unlikely Cave *Myotis lucifugus*
Muscatine –Bobcat Cave *Procyon lotor*
Winneshiek –Angry Hornet Cave *Perimyotis subflavus* –Carnolian Cave *Perimyotis subflavus* –Cave with Dome *Perimyotis subflavus* –Coldwater Cave *Perimyotis lucifugus* –Highlandville Cave #8 *Perimyotis subflavus* –Petersons's Cave *Myotis lucifugus* –Skunk Cave *Myotis lucifugus Perimyotis subflavus* –Wonder Cave, *Myotis lucifugus Perimyotis subflavus* –Yankee Blowhole *Perimyotis subflavus*

Webb et al. (1993) give data on vertebrates in Illinois caves by county as in Table 9.3.

Iowa. Five species of bats have been known to hibernate in the caves of Iowa (Harvey et al. 2011; Dixon 2010; and from Iowa Biological Cave Survey), Little brown bat, Northern long-eared bat, Indiana bat, Eastern pipistrelle, and the Big brown bat, although only two Indiana bats were found and they were found together. Table 9.4 lists the mammals of Iowa caves by county and species.

Dixon (2010) summarized data on bats of caves of Iowa by counties as follows:

Clayton—*Myotis lucifugus, Perimyotis subflavus*; Dubuque—*Perimyotis subflavus*; Delaware—*Myotis lucifugus*; Fayette—*Eptesicus fuscus, Myotis lucifugus, Perimyotis subflavus*; Floyd—*Myotis lucifugus, Perimyotis subflavus*; Jackson—*Eptesicus fuscus, Myotis lucifugus, Perimyotis subflavus*; Jones—*Eptesicus fuscus, Myotis lucifugus, Myotis septentrionalis, Perimyotis flavus*; Winneshiek—*Eptesicus fuscus, Myotis lucifugus, Perimyotis subflavus.*

Dixon (2011) also summarized information on smaller caves of Iowa (not more than 50 m in length). *Pipistrellus subflavus* occurred in 68% of those caves and *Myotis lucifugus* in 24%. Kunz and Schlitter (1968) presented information on bats of Iowa. They found *Lasionycteris noctivagans* in only eight counties, but thought it to be state wide in distribution, and did not find it in any caves. The earliest spring record was April 15 and latest fall record was October 2.

Minnesota. The Big brown bat, Eastern pipistrelle, Little brown bat, and Northern long-eared bat are found in Minnesota. Also, there is one old cave record of the silver-haired bat but Gerda Nordquist has never found it in winter (Gunderson and Beer 1953; Nordquist 1987, 2000, pers. comm., 2016).

Wisconsin. Six species of cave hibernating bats have been found in Wisconsin, the Big brown bat, Eastern pipistrelle, Silver-haired bat, Little brown bat, Northern long-eared bat, and the federally endangered Indiana bat (Jackson 1961) although only two of this latter species have been observed, Long 1976.

9.4.2 Mammals Other Than Bats

Eastern woodrats (*Neotoma floridana* (Ord) (Fig. 9.12) occur in caves, talus of limestone bluffs, associated woods, and other habitats. They are one of the most endangered mammals in Illinois. In the UMV region only four original populations remain. Three are in Jackson County (Fountain Bluff, Horseshoe Bluff, and Little Grand Canyon) and the largest population is found in Pine Hills, NW Union County (Hofmann 2010; Poole et al. 2013). Populations originally occurred at a number of sites in southern Illinois in Jackson, Union, Johnson, Pope, Gallatin, and Hardin counties (Nawrot and Klimstra 1976). More recently reintroductions were used in an attempt to establish populations in previously occupied sites (Poole et al. 2013). These are at Garden of the Gods in Saline County, Buzzards Point, High Knob, and Rim Rock/Pounds in Gallatin County, and Lusk Creek in Pope County.

It is of note that Eastern woodrats were very abundant in Meyer Cave deposits of some 5000 yr BP (Parmalee 1967). They are now restricted some 105 km to the southeast in Illinois at the Jackson and Union county sites. The former more northerly range may have been acquired in the post-glacial warm-dry xerothermic period with the eastward spread of the grasslands of the "Prairie Peninsula" at that time (Geis and Boggess 1968; Parmalee 1968).

Fig. 9.12 The Eastern woodrat, *Neotoma floridana*, is widespread and common in karst areas. They range from Texas to South Dakota and east to New York and south to Florida. In Illinois they are restricted to two extreme southern counties. They often make nests of shredded cedar bark in caves.

Mammals other than bats may occur in caves for various reasons. Caves often serve as dens or retreats for raccoons and these often forage in caves. Their droppings can serve as food for cave invertebrates. Various rodents and insectivores, particularly White-footed mice (*Peromyscus leucopus*) (Rafinesque) and short-tailed shrews (*Blarina brevicauda* (Say)) also commonly live in caves. Housecats (*Felis catus* L) have been known to enter caves and kill bats. Beavers have been observed in the lake of Carver's Cave in Minnesota (Brick 2009a) and in Lackey Cave, Illinois.

9.4.3 Remains of Extinct Mammals from Caves of the UMV Region

The fossil mammal records obtained from the Faunmap (2016) dataset indicate several mammal species in the UMV region which are now extinct or have vastly changed their geographic ranges. Data from Iowa indicate that Bogus Cave, Jones County had bones of Caribou (*Rangifer tarandus*), and several Mastodons have been found in Couch Cave, Monroe County, Illinois. LeConte (1848a, b) reported the first Flat-headed peccary (*Platygonus*) and other remains from a lead-lined cave-fissure fill in the Galena area of northwestern Illinois. Leidy (1860) reported more Flat-headed peccary remains from northwestern Illinois caves. Parmalee (1967) and Fowler (1959) reported more of these peccaries from rock shelters from Modoc, Randolph County, Illinois. Wyman and Leidy in Hall and Whitney (1862) reported on mammal bones from caves of Illinois and Wisconsin, including extinct Mastodon and the Giant ground sloth (*Megalonyx*). Mather (2009) and Widga et al. (2012) reported Scimitar-toothed cat (*Homotherium serum*) and Stag moose (*Cervalces sp.*) bones from Tyson Spring Cave, Fillmore County, Minnesota. These were the first records of these species from Minnesota. Extensive data on vertebrates and invertebrates from a deposit from Meyer Cave fissure in Monroe County, Illinois can be found in Parmalee (1967, 1968). This rich deposit dates from early postglacial times to present, and included the remains of several species of small mammals now with boreal or northern distributions suggesting a cool/moist post-Pleistocene (10,000–8000 yr BP) climate and others of a later (4000–1000 yr BP) warm/dry environment.

9.4.4 Other Vertebrates

Fish occurring in caves include various minnows, sculpins, suckers, and trout, which come into the caves from outside waters. There are false reports of cavefish in Cold Water Cave, Iowa (Anonymous 2011). Eyeless *Amblyopsis* cavefish (Amblyopsidae) do occur in Indiana and Missouri, but have not been found in southern Illinois. Another species in the same family, the Spring cavefish *Forbesichthys* (formerly *Chologaster*) *agassizii* (Putnam), with pigmentation and small eyes, occurs in springs in southern Illinois, and emerges at dusk to feed in surface waters. Webb et al. (1993) list fishes in caves in Monroe County, Illinois: Fogelpole Cave (Lemonade Cave), Banded sculpin *Cottus carolinae* Gill; Krueger Dry-Run Cave, *Ameiurus melas* (?) Black bullhead Rafinesque; Weeping Buddha Cave, Green sunfish *Lepomis cyanellus* Rafinesque.

Reptiles generally do not tolerate the colder temperatures of caves in the UMV region. Snakes seen (infrequently) in the caves of the area are copperheads *Agkistrodon contortrix* (L), blacksnakes *Elaphe obsoleta* (Say), and water snakes, *Nerodia sp.* Eastern timber rattlesnakes *Crotalus horridus* L. were reported by the fur-trader Pierre-Charles Le Sueur in 1700 from what is probably Snake Cave, near Potosi, Wisconsin (Reeder and Day 1990) and some up to 2 m long in the saltpeter caves on the west side of Lake Pepin of the Mississippi River in Goodhue County, Minnesota (Brick 2012). These snakes also occur in dens in the talus and talus caves of the limestone Mississippi River bluffs in southern Illinois in the Little Grand Canyon area and adjacent Pine Hills.

In contrast, some amphibians do frequent caves. These include the Cave salamander *Eurycea lucifuga* Rafinesque (Fig. 9.13) and Longtailed salamander *Eurycea longicauda* (Green) which live as larvae in cave streams in southwestern and southern Illinois and mature to terrestrial adults. They are usually found in the twilight zone of the caves. Also occurring in Illinois are Slimy salamanders *Plethodon glutinosus* (Green) with terrestrial juveniles. The latter two species feed in caves as well as outside (Peck 1974; Peck and Richardson 1976). Frogs often found in caves are the Pickerel frog *Lithobates palustris,* Leopard frog *Lithobates sphenocephalus* (Cope) and Green frog *Lithobates clamitans* (Latreille). Adult frogs overwinter in the spring-fed lake of Carver's Cave, Minnesota (Greg Brick, pers. comm., 2016). Webb et al. (1993) give data on amphibians and reptiles in caves by Illinois counties as in Table 9.5.

Birds are occasionally seen around the entrances to caves but the phoebe (*Sayornis phoebe*) is the only bird commonly found at caves of this area. It often makes its nests (mostly of mosses) on ledges near the cave entrance. Webb et al. (1993) give data on birds in caves by Illinois counties in Table 9.6.

9.5 Discussion

9.5.1 The Dynamics of Pleistocene-Recent Environmental Changes

In the past 2 million years the UMV region has experienced dramatic and recurrent environmental changes. These were repeated proximity or cover by continental glacial ice for long periods and then its retreat and disappearance for shorter intervals, accompanied by major fluctuations in average annual temperatures. Because the wholesale uptake of water by the great glaciers lowered ocean levels, there was great erosional deepening of river valleys during the glacial advances and concomitant cave enlargement, while the overriding ice created major changes in continental drainage patterns (e.g., the shifting of the positions of the Mississippi and Ohio Rivers). During retreat of the glaciers there was extensive valley and cave in-fill by outwash gravel and sediment. Many immense lakes were formed behind dams of glacial ice or outwash sediments. Huge dust storms deposited great blankets of loess. During all this, the regional animals and plants adjusted their presence or absence over geographic space and through evolutionary time according

Fig. 9.13 The Cave salamander *Eurycea lucifuga* is widely distributed from eastern Oklahoma, through southern Illinois and to Virginia and northern Alabama. © S. Peck 2019. All rights Reserved

Table 9.5 Records for amphibians and reptiles in caves in Illinois by county, cave, and species from Webb et al. (1993)

Adams –Bobtail Salamander Cave *Eurycea longicauda* –Lost Creek Cave (Pearl Cave) *Bufo americanus* –Slick Crawl Cave *Eurycea longicauda, Plethodon glutinosus*
Jackson –Ava Cave *Eurycea lucifuga*
Johnson –Cedar Bluff Cave *Eurycea lucifuga* –Mason Cave #1 *Eurycea lucifuga* –Pipistrellus Pit Cave *Eurycea lucifuga* –Procyon Cave (Unnamed Cave) *Eurycea lucifuga, E. longicauda, Plethodon glutinosus* –Sink Joint Cave *Eurycea lucifuga* –Teal's Cave *Eurycea lucifuga, Plethodon glutinosus*
Monroe –Fogelpole Cave (Lemonade Cave) *Ambystoma tigrinum, Lithobates palustris* –Illinois Cave (several other names) *Eurycea longicauda* –Kelly Spring Cave *Lithobates palustris* –Krueger Dry-run Cave *Plethodon dorsalis* –Terry Spring Cave *Eurycea lucifuga, Lithobates palustris* –Unnamed Cave near Collier Spring *Eurycea lucifuga* –Weeping Buddha Cave *Lithobates catesbiana, L. palustris*
Pope –Big Grand Pierre Creek Cave *Plethodon glutinosus* –Brasher Cave *Eurycea lucifuga, Plethodon dorsalis* –Simmons Creek Cave #2 *Eurycea lucifuga, E. longicauda, Plethodon dorsalis, P. glutinosus, Hyla triseriata* –Tube Cave *Plethodon dorsalis, P. glutinosus*
Pulaski –Boiling Spring *Eurycea longicauda*
Randolph –Indian Cave *Eurycea lucifuga, E. longicauda, Plethodon glutinosus, Lithobates palustris*
Randolph –Indian Cave *Eurycea lucifuga, E. longicauda, Plethodon glutinosus, Lithobates palustris*
St. Clair –Dieciseis Tigrinum Pit *Ambystoma tigrinum, Plethodon dorsalis, Hyla triseriata Lithobates catesbeiana*
Union County –Apis Annex *Eurycea lucifuga, Terrapene carolina* –Graig Cave *Ambystoma tigrinum* –Honeycomb Hole *Eurycea lucifuga* –Migrant Camp Cave *Eurycea lucifuga* –Rich's Cave *Eurycea longicauda, E. lucifuga, Lithobates palustris,* –Saratoga Cave *Plethodon glutinosus* –Shilly-Shally Cave *Eurycea lucifuga*

Table 9.6 Records for birds in Illinois caves by county, cave, and species (from Webb et al. 1993)

Johnson –Cedar Bluff Cave (Cedar Grove Cave) *Cathartes aura* (Turkey vulture) –Mason Cave #1 *Cathartes aura* –Simmons Creek Cave #2 *Sayornis phoebe* (Phoebe)
LaSalle –Mathiessen Park Cave *Columba livia* (Common pigeon)
Monroe –Fogelpole Cave (Lemonade Cave) *Sayornis phoebe* nest containing *Molothrus ater* (Brown headed cowbird) *egg* –Fults-Saltpeter Cave *Columba livia* –Weeping Buddha Cave *Sayornis phoebe*
Pike –Boat Ramp Cave *Sayornis phoebe* nest containing *Molothrus ater* egg
Randolph –Indian Cave *Sayornis phoebe*

to their needs and tolerances. This was all a prelude to the present make-up of the region's biotic communities and the distributions of their members.

We make the simple assumption that the vertebrates presently associated with caves in the UMV region have made this association sometime during recent times. The previous listing of vertebrate fossils from the region's caves show something of the dynamics of the range changes and the dynamics of the different species compositions of past times. There is however a more complex set of distributional patterns to consider for the invertebrates.

9.5.2 Invertebrates

9.5.2.1 Present Invertebrate Distributions

Most of the over 300 species of invertebrate animals found in the caves of the UMV region are not obligate cave inhabitants. The majority have distributions throughout the region (and sometimes far beyond) and also occur in non-cave habitats. In contrast, the cave-limited troglobites and stygobites are more apt to have smaller geographic ranges and some are single cave-site endemics, like the cave beetle *Pseudanophthalmus illinoisensis.*

At the other extreme, some troglobites have ranges that go far beyond the geographic limits of the UMV region. Examples with broad ranges of ten or more states are the dung fly *Spelobia tenebrarum* or the sheet-web spider *Phanetta subterranea.* As previously noted, other widespread troglobites, like the collembolan *Tomocerus missus* (reported from caves in Missouri, Indiana, Illinois, Kentucky, Colorado, Arizona) are likely complexes of closely related species (Christiansen and Bellinger 1980).

Both extremes in distribution exist among the stygobites as well. The two species of undescribed *Stygobromus* amphipods in southern Illinois are site-specific endemics. On the other hand, the isopod *Caecidotea stygia* as historically identified is reported from southwestern Ohio, across southern Indiana, through Kentucky into northern Tennessee, then west through the Shawnee Hills of southern Illinois to eastern Missouri. This species probably occurs in thousands of caves across this broad area, but has penetrated to the north into the glaciated regions by only a few kilometers, where it has been recorded from a limited number of collections found inhabiting the saturated interstices of glacial sediments. Ongoing evaluation of this isopod with molecular genetic techniques indicates that this widespread "species" is more likely a cluster of morphologically similar, but genetically distinct sibling species (Lewis, in progress).

Our knowledge of a few groundwater organisms from the till plains of the UMV region is from wells and is mostly owed to farmers who have placed unmeasured kilometers of drainage systems in their flat agricultural fields. Typically buried 1–2 m below the ground surface, these "drain tiles" collect and discharge both water and the groundwater animals that live there. Most of the fauna known to inhabit the saturated soil interstices tapped by these field tiles are crustaceans. Some of these are known from dozens of collections spanning large areas of the UMV region and beyond. These include the amphipod *Bactrurus mucronatus* and the isopods *Caecidotea kendeighi* and *C. beattyi* (Koenemann and Holsinger 2001; Lewis and Bowman 1981; Lewis 2015). Other species are known from one (e.g., *Caecidotea lesliei*) or a few (*Stygobromus iowae, S. putealis*) localities (Lewis and Bowman 1981; Holsinger 1972).

9.5.2.2 Achieving Present Invertebrate Distributions

The UMV region has had dynamic environments over the past 2 million years or so to which the animals have responded. Over this large geographic area there have been distributional and evolutionary adjustments. The observations above prompt the question: have the cave and groundwater crustaceans moved into glaciated regions since deglaciation, or did they survive in groundwaters for tens of thousands or more years while covered by glacial ice?

Two schools of thought exist on this subject. Both are predicated on the fact that an obligate groundwater fauna exists in many areas in glaciated North America. The first hypothesis is that the fauna was already present before glaciation and persisted under Pleistocene glacial ice cover. It is a corollary that these animals are thought to have relatively low dispersal capabilities and could not have moved into glaciated regions. Besides the fauna of the UMV region cited above, numerous species of the amphipod genus *Stygobromus* occur in formerly glaciated country in caves in New York, Vermont, Alberta, British Columbia, and southeastern Alaska (Wang and Holsinger 2001). Among the isopods, *Salmasellus steganothrix* Bowman is also known to occupy Castleguard Cave under the present Columbian Ice Field glacier in Alberta. Holsinger (1967) was a proponent of subglacial refugia for the species of *Stygobromus* amphipods because of the distance of their localities from Pleistocene glacial maxima. He believed that crustaceans could not travel the distances involved in the interval since the last glaciation, citing the presence of *Stygonectes allegheniensis* Holsinger (1967) restricted to caves from Maryland and Pennsylvania to central New York (Albany, Schoharie, Onondaga, and Herkimer counties) almost 400 km to the north of the Wisconsinan glacier's maximum extent (Holsinger 1967) and now known from a cave still farther north at Watertown, Jefferson County, New York, not far from the Canadian border (Peck unpublished).

The second hypothesis is that the groundwater fauna moved north into glaciated country after deglaciation. Dispersal may not have been a major challenge. In the above case of *Stygobromus* Lewis and Bowman (1981) calculated that over the course of the minimum amount of time available, some 10,000 years, this involved traveling only about 11 cm/24 h. As the glaciers receded vast amounts of meltwater were released and it seems reasonable to think that aquatic animals could disperse northward with the retreating ice fronts. Unlike *Stygobromus* and *Salmasellus,* some of the isopods of the glaciated areas, like *Caecidotea kendeighi*, retain significant pigmentation and eyes, suggesting the idea that their invasion of groundwater was relatively recent. Molecular genetic evidence in Europe supports the supposition of postglacial recolonization (Hewitt 1999) and it seems likely that this applies to North American faunas as well.

Regardless of which hypothesis one adheres to, the aquatic habitat under a massive continental glacier must have been a difficult one, to say the least. It is arguable that both caves and the spaces in unconsolidated sediments alike would have been crushed by the weight of the overlying ice, and it seems likely that caves under the ice would be filled by meltwater-carried sediments. Environmental conditions would be very cold with little, if any, possibility of food input after use or decay of the buried organic material overrun by the glaciers. Dating of cave sediments and speleothems relates these infills to Wisconsinan-Recent depositional-climatic events (see previous chapters).

9.5.2.3 Origins of Troglobites and Stygobites

Many different groups of animals have become specialized to subterranean life worldwide and this is also the case for the UMV region. That these different groups have almost all acquired some or many of the hallmark characters, or "troglomorphisms," of cave life shows how the environment of caves has caused a similarity in morphology through the process of convergent evolution.

Important questions are: when and where did this evolution occur, what selective pressures were involved, and how have the animals been distributed through time? For many years, it was hypothesized by many that the prime driver or selective pressure for entry into cave habitats and specialization to them was the cyclical changes in climates of the Pleistocene, with multiple periods of cool-moist (glacial) and warm-dry (interglacial) conditions. However, some recent studies of European cave faunas, using genetic analysis and molecular clock assumptions, have shown that at least some lineages invaded subterranean habitats much earlier, in mid or late Tertiary times, long before the climatic fluctuations of the Pleistocene (Cieslak et al. 2016). Another molecular genetic study proposes a similar pre-Pleistocene origin for a group of southeastern *Ptomaphagus* cave beetle species (Leray et al. 2019). Similar work with subterranean asellid isopods indicates divergence dating to the Miocene and before (Lewis et al. in progress). This pre-Pleistocene origin could also be true for a limited part of the cave faunas of the UMV region, but no studies have been completed on components of this particular fauna. We do think that this earlier pre-Pleistocene cave entry is possible in the UMV region at least for the aquatic faunas. We are less willing to accept this for the terrestrial fauna since it is difficult to believe that terrestrial animals could exist in caves probably filled by sediment and water under a glacier. Future molecular studies can test this.

Certainly, the Pleistocene must have been a time of some (small and/or large) changes in species distributions. Caves in southwestern and southern Illinois were never glaciated, and it is unknown what their periglacial environmental conditions were during glacial intervals when glacial ice fronts were not far off.

9.5.2.4 Cave Occupation During the Pleistocene?

It is now generally recognized that the "Driftless Area" was ice covered in the early (pre-Illinoian) Pleistocene but not the later Pleistocene (Illinoian and Wisconsinan glaciations). Stygobites may have been present and persisted during early glaciations but it seems unlikely than any troglobites did so. It is noteworthy that no stygobites have been found in the faunal surveys in the largest northern stream system caves such as Coldwater Cave, Iowa, and Mystery Cave, Minnesota.

The troglobites of the "Driftless Area" must have colonized the caves only in the later Pleistocene, sometime after the last glacial cover of the region during the past 790,000 years. This would be a maximum time for some or all of the cave occupation in the Driftless Area. It was believed that the local environment in the Last Glacial Maximum (Wisconsinan) was tundra or taiga-like vegetation and there is now evidence that some deciduous forests of oaks (*Quercus*) may have persisted in the Driftless Area while surrounded by ice in the Wisconsinan (Rowe et al. 2004). This supports the idea that the Driftless Area was a possible refugium for the survival of invertebrates during the periglacial climate of earlier Pleistocene glaciations.

Peck and Christiansen (1990) proposed differing times of cave occupation for 16 troglobitic species in the Driftless Area during the classic (but now outmoded) interglacials of Yarmouth, Sangamon, and Recent times, based on differences in the degree of morphological specialization and the extent of their geographic distribution. All in all, the troglobites of the Driftless Area are less diverse and appear less morphologically specialized than the assemblages of more southwestern and southern Illinois troglobites.

9.5.2.5 Post-Pleistocene Cave Occupation

In contrast to the scenarios for the invertebrate troglobites and stygobites, all trogloxenes and troglophiles (both vertebrates and invertebrates) have undoubtedly moved into previously glaciated regions from more southerly refugial regions and occupied the caves in which they now occur in post-Pleistocene times.

Additionally, after the last glaciation, in the recent, some stygobites may have moved from non-glaciated regions through water-filled soil, bedrock spaces, and crevices into glaciated areas to achieve their sometimes wide distributions. This is in addition to any of those which survived under Illinoian ice cover. Some less troglomorphic terrestrial troglobites such as the fly *Spelobia tenebrarum* and the spiders *Porrhomma* and *Phanetta subterranea* (*Porrhomma* is more troglomorphic than *Phanetta*) have wide distributions and occur in multiple areas not connected by continuous limestones that facilitate underground dispersal. Thus, subterranean movement into glaciated regions through air-filled cave spaces as dispersal routes seems highly unlikely for terrestrial species. The wide distributions of some species must have been achieved by epigean (above-ground) dispersal, possibly aided by cooler and moister climatic conditions shortly after glaciation.

In summary, the caves of the UMV region provide habitats for a wide array of vertebrates and invertebrate animals with a diversity of morphological properties and ecological requirements. The caves are therefore of biological as well as historical, geological, and geographical interest and value.

9.6 Cave Protection and Faunal Conservation

9.6.1 Federal and State Protection

The UMV region contains a number of caves that have been protected in various natural areas. In Illinois these include Cave-in-Rock State Park, Fogelpole, Pautler, Twin Culvert, and Stemler caves nature preserves and Illinois Caverns State Natural Area. With the exception of Cave-in-Rock, none of these sites is open to the public (some of these dedicated state nature preserves are privately owned). At Larue-Pine Hills Ecological Area there are no caves, but the Spring cavefish (*Forbesichthys agassizii*) is present in large springs emerging near the largest limestone cliffs in the state. In Iowa, Maquoketa Caves State Park features over a dozen caves open to the public (Brick 2004).

At 16 miles (25.7 km) in length, Fogelpole Cave is the longest cave in Illinois and second in the UMV region, topped only by Coldwater Cave in Iowa (a National Natural Landmark) with 17 miles (27.3 km) of passages. Mystery Cave in Mystery Cave State Park, Minnesota follows at 13 miles (20.9 km).

Perhaps the most notable obligate subterranean species in the UMV region is the Illinois cave amphipod *Gammarus acherondytes*. This stygobitic amphipod was described by Hubricht and Mackin (1940) from Illinois Caverns (Monroe County) and Stemler Cave (St. Clair County). At that time the caves were in rural settings. By the time that Peck and Lewis (1978) added three more cave records in Monroe County the suburbs of the St. Louis metropolitan area had begun to spread across the area. Due to the decline of the species it was listed as a federal endangered species in 1998. The only population in St. Clair County had been extirpated by septic system pollution from subdivisions occupying the recharge area of the cave. The project of Lewis et al. (2003) was initiated by The Nature Conservancy to further evaluate the situation with the Illinois cave amphipod and added several new cave localities. By 2003 the amphipod was known to continue to survive in six groundwater basins, and the number was increased to eight by Lewis (2003). Conservation efforts have focused on land acquisitions to secure the karstlands over the caves inhabited by *Gammarus acherondytes*, including purchase of relatively large tracts in the Illinois Caverns, Annbriar, and Fogelpole groundwater basins. Despite this, populations continue to decline due to ongoing problems with septic pollution, nutrient enrichment, and sedimentation.

9.6.2 Karst Conservancies

Another source of cave and karst protection are state karst conservancies affiliated with the National Speleological Society.

The Karst Conservancy of Illinois (KCI) owns the Pautler Nature Preserve. Although small in size at about three acres (1.2 ha), this preserve protects the main entrance to the Pautler Cave System, one of the longest caves in Illinois (survey incomplete but perhaps as much as 5 miles (8 km) in length). Pautler Cave is inhabited by one of the few remaining populations of the endangered Illinois cave amphipod as well as a diverse community of other subterranean invertebrates. The KCI also manages the Twin Culverts Nature Preserve, owned by The Nature Conservancy, where a cave is inhabited by the endangered Gray bat as well as an assemblage of troglobitic invertebrates.

Although outside the UMV, the Michigan Karst Conservancy significantly owns the 480 acre (194 ha) Fiborn Karst Preserve, on the site of a former limestone quarry, but more notably containing a variety of karst features including sinkholes, springs, and caves. Perhaps the most important feature of the preserve is Hendrie River Water Cave, at 1500 feet (457 m) the longest cave in the state. This cave is inhabited by hibernating Northern long-eared and Little brown bats. The Thunder Bay Karst Preserve is comprised of three units: Stevens Twin Sinks (31 acres, or 12.5 ha), Bruski Sink (2.5 acres, or 1 ha), and Mystery Valley (76 acres, or 31 ha). These preserves showcase numerous sinkholes, which at depths of 90 feet (27 m) contain floor microclimates inhabited by plant communities, particularly ferns, that are more characteristic of more northern latitudes. Mystery Valley is one of the largest collapse sinkholes in the Great Lakes region, the valley consisting of a 1.5 mile (2.4 km) long collapse of a former labyrinth of underground passages. The valley reaches a width of 500 yards (457 m) and a depth as much as 150 feet (46 m) (Michigan Karst Conservancy 2014).

Acknowledgements We thank Greg Brick and E. Calvin Alexander, Jr., for inviting us to prepare this faunal summary of the animals of caves in the UMV region and J.W. Dixon of the Iowa Biological Survey for supplying vertebrate information for Iowa caves. We thank William Shear and Olga Kukal for reviewing early versions of the manuscript. Angela Chamberlain helped obtain literature and in editing the vertebrate sections of the manuscript. Danté Fonolio and Adam Mann kindly provided photos.

References

Alexander CS, Prior JC (1968) The origin and function of the Cache Valley, southern Illinois. In: Bergstrom RE (ed) The Quaternary of Illinois: a symposium in observance of the centennial of the University of Illinois, pp 19–26. University of Illinois College Agri Spec Pub 14

Anonymous (1913) Blind crayfish in Carver's Cave. St Paul Pioneer Press, Nov 13

Anonymous (2011) Cold Water Cave's blind cave fish. https://therousedbearwordpress.com/2011/05/14/cold-water-cave%E2%80%99s-blind-cave-fish/

Brick GA (2000) Amphipods in Minnesota caves. J Cave Karst Stud 62 (1):31 (abstract)

Brick GA (2004) Iowa Underground: A Guide to the State's Subterranean Treasures. Trails Books, Black Earth, WI, 223p

Brick GA (2006) White Bugs & Black Holes: My Experience with Folsomia candida. Unpublished typescript, 2 p

Brick GA (2009a) History and geology at Carver's Cave. Ramsey County History 44(1):16–20

Brick GA (2009b) Subterranean Twin Cities. University of Minnesota Press, Minneapolis, 226p

Brick GA (2009c) Guanophile ecosystem of an urban cave polluted with raw sewage. In: Proceedings of 15th international congress speleology, Kerrville Texas, 19–26 July 2009, Greyhound Press 3: 1280

Brick GA (2012) Le Sueur's saltpeter caves at Lake Pepin, Minnesota, and wilderness gunpowder manufacture. Minnesota Archaeol 71:7–20

Christiansen KA, Bellinger PF (1980) The Collembola of North America north of the Rio Grande, a taxonomic analysis, 1st edn. Grinnell College, Grinnell IA

Christiansen KA, Willson M, Tecklin J (1961) The collembola of Hunter's Cave with a preliminary study of the microarthropod ecology of Hunter's Cave. NSS Bull 23:59–70

Cieslak A, Fresneda J, Ribera I (2016) Life-history specialization was not an evolutionary dead-end in Pyrenean cave beetles. Proc R Soc B 281: 20132978. https://doi.org/10.1098/rspb.2013.2798

Culver DC, Master l, Christman MC, Hobbs H III (2000) Obligate cave fauna of the 48 contiguous United States. Cons Biol 14:386–401

Dixon JW (2010) Mammalia, Chiroptera, Vespertilionidae: filling hibernacula distribution gaps for cave roosting bats from Iowa. Check List. Iowa Grotto National Speleological Society, P. O. Box 2285224, Iowa City, USA. 6:511–514

Dixon JW (2011) The role of small caves as bat hibernacula in Iowa. J Cave Karst Stud 73:21–27

Faunmap (2016). http://www.neotomaDB.org

Feldhamer GA, Hofmann JE, Carter TC, and Kath JA (2015) Bats of Illinois. ISU center for bat research, outreach, and conservation. Indiana State University, Terre Haute, Indiana

Flint RF, Colton R, Goldthwait R, Willman H (1959) Glacial map of the United States east of the Rocky Mountains. Geol Soc Am Scale 1:1,750,000

Fowler ML (1959) Modoc Rock Shelter: an early Archaic site in southern Illinois. Am Antiq 24:57–270

Gardner JE (1986) Invertebrate fauna from Missouri caves and springs. Missouri Dept Conser Nat Hist Ser no 3: 1072

Geis JW, Boggess WR (1968) The Prairie peninsula: its origin and significance in the vegetational history of Central Illinois. In: Bergstrom RE (ed) The Quaternary of Illinois: a symposium in observance of the centennial of the University of Illinois, University of Illinois College Agri Spec Pub 14, pp 89–103

Gunderson HL, Beer JR (1953) The Mammals of Minnesota. University of Minnesota Press

Harvey MJ, Altenbach JS, Best TL (2011) Bats of the United States and Canada. Johns Hopkins University Press, Baltimore

Hershler R, Holsinger JR, Hubricht L (1990) A revision of the North American freshwater snail genus *Fontigens* (Prosobranchia: Hydrobiidae). Smith Cont Zool 509:1–49

Hewitt GM (1999) Post-glacial re-colonization of European biota. Biol J Lin Soc 68:87–112

Hoffman RL (1999) Checklist of the millipeds of North and Middle America. Spec Publ No 8 Virginia Mus Nat Hist, Martinsville Virginia

Hofmann JE (2010) Field manual of Illinois Mammals. Illinois Nat Hist Surv Champaign, Manual 12

Holsinger JR (1967) Systematics, speciation and distribution of the subterranean amphipod genus *Stygonectes* (Gammaridae). Bull US Natl Mus 259

Holsinger JR (1972) The freshwater amphipod crustaceans (Gammaridae) of North America, Water Pollution Control Research Series. Biota Freshw Ecosyst Identif Manual 5:1–89

Hubricht L (1943) Studies in the nearctic freshwater Amphipoda III: notes on the freshwater Amphipoda of the eastern United States with descriptions of ten new species. Amn Midl Natur 29(3):683–712

Hubricht L, Mackin JG (1940) Descriptions of nine new species of fresh-water amphipod crustaceans with notes and new localities for other species. Am Midl Natur 23:187–218

Iannicelli M (2010) Evolution of the driftless area and contiguous regions of midwestern USA through Pleistocene periglacial processes. Open Geol J 4:35–54

Jackson HHT (1961) Mammals of Wisconsin. The University of Wisconsin Press, Madison, WI

Jass JP (1994) Investigation of crustacean amphipods in the family Crangonyctidae. Grant proposal for Lois Almon small grants program Wisconsin Academy of Sciences Madison WI 6 pp [from Natureserve Explorer comprehensive species report on *Stygobromus putealis*]

Klimaszewski J, Peck SB (1986) A review of the cavernicolous Staphylinidae of North America: part I Aleocharinae. Quaest Entomol 22:51–113

Koch DL, Case JC (1974) Fauna and flora. In: Report on Cold Water Cave Iowa. Geological Survey Unpublished report, pp 25–36.

https://s-iihr34iihr.uiowa.edu/publications/uploads/2015-12-17_15-12-33_mp-9.pdf

Koenemann S, Holsinger JR (2001) Systematics of the North American subterranean amphipod genus *Bactrurus* (Crangonycidae). Beaufortia 51(1):1–56

Kunz TH, Schlitter DA (1968) An annotated checklist of bats from Iowa. Trans Kansas Acad Sci 71:166–175

Layne JN (1958) Notes on mammals of southern Illinois. Am Midl Natur 60:219–254

LeConte JL (1848a) Notice of five new species of fossil mammalia from Illinois. Am J Sci Arts, 2nd ser 5:102–106

LeConte JL (1848b) On *Platygonus compressus*: a new fossil Pachyderm. Mem Am Acad Arts Sci 3:257–275

Leidy J (1860) Observations on the extinct Peccary of North America; Being a sequel to "A memoir on the extinct Dicotylinae of America." Trans Am Philos Soc 11:97–105

Leray VL, Caravas J, Friedrich M, Zigler KS (2019) Mitochondrial sequence data indicate "vicariance by erosion" as a mechanism of species diversification in North American *Ptomaphagus* (Coleoptera, Leiodidae, Cholevinae) cave beetles. Subterr Biol 29:35–57

Lewis JJ (1982) Systematics of the troglobitic *Caecidotea* (Crustacea: Isopoda: Asellidae) of the southern interior low plateaus. Brimleyana 8:65–74

Lewis JJ (2000) Five new species of *Pseudotremia* from caves in the Blue River area of southern Indiana (Diplopoda: Chordeumatida: Cleidogonidae). Myriapodologica 6(6):55–68

Lewis JJ (2002) *Chaetaspis aleyorum* a new species of milliped from Tumbling Creek Cave Missouri with a synopsis of the cavernicolous species of *Chaetaspis* (Diplopoda: Polydesmida). Myriapodologica 7(11):101–111

Lewis JJ (2003) The 2003 non-lethal censusing of *Gammarus acherondytes* communities in southwestern Illinois. Final report, US Fish and Wildlife Service and Endangered Species Program, Illinois Department of Natural Resources

Lewis JJ (2012) Chapter 8 Subterranean Ecosystems. In: Whitaker JO Jr, Amlaner CJ Jr (eds) Habitats and ecological communities of Indiana. Indiana University Press, Bloomington, IN, pp 156–175

Lewis JJ (2015) The subterranean asellids of Indiana (Isopoda) with the description of *Caecidotea dunlaporum* new species. J Cave Karst Stud 77(2):99–107

Lewis JJ, Bowman TE (1981) The subterranean asellids (Caecidotea) of Illinois (Crustacea: Isopoda: Asellidae). Smiths Contr Zool 335:1–66

Lewis JJ, Lewis SL (2014) The 2014 census of *Gammarus acherondytes* communities in southwestern Illinois Final Report. Endangered Species Program U S Fish and Wildlife Service and Endangered Species Program Illinois Department of Natural Resources

Lewis JJ, Moss P, Tecic D, Nelson ME (2003) A conservation focused inventory of subterranean invertebrates of the southwestern Illinois karst. J Cave Karst Stud 65:9–21

Lewis JJ, Reid JW (2007) Patterns and processes of groundwater invasion by Copepods in the interior low plateaus of the United States. Acta Carsologica 36(2):79–289

Long CA (1976) The occurrence status and importance of bats in Wisconsin with a key to the species. Wisconsin Acad Sci Arts Lett 64:62–82

Mackin JG, Hubricht L (1940) Descriptions of seven new species of *Caecidotea* (Isopoda Asellidae) from the central United States. Trans Am Micros Soc 59(3):383–397

Marshall SA, Peck SB (1984) Distribution of cave-dwelling Sphaeroceridae (Diptera) of eastern North America. Proc Entomol Soc Ontario 115:3–41

Mather D (2009) A new twist in Minnesota Megafauna History. Minnesota Conservation Volunteer. March-April, pp 21–29

Michigan Karst Conservancy (2014). http://caves.org/conservancy/mkc/preserve_mv.html

Montz GR (1993) The aquatic invertebrates of Mystery Cave Forestville State Park Minnesota. Unpublished report to the Minnesota Depart Nat Res, Ecological Services Section

Muck JA, Newman RM (1992) The distribution of amphipods in southeastern Minnesota and their relation to water quality and land use. J Iowa Acad Sci 99(1):34–39

Nawrot JR, Klimstra WD (1976) Present and past distribution of the endangered Southern Illinois Woodrat (*Neotoma floridana illinoensis*). Nat Hist Misc, Chicago Acad Sci, Chicago, Il. 12 pp

Nordquist GE (1987) Bats in Minnesota. Natural History Leaflet 8 James Ford Bell Mus Nat Hist, Minneapolis MN 8 p

Nordquist GE (2000) Winter use of subterranean cavities by bats in and near Minnesota. MSc thesis, University of Minnesota 84 p

Panno SV, Hackley KC, Kelly WR, Hwang HH, Wilhelm FM, Taylor SJ, Stiff BJ (2006) Potential effects of recurrent low oxygen conditions on the Illinois cave amphipod. J Cave Karst Studies 68:55–63

Parmalee PW (1967) A recent cave bone deposit in southwestern Illinois. Bull Nat Speleol Soc 29:119–147

Parmalee PW (1968) Cave and archaeological faunal deposits as indicators of post-Pleistocene animal populations and distribution in Illinois. In: Bergstrom RE (ed) The Quaternary of Illinois: a symposium in observance of the centennial of the University of Illinois, pp 104–114. University of Illinois College Agri Spec Pub 14

Pearson EW (1962) Bats hibernating in silica mines in southern Illinois. J Mammal 43:27–33

Peck SB (1974) The food of the salamanders *Eurycea lucifuga* and *Plethodon glutinosus* in caves. NSS Bull 36:7–10

Peck SB (1982a) New records of poorly known *Ptomaphagus* from Texas and the Ozarks (Coleoptera: Leiodidae: Cholevinae). Coleop Bull 36:109–111

Peck SB (1982b) The occurrence of *Ptomaphagus cavernicola* in forests in Florida and Georgia. Fla Entomol 65:378–379

Peck SB (1984) The distribution and evolution of cavernicolous *Ptomaphagus* beetles in the southeastern United States (Coleoptera: Leiodidae: Cholevinae) with new species and records. Can J Zool 62:30–740

Peck SB (1988) A review of the cave fauna of Canada and the composition and ecology of the invertebrate fauna of caves and mines in Ontario. Can J Zool 66:1197–1213

Peck SB (1998) A summary of diversity and distribution of the obligate cave-inhabiting faunas of the United States and Canada. J Cave Karst Stud 60:18–26

Peck SB, Christiansen K (1990) Evolution and zoogeography of the invertebrate cave faunas of the driftless area of the upper Mississippi River Valley of Iowa, Minnesota, Wisconsin, and Illinois USA. Can J Zool 68:73–88

Peck SB, Lewis JJ (1978) Zoogeography and evolution of the subterranean invertebrate faunas of Illinois and southeastern Missouri. NSS Bull 40:39–63

Peck SB, Richardson BJ (1976) Feeding ecology of the salamander *Eurycea lucifuga* in the entrance, twilight, and dark zones of caves. Ann Speleol 31:175–182

Peck SB, Thayer M (2003) The cave-inhabiting rove beetles of the United States (Coleoptera: Staphylinidae: excluding Aleocharinae and Pselaphinae): diversity and distributions. J Cave Karst Stud 65:37–42

Poole AK, Novosak BA, Gooley AC, Ing DM, Bluett RD, Carter TC, Feldhamer GA (2013) Reintroduction of the eastern woodrat (*Neotoma floridana*) in southern Illinois. Southeast Nat 12:1–10

Reeder PP, Day MJ (1990) Lead mining from the caves in the UMV region: an historical perspective. J Spelean Hist 24(1):14–22

Reid JW (2004) New records and new species of the genus *Diacyclops* (Crustacea: Copepoda) from subterranean habitats in southern Indiana USA. Jeffersoniana 12:1–65

Rowe JC, Heske EJ, Brown WP, Paige KN (2004) Surviving the ice: northern refugia and postglacial colonization. Proc Natl Acad Sci 101:10355–10359

Shear WA (1971) The millipede family Conotylidae in North America with a description of the new family Adritylidae (Diplopoda: Chordeumida). Bull Mus Comp Zool 141:55–98

Shear WA (1972) Studies in the millipede order Chordeumida (Diplopoda): a revision of the family Cleidogonidae and a reclassification of the order Chordeumida in the new world. Bull Mus Comp Zool 144:151–352

Shear WA, Lewis JJ, Farfan M (2007) Diplopoda Chordeumatida Cleidogonidae *Pseudotremia salisae* Lewis: distribution extension north of the Ohio River in Ohio and Illinois USA. Check List 2007:1–3

Shear WA, Shelley R (2007) *Tingupa tlingitorum* n sp a new milliped from Haines Alaska USA with notes on the generic distribution and a revised key to species (Chordeumatida: Tingupidae). Zootaxa 1393:53–59

Shelley RW (1994) The milliped family Nearctodesmidae in northwestern North America with accounts of *Sakophallus* and *S simplex* Chamberlin (Polydesmida). Can J Zool 72:470–495

Soto-Adames FN, Taylor SJ (2013) New species and new records of springtails (Hexapoda: Collembola) from caves in the Salem Plateau of Illinois USA. J Cave Karst Stud 75:146–175

Tilly LJ (1968) The structure and dynamics of Cone Spring. Ecol Monog 38:169–197

Wang D, Holsinger JR (2001) Systematics of the subterranean amphipod genus *Stygobromus* (Crangonyctidae) in western North America with emphasis on the hubbsi group. Amphipacifica 3:39–147

Webb DW, Page LM, Taylor SJ (1998a) The current status and habitats of the Illinois cave amphipod *Gammarus acherondytes* Hubricht and Mackin (Crustacea: Amphipoda). J Cave Karst Stud 60:172–178

Webb DW, Taylor SJ, Krejca JK (1993) The Biological Resources of Illinois Caves and other subterranean environments. Tech Rep 1993 (8) Illinois Natural History Survey Center for Biodiversity. 143 p

Webb DW, Wetzel MJ, Phillippe LR (1998b) The aquatic biota and groundwater quality of springs in the Lincoln Hills Wisconsin driftless and northern till plains sections of Illinois. Technical report 1998 (6). Illinois Natural History Survey Center for Biodiversity

Webb DW, Wetzel MJ, Reed PC, Phillippe LR, Harris MA (1995) Aquatic biodiversity in Illinois springs. J Kansas Entomol Soc 68 (suppl):93–107

Webb DW, Wetzel MJ, Reed RC, Phillippe LR, Young TC (1996) Karst springs in the sinkhole plain of Illinois; their community diversity and hydrogeology. Technical report 1996 (11). Illinois Natural History Survey Center for Biodiversity

Wetzel MJ, Taylor SJ (2001) First records of freshwater oligochaetes (Annelida Clitellata) from caves in Illinois and Missouri USA. J Cave Karst Stud 63:99–104

Whitaker JO Jr (1975) Bats of the caves and mines of the Shawnee National Forest of southern Illinois with particular emphasis on *Myotis sodalis* the Indiana Bat. In: Distributional studies of the Indiana bat (*Myotis sodalis)* on three national forests of the eastern region Eastern Region Forest Service USDA, pp 26–64

Whitaker JO Jr, Hamilton WJ Jr (1998) Mammals of the eastern United States, 3rd edn. Cornell University Press, Ithaca NY

Whitaker JO Jr, Kurta A, Carter TC (2011) Retention of the common name Eastern pipistrelle for *Perimyotis subflavus*. Bat Res News 52:53–54

Widga C, Fulton TL, Martin LD, Shapiro B (2012) *Homotherium serum* and *Cervalces* from the Great Lakes Region, USA: geochronology, morphology and ancient DNA. Boreas 41:546–556

Wilhelm FM, Taylor SJ, Adams GL (2006) Comparison of routine metabolic rates of the troglobite, *Gammarus acherondytes* (Amphipoda: Gammaridae), and the troglophile *Gammarus troglophilus*. Freshw Biol 51:1162–1174

Wilson DE, Reeder DM (2005) Mammal species of the world—A taxonomic and geographic reference. Third Ed. vol 1, 404 p

Wyman J, Leidy J (1862) Observations upon the remains of extinct and existing species of Mammalia found in the crevices of the lead-bearing rocks, and in the superficial accumulations within the lead region of Wisconsin, Iowa and Illinois. In: Hall J, Whitney JD (ed) Report on the Geological Survey of the State of Wisconsin. vol 1, pp 421–424. Albany, NY

Zacharda M, Puglsey CV (1988) *Robustocheles occulta* sp n, a new troglobitic mite (Acari: Prostigmata: Rhagidiidae) from North American caves. Can J Zool 66:646–650

Zeppelini D, Christiansen KA (2003) *Arrhopalites* (Collembola: Arrhopalitidae) in U.S. caves with description of seven new species. J Cave Karst Stud 65:36–42

Zeppelini D, Taylor SJ, Slay ME (2009) Cave *Pygmarrhopalites* Vargovitsch, 2009 (Collembola, Symphypleona, Arrhopalitidae) in United States. Zootaxa 2204:1–18

Zhang J, Holsinger JR (2003) Systematics of the freshwater amphipod genus *Crangonyx* (Crangonyctidae) in North America. Virginia Mus Nat Hist Mem 6:1–274

Intersection of Agriculture and Karst Terrains, Risks, and Solutions

10

Kevin Erb and Benjamin J. Maas

Abstract

Agriculture in Iowa, Illinois, Minnesota, and Wisconsin (the Upper Mississippi Valley region, or UMV) is an important component of each state's economy. Karst landscapes though in the UMV can be locally common and can complicate agricultural activities. Complications include thin and rocky soil, nutrient and fertilizer management concerns, and effects on groundwater resources. Best Management Practices (BMPs) for agricultural activities have historically been written for surface water concerns, with minimal consideration for the interconnection between surface and groundwater in karst areas. This chapter documents the concerns of higher livestock concentrations in karst landscapes and presents examples (and case studies) of approaches and innovative techniques to minimize agriculture's environmental impacts in karstic areas. The goal of this chapter is to provide helpful information that will facilitate changes in agricultural practices in karstic landscapes of the UMV that will aid in protecting groundwater and soil resources.

10.1 Introduction

Karst covers approximately 20% of the United States and karst aquifers supply about 40% of the drinking water to the residents of the United States (Green et al. 2006). A common characteristic of karstic landscape, and carbonate landscapes in general, is a thin layer of soil (Howell et al. 1995; Stiles and Stensvold 2008) (Fig. 10.1). This thin layer of soil makes karst aquifers susceptible to anthropogenic contamination from agricultural practices due to the dense network of conduits and fracture macropores that develop in the underlying shallow bedrock aquifers. This network of conduits and fractures (Fig. 10.2) makes it easy for water to infiltrate the shallow groundwater easily.

In these locations with shallow soils, direct conduits/sinkholes can develop over buried fractures, which can be indicated by visible differences (fracture traces) in plant growth (Fig. 10.2), allowing direct movement of surface water, contaminants, and sediment via the karst features into the underlying aquifers. The flow of water through sinkholes and fractures can be rapid, covering great distances in short periods of time (Luhmann et al. 2011; Wiersma et al. 1984). Agricultural operations, particularly tillage, fill in small sinkholes (<20 cm), making documentation and mitigation a challenge (Fig. 10.3). Sinkhole development and the movement of runoff and soil water into conduits varies significantly from year to year based on precipitation amounts, intensity, and climate factors. In some instances, transport of sediment from agricultural activities into karst can be impressive, such as at Big Spring and Spook Cave, in Iowa (Brick 2004, pp. 12–13, 65). However, understanding the impact agricultural activities have on a karstic landscape is difficult to discern as each karst terrain in the UMV can have different compositions and ages. Additionally, there are different types of karst patterns present in a karst terrain (Fig. 10.2). In the UMV, karst features are commonly found in carbonate bedrock, but features can develop in other rock types. Even in some non-karstic areas, artificial pseudokarst is present in the form of tile drains (Schilling and Helmers 2008) and agricultural drainage wells, which alter the hydrologic cycle (Baker et al. 1985; Capel et al. 2018). For the purposes of this chapter, karst is defined as carbonate bedrock overlain with less than 50 feet (15 m) of soil or glacial drift (Adams et al. 2016).

K. Erb (✉)
Division of Extension, Natural Resources Institute, University of Wisconsin-Madison, 2019 Technology Way, Green Bay, Wisconsin, 54311, USA
e-mail: kevin.erb@wisc.edu

B. J. Maas
Department of Environmental Science, Buena Vista University, 610 W. 4th Street, Storm Lake, Iowa, 50588, USA

G. A. Brick and E. C. Alexander Jr. (eds.), *Caves and Karst of the Upper Midwest, USA*, Cave and Karst Systems of the World, https://doi.org/10.1007/978-3-030-54633-5_10

Fig. 10.1 A, B Thin soil covering carbonate bedrock, common in karst areas. Photos courtesy Calumet County (Wisconsin) Land and Water Conservation Department

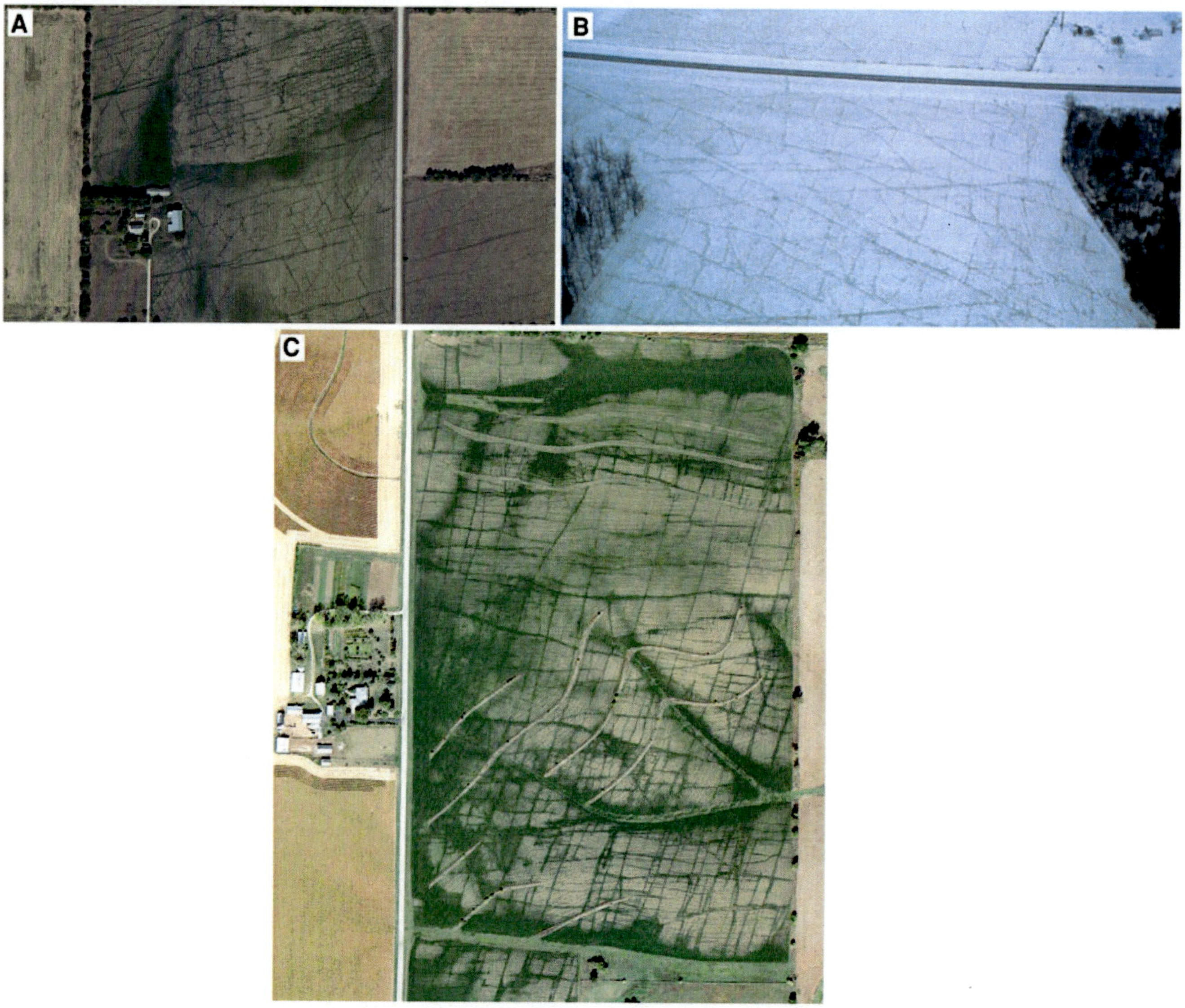

Fig. 10.2 A Examples of fracture trace pattern in karstic terrain in Door County, Wisconsin showing both multiple angles and fractures extending past the local road in the center (aerial photo courtesy of Bing maps), B Door County, winter (aerial photo courtesy University of Wisconsin-Green Bay) and C fracture traces (green pattern) in a field with waterways and erosion control terraces. Jo Daviess County, IL (photo from Panno et al. 2015)

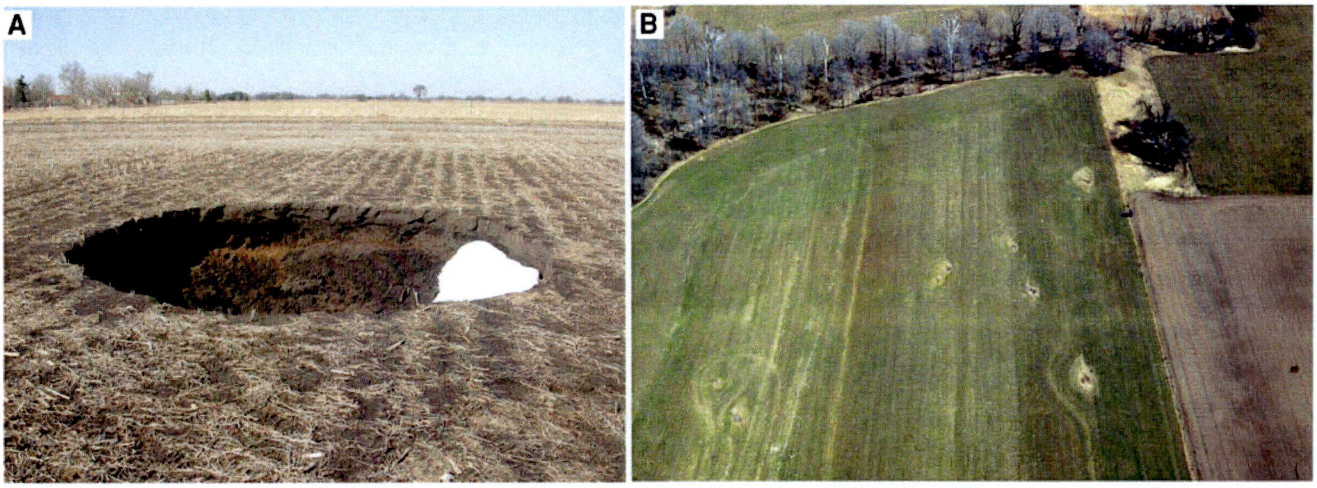

Fig. 10.3 Examples of sinkholes in Wisconsin agricultural fields indicating variation in size and related difficulty in associated with locating sinkholes. **A** A sinkhole in Waukesha County, Wisconsin that collapsed less than 24 h after the last equipment passed over the site, courtesy Wisconsin Geological and Natural History Survey (white area is unmelted snowfall). **B** Numerous smaller sinkholes that have developed over time, with visual evidence of the impact of continuing agricultural operations (harvest patterns). Courtesy Calumet County Land and Water Conservation Department

In addition to sediment pollution, nitrate as nitrogen (NO_3-N) is another groundwater pollutant of concern across the US in shallow aquifers as they are influenced by anthropogenic activities (Burow et al. 2010). These shallow aquifers are both karstic and non-karstic. Nitrate as nitrogen has been linked to human health risks (US EPA). Nitrate pollution in groundwater wells has been a concern nationally and was summarized nicely in the classic paper by Madison and Brunett (1984). Examples of elevated NO_3-N concentrations in groundwater also can be found in sandy aquifers, such as the sand plain region of Wisconsin and the sandhills in Nebraska, which can have NO_3-N concentrations in excess of 20 ppm (Spalding et al. 1978; Stites and Kraft 2001; Zaporozec 1983).

Diverse agricultural systems, different geological and climate across the UMV precludes a uniform management strategy. Variations in livestock type (cattle, pigs and hogs, chickens, and turkeys) (USDA NASS 2012), density and livestock waste management strategies interact with diverse cropping systems/rotations (corn, soybeans, oats, winter wheat, spring wheat, alfalfa, hay, and specialty crops (peas, beans, barley, sunflower, potatoes, sweet corn, and sugar beets). According to USDA NASS 2012 Census of Agriculture data, Iowa produces the most pigs and hogs, chickens, corn for grain, and soybeans (USDA NASS 2012). The same source indicates that Minnesota produces the most turkeys, and Wisconsin produces the most corn for silage, cranberries, and is second nationally in dairy production (USDA NASS 2012).

Due to the susceptibility of UMV karst aquifers to contamination, numerous studies have examined the interactions between animal and cropping agriculture and the karst aquifer (Angel and Peterson 2015; Gupta et al. 2004; Kresic et al. 1992; Long et al. 2008; Mooers and Alexander 1994; Nguyet and Goldscheider 2006; Panno and Kelly 2004; Peterson et al. 2002). The contamination risk is further exacerbated by macropore development in highly elastic clay soils and other soils under no-till farming systems (Erb et al. 2015; Hoorman and Shipitalo 2006) (Fig. 10.4). Another route for contaminants is tile drains, which act like pseudokarst in that they rapidly move water from agricultural fields to creeks and streams (Schilling and Helmers 2008) (Fig. 10.5). In areas with both shallower and deeper karst, these underground tile drainage systems and macropore development may intersect shallow karst areas, creating additional indirect or direct connections. In the past, some tile drain systems were designed to discharge directly to sinkholes or other karst features. These buried drainage systems are similar to karst features in that they discharge to surface water (and occasionally karst features), transporting water directly to discharge points with little to no filtering. Despite the fact there are many instances of rapid transport of surface water through the subsurface, in karstic and non-karstic systems, many agricultural landscapes have been classified as slow-flow or mixed-flow, a combination of slow-flow and fast-flow (Capel et al. 2018). However, in karst areas, travel times to aquifers can be on the order of minutes to hours depending on the soil thickness and the depth to the water table (Erb et al. 2015; Wiersma et al. 1984).

The rapid transport of water from the surface to an aquifer is a concern because the application of fertilizer, both animal manures and synthetic, can be a source of nitrogen (Maticic 1999), chloride (Link and Inmann 2003; Mooers and Alexander 1994; Panno et al. 2006) phosphorus, and pathogens (Muldoon et al. 2018) to shallow aquifers.

Fig. 10.4 Dye test showing marcroporosity in soil cracks (right), and resultant dye movement along the base of the plow layer (left), and downward below the plow layer through earthworm burrows acting as macropores. (Photo courtesy: Shipatalo, USDA-ARS)

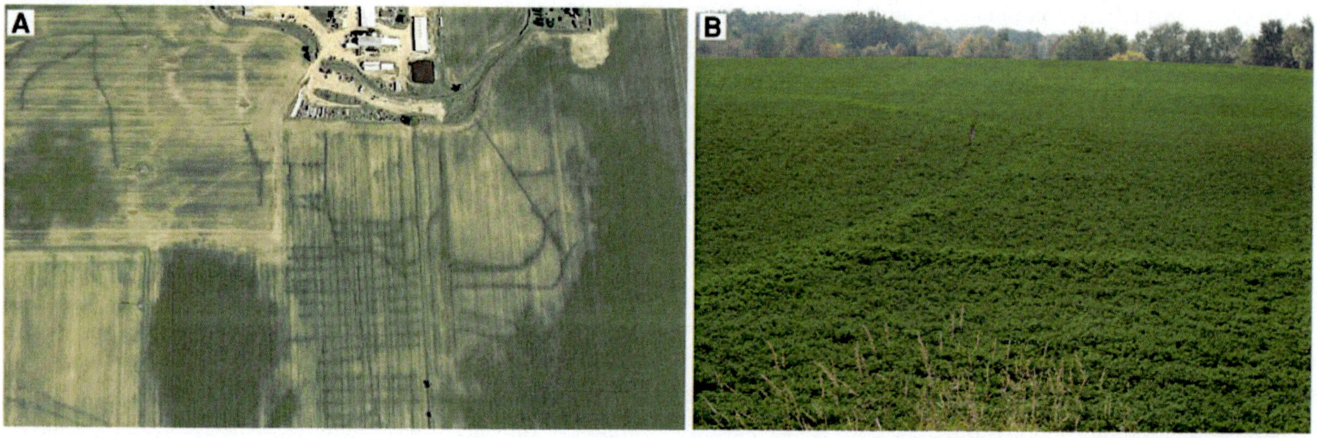

Fig. 10.5 Examples of artificial pseudokarst **A** Google Earth image from Calumet County, Wisconsin and **B** ground-level photo in alfalfa in Brown County, Wisconsin. The improved drainage created by buried tile systems can positively impact crop growth, resulting in similar crop impacts to karst fracture traces in certain seasonal growing conditions. (Images: Google Earth (left) Kevin Erb, UW-Madison Division of Extension (right))

Manure can also be transported into streams during precipitation events and then infiltrate in the groundwater through cracks, conduits, and sinkholes (Gupta et al. 2004). Complicating the management of animal manure is the different abundances of livestock and the different concentrations of nitrogen and phosphorus compounds, which can result in either an abundance or deficit of nitrogen and phosphorus for plant fertilization and an excess of nitrogen to aquifers (Capel and Hopple 2018; Capel et al. 2018; Maticic 1999; Smith and Joern 2012). Additionally, liquid animal manure impacts karst areas more than solid animal manure as the liquid animal manure is mobilized more rapidly (Czymmek et al. 2011). Synthetic fertilizers and pesticides also impact karstic aquifers. The use of animal waste on fields introduces bacteria, virus, pathogens, and hormones into both surface water and groundwater (Davis et al. 2005; Gupta et al. 2004; Peterson et al. 2000). The presence of bacteria, viruses, and pathogens is a concern because there is limited soil development in karstic terrains (Howell et al. 1995; Stiles and Stensvold 2008). In areas with high agricultural activity the dominant source of bacteria, fecal coliform and *Escherichia coli* (*E. coli*), is from livestock (Davis et al. 2005; Howell et al. 1995; Muldoon et al. 2018). Elevated fecal coliform concentration in water is a source of concern because of associated health risks and to trapping of bacteria in the epikarst (Massei et al. 2002). The trapping of fecal coliform in epikarst is a concern because soils in agriculturally active areas can support the survival of fecal coliform for up to six months after spring application of manure and two months after fall applications (Stoddard et al. 1998). Additional anthropogenic potential contaminants in rural areas including road salt, septic effluent, synthetic fertilizers, and pesticides have been documented to impact surface water and karst springs (Angel and Peterson 2015; Crain 2002; Kelly 2008; Lax and Peterson 2009; Lax et al. 2017; Panno et al. 2006; Panno and Kelly 2004; Taraba et al. 1997).

One challenge to agricultural (livestock and non-livestock) operations in karst areas is the difficulty in determining where direct conduits to groundwater exist (Fig. 10.6). Normal soil-moving farming operations often cover up or fill in small sinkholes, resulting in smaller features being hidden by the soil. Some features show up as fracture traces for a few weeks per year (combination of dry weather and when certain crops are grown) (Panno et al. 2015) or are determined when ponded water infiltrates over a short period in late winter (the frost leaving the soil allows ponded water to quickly infiltrate into direct conduits without creating a surface sinkhole). Pseudokarst, Fig. 10.3, (underground drainage systems) can mimic the crop growth impacts of karst features, but often appear as a grid pattern or follow the low spots in the field rather than the "cracked glass" pattern typical of karst features (Fig. 10.2).

Because of the susceptibility of karst aquifers to agricultural activities, county and state governments, along with conservation groups, have taken an active role in implementing new regulations to decrease effects of agricultural practices. In response to these and other developments, locally-based environmental organizations such as the Crawford Stewardship Project have become increasingly prominent, enlisting professional assistance and organizing networks of volunteers to monitor variables such as air and water quality and disseminate information (CSP 2016).

Regional karst science has yet to be incorporated adequately into planning mechanisms, and there have been calls for regional karst studies similar to that carried out and applied to land use planning in northeastern Wisconsin (Erb and Stieglitz 2007). This paper will highlight the variability of animal density across the UMV and present two case studies from the eastern side of Wisconsin, an area with karstic terrains and dense agricultural activity. These case studies will highlight how these new regulations have positively affected both groundwater quality and environmental quality, along with limitations of current efforts. Regardless of the outcomes of the current conservation methods, there has been an increase public awareness of southwestern Wisconsin's karst, which is now a central theme for local non-government organizations (NGOs) and the regional press.

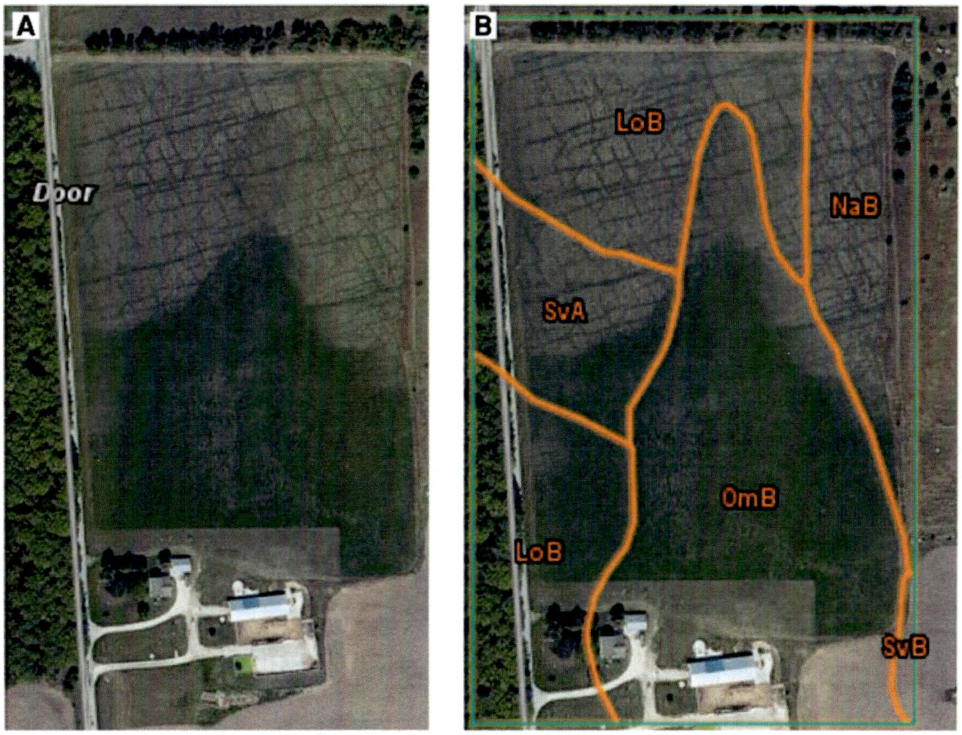

Fig. 10.6 **A** Fracture trace image from Door County, Wisconsin (left, courtesy Bing Maps), showing both the "cracked glass" pattern typical of fracture traces, and the impact of soil depth on feature visibility and the soil date, right. **B** The top half of the field has shallower soils than the bottom half, resulting in the fractures being more visible in dry years (courtesy USDA-NRCS Web Soil Survey)

10.2 Methods

To better show the variations in livestock populations across the UMV, maps showing areas with both agriculture (cropland acres, livestock density) and karst are shown in Figs. 10.7, 10.8, 10.9 and 10.10. Land used for agricultural activities was identified from data published by the USDA: National Agricultural Statistics Service (Han et al. 2014). After identifying the land used for agriculture, the map data were reclassified from a raster data model to vector data model using ESRI ArcMap. Converting from a raster data model allowed for locations that had both agriculture and karst on a landscape to be identified and extracted from the larger datasets into a new map dataset. The locations with karst are shown using the karst map, available from the USGS (Weary and Doctor 2014). Karst was defined as carbonate bedrock less than 50 feet (15 m) from the surface, similar to other classification methods (Adams et al. 2016). Locations with karst within 50 feet (15 m) of the surface in agricultural areas were then determined by merging two datasets and selecting the locations where both layers overlapped. These data layers were then merged into one layer and displayed in a gray color (Fig. 10.7), which highlights only the agricultural areas that are on top of

Fig. 10.7 Locations of karst and agriculture are shown in gray, at the county level for the UMV

karstic areas. Due to scale, resolution, and variations in known soil depth, these maps should be considered generalizations and not used to guide site-specific management.

Once the karst and agricultural areas were identified (locations susceptible to agricultural practices), the next goal was to display the variation in agricultural animal density across the UMV. The agricultural animal density maps were achieved by determining the number of animals in each county using publicly available data from the USDA for the years 2012 and 2017, the most recent data that are available (Quick Stats 2018). The animals investigated were cattle (2017 data), swine (2012 data), and poultry, a combination of turkey and chickens (2012 data). Crop data used were also accessed using publicly available data for the year 2017 (Quick Stats 2018). The crop data at the county level include corn, soybeans, oats, winter wheat, spring wheat, alfalfa, hay, beans, barley, sunflower, and sugar beets. The individual crop data, as acres, were added together for a total number of crop land in each county. Numbers of animals in each county were then divided by the amount of cropland in each county. The result of the animal density at the county level are shown in Figs. 10.8, 10.9 and 10.10 with light red colors indicating lower densities and darker red, higher densities. This cropland density approach removes variability caused by different percentages of total land in each county with different land uses (natural areas, urban, transportation, etc.). Each state has been scaled so that they cover the same numeric range for each animal density.

10.3 Results and Discussion

The locations with both karst, defined as shallow carbonate bedrock within 50 feet (15 m) to the surface (Adams et al. 2016) and agricultural activity are indicated for the UMV (Fig. 10.7). Urban areas and locations with more than 50 feet (15 m) of material covering carbonate bedrock were removed from the karst base map published by the USGS (Weary and Doctor 2014) as these are locations that are less susceptible to agricultural activities.

Agricultural land use is variable across the UMV as indicated by analyses of the USDA data, presented in Figs. 10.8, 10.9 and 10.10. The analyses of data presented in this paper indicate there are different densities of livestock production across both the UMV and across each state in both karstic and non-karstic areas. Iowa has some of the highest densities (swine, poultry), but not necessarily in karstic areas. While elevated swine densities are found in the karstic areas of Iowa, poultry is mostly located in non-karstic areas and concentrated in only a few counties (Figs. 10.8 and 10.9). Wisconsin has some counties with high densities of cattle in karstic areas, but also has elevated cattle densities outside karst areas (Fig. 10.10). The other UMV states have significant of cattle production, but at densities lower than those found in Wisconsin (Fig. 10.10).

As indicated in Figs. 10.8, 10.9 and 10.10, there are several counties with elevated animal densities, which present management challenges to land managers. These challenges include both large amounts of animal manure and different types of animal manure, water resource demands, pathogens, and the application of synthetic fertilizers (Capel et al. 2018; Czymmek et al. 2011; Gupta et al. 2004; Smith and Joern 2012). Below are two case studies in the management of these risks, the first examining the impact of regulations in Northeast Wisconsin and the second the grassroots solutions and approaches in both Illinois and Wisconsin.

10.4 Case Studies

10.4.1 Case Study 1—Impact of Regulations

The counties alongside the western shore of Lake Michigan in Wisconsin have areas of shallow soils (<5 m) underlain by a highly fractured Silurian dolomite. Surface water runoff enters the aquifer system both through sinkholes and direct infiltration—reports of Brown Water Incidents (BWIs) in local wells are not uncommon, and usually occur within a few days of precipitation or a snowmelt runoff event. The runoff entering the aquifer during BWIs carries soil, nutrients, animal manure, and pathogens directly into the groundwater system (Erb and Stieglitz 2007; Muldoon et al. 2018).

Clusters of BWIs, bacterial and nutrient contamination in the early spring (February–March) occurred in a five-county area in 2004, 2005, and 2006, with the 2006 event resulting in 78 contaminated wells in a single township (Morrison, Brown County, Wisconsin, 93 km^2). The incidents led to the formation of the Northeast Wisconsin Karst Task Force (KTF), which studied the issue and created a set of recommendations for landowners, local governments (in five counties), and the state government to implement to reduce the risk of BWIs (Erb and Stieglitz 2007).

Within a few months of the KTF recommendations in 2007, two Wisconsin counties, Brown and Manitowoc, implemented regulations on local livestock farms to regulate spreading of manure on frozen and snow-covered soils. While both counties have actively mapped karst features, each county is managing its karst features differently. Brown County required land managers to prioritize fields for application based on risk, apply animal manure to lowest risk fields first, and follow a winter manure spreading plan. Manitowoc County restricted applications of animal manure to flatter fields and required incorporation of manure within

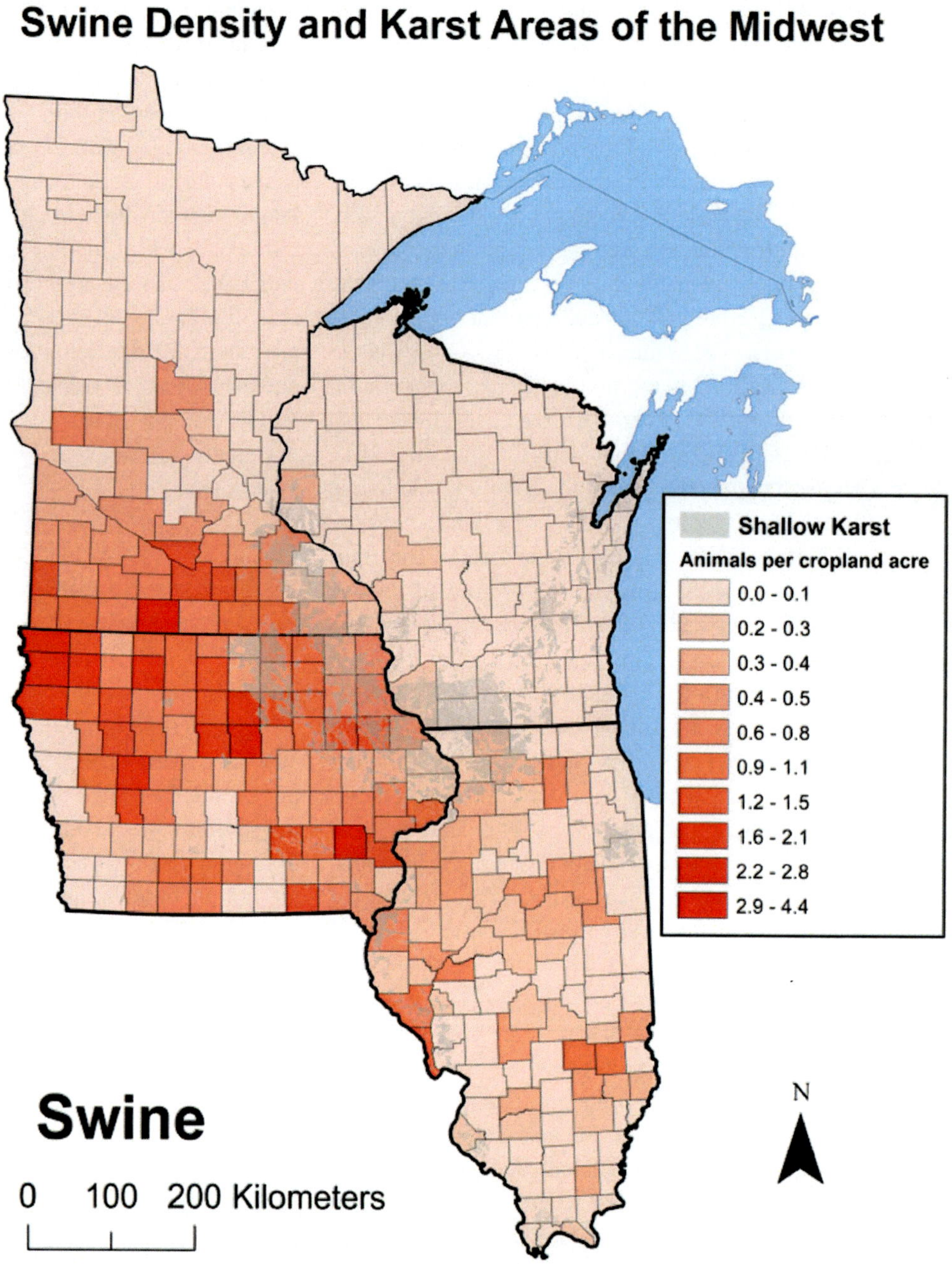

Fig. 10.8 Locations of karst and agriculture (gray) overlying swine density at the county level for the UMV (shades of red). Darker colors of red indicate elevated densities. See text for explanation of how densities were determined

48 h. The remaining counties did not change local regulations but did increase education and outreach to land managers.

Researchers from the University of Wisconsin-Extension (UWEX), UW-Green Bay, and UW-Madison, inventoried independently documented or verified well-contamination reports in Brown and Manitowoc counties, which implemented regulations, and two adjacent counties (Calumet, Kewaunee) that did not. Contamination incident data were compared in the five years prior to the regulations being implemented (2002–2006) and the seven years afterward (2008–2014). Impacts examined included BWIs, presence of manure odors, and documented pathogen contamination (Erb et al. 2015).

The statistical analysis showed that the implementation of restrictions on winter spreading (frozen ground) of animal waste has likely had an impact on the documented instances of well contamination. In the counties with new regulations, almost every impact variable showed a statistically significant difference between the pre-regulation period (2002–2006) and the post-regulation period (2008–2014) in one of the two counties.

The data also indicate an overall decrease in the number of documented wells with pathogenic indicator bacteria in the counties that implemented regulations, with 28 recorded incidents in the pre-2007 period and eight in the post-2007 period. Reports of pathogenic indicator bacteria remained constant (7) in the two counties that did not implement

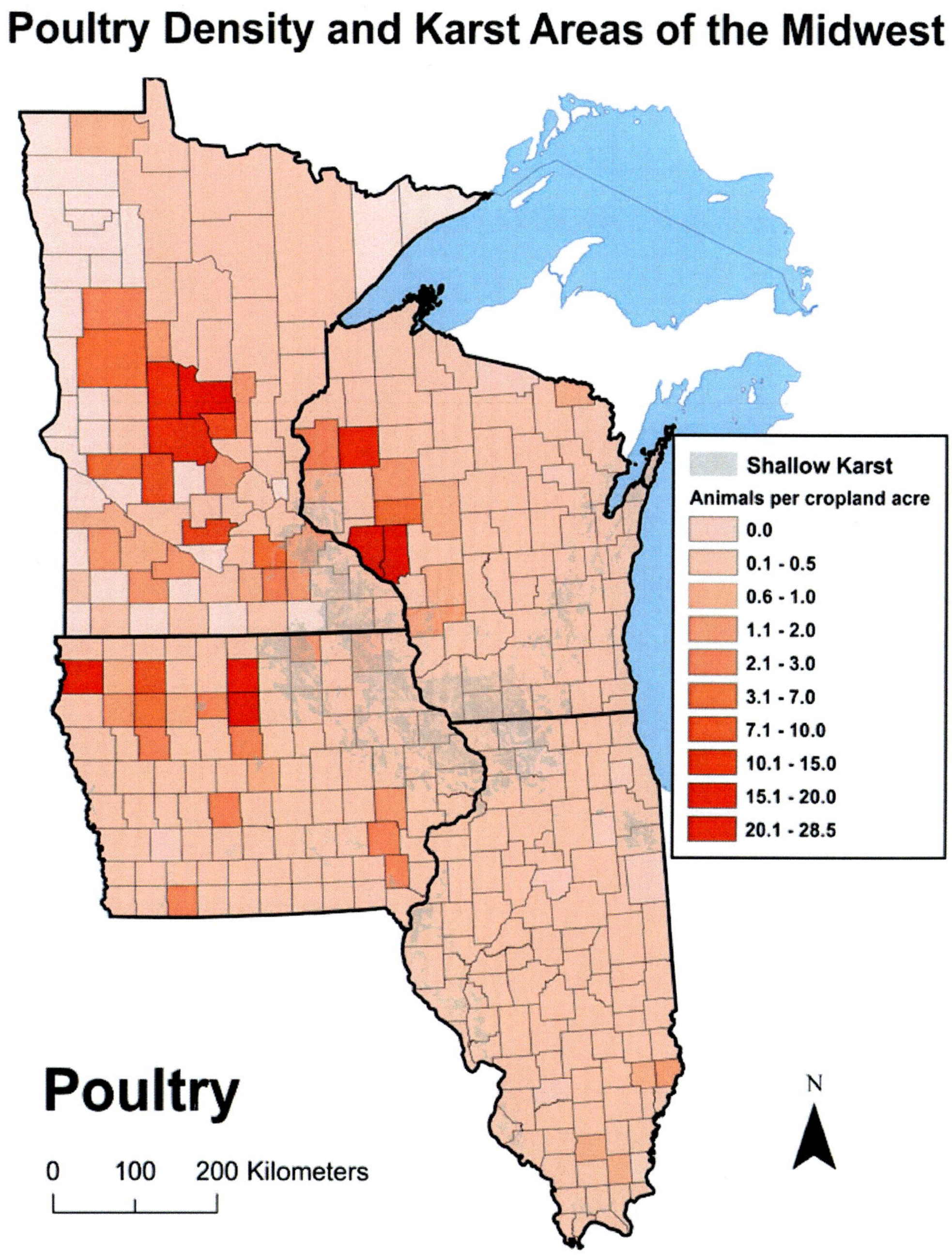

Fig. 10.9 Locations of karst and agriculture (gray) overlying poultry density at the county level for the UMV (shades of red). Darker colors of red indicate elevated densities. See text for explanation of how densities were determined

regulations. BWIs showed a lower rate of increase (38%) in regulated counties (5 pre/8 post) *versus* 70% in non-regulated counties (10 pre/17 post).

Livestock numbers increased in all four counties, pre-regulation to post-regulation, resulting in more animal waste applied in the target area. Since the publication of the study, Kewaunee County has implemented winter spreading regulations, the State of Wisconsin has revised state regulations (NR151) to reduce risk in 15 Silurian dolomite counties, and several of the counties are considering incorporating the new state regulations into their local enforcement ordinances. In the first 30 months since the Kewaunee regulations were passed, the local agency staff has strongly felt the risk of groundwater contamination has been reduced (D. Bonness, pers. comm., 10/2/2018). Future studies will determine the actual impact.

By comparing the pre-regulation and post-regulation statistics, the counties that implemented even minor regulations had a statistically significant impact on the number of BWIs, odors, and pathogenic indicator bacteria in groundwater. This study verified that implementing the winter spreading practices reduced BWIs, odors, and pathogenic indicator bacteria in groundwater.

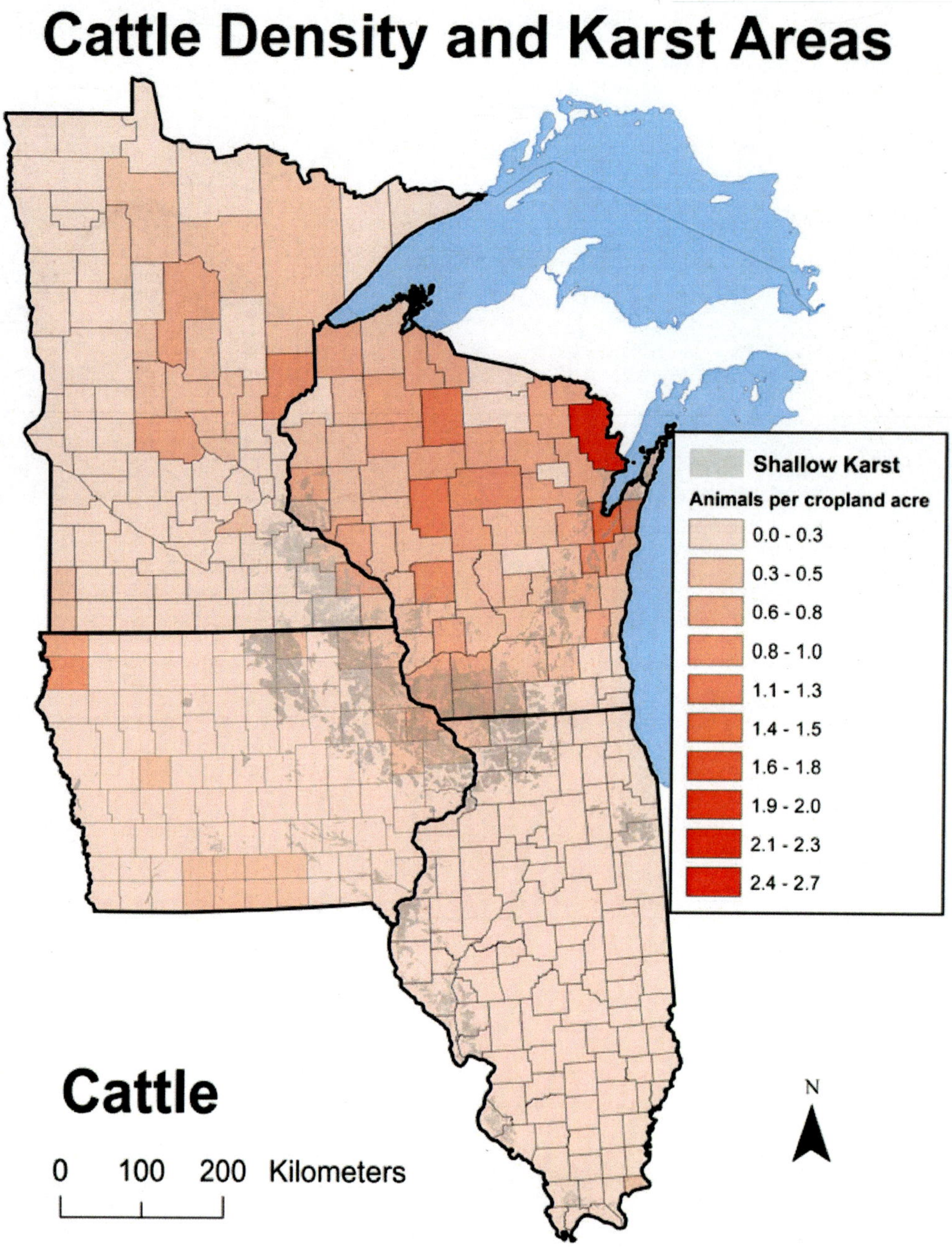

Fig. 10.10 Locations of karst and agriculture (gray) overlying cattle density at the county level for the UMV (shades of red). Darker colors of red indicate elevated densities. See text for explanation of how densities were determined

10.4.2 Case Study 2: Grassroots, State, and Regional Approaches

Regulations at the state and local level are a proven method to reduce groundwater contamination in agricultural areas (Church and Prokopy 2017; Erb et al. 2015). Erb and Stieglitz (2007) indicated there is no way to prevent 100% of all groundwater contamination in karst areas. Furthermore, some best management practices (BMPs) for reducing nutrients and pathogen movement to groundwater, such as leaving manure on the surface to reduce pathogen viability, can increase the risk of nutrients and pathogens entering surface water via runoff. Likewise, anaerobic digestion and other treatment systems for human, industrial, and livestock waste reduce particle size and total dry matter content of liquid manure, increasing the risk of downward movement in the soil profile via macropores (Hoorman and Shipitalo 2006).

Recognizing the risk, land managers, and resource agency staff in the UMV have created a variety of tools to identify and reduce contamination risk. At the same time, groups of land managers have come together to voluntarily create and implement risk reduction strategies. The long-term impact of both approaches has yet to be measured, but many hold promises to reduce contamination incidents. Examples of these approaches include focusing on developing risk reduction tools, implementing mitigation strategies, and changing land management practices. Provided below are descriptions of resources and website locations of tools available that are designed to help improve land management.

10.4.2.1 Weather

Groundwater contamination in karst systems is influenced by weather events, which include precipitation and snowmelt. These weather events directly contribute contaminants to groundwater as infiltration through shallow soils or as runoff to conduits, such as sinkholes. Partnering with the National Weather Service, six Midwestern states have implemented a Runoff Risk Advisory System. Originally developed for surface water protection, the system can be used by land managers to assess risk for groundwater as well. It integrates soil moisture measurements, soil temperature, flood prediction, and weather forecasting models to create a risk index of surface runoff leaving the landscape. The January 2019 release provides a resolution of 2 km by 2 km and gives those who apply manure and other products to the landscape a tool to estimate the risk of applying on different days. Information on the Runoff Risk Advisory System can be found at https://www.mda.state.mn.us/protecting/cleanwaterfund/toolstechnology/runoffrisk.

10.4.2.2 Pathogen/Endocrine Disrupter/ Pharmaceutical Risk

Increasing concerns with odor and greenhouse gas emissions led livestock producers to try different manure management/treatment systems (e.g., anaerobic digestion, composting) to address odor and greenhouse gas emission concerns. This resulted in significant federal, state, and private funding invested in digesters (AgSTAR 2017). Operational challenges have limited the use and lifespan of anaerobic digestion systems, which has decreased interest in the technology. State regulatory rules implemented in Wisconsin in 2018 (NR151) granted greater land application flexibility to livestock farms that could reduce pathogens below statutory thresholds (500,000 Colony Forming Units (CFU) per 100 ml for liquid manure or per gram total solids for solid manure). However, meeting the pathogen thresholds is a new concept for digester operators. Similar digesters used in the management of human waste are not effective at removing endocrine disrupters and pharmaceutical products that enter the treatment facility (Samaras et al. 2013). Source control (reducing pathogen load pre-treatment) may have significant potential to reduce loading from land application by reducing the need for treatment.

Changing how the livestock are fed can reduce the E. coli O157:H7 excreted in animal manure, thereby decreasing the loading onto sensitive soils (Callaway et al. 2003). Reductions in grain feeding and increases in forage feeding hold more promise in the reduction of pathogens with beef production, as most dairy farms are already feeding high levels of forage.

Taking advantage of natural solar UV radiation to decrease pathogen survival by leaving animal wastes on the surface instead of incorporating them can reduce pathogens risk; however, not incorporating the waste into the soil may increase the risk of runoff carrying nutrients and pathogens into surface water or into groundwater (Nicholson et al. 2005).

10.4.2.3 Managing the Land Application Risk

One of the more cost-effective approaches is to reduce the risk during application. Each of the four states in the UMV has state-specific regulations on livestock operations that require either setbacks or mitigation practices, such as vegetated buffers and incorporation, around karst features (Erb and Stieglitz 2007). Identification and mapping of these features is the responsibility of the landowner or land manger and their consultant who develops the nutrient management plan. While some states and counties have tried to inventory all features, good maps only exist in a few areas (D. Bonness, pers. comm., 10/2/2018).

A common approach has developed independently across the UMV to identify locations with karst. The map developer, consultant, or agency resource manager start with marking visible and known karst features on an aerial photo of the field and surrounding area. Often multiple years of aerial photos are used to try to identify features that may only be visible certain times of the year or while specific crops are being grown. Soil depth mapping data is added, as greater soil depths can hide fractures and sinkholes. These maps are then examined and updated on an annual basis, and new features added as they are observed.

These efforts to identify karstic features are often aided by county-level depth to bedrock mapping research. Newer techniques, including LiDAR mapping of closed depressions, ground-penetrating radar, and seismic analysis can be useful in the field map development process (Cooley et al. 2013). Consultants and agencies are exploring other options, such as electrical conductivity measuring and remote sensing in shallow bedrock zones.

Mitigation strategies during land application include maximizing the soil/waste interface (mixing of soil and waste during application/avoiding injecting larger volumes in limited bands), maximizing the distance between the karst interface by prohibiting deeper injections/incorporation, and pre-tilling fields before application to eliminate macropores and preferential flow pathways that bypass natural soil filtering. Rates are often limited based on soil moisture to ensure oversaturation does not occur, or applications are split and occur a week or more apart. Seasonal restrictions have also proven effective (Erb et al. 2015). Farms in the UMV are experimenting with in-season application, timed to greatest crop uptake/greatest evapotranspiration of soil moisture, to reduce the risk of loss as well.

10.4.2.4 Producer Led Innovations

In several cases, local landowner-led or agency and landowner workgroups have made significant progress in cataloging or creating local solutions and implementing landscape management changes. Examples include the Southwest Illinois Sinkhole Plain BMP Manual (Committee 1997), the Northeast Wisconsin Karst Task Force Report, Morrison Farmers Group (Erb and Stieglitz 2007), and Peninsula Pride Farmer Led Watershed Group in Kewaunee/Door Counties, Wisconsin. The former two groups focused on documenting research and BMPs, while the farmer-led groups have focused on creating localized resources and solutions. The Morrison group created a list of effective, feasible, and practical BMPs, and worked together to implement them in Brown County, Wisconsin. The Peninsula Pride group secured funding to assist landowners in mapping depth to bedrock in 60% of the sensitive areas in Kewaunee County in less than one year, accurate to within one inch (2.5 cm) where soil depths were less than 3 feet (1 m). They are also working with local agencies to increase adoption when the regulations require or recommend BMPs such as split manure/fertilizer applications, and the establishment of buffers around sinkholes and known karst features.

10.4.2.5 Paradigm Shifts

Going forward, wholesale management changes will be needed to fully address the impact of livestock agriculture in karst areas. Local stakeholder workgroups in Northeast Wisconsin have identified several shifts in thinking that would reduce the risk of pollution, including

- Converting from managing manure as a liquid to managing manure as a solid.
- Shifting a greater percentage of manure application to in-season rather than pre-season applications.
- Adapting a managed grazing system.
- Growing shorter season varieties to widen the application window during the growing season.
- Relocating livestock operations from high-risk areas to lower risk areas. This has already been successfully done for surface water reasons under Wisconsin's former Priority Watershed program.

Any paradigm shift must be fully analyzed to document and address any unintended consequences. In Northeast Wisconsin, documenting depth to bedrock using soil disturbing methods (backhoe, probe) has led to the development of sinkholes where they did not previously exist, so the focus has changed to non-invasive methods of soil depth mapping (D. Bonness, pers. comm., 10/2/2018). Likewise, implementing certain soil health practices, such as no-till, increase water infiltration and are documented to increase contamination risk to both groundwater and subsurface drainage systems, and must be carefully considered before implementation (D. Hart, pers. comm., 8/31/2018; Hoorman and Shipitalo 2006).

10.5 Summary

Agricultural activities and shallow karst commonly intersect in the UMV, which can lead to multiple risks to the residents of the UMV and the environment, especially to water quality. This risk to water quality can result in health concerns due to the presence of excess nitrogen as nitrate and bacteria, pathogens, and viruses. Data presented in this paper indicate the density of livestock varies within each state and in some instances elevated densities occur in karstic areas. However, agriculture is an important component of each UMV state's economy. In attempts to better manage karst landscapes, especially the application of animal manure, local conservation efforts have been implemented in Wisconsin and Illinois, with positive results. Plans for the expansion of these voluntary and regulatory approaches are in place and suggests that with proper management, at the county level, environmental quality can be improved even in areas with elevated agricultural activities. Regional karst science has yet to be incorporated adequately into planning mechanisms, and there have been calls for a regional karst study similar to that carried out and applied to land use planning in northeastern Wisconsin. Regardless of the outcomes of the current conservation methods, there has been an increased public awareness of Wisconsin's karst, which is now a central theme for local Non-Government Organizations (NGOs) and the regional press.

References

Adams R, Barry J, Green J (2016) Minnesota regions prone to surface karst feature development Minnesota Department of Natural Resources, Ecological and Water Resources Division Series GW-01:1–3

AgSTAR (2017) EPA, Biogas Recovery in the Agriculture Sector

Angel JC, Peterson EW (2015) Nitrates in karst systems: comparing impacted systems to a relatively unimpacted system. J Geogr Geol 7:65–76

Baker JL, Kanwar RS, Austin TA (1985) Impact of agricultural drainage wells on groundwater quality. J Soil Water Conserv 40:516–520
Brick GA (2004) Iowa Underground: A Guide to the State's Subterranean Treasures. Trails Books, Black Earth, WI, 223 p
Burow KR, Nolan BT, Rupert MG, Dubrovsky NM (2010) Nitrate in groundwater of the United States, 1991–2003. Environ Sci Technol 44:4988–4997
Callaway T, Elder R, Keen J, Anderson R, Nisbet D (2003) Forage feeding to reduce preharvest Escherichia coli populations in cattle, a review. J Dairy Sci 86:852–860
Capel PD, Hopple JA (2018) Data and citations describing the connections between agriculture and water quality in the United States. USGS data release
Capel PD, McCarthy KA, Coupe RH, Grey KM, Amenumey SE, Baker NT, Johnson RL (2018) Agriculture—A river runs through it —The connections between agriculture and water quality USGS circular 1433:1–201
Church SP, Prokopy LS (2017) The influence of social criteria in mobilizing watershed conservation efforts: a case study of a successful watershed in the Midwestern US. Land Use Policy 61:353–367
Committee (1997) Southwestern Illinois sinkhole plain; best management practices manual: practices that help protect the fragile karst environment
Cooley E, Radatz A, Frame D (2013) Executive summary: mapping bedrock University of Wisconsin-Extension discovery farms: 2
Crain AS (2002) Pesticides and nutrients in karst springs in the Green River basin, Kentucky, May-September 2001 USGS 133-01
CSP (2016) Crawford Stewardship Project. https://www.crawfordstewardship.org/. Accessed 20 Jan 2019
Czymmek K, Geohring L, Lendrum J, Wright P, Albrecht G, Brower B, Ketterings Q (2011) Manure management guidelines for limestone bedrock/karst areas of Genesee County. New York: Practices for Risk Reduction. Animal Science Publication Series No. 240:1–9
Davis RK, Hamilton S, Brahana JV (2005) Escherichia coli survival in mantled karst springs and streams, northwest Arkansas Ozarks, USA. J Am Water Resour Assoc 41:1279–1287
Erb K, Ronk E, Koundinya V, Luczaj J (2015) Groundwater quality changes in a karst aquifer of northeastern Wisconsin, USA: Reduction of brown water incidence and bacterial contamination resulting from implementation of regional task force recomendations Resources 4:655–672
Erb K, Stieglitz R (2007) Final report of the Northeast Wisconsin Karst Task Force, 51 p
Green RT, Painter SL, Sun A, Worthington SRH (2006) Groundwater contamination in karst terranes. Water Air Soil Pollut 6:157–170
Gupta S, Munyankusi E, Moncrief J, Zvomuya F, Hanewall M (2004) Tillage and manure application effects on mineral nitrogen leaching from seasonally frozen soils. J Environ Qual 33:1238–1246
Han W, Yang Z, Di L, Yue P (2014) A geospatial web service approach for creating on-demand cropland data layer thematic maps. Trans Am Soc Agri Biol Eng 57:239–247
Hoorman JJ, Shipitalo MJ (2006) Subsurface drainage and liquid manure. J Soil Water Conserv 61:94A-97A
Howell JM, Coyne MS, Cornelius P (1995) Fecal bacteria in agricultural waters of the bluegrass region of Kentucky. J Environ Qual 24:411–419
Kelly WR (2008) Long-term trends in chloride concentrations in shallow aquifers near Chicago. Ground Water 46:772–781
Kresic N, Papic P, Golubovic R (1992) Elements of groundwater protection in a karst environment. Environ Geol Water Sci 20:157–164
Lax SM, Peterson EW (2009) Characterization of chloride transport in the unsaturated zone near salted road. Environ Geol 58:1041–1049
Lax SM, Peterson EW, Van der Hoven S (2017) Quantifying stream chloride concentrations as a function of land-use. Environ Earth Sci 76:12
Link ML, Inmann DL (2003) Ground water monitoring at livestock waste control facilities in Nebraska ground water unit, water quality assessment section, Nebraska Department of Environmental Quality
Long AJ, Sawyer JF, Putnam LD (2008) Environmental tracers as indicators of karst conduits in groundwater in South Dakota, USA. Hydrogeol J 16:263–280
Luhmann AJ, Covington MD, Peters AJ, Alexander SC, Anger CT, Green JA, Runkel AC, Alexander EC Jr (2011) Classification of thermal patterns at karst springs and cave streams. Ground Water 49 (3):324–334
Madison RJ, Brunett JO (1984) Overview of the occurrence of nitrate in ground water of the United States National Water Summary 1984—Water-Quality Issues 2275:93–105. https://doi.org/10.3133/wsp2275
Massei N, Lacroix M, Wang HQ, Dupont JP (2002) Transport of particulate material and dissolved tracer in a highly permeable porous medium: comparison of the transfer parameters. J Contam Hydrol 57(1–2):21–39
Maticic B (1999) The impact of agriculture on ground water quality in Slovenia: standards and strategy. Agric Water Manag 40(2):235–247. https://doi.org/10.1016/S0378-3774(1098)00124-00123
Mooers HD, Alexander EC Jr (1994) Contribution of spray irrigation of wastewater to groundwater contamination in the karst of southeastern Minnesota, USA. Appl Hydrogeol 2:34–44
Muldoon MA, Borchardt MA, Spencer SK, Hunt RJ, Owens D (2018) Using enteric pathogens to assess sources of fecal contamination in the silurian dolomite aquifer: preliminary results. Karst groundwater contamination and public health. Springer, Cham, pp 209–213
NASS U (2012) Census of Agriculture, Ag Census Web Maps
Nguyet VTM, Goldscheider N (2006) Tracer tests, hydrochemical and microbiological investigations as a basis for groundwater protection in a remote tropical mountainous karst area. Vietnam Hydrogeol J 14:1147–1159
Nicholson FA, Groves SJ, Chambers BJ (2005) Pathogen survival during livestock manure storage and following land application. Biores Technol 96:135–143
Panno SV, Hackley KC, Hwang H, Greenberg S, Krapac I, Landsberger S, O'Kelly D (2006) Characterization and identification of Na-Cl sources in ground water. Groundwater 44:176–187
Panno SV, Kelly W (2004) Nitrate and herbicide loading in two groundwater basins of Illinois' sinkhole plain. J Hydrol 290:229–242
Panno SV, Luman DE, Kolata DR (2015) Characterization of karst terrain and regional tectonics using remotely sensed data in Jo Daviess county, Illinois. Illinois State Geol Surv Circ 589:29
Peterson E, Davis R, Brahana J, Orndorff H (2002) Movement of nitrate through regolith covered karst terrane, northwest Arkansas. J Hydrol 256:35–47
Peterson E, Davis R, Orndorff H (2000) 17b-estradiol as an indicator of animal waste contamination in mantled karst aquifers. J Environ Qual 29:826–834
Quick Stats (2018) Survey & Census Data. https://www.nass.usda.gov/Quick_Stats/
Samaras VG, Stasinakis AS, Mamais D, Thomaidis NS, Lekkas TD (2013) Fate of selected pharmaceuticals and synthetic endocrine disrupting compounds during wastewater treatment and sludge anaerobic digestion. J Hazard Mater 244:259–267
Schilling KE, Helmers M (2008) Tile drainage as karst: Conduit flow and diffuse flow in a tile-drained watershed. J Hydrol 349:291–301
Smith D, Joern B (2012) Dietary phytase to reduce phosphorus losses from animal manure SERA-17 Report:2

Spalding RF, Gormly JR, Curtiss BH, Exner M (1978) Nonpoint nitrate contamination of ground water in Merrick County, Nebraska. Ground Water 16:86–95. https://doi.org/10.1111/j.1745-6584.1978.tb03207.x

Stiles CA, Stensvold KA (2008) Loess contribution to soils forming on dolostone in the driftless area of Wisconsin. Soil Sci Soc Am J 72:650–659

Stites W, Kraft GJ (2001) Nitrate and chloride loading to groundwater from an irrigated north-central US sand-plain vegetable field. J Environ Qual 30(4):1176–1184

Stoddard CS, Coyne MS, Grove JH (1998) Fecal bacteria survival and infiltration through a shallow agricultural soil: timing and tillage effects. J Environ Qual 27:1516–1523

Taraba J, Dinger J, Sendlein L, Felton G (1997) Land use impacts on water quality in small karst agricultural watersheds. Proceedings Karst-Water Environment Symposium 1997, 127 p

Weary DJ, Doctor DH (2014) Karst in the United States: a digital map compilation and database. USGS Open-File Rep 2014–1156:23

Wiersma JH, Stieglitz RD, Cecil DL, Metzler GM (1984) Characterization of the shallow groundwater system in an area with thin soils and sinkholes (Door Co., WI). In: Multidisciplinary conference on sinkholes 1, pp 305–310

Zaporozec A (1983) Nitrate concentrations under irrigated agriculture. Environ Geol 5(1):35–38

11 Paleoclimate Studies from Midwestern Caves

Jeffrey A. Dorale

Abstract

Caves of the Midwestern U.S.A. have a history of paleoclimate studies going back more than 50 years. The most valued paleoclimate archives are speleothems, especially stalagmites, which commonly contain a straightforward internal stratigraphy, grow over many thousands of years in some cases, and can be readily dated using uranium-series techniques. Some of the more unique contributions that have come from Midwestern caves include: using speleothem carbon isotopes as a vegetation proxy, reconstructing the episodic nature of cave flooding from detrital layers in speleothems, and using broken speleothems and altered speleothem growth to identify the timing of past seismic events. Because speleothems contain high-quality paleoclimate information, the pace of cave paleoclimate studies has intensified in recent years. This increased demand on a fragile resource highlights the need for a strong conservation ethic among scientists working in caves.

11.1 Introduction

Caves are uniquely protected from many of the destructive processes that operate at the Earth's surface, yet are still in close communication with surface conditions via hydrologic, biologic, and geochemical processes. Caves may therefore exist in stable configurations for many thousands of years, providing valuable sites for capturing long-term records of climatic and environmental changes. Paleoclimate studies from caves have greatly increased in number over the past two decades, from locations around the globe, but some of the earliest history of the science can be traced back to Midwestern settings. The states of Iowa, Illinois, Minnesota, and Wisconsin, as well as neighboring Missouri, Arkansas, and Indiana, all host significant karst landscapes and have played an important role in piecing together climatic histories of the region.

The range of cave-based topics underneath the broad umbrella of *paleoclimate* spans the spectrum from temperature and vegetation reconstructions, to landscape evolution, to paleohydrology, and even paleoseismicity, with relevance at various scales, from local to global. Speleothem records have great potential to contribute fundamentally to the field of paleoclimatology because they can exhibit a suite of favorable characteristics found in few other single archives. These characteristics include (1) the ability to date small carbonate sub-samples precisely and accurately by U-series methods (only rivaled by records that have truly annual layers), (2) high resolution (approaching that in ice-core records), and (3) continuous or near-continuous deposition over thousands of years (Fig. 11.1). Here I provide a brief overview of some of the history of paleoclimate studies in Midwestern caves. I do not attempt to cover the full range of topics that have been considered in these studies, but instead I highlight what I believe are some of the more unique and interesting contributions to paleoclimate research that have come from Midwestern caves.

11.2 Early Work in Midwestern Caves

A seminal paper by Hendy and Wilson (1968) set the stage for paleoclimate studies in caves. Around this same time, Derek Ford and Henry Schwarcz from McMaster University began working on speleothems from North American caves. Some of this early work included uranium-series dating and oxygen isotope analysis on speleothems from Cold Water Cave in northeastern Iowa (Figs. 11.2 and 11.3) which had been discovered in 1967 (Koch and Case 1974). Similar age-dating was carried out some years later on speleothems

J. A. Dorale (✉)
Department of Earth and Environmental Sciences, University of Iowa, Trowbridge Hall, Iowa City, Iowa 52242, USA
e-mail: jeffrey-dorale@uiowa.edu

G. A. Brick and E. C. Alexander Jr. (eds.), *Caves and Karst of the Upper Midwest, USA*, Cave and Karst Systems of the World, https://doi.org/10.1007/978-3-030-54633-5_11

Fig. 11.1 A polished 23-cm-long stalagmite segment from Crevice Cave, Missouri, displaying some of the features inherent to high-quality speleothem records. The growth layers are consistently near-horizontal in the center of the stalagmite, and the calcite is dense and largely immune to post-depositional alteration, which allows for accurate dating. The small, closely-spaced series of drill holes along the central axis are sub-samples analyzed for oxygen and carbon isotopes, and used to generate a time-series of temperature and vegetation changes with decadal-scale resolution. The larger holes offset from the central axis were dated by U-Th techniques. The bottom and top dates are 112,180 ± 640, and 76,230 ± 260 years old, respectively, revealing that this 23 cm segment covers about 36,000 years of Earth history assuming there are no hiatuses in growth. Note the variability in colors and the discernible brown layers within the whiter calcite—these are detrital layers introduced by floodwaters, and they show an episodic history. Note how these detrital layers in some places (especially in the upper part of the stalagmite) fade along the central axis of the stalagmite, presumably because the falling drip water (after flood waters receded) flushed some of the detritus away from the center. The red background is colored epoxy, used to encase the stalagmite before sawing, which protects the sample and provides a stable working surface. This stalagmite was found naturally broken, lying in the cave stream, and although corroded on its exterior, the interior is pristine due to its dense crystal structure. Broken formations may have some drawbacks from a scientific viewpoint, but are an obvious choice from a conservation viewpoint, and can still provide extremely high-quality paleo environmental information

from Mystery Cave in southeastern Minnesota (Lively 1983). The results of these early studies showed promise for understanding paleoclimate for parts of the Quaternary (e.g., Thompson et al. 1974; Harmon et al. 1979; Lively 1983). It is interesting that the basic approach of these early studies remains much the same in modern speleothem studies—establish a chronology by U-series dating, which may be useful in and of itself, and which also provides the timescale for isotopic or other internal proxies of paleoclimatic conditions.

11.3 Age Dating

The ability to accurately and precisely date speleothem carbonate by U-series methods has proven to be one of the greatest strengths of the science. The extensive number of published uranium–thorium (U-Th) dates on speleothems has made it clear that well-chosen speleothem sub-samples are nearly ideal candidates for reliable age-dating. The first U-Th measurements on carbonates were made on corals (Barnes et al. 1956), followed a few years later by similar measurements made on cave calcites (Rosholt and Antal 1962), using decay-counting techniques. The routine application of U-Th dating to speleothems gained prominence in the early 1970s with the studies of the McMaster group (Harmon et al. 1975). Dating capabilities advanced significantly in the late 1980s with the development of mass spectrometric techniques for measuring U-series isotopes (Edwards et al. 1987). These new techniques improved the precision with which U-Th ages could be determined, decreased sample size requirements, and extended the range of U-Th dating to both younger and older times. Given the low deposition rates of many speleothems, these advances have significantly improved the science in the modern era compared to the earlier era of decay-counting, even though there has been little advance in the actual concept. By applying these modern techniques to appropriate materials, one can obtain precise, and potentially accurate ages over the past half million years. This range extends well beyond that of the radiocarbon method (~45,000 years) and therefore U-Th dating plays an important role in elucidating earlier parts of Quaternary climate history. Richards and Dorale (2003) and Dorale et al (2004) provide reviews of U-series dating applied to speleothems.

11.4 Approaches to Paleoclimatology

Carbon and oxygen isotopic variations in speleothems have been studied for more than five decades, commonly with the central goal of reconstructing past climatic conditions. In a seminal paper, Hendy (1971) outlined the various

Fig. 11.2 Cold Water Cave, Iowa illustrating beautiful speleothems as well as the strong bedding plane control of the cave's main passage morphology. The cave was discovered in 1967 and has received scientific interest ever since that time. Access to the cave is closely regulated today. Photo courtesy of www.scottdankofphoto.com

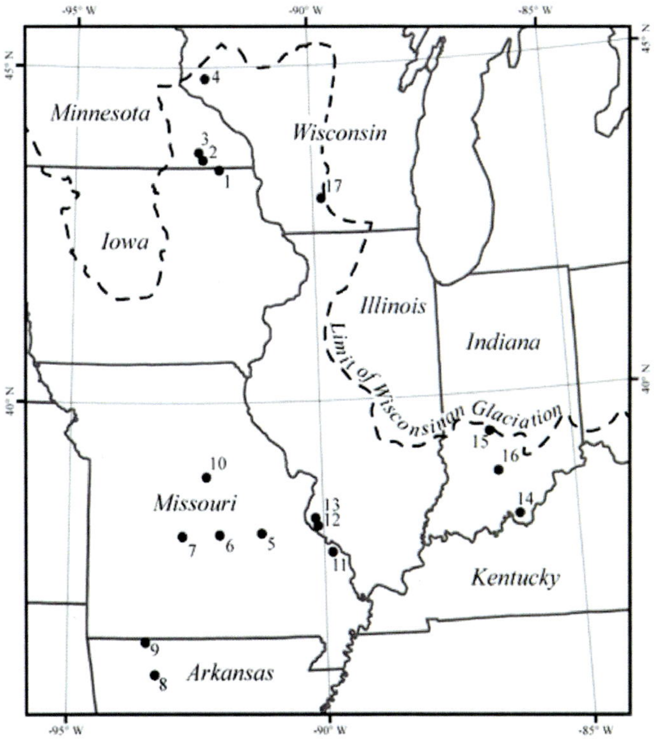

Fig. 11.3 Map of the Midwestern U.S.A. showing numbered locations of cave study sites referenced in this chapter. The caves are: (1) Cold Water Cave, (2) Mystery Cave, (3) Spring Valley Caverns, (4) Crystal Cave, (5) Onondaga Caverns, (6) Ozark Caverns, (7) Bridal Cave, (8) Beckham Creek Cave, (9) Cosmic Caverns, (10) Devil's Icebox Cave, (11) Crevice Cave, (12) Fogelpole Cave, (13) Illinois Caverns, (14) Indiana Caverns, (15) Porter Cave, (16) Donahue Cave, and (17) Cave of the Mounds. The dashed line shows the maximum extent of Wisconsinan glaciation. Note that all of the caves are outside the glacial margin

equilibrium and non-equilibrium processes that govern the distribution of oxygen isotopes ($\delta^{18}O$) and carbon isotopes ($\delta^{13}C$) during calcite ($CaCO_3$) speleothem formation, and discussed means of recognizing whether speleothem isotopic signatures might serve as appropriate paleoclimatic indicators. The delta notation (δ) for oxygen and carbon isotope ratios refers to the $^{18}O/^{16}O$ and $^{13}C/^{12}C$ ratios in a sample compared to those ratios in a standard (e.g., VSMOW for $\delta^{18}O$, and VPDB for $\delta^{13}C$), with the difference between sample and standard reported in per mil (‰, or parts per thousand).

Since the early 1970s, many fundamental aspects of speleothem research besides stable isotopic work have expanded and improved enormously, including luminescent and optical banding (e.g., Shopov et al. 1994), growth rates (e.g. Baker et al. 1998a, b), crystallography and mineralogy (e.g., Frisia et al. 2000), and trace element incorporation (e.g. Fairchild et al. 2000). In addition, high-resolution ice-core records, marine sediment records, pollen records, and numerous other paleoclimate proxy records have added tremendously to our overall knowledge of Quaternary climate history. Collectively, this knowledge provides a more informed regional and global perspective for interpreting speleothem $\delta^{18}O$ and $\delta^{13}C$ records than was possible four decades ago.

11.5 Stable Isotopes as Paleoclimatic Indicators

The $\delta^{18}O$ and $\delta^{13}C$ values of speleothem $CaCO_3$ are related to the primary sources of oxygen and carbon in the cave seepage water. In the case of oxygen, this is rain and snow. In the case of carbon, this is soil carbon dioxide and carbonate bedrock. The process of speleothem deposition can be traced back to the soil horizon where biological activity produces high levels of CO_2. This soil CO_2 acidifies seepage waters, which in turn dissolve carbonate bedrock enroute to underlying caves. Upon entering a cave passage of lower CO_2 concentration (relative to the soil atmosphere) the seepage water releases CO_2, and $CaCO_3$ deposition takes place (Holland et al. 1964). Because bicarbonate concentrations of karst groundwaters are typically in the parts per thousand range, the $\delta^{18}O$ composition of the water and the dissolved carbonate species are dominated by the water molecules themselves, which originated as meteoric precipitation. Therefore, the $\delta^{18}O$ values of speleothems are generally not significantly influenced by the bedrock $\delta^{18}O$ value. Speleothem $\delta^{13}C$ values, however, are significantly influenced by the isotopic composition of the bedrock, and the soil CO_2. Soil CO_2 is related to the vegetation overlying the cave, and vegetation at the regional scale is correlated to climate.

We can therefore see that speleothem $\delta^{18}O$ tells us something about atmospheric conditions and precipitation (and possibly temperature), and speleothem $\delta^{13}C$ traces processes from the vegetation and soil on down to the cave. Temperature is always a climatic variable of interest, and deep caves have a unique characteristic in that they generally reflect mean conditions, that is to say, the near-steady temperature of caves reflects the mean annual average. Thus one of the great attributes of caves for paleoclimate studies is they are a very stable environment and are well-positioned to average out high-frequency fluctuations and capture long-term changes. The link of $\delta^{18}O$ to temperature stems from the fact that there is temperature-dependent oxygen isotopic fractionation between water and calcite during calcite deposition, such that fractionation is greater at colder temperatures. There are, however, complications in estimating accurate paleo-temperatures from speleothem $\delta^{18}O$, due to variable behavior between rainwater $\delta^{18}O$ values and atmospheric temperature, but methods have developed to at least use speleothem $\delta^{18}O$ as a qualitative measure of temperature change (e.g., Dorale et al. 1992; Denniston et al. 1999a).

A number of studies from Midwestern caves (Fig. 11.3) have used $\delta^{18}O$ and $\delta^{13}C$ changes across time to elucidate climate histories. One of the earliest was that of Harmon et al. (1979), who inferred temperature changes from $\delta^{18}O$ variations in a Cold Water Cave stalagmite reported to have grown continuously from 25,000 to 6,000 years ago. This early work brought widespread attention to the great potential of speleothems to provide long climate records; unfortunately more recent age-dating work has placed some of these early results in question (Alexander et al. 2001). More work is needed to better understand the discrepancies reported by Alexander et al. (2001). Cold Water Cave was studied in more recent times by students of Luis González at the University of Iowa. The first of this new era was the work of Dorale et al. (1992), who studied several stalagmites and reported temperature and vegetation changes during the Holocene. Other students of González continued more research on Cold Water Cave through the 1990s (e.g. Suzuki 1998; Denniston et al. 1999a).

11.6 Speleothem $\delta^{13}C$ as Vegetation Proxy

One of the significant contributions of the new era of research at Cold Water Cave was the development of speleothem $\delta^{13}C$ as a vegetation proxy through a comparative approach of speleothem data with pollen and plant macrofossil data. Although a host of factors can potentially affect the $\delta^{13}C$ values of speleothems, vegetation is a major factor because soil CO_2 is generated largely by the microbial oxidation of soil organic matter, which is derived from vegetation. C_3 and C_4 photosynthetic pathways produce large differences in the $\delta^{13}C$ values of plant tissue. C_3 plants have $\delta^{13}C$ values averaging ca. –26 ‰, whereas C_4 plants average ca. –12 ‰. C_4 plants are mostly warm-season grasses found in tropical and temperate grasslands (Teeri and Stowe 1976). In Midwestern prairies, this includes common, biomass-dominant grass species such as big bluestem (*Andropogon gerardi*) and Indian grass (*Sorghastrum nutans*). C_3 plants, on the other hand, are mostly trees, shrubs, cool-season grasses, and herbs. Speleothems are thus capable of recording the proportion of C_3 to C_4 plant biomass through time in their $\delta^{13}C$ values, and are particularly useful in regions where C_3 versus C_4 dominated biomes may have traded prevalence through time, such as in the Midwestern U.S.A.

Cold Water Cave and Robert's Creek (located ~60 km SE of Cold Water) provide one of the best examples of the

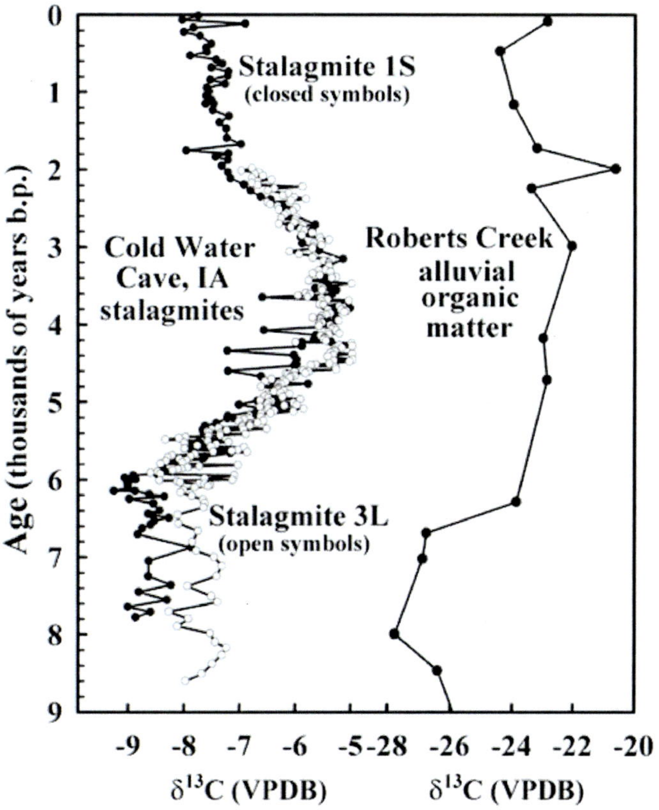

Fig. 11.4 Holocene $\delta^{13}C$ profiles for stalagmites 1S and 3L from Cold Water Cave, Iowa and alluvial sedimentary organic matter from Roberts Creek. Note the general similarity in timing and trend of the different $\delta^{13}C$ records. The $\delta^{13}C$ values of the stalagmites reflect the sources of carbon, namely, soil CO_2 ultimately derived from the vegetation, for which the Roberts Creek record provides representative data, and the overlying limestone, which typically has $\delta^{13}C$ values close to 0 ‰ VPDB. In principle, these two sources contribute 50:50 to the dissolved carbonate species, although complex isotopic exchange scenarios are possible that may modify this ratio by the time seepage waters reach the caves. Nonetheless, the carbon which originates as soil CO_2 undergoes isotopic fractionation as it converts from dissolved CO_2 to aqueous CO_2 to HCO_3^- to $CO_3^=$ to solid $CaCO_3$; under equilibrium conditions, this sequence represents an approximate 10 ‰ enrichment. Thus, when soil CO_2 with a $\delta^{13}C$ value of –28 ‰ (representing a typical C_3 plant value) fractionates approximately 10 ‰ and mixes 50:50 with bedrock carbonate with a $\delta^{13}C$ value of ~0 ‰, the resultant speleothem carbonate would have a $\delta^{13}C$ value of ~−9 ‰, consistent with the observations at Cold Water Cave and Roberts Creek. Because of the 50:50 contribution and the invariant nature of the bedrock component, shifts in the vegetation $\delta^{13}C$ values should cause speleothem shifts that are roughly half the magnitude of the vegetation shifts, again consistent with the observations. Note the factor of two difference in the $\delta^{13}C$ scales. Figure modified from Baker et al. (1998a, b)

relationship between speleothem $\delta^{13}C$ values and the C_3/C_4 biomass ratio (Baker et al. 1998a, b). Pollen, plant macrofossils, and sedimentary organic matter $\delta^{13}C$ values from well-dated alluvial deposits along Roberts Creek were compared to stalagmite $\delta^{13}C$ values from Cold Water Cave. The pollen and plant macrofossil evidence (which in many cases allowed species-level plant identification) showed that a C_4-inclusive prairie replaced a C_3-rich deciduous forest around 6,300 years ago, which in turn was replaced by oak savanna (and a more intermediate C_3-C_4 mixture) approximately 3,500 years ago. The sedimentary organic matter $\delta^{13}C$ values track the vegetation changes. The similarity of both the timing and isotopic trend of the two speleothem records from Cold Water Cave with the vegetation record from Roberts Creek argues that the speleothems are recording long-terms changes in the isotopic composition of the soil organic matter resulting from regional changes in the vegetation (Fig. 11.4).

More detail on the movement of the prairie-forest border in the Upper Midwest was provided by the work of Denniston et al. (1999b), who expanded speleothem work to Mystery Cave and Spring Valley Caverns in Minnesota, and Crystal Cave in Wisconsin. This work revealed a sharp prairie-forest border in the region that persisted for several thousand years during the middle Holocene, as had been suggested by Baker et al. (1996).

Similar tracing of the prairie-forest ecotone during the Holocene was carried out by Denniston et al. (1999c, 2000) in the Lower Midwest, by looking at speleothem records from caves in Missouri and Arkansas. Beyond the last glacial margin (Fig. 11.3), vegetation reconstructions are hampered by a lack of natural lakes and wetlands that typically yield this type of information in the glaciated country to the north, so the cave reconstructions help significantly with filling in the regional picture. The state of Missouri is reported to contain more than 6,000 known caves, and the Ozark region of Arkansas is also home to abundant karst and caves; thus there is tremendous opportunity for speleothem work in this region. The work of Denniston et al. (2000, 2007) utilizing speleothem $\delta^{13}C$ revealed a number of vegetation oscillations in the region during the Holocene, and inferred several periods of drying climate.

11.7 Long Climate Records from Midwestern Caves

Only a handful of published speleothem records from Midwestern caves reach into the last glacial period and beyond, although the potential certainly exists for long records in this region. One of the better known records is that from Crevice Cave in southeastern Missouri (Dorale et al. 1998). This publication describes a record spanning the period from 25,000 to 75,000 years ago, with temperature oscillations interpreted from variations in stalagmite $\delta^{18}O$, and vegetation changes inferred from stalagmite $\delta^{13}C$, in much the same way as was done at Cold Water Cave. One of the larger climatic transitions was noted to have taken place around 55,000 years ago, as the Laurentide Ice sheet built in size during the last glacial cycle. One of the more important contributions of this study was the demonstration that speleothem isotope histories could be replicated in multiple speleothems over very long time periods, which provides some measure of confidence that the proxies are directly recording climatic variables of interest. The topic of replication is reviewed by Dorale and Liu (2009).

Just north of Crevice Cave, Illinois Caverns and Fogelpole Cave in southwestern Illinois have also produced climate records from the last glacial and interglacial periods (Panno et al. 2004, 2012). These studies used speleothem ages as well as cave sediments to infer a history of cave development and subsequent infilling, with regional climatic implications. Based on the constraints provided by dated speleothems and calculated rates of incision and passage dimensions, the age of speleogenesis for Fogelpole Cave and Illinois Caverns was estimated to range from 120,000 and 164,000 years ago.

Several cave sites in southern Indiana have recently produced notably long records. The studies of Chirienco (2010) and Panno et al. (2016) at Donahue Cave describe speleothem growth extending, intermittently, back to 300,000 years ago, and interpretations of paleoclimatic and paleoseismic conditions have also been inferred. Akers (2016) describes several speleothems from Indiana Caverns with ages as old as 90,000 years, and with intermittent growth thereafter, presumably reflecting significant changes in hydrologic conditions. Moellers et al. (2006) describe a speleothem record from Cosmic Caverns in northern Arkansas from 141,000 to 128,000 years ago, which covers the penultimate transition from glacial to interglacial conditions.

In the Upper Midwest, speleothems have also been reported that reach back several hundred thousand years. Dasgupta (2008) dated a large number of old speleothems from several Minnesota caves including Spring Valley Caverns (104,000 years), Echo Chamber (297,000 years), and Mystery Cave. The oldest sample from Mystery Cave was 128,000 years old, which is consistent with the earlier work of Lively (1983), who also found speleothems of that age, reflecting growth during the previous interglacial period. More recently, Batchelor et al. (2018) have dated a large number of speleothems from Cave of the Mounds in southwestern Wisconsin, and report speleothem growth occurring throughout much of the past 257,000 years, which has important implications for climatic boundary conditions (such as permafrost) over the past several glacial/interglacial cycles.

Long speleothem records from Midwestern caves that span multiple glacial cycles are important for a number of reasons. First, the oldest speleothem age for a given cave provides a simple, yet powerful constraint on the minimum age of the cave system, thereby providing evidence for rates of karst development and landscape evolution. Second, speleothems are some of the most valuable archives of climatic conditions on the continents. Such records extend back into parts of the Quaternary that are well beyond the reach of radiocarbon and provide one of the very few high-resolution archives of terrestrial paleoclimatic conditions that can be reliably dated over multiple glacial/interglacial cycles; therefore allowing comparison to long reference climate records from ice cores or ocean sediments.

What's the oldest reported speleothem from a Midwestern cave? As noted above, a stalagmite from Cave of the Mounds reaches back to 257,000 years and Echo Chamber in Minnesota and Donahue Cave in Indiana both extend to 300,000 years. In my collection from Crevice Cave, Missouri, the oldest U-Th date, from the base of an 87-cm-long stalagmite, is 416,450 ± 5,000 years old (unpublished data). This particular stalagmite, which experienced numerous changes in growth patterns and a series of hiatuses, finally stopped growing at 121,600 ± 550 years ago. When better developed and fully integrated, these records will reveal the previously unknown, complex, and exciting history of midcontinent climatic variability over several glacial/interglacial cycles.

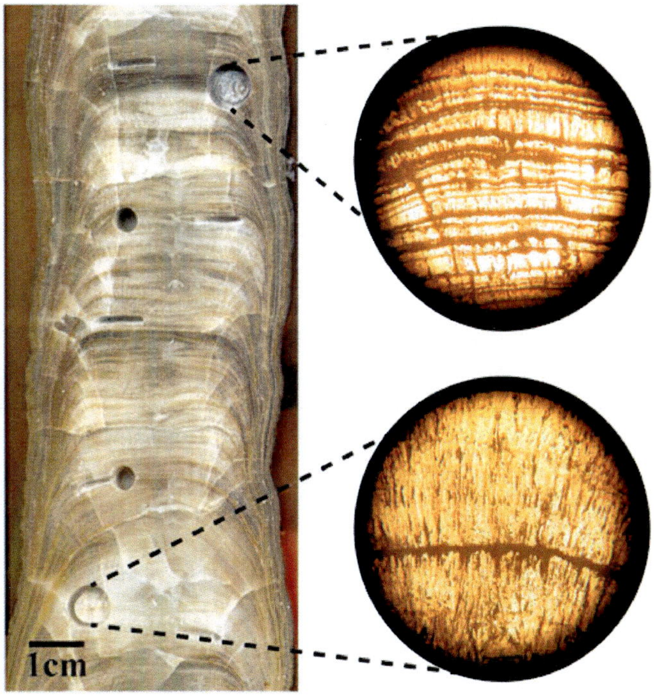

Fig. 11.5 A Holocene stalagmite from Crevice Cave, Missouri, showing detrital flood layers preserved in calcite. Although detrital layers can be observed visually in reflected light on polished surfaces, thick-section photomicrographs allow for careful identification of individual flood layers. The lower photomicrograph shows a single detrital layer, while the upper photomicrograph shows more than a dozen, and illustrates the challenge in making accurate counts

11.8 Flood Records from Midwestern Caves

Because stalactites and stalagmites form from dripping water, they are sensitive to any variations in the hydrologic cycle that alter the flow of water from the ground surface to the drip site in the cave. These types of hydrologic variations may express themselves in a number of ways, ranging from deposition versus non-deposition at the extreme case, to more subtle variations in speleothem growth rates or axis shifts, banding features, or trace element concentrations.

One of the more dramatic types of hydrologic phenomena that may affect speleothems growing in caves is flooding of the cave. The nature of flooding in caves is highly variable depending on the specific characteristics of the cave hydrology, and the nature of the external events creating the floodwaters. Wood et al. (2010), for example, interpreted cave sediment sequences in Porter Cave in Indiana to indicate catastrophic flooding by glacial meltwaters around 40,000 years ago. Most cases of cave flooding are caused by much less extreme phenomena than glacial meltwater surges, however, and instead generally reflect episodes of heavy rainfall. Shallow groundwater flow in a typical karst terrain is commonly regulated by numerous recharge points throughout the epikarst, but with a limited number of outlets, creating a condition of slow backfilling in cave systems during high recharge events. Detrital particles are likely to become suspended within the cave floodwaters, and these particles get deposited on bedrock surfaces and speleothems that are inundated by the flooding. After floodwaters recede, thin clay-rich deposits may be preserved by the subsequent resumption of speleothem growth, preserving the deposits in calcite (Fig. 11.5), recording the flood event.

One of the earliest mentions of this phenomenon was made by White (1976), who recognized that mud layers within speleothems may have an origin tied to flooding of the cave. Despite such early recognition, there was no focused effort to utilize detrital layers in speleothems as a proxy for episodic flooding until Lepley (2004) and Dorale et al. (2005) reported Holocene records of cave flooding for Crevice Cave in Missouri. Knight (2006) extended this work back to 130,000 years ago for Crevice Cave, and Dasgupta et al. (2010) described a Holocene flood record from Spring Valley Caverns in Minnesota. The technique has since gained widespread attention, and speleothem flood records have now been reported from around the globe, including those from Central America (Frappier et al 2014), Europe (Gonzalez-Lemos et al. 2015), and Australia (Denniston et al. 2015). Denniston and Luetscher (2017) provide an excellent recent review on the topic of flood records preserved in speleothems.

11.9 Paleoseismic Events Recorded in Midwestern Caves

Large earthquakes are not common to the Midwest in the same way they are in more tectonically active regions of Earth. But large earthquakes do occasionally occur in the region, and sometimes with historic consequences. The famous New Madrid event of 1811–1812 was comprised of more than a thousand small earthquakes but included four

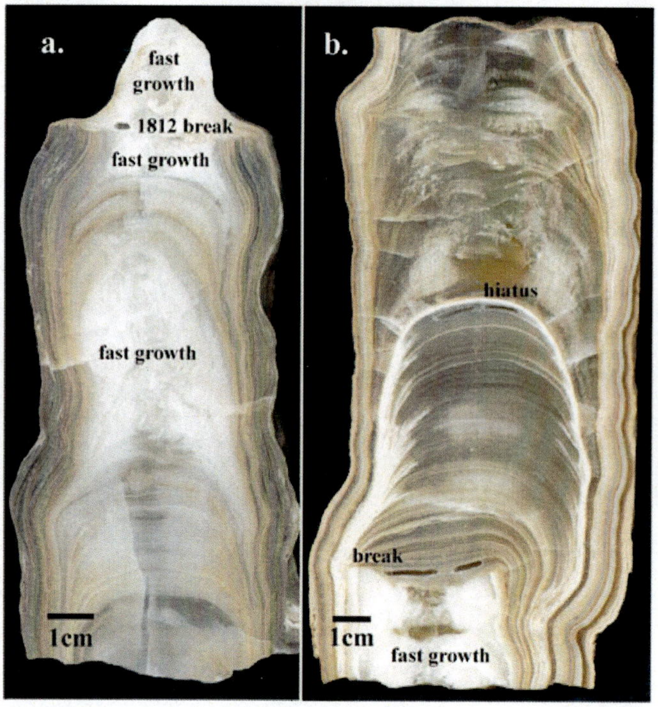

Fig. 11.6 Two stalagmites from Crevice Cave, Missouri, revealing evidence of past seismic activity. The stalagmite in panel a shows a clear break and subsequent deposition of porous, white calcite. The U-Th date at the onset of this new calcite growth is consistent with the 1811–1812 New Madrid seismic event. Two earlier periods of rapid growth, again whiter in color and more porous, can be observed, possibly also indicating past seismic influences. The stalagmite in panel b shows a break right above a period of rapid, white calcite deposition, indicating, perhaps, dilation of hydraulic pathways prior to the breakage event. The base of this stalagmite, found as a separate piece, has been determined by U-Th methods to be 416,450 ± 5,000 years old, and may be the oldest known speleothem from the Midwestern U.S.A

large individual earthquakes with moment magnitudes in the 7.3–7.7 range. The strongest earthquake occurred on February 7, 1812. This famous earthquake caused massive landslides, snapped trees, and caused the Mississippi river to flow backwards—a so-called fluvial tsunami.

Caves are potential repositories for records of seismic events based on damage to the cave and speleothems. In the Midwestern U.S.A, Sam Panno and colleagues have championed the technique of extracting past seismic information from caves (e.g. Panno et al. 2009, 2016). Notable effects to speleothems include breakage, changes in the orientation of the growth axis, and cessation or initation of growth, or a different style of carbonate precipitation (Fig. 11.6). The change of growth pattern is presumably due to dilation or contraction of hydraulic pathways, and/or sedimentation within their pathways during seismic activity (Panno et al. 2016). The alteration of hydraulic pathways in the epikarst might be expected to significantly change flow dynamics above the cave, leading to changes in the drip rate and possibly the rate of calcite deposition. Panno et al. (2009) have shown that the onset of relatively fast speleothem growth (confirmed by U-Th dating) seems to be one of the common characteristics associated with seismic activity. In some cases, too, this fast new calcite growth is conspicuously white in color, which possibly reflects fast transit of the vadose waters through the epikarst with little time to pick up soil organic acids that typically color speleothem calcite.

The value of speleothem-based studies of earthquake occurrence is they can help establish the long-term recurrence intervals of large earthquakes associated with dangerous seismic zones. Information on recurrence intervals can be used for hazard evaluation and can lead to a better understanding of the physical nature of the seismic zone itself.

11.10 Conclusions

Caves are unique environments, simultaneously shielded from, and connected to, the world above. Whether it is using carbon isotopes to reconstruct the vegetation that once grew over the cave, identifying ancient seismic events from broken speleothems, or piecing together a history of cave flooding from fine layers of mud tapped in their fabrics, speleothem-based paleoclimate reconstruction techniques continue to evolve, revealing the secrets of a former world. Many mysteries remain, and much work remains to be done. The earliest paleoclimate work in caves, some of which can be traced back to Midwestern settings, was perhaps a bit underappreciated at the time, but it set the stage for an explosive contemporary scene in cave studies. Caves are without question fruitful areas for scientific inquiry, but in this modern era of fast-paced research I believe it is worth keeping in mind that caves, in general, are delicate settings and deserve our utmost respect. Cavers have long abided by the rules of respect, and I urge the new generation of scientists to follow suit as they pursue their research in caves.

Acknowledgments I would like to thank Phil Kerr for help with the figures, and Scott Dankof for sharing his fine photograph of Cold Water Cave. Rhawn Denniston provided suggestions that improved the manuscript. I would like to thank the editors for their patience as they waited for this chapter, and worked through its editing.

References

Akers P (2016) Paleoclimate change in southern Indiana determined from speleothem climate proxies and analysis of modern precipitation oxygen isotope variations. PhD dissertation, University of Georgia

Alexander EC Jr, Dorale JA, Shade BL, Edwards RL, Lively RS, Worthington SRH, Serefiddin F, Ford DC (2001) A major revision of U/Th speleothem dates from Cold Water Cave, Iowa. Geol Soc Am Abstracts Prog 33(6):71

Baker A, Genty D, Dreybrodt W, Barnes WL, Mockler NJ, Grapes J (1998) Testing theoretically predicted stalagmite growth rate with recent annually laminated samples; implications for past stalagmite deposition. Geochim Cosmochim Acta 62:393–404

Baker RG, Bettis EA III, Schwert DP, Horton DG, Chumbley CA, González LA, Reagan MK (1996) Holocene paleonvironments of northeast Iowa. Ecol Monogr 66:203–234

Baker RG, González LA, Raymo M, Bettis EA III, Reagan MK, Dorale JA (1998) Comparison of multiple proxy records of Holocene environments in the Midwestern United States. Geology 26:1131–1134

Barnes JW, Lang EJ, Potratz HA (1956) The ratio of ionium to uranium in coral limestone. Science 124:175–176

Batchelor CJ, Orland IJ, Marcott SA, Slaughter R, Edwards RL (2018) A high-precision U-Th chronology of calcite deposition at Cave of the Mounds, Wisconsin, and its implications for climate and permafrost in the late Pleistocene. Geol Soc Am Abstracts Prog 50 (4)

Chirienco MI (2010) Interpretations of paleoclimate and speleogenesis from speleothems in Donnehue's Cave, Indiana. MS thesis, University of Illinois

Dasgupta S (2008) High-resolution speleothem record of late Quaternary climate change from the upper Midwest, USA. PhD dissertation, University of Minnesota

Dasgupta S, Saar MO, Edwards RL, Shen CC, Cheng H, Alexander EC Jr (2010) Three thousand years of extreme rainfall events recorded in stalagmites from Spring Valley Caverns, Minnesota. Earth Planet Sci Lett 300:46–54

Denniston RF, DuPree M, Dorale JA, Asmerom Y, Polyak V, Carpenter SJ (2007) Episodes of late Holocene aridity recorded by stalagmites from Devil's Icebox Cave, central Missouri, USA. Quatern Res 68:45–52

Denniston RF, González LA, Asmerom Y, Baker RG, Reagan MK, Bettis EA III (1999) Evidence for increased cool season moisture during the middle Holocene. Geology 27:815–818

Denniston RF, González LA, Asmerom Y, Reagan MK, Recelli-Snyder H (2000) Speleothem carbon isotopic records of Holocene environments in the Ozark Highlands, USA. Quatern Int 67:21–27

Denniston RF, González LA, Baker RG, Asmerom Y, Reagan MK, Edwards RL, Alexander EC Jr (1999) Speleothem evidence for Holocene fluctuations of the prairie-forest ecotone, north-central USA. Holocene 9(6):671–676

Denniston RF, González LA, Semken H, Asmerom Y, Baker RG, Recelli-Snyder H, Bettis EA III (1999) Integrating stalagmite, vertebrate, and pollen sequences to investigate Holocene vegetation and climate change in the southern Midwestern United States. Quatern Res 52(3):381–387

Denniston RF, Luetscher M (2017) Speleothems as high-resolution paleoflood archives. Quatern Sci Rev 170:1–13

Denniston RF, Villarini G, Gonzales AN, Wyrwoll K-H, Polyak VJ, Ummenhofer CC, Lachniet MS, Wanamaker AD Jr, Humphreys WH, Woods D, Cugley J (2015) Extreme rainfall activity in the Australian tropics reflects changes in the El Nino/southern oscillation over the last two millennia. Proc Natl Acad Sci 112:4576–4581

Dorale JA, Edwards RL, Alexander EC Jr, Shen CC, Richards DA, Cheng H (2004) Uranium-series dating of speleothems: current techniques, limits & applications. In: Sasowsky ID, Mylroie JE (eds) Studies of cave sediments: physical and chemical records of paleoclimate. Springer, Dordrecht, Netherlands, pp 177–197

Dorale JA, Edwards RL, Ito E, González LA (1998) Climate and vegetation history of the midcontinent from 75 to 25 ka: a speleothem record from Crevice Cave, Missouri, USA. Science 282:1871–1874

Dorale JA, González LA, Reagan MK, Picket DA, Murrell MT, Baker RG (1992) A high-resolution record of Holocene climate change in speleothem calcite from Cold Water Cave, Northeast Iowa. Science 258:1626–1630

Dorale JA, Lepley SW, Edwards RL (2005) The ultimate flood recorder: flood deposited sediments preserved in stalagmites. European Geosciences Union, Vienna, Austria

Dorale JA, Liu Z (2009) Limitations of Hendy test criteria in judging the paleoclimatic suitability of speleothems and the need for replication. J Cave Karst Stud 71:73–80

Edwards RL, Chen JH, Wasserburg GJ (1987) ^{238}U–^{234}U-^{230}Th-^{232}Th systematics and the precise measurement of time over the past 500,000 y. Earth Planet Sci Lett 81:175–192

Fairchild IJ, Borsato A, Tooth AF, Frisia S, Hawkesworth CJ, Huang Y, McDermott F, Spiro B (2000) Controls on trace elements (Sr-Mg) compositions of carbonate cave waters: implications for speleothem climatic records. Chem Geol 166:255–269

Frappier AB, Pyburn J, Pinkey-Drobnis AD, Wang X, Corbett DR, Dahlin BH (2014) Two millennia of tropical cyclone-induced mud layers in a northern Yucatan stalagmite: multiple overlapping climatic hazards during the Maya Terminal Classic "megadroughts." Geophys Res Lett 41(14):5148–5157

Frisia S, Borsato A, Fairchild IJ, McDermott F (2000) Calcite fabrics, growth mechanisms, and environments of formation in speleothems from the Italian Alps and southwestern Ireland. J Sediment Res 70:1183–1196

Gonzalez-Lemos S, Müller W, Pisonero J, Cheng H, Edwards RL, Stoll HM (2015) Holocene flood frequency reconstruction from speleothems in northern Spain. Quatern Sci Rev 127:129–140

Harmon RS, Schwarcz HP, Ford DC, Koch DL (1979) An isotopic paleotemperature record for late Wisconsinan time in northeast Iowa. Geology 7(9):430–433

Harmon RS, Thompson P, Schwarcz HP, Ford DC (1975) Uranium-series dating of speleothems. Nat Speleol Soc Bull 37:21–33

Hendy CH (1971) The isotopic geochemistry of speleothems, 1. The calculation of the effects of different modes of formation on the isotopic composition of speleothems and their applicability as paleoclimatic indicators. Geochim Cosmochim Acta 35:801–824

Hendy CH, Wilson AT (1968) Palaeoclimatic data from speleothems. Nature 216:48–51

Holland HD, Kirsipu TV, Huebner JS, Oxburgh UM (1964) On some aspects of the chemical evolution of cave waters. J Geol 72:36–67

Knight CW (2006) An interglacial/glacial flood record from stalagmites from Crevice Cave, Missouri. MS thesis, University of Iowa

Koch DL, Case JC (1974) Report on Cold Water Cave. Iowa Geological Survey Report 80

Lepley SW (2004) Holocene ENSO cyclicity from U-Th dated sedimentation events in stalagmites from Crevice Cave, Missouri. MS thesis, University of Missouri-Columbia

Lively RS (1983) Late-Quaternary U-series speleothem growth record from southeastern Minnesota. Geology 11:259–262

Moellers L, Denniston RF, Carpenter SJ, Dorale JA, Polyak V, Asmerom Y (2006) A stalagmite record of the penultimate deglaciation from northwestern Arkansas. Geol Soc Am Abstracts Prog 38(4):27

Panno SV, Chierinco MI, Bauer RA, Lundstrom CC, Zhang Z, Hackley KC (2016) Earthquakes recorded in stalagmites from a cave in south-central Indiana. Bull Seismol Soc Am 106(5):1–12

Panno SV, Curry BB, Wang H, Hackley KC, Liu C-L, Lundstrom C, Zhou J (2004) Climate change in southern Illinois, USA, based on the age and δ13C of organic matter in cave sediments. Quatern Res 61(3):301–313

Panno SV, Curry BB, Wang H, Hackley KC, Zhang Z, Lundstrom C (2012) The effects of climate change and glaciation on speleogenesis and karstification in southwestern Illinois' sinkhole plain. Carbonates Evaporites 24:87–94

Panno SV, Lundstrom C, Hackley KC, Curry BB, Fouke B, Zhang Z (2009) Major earthquakes recorded by speleothems in Midwestern U.S. caves. Bull Seismol Soc Am 99(4):2147–2154

Richards DA, Dorale JA (2003) Uranium-series chronology and environmental applications of speleothems. Rev Mineral Geochem 52:407–460

Rosholt JN, Antal PS (1962) Evaluation of the ^{231}Pa/U-^{230}Th/U method for dating Pleistocene carbonate rocks. US Geological Survey Prof. Paper 450-E:108

Shopov YY, Ford DC, Schwarcz HP (1994) Luminescent microbanding in speleothems: High-resolution chronology and paleoclimate. Geology 22:407–410

Suzuki AT (1998) The geochemical evolution of vadose fluids and gases above Cold Water Cave, Northeast Iowa; transportation and preservation of seasonal climatic signals. MS thesis, University of Iowa, p 1–64

Teeri JA, Stowe LG (1976) Climatic patterns and distribution of C_4 grasses in North America. Oecologia 23:1–12

Thompson P, Schwarcz HP, Ford DC (1974) Continental Pleistocene climatic variations from speleothem age and isotopic data. Science 184:893–895

White WB (1976) Cave minerals and speleothems. In: Ford TD, Cullingford CHD (eds) The science of speleology. Academic Press, London, pp 267–328

Wood JR, Forman SL, Everton D, Pierson J, Gomez J (2010) Lacustrine sediments in Porter Cave, central Indiana, USA and possible relation to Laurentide ice sheet marginal positions in the middle and late Wisconsinan. Palaeogeogr Palaeoclimatol Palaeoecol 298(3–4):421–431

Index

G. A. Brick and E. C. Alexander Jr. (eds.), *Caves and Karst of the Upper Midwest, USA*,
Cave and Karst Systems of the World, https://doi.org/10.1007/978-3-030-54633-5

T

U

V

W